Study Guide and Solutions Manual to Accompany

Core Organic Chemistry

Study Guide and Solutions Manual to Accompany

Core Organic Chemistry

with

Nucleophile/Electrophile
Reaction Guide

Dr. Donna Nelson

Marye Anne Fox

James K. Whitesell

The University of Texas
Austin, Texas

Jones and Bartlett Publishers
Sudbury, Massachusetts

Boston   London   Singapore

Editorial, Sales, and Customer Service Offices

Jones and Bartlett Publishers
40 Tall Pine Drive
Sudbury, MA 01776
info@jbpub.com
http://www.jbpub.com

Jones and Bartlett Publishers International
Barb House, Barb Mews
London W6 7PA
UK

ISBN: 0-7637-0440-7

Printed in the United States of America
00 99 98 97   10 9 8 7 6 5 4 3 2 1

Contents

Suggested Study Methods   vi

1 Structure and Bonding in Alkanes
    Key Concepts   1
    Answers to Exercises   1
    Answers to Problems   11
    Important New Terms   15

2 Alkenes, Dienes, Aromatic Hydrocarbons, and Alkynes
    Key Concepts   19
    Answers to Exercises   19
    Answers to Problems   29
    Important New Terms   34

3 Functional Groups Containing Heteroatoms
    Key Concepts   37
    Answers to Exercises   37
    Answers to Problems   47
    Important New Terms   53

4 Chromatography and Spectroscopy: *Purification and Structure Determination*
    Key Concepts   57
    Answers to Exercises   57
    Answers to Problems   64
    Important New Terms   73

5 Stereochemistry
    Key Concepts   77
    Answers to Exercises   77
    Answers to Problems   85
    Important New Terms   89

6 Understanding Organic Reactions
    Key Concepts   93
    Answers to Exercises   93
    Answers to Problems   105
    Important New Terms   111

7 Mechanisms of Organic Reactions
    Key Concepts   115
    Answers to Exercises   115
    Answers to Problems   124
    Important New Terms   131

8 Substitution by Nucleophiles at sp^3-Hybridized Carbon
    Key Concepts   135
    Answers to Exercises   135
    Answers to Problems   144
    Important New Terms   152

9 Elimination Reactions

Key Concepts   153

Answers to Exercises   153

Answers to Problems   162

Important New Terms   171

10 Addition to Carbon—Carbon Multiple Bonds

Key Concepts   173

Answers to Exercises   173

Answers to Problems   188

Important New Terms   202

11 Electrophilic Aromatic Substitution

Key Concepts   205

Answers to Exercises   205

Answers to Problems   218

Important New Terms   227

12 Nucleophilic Addition and Substitution at Carbonyl Groups

Key Concepts   229

Answers to Exercises   229

Answers to Problems   243

Important New Terms   251

13 Substition Alpha to Carbonyl Groups:
Enolate Anions and Enols as Nucleophiles

Key Concepts   253

Answers to Exercises   253

Answers to Problems   261

Important New Terms   267

14 Skeletal-Rearrangement Reactions

Key Concepts   269

Answers to Exercises   269

Answers to Problems   276

Important New Terms   285

15 Multistep Syntheses

Key Concepts   287

Answers to Exercises   287

Answers to Problems   295

Important New Terms   298

16 Polymeric Materials

Key Concepts   299

Answers to Exercises   299

Answers to Problems   308

Important New Terms   314

SUGGESTIONS FOR STUDYING ORGANIC CHEMISTRY

Mastering organic chemistry is likely to be one of the most stimulating and exhilarating learning experiences you will encounter as part of your undergraduate studies. Being able to address the structures and function of new and of naturally occurring molecules makes it possible for you to understand much of the excitement in many of the most fascinating areas of science. A command of the principles of organic chemistry, for example, is absolutely critical for further work not only in chemistry, but also in biology, the health sciences, materials science, and solid state physics.

Nonetheless, because of its unfairly earned reputation, organic chemistry is sometimes regarded with alarm, verging on dread, by many students when they first enroll in the course. This is an unfair and inaccurate characterization, for although the course will cover a great deal of material, a hard-working student will find that organic chemistry has a intellectually appealing logical basis that makes it possible to generalize about newly encountered reactions from a smaller group of reactions that have been thoroughly studied before. But it would be quite wrong to think that this course can be treated casually or that it is possible to cram before an examination and expect to assimilate the key ideas adequately. This course requires dedication: a scheduled, regular, firm commitment of both time and effort. It will require the kind of discipline you may have experienced in studying a foreign language (to learn the terminology associated with this subject), combined with that encountered in other survey courses where you are asked to deal with a large number of facts and integrate them, deductively and inductively, into a coherent framework.

Learning is an individual activity. It is therefore difficult to say exactly what method will work best for each student, but faculty who have taught this course for many years often find that the following study techniques are generally most effective for most students. They include:

- Adequately preparing for lectures. This means reading the material in the text that will be covered in the lecture *before* it is presented. This is crucial if you are to be able to ask intelligent questions about topics that require further clarification.

- Regularly attending lectures, with a clear determination to extract from your instructor an appreciation of the relative importance of topics being covered. You should take good notes, writing down key points as they are presented and should review (or, in some cases, even transcribe) the notes, possibly comparing your notes with those of another student, after the lecture.

- Conscientiously working the in-chapter exercises after the material in each section has been covered in lecture to be sure that critical concepts have been adequately understood. You should work the exercises on your own and should not skim back to the text or to the Study Guide to find an answer. Nor should you skip over questions because they look obvious or easy – if you do so, you can't be sure that what you thought simple really was. You should write down the answer you think is correct and *then* compare that given in the Study Guide. If your answer doesn't match, check the discussion of that relevant material in the text or your notes and if a mistake is not obvious, get some help from your instructor, a teaching assistant, or another student.

- Promptly working the end-of-chapter problems when a chapter has been completed. This assures you that the material covered in the individual sections has been integrated at a level required for generalization in later parts of the course. This activity, together with due diligence on the in-chapter exercises, is **the most effective way** to be sure that you have mastered organic chemistry.

- Using the answers provided in the Study Guide for the exercises and problems only to *check* the answers you have independently written down. Students who simply peruse the answers to find out how to answer the questions without having dealt with them independently will *not* assimilate the applicable concept well enough to be able to recall it later.

- Availing yourself of the learning aids available in the text and Study Guide. Among these are the Summary of New Reactions and Conclusions sections that appear at the end of each Chapter, the Tables of reactions, pK_as, bond energies, etc. that appear in many chapters (some of which are reproduced on the endpapers inside the back cover), the comprehensive glossary with cross-references to the section of the text where the topic is developed, and the index at the end of the book. The boxed material that appears in each chapter may also serve as a mnemonic device to help some students recall the relevance of some of the material.

- Designing self-selected learning aids, perhaps by routinely using a molecular model set to visualize reactions and by constructing a set of index cards summarizing important reactions as they are introduced, and using them as reinforcing drill.

- Seeking out additional local assistance designed to help you. For example, some students derive unexpected benefits by taking advantage of their instructor's or a teaching assistant's office hours, by actively participating in recitation sections and/or help sessions, by seeking out any additional supporting materials (handouts, sample tests, computer programs for drill or visualization), by reviewing audio or video tapes of lecture discussions, and by forming regularly scheduled study groups with friends or other students in the class.

- Making the most of the laboratory experience that is associated with most organic courses. Actually working with organic compounds when you have prepared sufficiently well for the experiment will reinforce the utility of the reactions you learn in the lecture and from the text.

Each chapter of this Study Guide provides a list of Key Concepts from each chapter (making it unnecessary to ask your instructor what will be on the exam). This is followed by the Answers to the Exercise and the Problems, as well as a glossary of Important New Terms that are relevant to that chapter.

Study frequently, without skipping sections or chapters because of other time commitments. A student who actively and conscientiously incorporates each of these activities into his or her study program *will succeed*. We apologize if we have made any mistakes in compiling the Study Guide, and would welcome any comments from you about the text, the Study Guide, or any advice you would like to offer to future students.

Marye Anne Fox
James Whitesell
Department of Chemistry
University of Texas at Austin
Austin, TX 78712

Structure and Bonding in Alkanes

1

Key Concepts

Hybridization

Covalent and ionic bonding

Sigma (σ) bonds

Geometry at sp^3-hybridized (tetrahedral) carbon atoms

Group characteristics of alkanes

Free rotation

Positional isomerism

Drawing structures in three dimensions

Ring strain in cycloalkanes

Naming alkanes

Measuring relative stabilities of isomeric alkanes

Relationship between structure and chemical stability

Answers to Exercises

Ex 1.1 (a) Boron has an atomic number of 5 and is in the third column and second row of the periodic table. Its electronic configuration has a completed first shell ($1s$) and three electrons in the second shell. Its configuration is thus $1s^2, 2s^2, 2p^1$.

(b) Metallic magnesium has an atomic number of 12. It is in the second column of the third row of the periodic table and, in its neutral state, has the same number of electrons as protons in its nucleus, hence 12. The electronic configuration is $1s^2, 2s^2, 2p^6, 3s^2$.

(c) Divalent magnesium ion (Mg^{2+}) has two fewer electrons than metallic magnesium. In accord with the reasoning in part b, its electronic configuration is $1s^2, 2s^2, 2p^6$.

(d) Elemental phosphorus appears in the third row and fifth column of the periodic table, below nitrogen, and has an atomic number of 15. Its electronic configuration has filled first and second shells, with five electrons in the third shell. It is therefore $1s^2, 2s^2, 2p^6, 3s^2, 3p^3$.

(e) Sulfur appears in the third row of the periodic table beneath oxygen and has an atomic number of 16. S^{2-} has two more electrons than required for neutrality and thus a total of 18 electrons. With this number of electrons, the first three shells are filled, with an electronic configuration equivalent to that of argon: $1s^2, 2s^2, 2p^6, 3s^2, 3p^6$.

Ex 1.2 To solve this Exercise, you must determine how many electrons are required for a completely filled valence shell in each ion. For the cations, one or more electrons must be removed to obtain a configuration with a filled shell, whereas for the anions, one or more electrons must be added to the neutral molecule in order to fill the valence shell.

(a) The configuration of H^- is $1s^2$ and the valence shell is already filled.

(b) The configuration of Ca^{2+} is $1s^2, 2s^2, 2p^6, 3s^2, 3p^6$, containing two fewer electrons than a neutral calcium atom.

(c) H$^+$ has one fewer electron than a neutral hydrogen atom. Hence, it has no electrons and does not have an incompletely filled valence shell. A completely filled valence shell for a first-row element (hydrogen or helium) would have two electrons. With hydrogen, the presence of two electrons would produce an anion H$^-$, reflecting the extra electron beyond that required to charge-balance the number of protons in the hydrogen nucleus.

(d) Mg$^+$ has one fewer electron than a magnesium atom. Hence, it has one s electron (compare with Exercise 1.1) more than required for a filled second-row valence shell ($1s^2$, $2s^2$, $2p^6$, $3s^1$). It is easier to lose one electron than to take on seven (to achieve a filled third-row valence shell). Losing this one electron will produce Mg^{2+}, which is the stable oxidation state for ionic magnesium.

(e) Chloride has one more electron than a chlorine atom: hence, $1s^2$, $2s^2$, $2p^6$. This is a filled second-row valence shell and it is necessary neither to add nor to remove any electrons.

Ex 1.3 The position of atoms in a molecule (and therefore its three-dimensional shape) is dictated by electron-electron repulsion between valence electrons. Thus, the atoms adopt positions so as to maximize separation of centers of electron density. With two substituents the bond angle is 180°; with three 120°; and with four, 109.5°. There are small deviations from these ideal angles when these substituents are different.

(a) There are two second-row atoms in CH$_3$OH (methyl alcohol), carbon and oxygen, and four atoms are attached to carbon (three hydrogen atoms and an oxygen atom). Thus, the bond angles will be approximately tetrahedral, and both the H—C—H and the H—C—O bond angles will be about 109.5°.

(b) The two carbon atoms in H$_2$C=CH$_2$ (ethene) are identical and each has three substituents. Thus, the bond angles are approximately 120°.

(c) The CH$_3$ group of H$_3$C—C≡CH (propyne) bears four substituents (three hydrogen atoms and one carbon atom) and therefore has tetrahedral geometry with bond angles (H—C—H and H—C—C) of about 109.5°. The two remaining carbon atoms are bound only to each other and one additional substituent; thus, the bond angles for these carbon atoms (C—C≡C and C≡C—H) are 180°.

Ex 1.4 Bond angles are determined primarily by the valence electrons, which includes both bonding and nonbonding electrons.

(a) Although the oxygen atom in water is attached to only two hydrogen atoms, there are four pairs of valence electrons on this atom—two pairs in covalent O—H bonds and two additional pairs of electrons that do not participate in bonding. Likewise, there are four pairs of electrons in the valence shell of nitrogen in ammonia—three pairs in the three N—H bonds and one additional pair of nonbonded electrons. All electrons in the valence shell (not just the bonding electrons) must be considered. With four electron pairs available to the second-row atom in both compounds, both molecules will have roughly tetrahedral geometry about the heteroatoms.

(b) The electrons in both water and ammonia are arranged in a tetrahedron to minimize electron-electron repulsion. Thus, the bond angles in both compounds are approximately 109.5°.

(c) The bond angles (H—N—H) in ammonia are smaller than 109.5° because the four pairs of electrons in the valence shell are not identical. The pair of electrons not used in bonding is closer to the nucleus and, as a result, repulsion between this pair and the three pairs of electrons used in bonding is greater than that between any two bonding pairs of electrons. The bonding pairs are thus pushed slightly closer to each other and further from the pair not involved in bonding.

The situation is similar for the oxygen atom of water: there is less repulsion between the two bonding pairs of electrons than either between one of the bonding pairs with one of the non-bonding pairs or between the non-bonding pairs with each other. The two pairs of electrons not used in bonding push the two hydrogen atoms closer together, making the H—O—H angle somewhat smaller than the tetrahedral angle.

Ex 1.5 (a) Because electronegativity increases in the progression from left to right along a row of the periodic table, oxygen is more electronegative than nitrogen. Oxygen therefore attracts the two electrons in the O—H covalent bond more strongly than nitrogen attracts those in an N—H bond. Hence, the O—H bonds in HOH are more polar than the N—H bonds in NH_3.

(b) Fluorine, on the right-hand side of the periodic table, is appreciably more electronegative than hydrogen on the left-hand side; thus, the carbon–fluorine bond is considerably more polar. In fact, carbon and hydrogen have very similar electronegativities and a C—H bond is therefore nonpolar.

(c) For the reasons given in part *a*, the carbon–oxygen bond is more polar than a carbon–nitrogen bond.

(d) Electronegativity decreases from top to bottom of a column of the periodic table, and oxygen is more electronegative than sulfur. A carbon–oxygen bond is correspondingly more polar than a carbon–sulfur bond. However, the larger sulfur atom is more polarizable and responds more to some polar reagents, despite its lower electronegativity.

(e) Electronegativity increases from left to right along a row in the periodic table and decreases from top to bottom along a column. Thus, although oxygen is found to the left of the halogens in the periodic table, it is still more electronegative than bromine, and, as a result, the carbon–oxygen bond is more polar.

Ex 1.6 In drawing a Lewis dot structure, the valence electrons of each atom are shared between atoms so as to provide access to a filled valence shell (two for hydrogen; eight for all second-row elements).

Ex 1.7 One calculates formal charge by comparing the number of valence electrons of the neutral element with the sum of the number of unshared electrons and half the number of shared electrons available to the atom.

(a) carbon at left as drawn: $4 - [0 + (8\div2)] = 0$; carbon in center: $4 - [0 + (8\div2)] = 0$; nitrogen: $5 - [2 + (6\div2)] = 0$; hydrogens: $1 - [0 + (2\div2)] = 0$.

(b) carbon: $4 - [0 + (8\div2)] = 0$; singly bound oxygens: $6 - [6 + (2\div2)] = -1$; doubly bound oxygen: $6 - [4 + (4\div2)] = 0$.

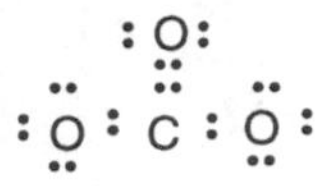

$$\overset{\displaystyle :O:}{\underset{}{:O:C:O:}}$$

(c) left oxygen: $6 - [4 + (4 \div 2)] = 0$; center oxygen: $6 - [2 + (6 \div 2)] = +1$; right oxygen: $6 - [6 + (2 \div 2)] = -1$.

$$\overset{..}{O}::\overset{..}{O}:\overset{..}{\underset{..}{O}}:$$

(d) methyl carbons: $4 - [0 + (8 \div 2)] = 0$; carbonyl carbon: $4 - [0 + (8 \div 2)] = 0$; carbonyl oxygen: $6 - [4 + (4 \div 2)] = 0$; hydrogens: $1 - [0 + (2 \div 2)] = 0$.

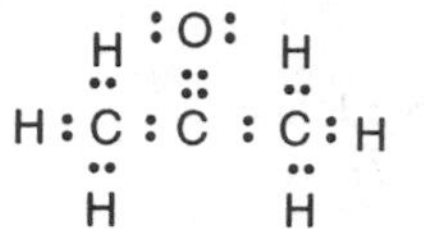

$$\begin{array}{ccc} & \text{H} & :\text{O}: & \text{H} \\ & .. & :: & .. \\ \text{H}:&\text{C}:\text{C}&:\text{C}:&\text{H} \\ & .. & & .. \\ & \text{H} & & \text{H} \end{array}$$

(e) nitrogen: $5 - [0 + (8 \div 2)] = +1$; carbon: $4 - [0 + (8 \div 2)] = 0$; hydrogens: $1 - [0 + (2 \div 2)] = 0$.

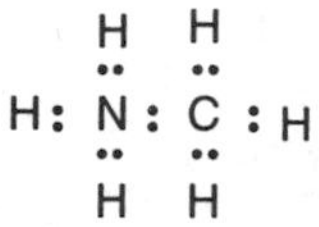

$$\begin{array}{cc} \text{H} & \text{H} \\ .. & .. \\ \text{H}:\text{N}:&\text{C}:\text{H} \\ .. & .. \\ \text{H} & \text{H} \end{array}$$

Ex 1.8   The hydrogen atoms on the two methyl groups of ethane change relative positions by rotation about the bond between the two carbon atoms. Two extremes are shown below. No such change is possible for methane.

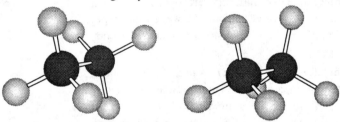

Ex 1.9   There are many possible correct answers to this exercise. One arbitrarily chosen set is shown.

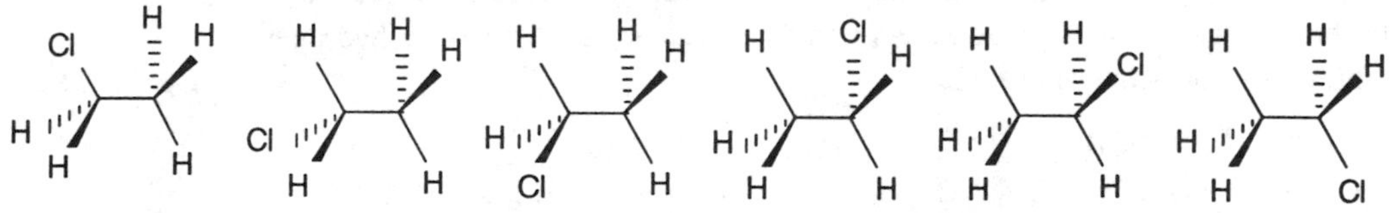

Ex 1.10 Although naming of compounds is given later in the chapter and names are not needed to complete the exercise, correct names for each of the isomeric hydrocarbons drawn in this exercise are given to help you when you review this material later.

(a) The formula C_5H_{12} corresponds to a formula C_nH_{2n+2}. Thus, this structure is a simple alkane. Construct the isomers in a systematic way by first arranging all five carbons in a line.

Then remove one carbon from the end and attach it to the second carbon of the chain. (Attaching it to the other end would produce the same linear chain of five carbon atoms.) Because carbons 2 and 3 of a linear four-carbon chain (that is, n-butane) are the same, exactly the same structure would be obtained by attaching the methyl group to C-3.

Pentane

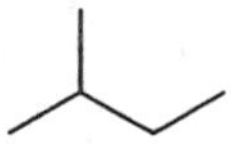

2-Methyl-
butane

The final isomer is obtained by removing two carbons from the five-carbon linear chain, leaving a three-carbon linear chain and two one-carbon fragments (methyl groups). These two carbons are then attached to C-2. If either of these methyl groups were added to C-1 or C-3, the C_4 linear arrangement would be obtained and we would have the same arrangement described just above. Instead of attaching two one-carbon fragments to C-2, we could have attached a single two-carbon fragment to C-2. However, this arrangement is identical with 2-methylbutane. (Use your model kit to convince yourself that this is so.)

2,2-Dimethyl-
propane

(b)   As in part *a*, the overall formula C_6H_{14} corresponds to the general formula for an alkane in which $n = 6$. This is an acyclic alkane. Begin by building a structure that has six atoms in a row. Then change the structure by moving one methyl group to C-2 or C-3 of the chain. Attachment of this methyl group to the first carbon of the chain reproduces the original C_6 straight-chain hydrocarbon, not a structural isomer.

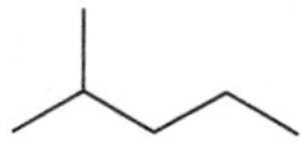

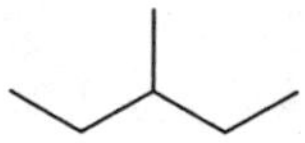

Hexame   2-Methylhexane   3-Methylhexane

Removing two carbons results in a linear C_4 hydrocarbon skeleton to which we add two methyl groups, either both at C-2 or one at C-2 and one at C-3. (Adding the two carbons as an ethyl group to C-2 of the butane skeleton would produce the third structure shown above, 3-methylpentane.)

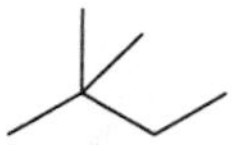

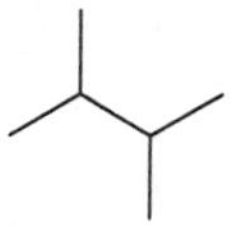

2,2-Dimethylbutane   2,3-Dimethylbutane

(c)   The formula C_7H_{16} also corresponds to an acyclic alkane. To find all the isomers, we follow the procedure described in part *b*. The first isomer has seven carbons arranged in a line. The next has six carbons arranged in a line with a single methyl group attached at various positions along the chain. However, the molecule is symmetrical and there are only two unique ways to add the additional carbon: to C-2 or to C-3 (C-4 is equivalent to C-3 and C-5 is equivalent to C-2).

Heptane   2-Methylhexane   3-Methylhexane

Removing an additional carbon from the chain results in a longest chain of five carbons, to which two methyl groups are attached in one of four ways: both to C-2 (the same as attaching both to C-4); one to C-2 and one to C-3; one to C-2 and one to C-4; or both to C-3. Attachment of a two-carbon fragment (an ethyl group) to C-3 of the five-carbon fragment results in a different structure.

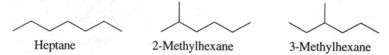

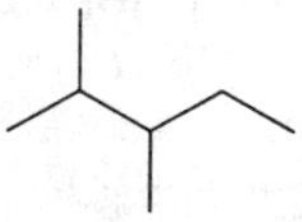

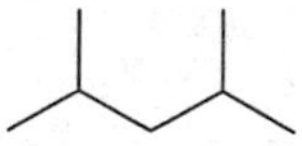

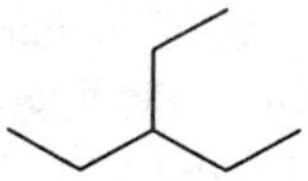

2,2-Dimethyl-   2,3-Dimethyl-   2,4-Dimethyl-   3,3-Dimethyl-   3-Ethyl-
pentane   pentane   pentane   pentane   pentane

Removing one more carbon leaves a four-carbon chain, to which three carbons are attached to C-2 and C-3. If these three carbons were attached as a three-carbon chain to either C-2 or C-3, we would obtain a six-carbon fragment with a single methyl bound to C-3. If a two-carbon fragment (ethyl group) were attached to C-2 or C-4 of the four-carbon backbone, a longer C_5 backbone would result in one of the isomers already described. The only new structure results from attaching two carbons to C-2 and one to C-3.

2,2,3-Trimethylbutane

Notice that as the number of carbon atoms in the alkane increases, the number of possible isomers also increases: there are three isomers for C_5H_{12}, five for C_6H_{14}, and nine for C_7H_{16}.

Ex 1.11 The number of rings in hydrocarbons containing only σ bonds can be deduced from the molecular formula by noting the difference between the given formula and the formula expected for a saturated acyclic hydrocarbon: C_nH_{2n+2}.

(a) For a C_5 hydrocarbon, an acyclic structure has a formula of C_5H_{12}. The formula C_5H_{10} lacks two hydrogens from that calculated, and is consistent with the presence of one ring. Because there are only five carbon atoms, the largest ring possible is cyclopentane and the smallest ring is cyclopropane. Cyclopentane accounts for all the carbon atoms present. With only four carbons in the ring, one more is attached as a substituent (as a methyl group) and there is only one way to do this.

Cyclopentane   Methylcyclobutane

We must add two additional carbons to a three-membered ring, and we can do this on the same carbon, on adjacent carbons, or as a two-carbon fragment at one carbon atom. The cyclopropane ring has two faces, and so the two methyl groups can be oriented either on the same side (the *cis* isomer) or on opposite sides (the *trans* isomer) of the ring.

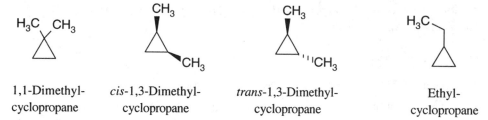

1,1-Dimethyl-   *cis*-1,3-Dimethyl-   *trans*-1,3-Dimethyl-   Ethyl-
cyclopropane   cyclopropane   cyclopropane   cyclopropane

(b) C_5H_8 differs from C_nH_{2n+2} by four hydrogen atoms. This tells us that there must be two rings present. (Nomenclature for most compounds containing more than one ring is beyond the scope of a one-year course.)

There are several ways to formulate two rings. First, it is possible to fuse two rings together, so that two adjacent atoms are common to each ring. In principle, the hydrogens at the bridge positions (atoms common to both rings) can be either *cis* or *trans*. These rings can be a four-membered ring fused to a three-membered ring or a three-membered ring fused to a three-membered ring bearing a single methyl substituent, with *cis* and *trans* isomers possible. It is extremely difficult to force a *trans* ring fusion in a three- or four-membered ring, however, and structures including a *trans* ring junction with fused three- and four-membered rings are omitted. It is also possible to form two rings with two nonadjacent atoms in common. Finally, it is possible to have two rings that have only one atom in common.

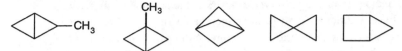

(c) C_6H_{12} differs from the formula expected for a C_6 acyclic alkane (C_6H_{14}) by two hydrogen atoms. Thus, C_6H_{12} contains one ring. Structures meeting this requirement can be an unsubstituted six-membered ring; a methyl-substituted five-membered ring; five different four-membered rings substituted with two methyl groups and one such ring substituted with a single ethyl group; three trimethyl- substituted three-membered rings; three isomeric ethyl, methyl-substituted three-membered rings; and two three-membered rings bearing either a branched three-carbon substituent (an isopropyl group) or a straight-chain three-carbon substituent (an *n*-propyl group).

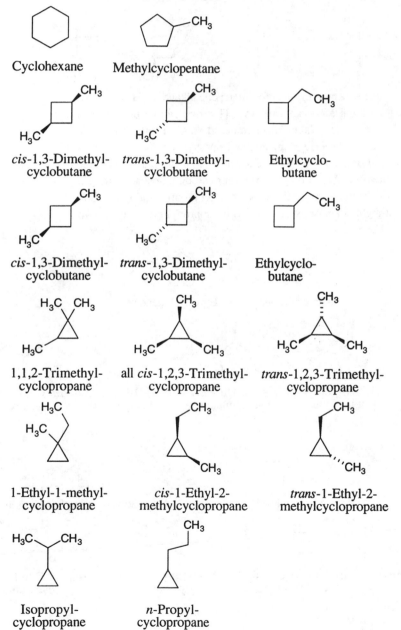

Cyclohexane   Methylcyclopentane

cis-1,3-Dimethyl-   *trans*-1,3-Dimethyl-   Ethylcyclo-
cyclobutane   cyclobutane   butane

cis-1,3-Dimethyl-   *trans*-1,3-Dimethyl-   Ethylcyclo-
cyclobutane   cyclobutane   butane

1,1,2-Trimethyl-   all *cis*-1,2,3-Trimethyl-   *trans*-1,2,3-Trimethyl-
cyclopropane   cyclopropane   cyclopropane

1-Ethyl-1-methyl-   *cis*-1-Ethyl-2-   *trans*-1-Ethyl-2-
cyclopropane   methylcyclopropane   methylcyclopropane

Isopropyl-   *n*-Propyl-
cyclopropane   cyclopropane

(d) C_6H_{10} differs from the expected formula for an acyclic alkane in lacking four hydrogens. This means that the compound must contain two rings. In addition to methylated analogs (including *cis-trans* isomers of the compound in part *b*), one can also have additional fused (a three-membered ring fused to a five-membered ring) structures and a bridged structure. As in part *b*, names are not included for these bicyclic compounds.

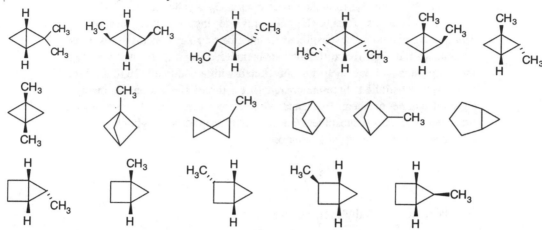

(e) C_7H_{14} differs from the expected formula for an acyclic alkane (C_7H_{16}) by two hydrogens, indicating the presence of one ring. This ring can be present as a seven-membered ring; a methylated six-membered ring; an ethylated or a dimethylated five-membered ring; a propylated or isopropylated four-member-ed ring; a four-membered ring containing one ethyl and one methyl group or three methyl groups; a three-membered ring bearing four carbon atoms as a straight chain or as various isomeric arrangements, as a methyl group and three carbon atoms either as a straight chain or as an *n*-propyl or isopropyl group, or as two ethyl groups.

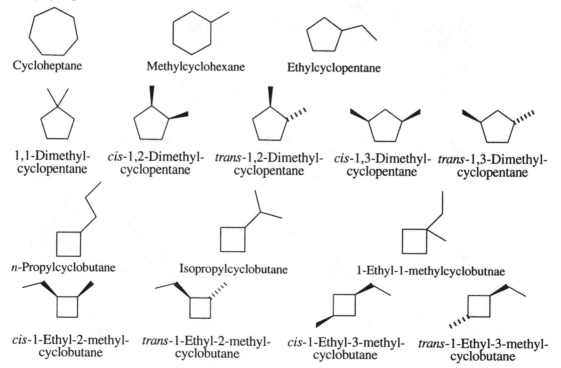

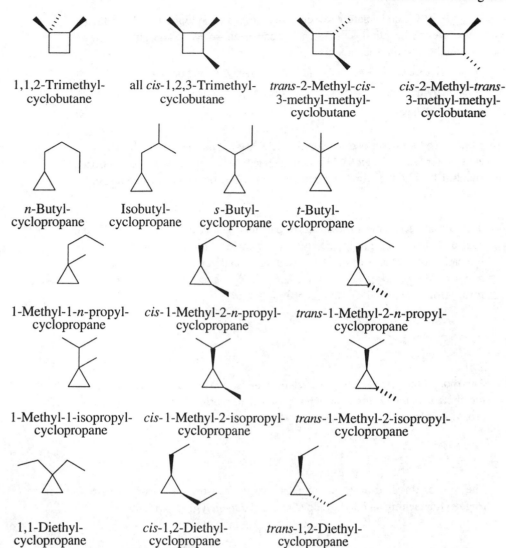

1,1,2-Trimethyl-cyclobutane   all *cis*-1,2,3-Trimethyl-cyclobutane   *trans*-2-Methyl-*cis*-3-methyl-methyl-cyclobutane   *cis*-2-Methyl-*trans*-3-methyl-methyl-cyclobutane

n-Butyl-cyclopropane   Isobutyl-cyclopropane   *s*-Butyl-cyclopropane   *t*-Butyl-cyclopropane

1-Methyl-1-*n*-propyl-cyclopropane   *cis*-1-Methyl-2-*n*-propyl-cyclopropane   *trans*-1-Methyl-2-*n*-propyl-cyclopropane

1-Methyl-1-isopropyl-cyclopropane   *cis*-1-Methyl-2-isopropyl-cyclopropane   *trans*-1-Methyl-2-isopropyl-cyclopropane

1,1-Diethyl-cyclopropane   *cis*-1,2-Diethyl-cyclopropane   *trans*-1,2-Diethyl-cyclopropane

Ex 1.12 To become more familiar with line notation for representing structures, specifically draw in all the C—H bonds implied in each of these structures. Then you can simply count the number of carbon and hydrogen atoms present. By comparing the formula obtained in this way with that expected for an acyclic alkane (C_nH_{2n+2}), you can obtain the number of rings as the difference between the number of hydrogen atoms divided by 2.

(a) C_6H_{12}: $(14 - 12) \div 2 =$ one ring.

(b) C_9H_{16}: $(20 - 16) \div 2 =$ two rings.

(c) C_7H_{12}: $(16 - 12) \div 2 =$ two rings.

(d) C_8H_8: $(18 - 8) \div 2 =$ five rings.

Ex 1.13 Recognizing that the strain energy of cyclopropane is constrained to three carbon atoms while that for cyclobutane is constrained to four, one can see that the strain energy per carbon is higher in cyclopropane (27.6 kcal/mole = 9.2 kcal/mole per carbon) than in cyclobutane (26.4 kcal/mole = 6.6 kcal/mole per carbon).

Ex 1.14 To use the IUPAC nomenclature system, we first locate the longest chain and name branching groups as alkyl substituents, numbering so as to assign the lowest possible numbers to the substituents.

(a) 3-methylhexane

(b) 4-isopropylheptane

(c) s-butylcyclohexane

(d) 4,5-dimethyl-6-ethyldecane

(e) t-butylcyclopropane

Ex 1.15 Using the procedure described in Exercise 1.14, we obtain the names shown in the answers to Exercises 1.10 and 1.11. As mentioned earlier, the naming of multi-ring compounds is beyond the scope of a one-year course and no names are given here for such structures.

Ex 1.16 (a) To draw an acceptable structure corresponding to an IUPAC name, identify the root and functional group—in this case, octane—and draw a straight chain indicative of that structure. Then place three methyl groups at positions 3, 3, and 4 on the chain. (Because the chain is end-to-end symmetrical, it doesn't matter from which end we start numbering.)

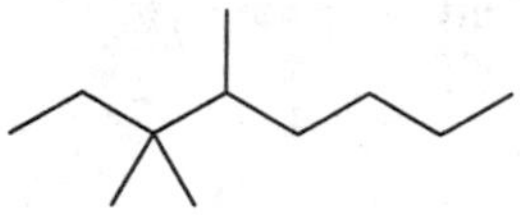

(b) The root of this hydrocarbon is cyclopentane, so we begin by drawing a five-membered saturated ring to which is attached an n-propyl group. Recall that n-propyl is a three-carbon fragment for which the attachment point is at C-1.

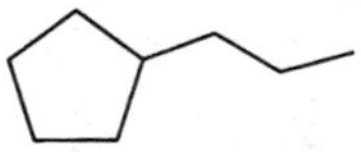

(c) The root of this hydrocarbon is hexane, so we construct a linear six-carbon fragment, attaching an ethyl group to C-3 and a methyl group to C-2.

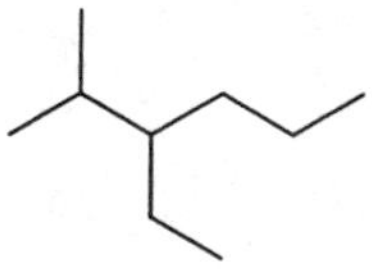

(d) As in part b, the root of this hydrocarbon is cyclopentane. In addition, there are two methyl groups located on adjacent carbons (1,2-) and on the same side of the ring (cis).

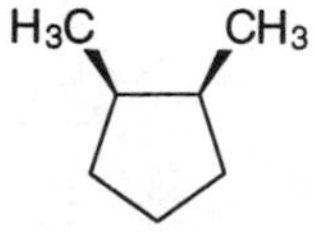

H_3C CH_3

(e) The root of this hydrocarbon is cyclohexane, to which two methyl groups are attached in a 1,4- relation and on opposite sides of the ring.

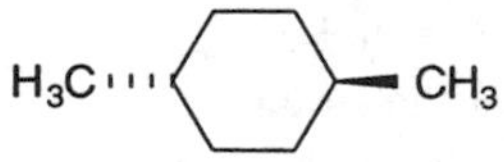

H_3C ''' CH_3

Ex 1.17 (a) The heat required to warm the water is obtained by multiplying 1000 g of water by its heat capacity. This amount of heat is equivalent to that released by burning the hydrocarbon. It can be calculated by multiplying the heat of combustion by the number of moles of hydrocarbon. This requires conversion of the 1 g to a molar equivalent (that is, by dividing by the molecular weight of C_3H_6 = 42 g/mole) and completion of the mathematics as follows:.

$$(1000 \text{ g}) \left(\frac{1 \text{ cal}}{\text{g °C}}\right) (12 \text{ °C}) = 12{,}000 \text{ cal}$$

$$12{,}000 \text{ cal} = \left(\text{heat of combustion } \frac{\text{kcal}}{\text{mole}}\right) \left(\frac{1 \text{ mole}}{42 \text{ g}}\right) \left(\frac{1000 \text{ cal}}{\text{kcal}}\right) (1 \text{ g})$$

$$\text{heat of combustion} = 12{,}000 \text{ cal} \left(\frac{1}{\text{g}}\right) \left(\frac{42 \text{ g}}{\text{mole}}\right) \left(\frac{1 \text{ kcal}}{1000 \text{ cal}}\right) = 504 \frac{\text{kcal}}{\text{mole}}$$

Note that the value calculated here (504 kcal/mole) has too many significant figures, given the precision of the data. To report a heat of combustion to three significant figures, an experimenter would have had to have measured the amount of hydrocarbon used (1.00 g), the water volume (1000 mL), and the temperature increase (12.0 °C) each to an accuracy of three significant figures (e.g., 0.1 °C). Such values require very careful measurements! Because one of the values (the number of grams of water) was measured to only one significant figure, the answer should be rounded to a value that also has only one significant figure; that is, to 500 kcal/mole. (Compare your answer with that in Table 1.5 for cyclopropane. Does this measurement allow you to say which C_3H_6 isomer was used in the experiment?)

(b) The analysis is identical with part *a*, with a revised molecular weight for C_6H_{12}.

$$(250 \text{ g}) \left(\frac{1 \text{ cal}}{\text{g °C}}\right) (45 \text{ °C}) = 11{,}250 \text{ cal}$$

$$11{,}250 \text{ cal} = \left(\text{heat of combustion } \frac{\text{kcal}}{\text{mole}}\right) \left(\frac{1 \text{ mole}}{84 \text{ g}}\right) \left(\frac{1000 \text{ cal}}{\text{kcal}}\right) (1 \text{ g})$$

$$\text{heat of combustion} = 11{,}250 \text{ cal} \left(\frac{1}{1 \text{ g}}\right) \left(\frac{84 \text{ g}}{\text{mole}}\right) \left(\frac{1 \text{ kcal}}{1000 \text{ cal}}\right) = 945 \frac{\text{kcal}}{\text{mole}}$$

Again, the same reservations about data precision as those in part *a* apply.

Answers to Review Problems

Pr 1.1 The identity of a chemical element is given by the number of protons in the nucleus. The magnitude of the positive charge on a cation defines the number of electrons missing from those required to offset the nuclear charge in a neutral atom, whereas the magnitude of the negative charge of an anion defines the number of extra electrons present. In a neutral atom, the number of protons is equal to the number of electrons. Thus, by counting the number of electrons present in each example and adjusting for charge, the atomic number of the atom or ion is defined.

(a) Adding the number of electrons in the $1s, 2s$, and $2p$ orbitals (that is, the superscripts in the problem) gives ten electrons. The +1 charge of a monocation requires one more electron for charge neutralization to produce a neutral atom: this element must therefore have an atomic number of 11. From the periodic table, we locate this element as Na^+.

(b) The same electron configuration with ten electrons as a dication requires an atomic number of 12: hence, Mg^{2+}.

(c) A dianion with 18 electrons must have two fewer protons in its nucleus. The element with atomic number 16 is sulfur: hence, S^{2-}.

(d) A neutral atom with ten electrons has an equal number of protons: hence, Ne.

Pr 1.2 A bond is polar when the two participating atoms have modestly different electronegativities. A bond is nonpolar when an electron pair is shared between atoms of very similar or identical electronegativities. A bond is ionic bond when the two associated atoms differ so significantly in electronegativity that the electron-rich atom (an anion) has no tendency to share its electrons with the electron-deficient atom (a cation), so that the resulting ions are associated through electrostatic attraction rather than a sharing of electrons.

(a)   Nonpolar: C and H have nearly identical electronegativities.

(b)   Polar: O is more electronegative than C.

(c)   Ionic: Na^+ and RO^- are stable ions whose electronegativities are quite different.

(d)   Nonpolar: the C—H bond is nonpolar (see part *a*), despite the fact that the molecule is polar by virtue of the presence of a polar C—O bond at another position in the molecule.

(e)   Polar: S is more electronegative than C.

(f)   Polar: Br is more electronegative than C.

Pr 1.3   To calculate formal charge, compare the number of valence electrons for the neutral element with the sum of half of the number of shared electrons and the number of unshared first electrons for the atom in the molecule. To obtain these latter numbers, draw a Lewis dot structure. A Lewis dot structure has the sum of the number of valence electrons for each element, adjusted for the net charge, distributed to provide access to an octet of electrons (or two electrons for each H) to each atom.

(a) BF_4^- has 32 electrons: seven from each F, three from B, and one from the (minus one) charge. Boron has three valence electrons. Its formal charge is:

$$3 - [0 + (8 \div 2)] = -1.$$

Fluorine has seven valence electrons. Its formal charge is:

$$7 - [6 + (2 \div 2)] = 0.$$

(b) NH_4^+ has eight electrons in a Lewis dot representation: five from nitrogen, one from each hydrogen, less one for the (plus one) charge. Nitrogen has five valence electrons. Its formal charge is:

$$5 - [0 + (8 \div 2)] = +1.$$

Each hydrogen has one valence electron. Hydrogen's formal charge is:

$$1 - [0 + (2 \div 2)] = 0.$$

(c) CH_4 has eight electrons in a Lewis dot representation: four from carbon and one from each hydrogen. Carbon has four valence electrons. Thus, its formal charge is:

$$4 - [0 + (8 \div 2)] = 0.$$

Each hydrogen has one valence electron. Hydrogen's formal charge is:

$$1 - [0 + (2 \div 2)] = 0.$$

(d) H_3O^+ has eight electrons in a Lewis dot representation: six from oxygen, one from each hydrogen, less one for the (plus one) charge. Oxygen has six valence electrons. Thus, its formal charge is:

$$6 - [2 + (6 \div 2)] = +1.$$

Each hydrogen has one valence electron. Hydrogen's formal charge is:

$$1 - [0 + (2 \div 2)] = 0.$$

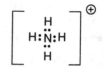

(e) H_2 has two electrons in a Lewis dot representation: one from each hydrogen. Hydrogen has one valence electron. Thus, its formal charge is:

$$1 - [0 + (2 \div 2)] = 0.$$

Pr 1.4 To draw a Lewis dot structure, begin by assembling the atoms in the correct bond sequence. Then place two electrons for each bond between the bonded atoms. Finish by adding lone pairs of electrons as needed so that each atom has an octet of electrons, if possible.

(a) Four valence electrons are available from carbon and six from oxygen, for a total of ten. (Carbon monoxide represents an unusual bonding arrangement as both carbon and oxygen have three bonds, one fewer for carbon and one more for oxygen than normal. The two atoms also bear unusual formal charge.)

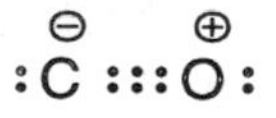

(b) Four valence electrons are available from carbon and six from each of two oxygen atoms, for a total of 16.

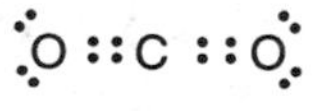

(c) Four valence electrons are available from each of two carbons and one from each of two hydrogen atoms, for a total of ten.

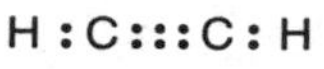

(d) Four valence electrons are available from carbon and six from each of three oxygen atoms, with two extra electrons to accommodate the negative charge, for a total of 24.

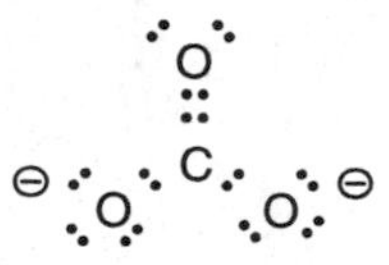

(e) Four valence electrons are available from carbon, one from each of the hydrogens, and six from oxygen, for a total of 12.

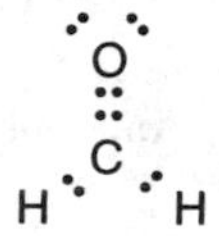

Pr 1.5 The same procedure is followed as in Problem 1.4.

(a) Each carbon contributes four valence electrons (3×4) and each hydrogen one valence electron (6×1), for a total of 18 electrons.

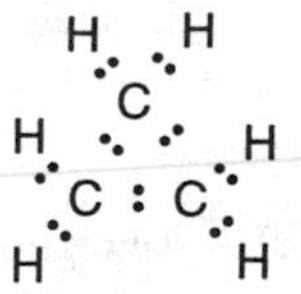

(b) Each carbon contributes four valence electrons (3×4) and each hydrogen one valence electron (8×1), for a total of 20 electrons.

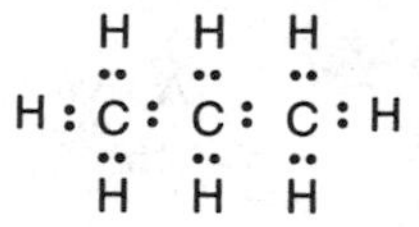

(c) Each carbon contribute four valence electrons (3×4), each hydrogen atom one valence electron (7×1), and the chlorine atom has seven, for a total of 26 electrons.

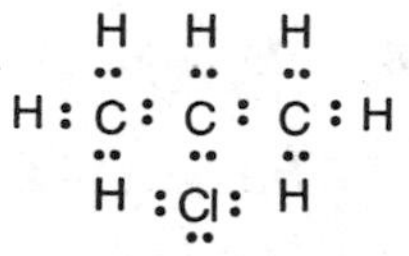

(d) Nitrogen contributes five valence electrons and each hydrogen one (4×1), with one electron having been removed to accommodate the +1 charge, for a total of eight electrons.

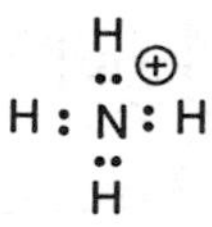

Pr 1.6 As in Problem 1.3, formal charge is calculated at each atom by comparing the number of valence electrons with the sum of the number of nonbonded electrons and half of the bonded electrons to which the atom has access.

(a) Each hydrogen, with one valence electron, has access to two bonded electrons:

H: $1 - (0 + 2 \div 2) = 0$

C: $4 - (0 + 8 \div 2) = 0$

O: $6 - (6 + 2 \div 2) = -1$

S: $6 - (0 + 8 \div 2) = +2$

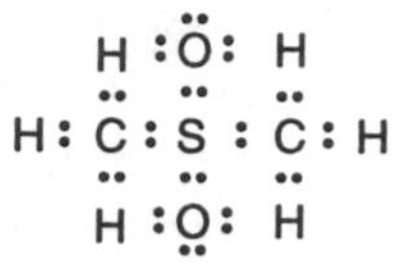

H : Ö : H
H : C : S : C : H
H : Ö : H

(b) H: $1 - (0 + 2 \div 2) = 0$

C: $4 - (0 + 8 \div 2) = 0$

O: $6 - (2 + 6 \div 2) = +1$

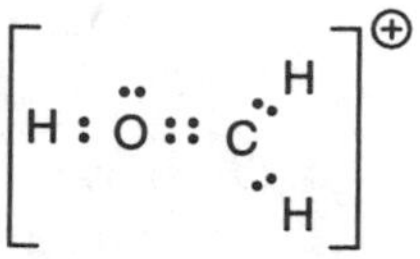

$$\left[\text{H : Ö :: C} \begin{smallmatrix} H \\ \\ H \end{smallmatrix} \right]^{\oplus}$$

(c) H: $1 - (0 + 2 \div 2) = 0$

C: $4 - (0 + 8 \div 2) = 0$

O: $6 - (2 + 6 \div 2) = +1$

$$\left[\begin{smallmatrix} & & H \\ \text{H : Ö : C : H} \\ & & H & H \end{smallmatrix} \right]^{\oplus}$$

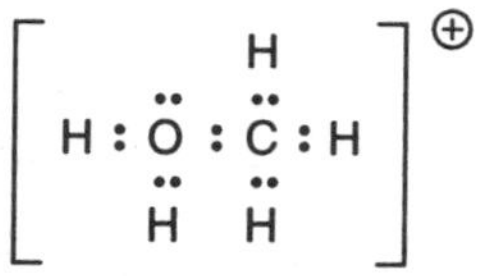

(d) H: $1 - (0 + 2 \div 2) = 0$

C: $4 - (0 + 8 \div 2) = 0$

N: $5 - (2 + 6 \div 2) = 0$

O: $6 - (2 + 6 \div 2) = +1$

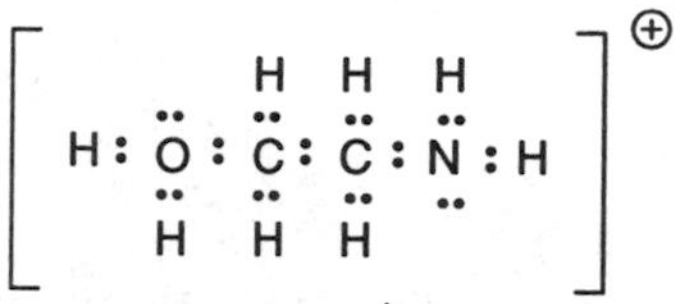

$$\left[\begin{smallmatrix} & H & H & H \\ \text{H : Ö : C : C : N : H} \\ & H & H & H \end{smallmatrix} \right]^{\oplus}$$

Pr 1.7 (a) An *s*-butyl group is bound to C-5 of a linear nine-carbon chain.

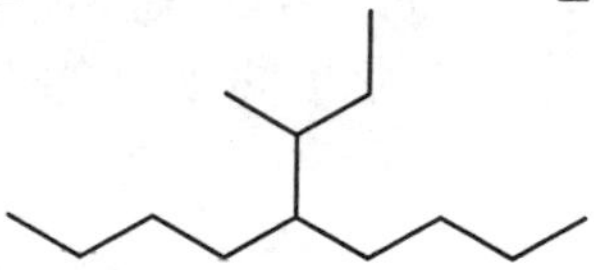

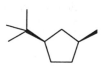

(b) A seven-membered ring is substituted by ethyl groups in a 1,3-relation and directed toward opposite sides of the ring.

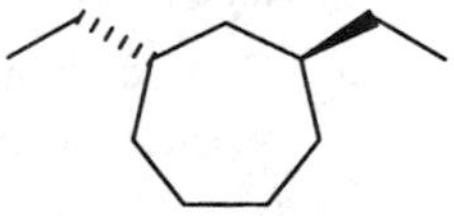

(c) An eight-carbon linear chain with an isopropyl group at C-4.

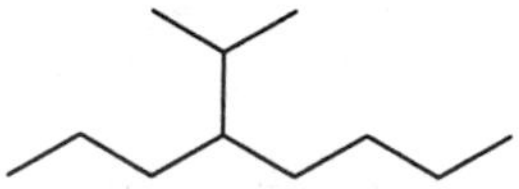

(d) A five-membered ring with *t*-butyl and methyl groups 1,3 and *cis* to each other.

Pr 1.8 The "squiggly" lines indicate that only part of a structure is specified.

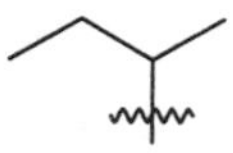

(a) 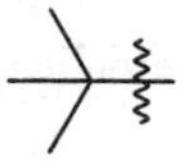   (b)   (c)   (d)

t-butyl   isopropyl   *s*-butyl   ethyl

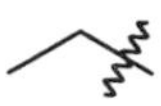

Pr 1.9 (a) *n*-pentane

(b) *s*-butylcyclohexane

(c) 2,2-dimethylbutane

(d) 2,2-dimethylbutane (a second representation of the same isomer as in part *c*)

(e) 3-methyl-4-isopropyl-5-ethylnonane

(f) *cis*-1,2-dimethylcyclobutane

(g) *trans*-1,3-dimethylcyclobutane

(h) *cis*-1-isopropyl-3-methylcyclohexane

(i) 4-*t*-butylheptane

Pr 1.10 Recall from Exercise 1.10, part (a), that there are three skeletal isomers with a formula C_5H_{12}. We begin with these hydrocarbons and consider whether isomers are obtained as each type of hydrogen in each skeleton is successively replaced by a fluorine atom.

   (a)   2,2-dimethylpropane: all hydrogens are equivalent, so there is but one monofluorinated isomer.

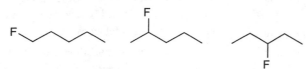

   (b)   *n*-pentane: 1-, 2-, and 3- fluoro derivatives.

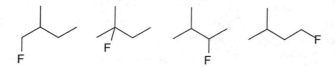

   (c)   2-methylbutane: 1-, 2-, 3-, and 4-fluoro derivatives.

Pr 1.11 The heat of combustion of a linear alkane is higher than that for a branched alkane. The greater the degree of branching, the lower is the heat released upon combustion. Thus, to answer this question it is only necessary to determine which isomer is less branched.

   (a)   heptane   (b)   2-methylbutane   (c)   octane

Pr 1.12 (a)   Compositionally different, with one fewer fluorine, one more carbon and iodine, and two more hydrogens.

   (b)   Structural isomers: here the two fluorines are bound to the same carbon atom instead of adjacent atoms.

   (c)   Same compound: three-dimensional representations differing only by rotations about a single bond.

   (d)   Structural isomers: here the two fluorines are bound to the first and third carbon of the chain instead of to adjacent carbon atoms.

Important New Terms

Acyclic: lacking rings (1.3)

Alkanes: a family of saturated hydrocarbons with the empirical formula C_nH_{2n+2} for acyclic members (1.3)

Alkyl group: a fragment derived from an alkane by removal of one hydrogen (1.5)

Angle strain: the destabilization caused by deformation from normal bonding angles for atoms in a cyclic compound (1.4)

Atomic orbitals: the probability surfaces associated with an atom within which an electron is likely to be found (1.2)

Ball-and-stick model: three-dimensional representation of a molecule in which bonds are indicated by lines and atoms by spheres

Bond angle: the angle formed by two bonds intersecting at an atom; varies with hybridization: typically about 109.5° at an sp^3-hybridized atom, 120° at an sp^2-hybridized atom, and 180° at an sp-hybridized atom (1.2)

Bond length: the equilibrium distance between two covalently bonded atoms; varies with hybridization: 1.54 Å between sp^3-hybridized carbon atoms in ethane (1.2)

***n*-Butyl group**: $-(CH_2)_3CH_3$, an unbranched four carbon alkyl group attached through the primary carbon (1.5) *s*-**Butyl group**: $-CH(CH_3)CH_2CH_3$, a four-carbon alkyl chain attached through the secondary carbon (1.5)

t-**Butyl group**: $-C(CH_3)_3$, a four-carbon alkyl group attached through the tertiary carbon (1.5)

Calorimeter: a device with which the heat released or consumed in a chemical reaction can be accurately measured (1.6)

Chemical bond: an energetically favorable interaction between two atoms induced by a pair of electrons mutually attracted to both nuclei or by electrostatic attraction between two ions (1.2)

cis **isomer**: designation of the relative positioning of two groups on the same side of a bond about which rotation is restricted (1.5)

Combustion: burning in air (1.6)

Conformational isomers: stereoisomers that are inter-converted by rotation about covalent bonds (1.3)

Connectivity: representation of the attachments of atoms in a molecule (1.2)

Cyclic: containing one or more rings (1.4)

Cycloalkanes: saturated hydrocarbons containing one or more rings with the empirical formula in which two hydrogens per ring are subtracted from the formula of an acyclic alkane (C_nH_{2n+2}) (1.4)

Cyclopropane: C_3H_6; the simplest cycloalkane (1.4)

Electron configuration: an atomic-orbital description of the electrons associated with a given atom, listing the principal quantum number, the hydrogenic orbital type, and the number of electrons occupying the suborbital (1.2)

Electronegativity: the tendency of an atom to attract electrons, thus polarizing a covalent bond (1.2)

Electrostatic attraction: the favorable interaction between two species of opposite charge (1.2)

Electrostatic repulsion: the unfavorable interaction between two species of like charge (1.2)

Empirical formula: quantitative description, in smallest whole numbers, of the relative proportion of elements present in a compound (1.3)

Ethane: C_2H_6; the simplest saturated hydrocarbon containing a C—C bond (1.3)

Formal charge: a construct used to describe electron distribution in a molecule by comparing the number of valence electrons in a neutral atom with the sum of the number of unshared electrons plus half the number of shared electrons available to that atom; difference between the number of electrons accessed by an atom in a molecule and in its elemental state (1.2)

Free rotation: the motion attained when orbital overlap is unaffected by rotation about the internuclear axis of a sigma bond (1.2)

Hatched line: a graphic representation in a three-dimensional structure indicating a group positioned away from the observer (1.2)

Heat of combustion: the heat released when one mole of a compound is completely oxidized to CO_2 and H_2O (1.6)

Heat of formation: a theoretical description of the energy that would be released if a molecule were formed from its component elemental atoms in their standard states (1.6)

Hund's rule: when possible, electrons singly occupy orbitals of identical energy (1.2)

Hybrid orbitals: the orbitals formed by mixing hydrogenic (*s,p,d*, etc.) atomic orbitals (1.2)

Hydrocarbons: compounds that contain only carbon and hydrogen (1.3)

Hydrogenic atomic orbitals: the atomic orbitals calculated precisely for hydrogen, including spherical *s* orbitals, propeller-shaped *p* orbitals, dumbbell-shaped *d* orbitals, more complex shaped *f* orbitals, etc. (1.2)

Inert gas: an atom that does not readily enter into chemical bonding with other atoms because the valence electron shell is filled; found at the far right column of the periodic table (1.2)

Ionic bond: an attractive electrostatic association between two oppositely charged ions (1.2)

Isobutyl group: $-C(CH_3)_3$, attached through the tertiary carbon (1.5)

Isomers: different structural arrangements constituted from the same atoms (1.3)

Isopropyl group: $-CH(CH_3)_2$, a branched three-carbon alkyl group attached through a secondary carbon (1.5)

IUPAC rules: a set of procedures for naming organic compounds; a root word describes the number of backbone carbon atoms, a suffix defines the functional group, and a prefix gives the position of each substituent (1.5)

Lewis dot structure: a representation in which electrons available to a given atom are indicated either as a non-bonded lone pair (by a pair of dots) or as a shared bonding pair (as a pair of dots between two atoms) (1.2)

Methane: CH_4, the simplest hydrocarbon, composed of carbon surrounded by four hydrogens (1.3)

Molecular formula: description of the number of each type of atom present in a molecule (1.3)

Nodal surface: a position in an atomic or molecular orbital at which electron density is zero (1.2)

Nonpolar covalent bond: a chemical bond characterized by the absence of appreciable partial charge separation because of nearly equal sharing of the electrons in the bond by the two bonded atoms (1.2)

Normal alkane: a straight-chain alkane (1.3)

Orbital: the probability surface describing the volume in which an electron is likely to be found (1.2)

Orbital overlap: the spatial intersection of atomic or hybrid atomic orbitals required for forming a chemical bond (1.2)

Organic chemistry: the chemistry of carbon compounds (1.1)

Pauli exclusion principle: a theoretical statement that each electron must be unique; that is, each must have a distinct set of principal, secondary, azimuthal, and spin quantum numbers; a statement that no more than two electrons, which must have opposite spins, can occupy the same orbital (1.2)

Periodic table: an orderly arrangement of the elements grouped according to their atomic number and electronic configuration (1.2)

Polar covalent bond: a chemical bond characterized by appreciable partial charge separation because of unequal

Solid wedges: the graphic representation in a three-dimensional structure to indicate a group positioned near the observer (1.2)

Space-filling model: a representation indicating van der Waals radii of all component atoms oriented in space (1.2)

sp^3-**Hybrid orbitals**: hybrid orbitals formed by mathematically mixing one s and three p atomic orbitals; the four hybrid orbitals so formed are tetrahedrally dispersed (separated by 109.5°) from the atom's nucleus (1.2)

Structural isomers: isomers in which the carbon backbones differ (1.3)

Tertiary carbon: a carbon atom chemically bonded to only three other carbon atoms (1.5)

Tetrahedral carbon: an sp^3-hybridized carbon atom bearing four substituents directed at 109.5° from each other (1.2)

trans **isomer**: designation of the relative positioning of two groups on opposite sides of a bond about which rotation is restricted (1.5)

sharing of the electrons in the bond between two bonded atoms (1.2)

Polarization: a partial charge separation induced by a difference in electronegativity between carbon and a heteroatom (1.2)

Primary carbon: a carbon atom chemically bonded to only one other carbon atom (1.5)

n-**Propyl group**: $-(CH_2)_2CH_3$, an unbranched three carbon alkyl group attached through the primary carbon (1.5)

Ring strain: the destabilization caused by angle strain and eclipsing interactions in a cyclic compound (1.4)

Saturation: the condition of a compound containing only sp^3-hybridized atoms (1.3)

Sawhorse representation: one employing solid wedges and dashed lines to represent three-dimensional structure (1.2)

Secondary carbon: a carbon atom chemically bonded to only two other carbon atoms (1.5)

Sigma (σ) bond: a covalent chemical bond in which electron density is arranged symmetrically along the axis connecting the two bonding atoms; results from direct overlap of hybrid orbitals having some s character) (1.2)

Valence electrons: electrons occupying an incompletely filled quantum level (1.2)

Valence shell: the outermost atomic shell that typically contains electrons (1.2)

van der Waals attraction: the energetically favorable force resulting from the interaction of the bonded electrons of one molecule and the nuclei of another (1.2)

van der Waals radius: the effective size of an atom (1.2)

van der Waals repulsion: the energetically unfavorable force resulting from the interaction of the bonded electrons of one molecule and those of another, or of the nuclei of one molecule with those of another; repulsive intermolecular dipole-dipole interaction (1.2)

Wedged line: the graphic representation in a three-dimensional structure indicating a group positioned near the observer (1.2)

Alkenes, Aromatic Hydrocarbons, and Alkynes

Key Concepts

Geometry at sp^2- and sp-hybridized carbon atoms

Multiple bonding

Pi bonds

Restricted rotation

Bonding and antibonding molecular orbitals

Cis-trans (Z/E) isomerism

Relative stabilities of isomeric alkenes

Group characteristics of alkenes and alkynes

Index of hydrogen deficiency

Naming alkenes, alkynes, and aromatic hydrocarbons

Classification as primary, secondary, tertiary carbons, alcohols, amines

Conjugation

Hyperconjugation

Resonance

Resonance hybrid structures

Aromaticity

Hückel's rule

Heteroaromatics

Relationship between physical properties and structure

Answers to Exercises

Ex 2.1 (a) This exercise asks that you determine possible isomeric structures with a given molecular formula. Although it does not ask you to name the structures you have drawn, names are given here so that you can return to this exercise later in the Chapter for practice in naming isomers. The formula C_5H_{10} differs by two hydrogen atoms from that of a saturated hydrocarbon and thus has either one double bond or one ring. We therefore begin by constructing possible *cis* and *trans* and positional isomers for a C_5 skeleton. We repeat the process for a methyl-substituted C_4 skeleton and then for structures containing rings.

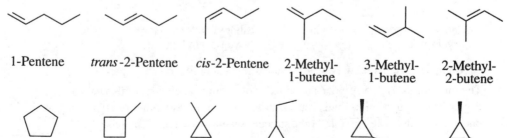

| 1-Pentene | *trans*-2-Pentene | *cis*-2-Pentene | 2-Methyl-1-butene | 3-Methyl-1-butene | 2-Methyl-2-butene |

| Cyclopentane | Methylcyclo-butane | 1,1-Dimethyl-cyclopropane | Ethylcyclo-propane | *cis*-1,2-Dimethyl-cyclopropane | *trans*-1,2-Dimethyl-cyclopropane |

(b) The formula C_5H_8 differs from a saturated hydrocarbon formula by four hydrogen atoms, indicating either two double bonds, one ring and one double bond, or two rings. On a C_5 fragment, we can locate two double bonds at the 1,2-, 1,3-, 1,4-, or 2,3-positions.

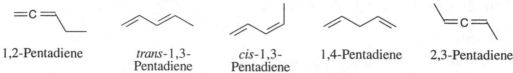

| 1,2-Pentadiene | *trans*-1,3-Pentadiene | *cis*-1,3-Pentadiene | 1,4-Pentadiene | 2,3-Pentadiene |

On a four carbon skeleton, we can put the double bonds in a 1,2- or a 1,3-relationship with a methyl group at C-2. No C_3 chain isomers are possible without providing a longer base chain with the necessary substitution.

3-Methyl-1,2-butadiene   2-Methyl-1,3-butadiene

Among the cyclic structures, we begin with a C_5 ring. With a C_4 ring bearing a double bond, the double bond can be endocyclic (inside the ring) or exocyclic (outside the ring) with a methyl on the double bond or at a position one atom removed.

Cyclopentene   1-Methyl-cyclobutene   Methylene-cyclobutane   3-Methyl-cyclobutene

For the three-membered rings with the double bond within the ring, the other two carbons of C_5H_8 can be attached as either two methyl groups bound, respectively, to the same atom or to adjacent atoms or as an ethyl group either on the sp^2- or sp^3-hybridized carbon.

1,2-Dimethyl-cyclopropene   1,3-Dimethyl-cyclopropene   3,3-Dimethyl-cyclopropene   1-Ethyl-cyclopropene   3-Ethyl-cyclopropene

There are three ways in which the double bond can be outside a three-membered ring, as shown below.

2-Methyl-methylene-cyclopropane   Ethylidene-cyclopropane   Ethenylcyclopropane (Cyclopropylethene)

Compounds containing two rings were discussed in Exercise 1-F, part *b*.

(c) The formula C_6H_{12} corresponds to an index of hydrogen deficiency (also called degree of unsaturation) of one. Thus, possible structures have either one ring or one double bond. The possible cyclic structures were discussed in Exercise 1.11, part *c*.

In addition, acyclic structures that contain one double bond also correspond to this formula. With a linear C_6 chain, the double bond can be at the 1-, 2-, or 3-position with *cis* and *trans* isomers being possible for the latter two.

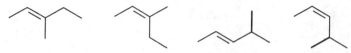

1-Hexene   *trans*-2-Hexene   *cis*-2-Hexene   *trans*-3-Hexene   *cis*-3-Hexene

In the C_5 linear chains, one begins by placing a double bond at the C-1 position, with the last carbon being added as a methyl group at C-2, C-3, or C-4.

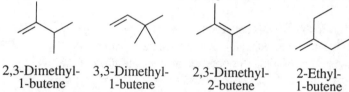

| 2-Methyl-
1-pentene | 3-Methyl-
1-pentene | 4-Methyl-
1-pentene | 2-Methyl-
2-pentene |

Another set of three isomers have the double bond at the C-2 position, with *cis* and *trans* isomers being possible when the additional methyl group is at C-3 or C-4 of the chain.

| *trans*-3-Methyl-
2-pentene | *cis*-3-Methyl-
2-pentene | *trans*-4-Methyl-
2-pentene | *cis*-4-Methyl-
2-pentene |

With the C_4 chain, the last two carbons can be present either as two methyl groups at C-2 or C-3 or as one ethyl group at C-2.

| 2,3-Dimethyl-
1-butene | 3,3-Dimethyl-
1-butene | 2,3-Dimethyl-
2-butene | 2-Ethyl-
1-butene |

Only a single isomer is possible in 2-ethyl-1-butene because attachment of the ethyl group at any other position produces a longer chain. There is no acyclic structure possible with a three-carbon alkene as the longest chain, with the formula C_6H_{12}.

Ex 2.2 (a) Locate the longest continuous carbon chain that contains the functional group, in this case, the double bond. This compound is a hexene, and we indicate that the double bond is between carbons-1 and -2 by calling it 1-hexene. Thus, the correct name is 3-methyl-1-hexene.

(b) 3-Methyl-1-pentene. The longest continuous chain of atoms containing the functional group is C_5, and this compound is a pentene with a double bond between C-1 and C-2. Thus, it is a 1-pentene, with a methyl group attached to carbon-3.

(c) This is a ring compound containing a double bond, so it is a cycloalkene. Because the ring contains six atoms, this compound is a cyclohexene. The carbons of the double bond are C-1 and C-2, the order of which is assigned so that any other substituents have the lowest possible numbers. Thus, this compound is 3-methylcyclohexene. It is unnecessary to call this compound 3-methyl-*1*-cyclohexene because the rules state that the double bond dictates where the numbering of the ring carbons starts, and so the position of the alkene functional group at carbon-1 is understood.

(d)   *trans*-3-Hexene. The longest continuous chain of carbon atoms containing the functional group is a hexene. Counting from either end of the chain, we find that the first carbon of the double bond is C-3, hence 3-hexene. There are *cis* and *trans* isomers of this alkene, and the structure in this exercise is the *trans* isomer, with the two alkyl groups directed toward opposite sides of the double bond. It is also called (*E*)-3-hexene.

Ex 2.3   (a)   Draw six carbons with a double bond between carbons 1 and 2 and with a cyclopropyl group attached to carbon 2.

(b)   Begin by drawing an eight-carbon skeleton with a double bond between carbons 1 and 2 and an ethyl group attached to C-3.

(c)   Construct a 1-heptene with an isobutyl group appended at C-2.

Ex 2.4   (a)   *cis*-2-Pentene ((*Z*)-2-pentene). This name indicates the double bond is at the second position along the longest chain with the two alkyl groups (of higher priority than the two hydrogen atoms) at C-2 and C-3 on the same side of the double bond.

(b)   (*E*)-3-Methyl-3-hexene. The double bond is between C-3 and C-4. An ethyl group has higher priority than a methyl group and also has higher priority than hydrogen. Because the two groups of higher priority are on opposite sides of the double bond, this is an *E*-isomer.

(c)   *trans*-3-Methyl-3-hexene ((*E*)-3-methyl-3-hexene). The position of the double bond is always assigned the lowest possible number counting either from the left or the right. The two alkyl groups on the double bond are on opposite sides.

(d)   4,5-Dimethyl-*cis*-4-octene (4,5-dimethyl-(*Z*)-4-octene). The longest octene chain has a double bond between C-4 and C-5. The three-carbon fragment has higher priority than the one-carbon fragment. Because the groups of highest priority are located on the same side of the double bond, this is a *cis* (or *Z*) isomer.

Ex 2.5   (a)   An eight-carbon chain with the double bond at the second possible position has a hydrogen and methyl at one terminus of the double bond, with methyl having higher priority. At the other terminus, the groups attached are hydrogen and a *n*-pentyl group: the alkyl group has higher priority. In the *E*-isomer, these alkyl groups are located in a *trans* relation.

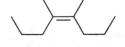

(b)   In a C_8 chain bearing a double bond at the 3-position, two hydrogen atoms at each terminus of the double bond have lower priority than the alkyl groups at the same positions and are on the same side of the double bond in the *Z*-isomer.

(c)   The same analysis as in part *b*, except that the position of the double bond is between C-2 and C-3.

(d)   The same analysis as in part *b*, except with the double bond in a different position and the alkyl groups in a *trans* (*E*) relation.

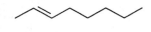
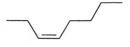

Ex 2.6 Combustion of a hydrocarbon converts it into carbon dioxide and water. With isomeric pairs, there is no need to correct for the number of moles of carbon dioxide produced. Therefore, the member of the isomeric pair with the higher heat of formation (less stable) has the higher heat of combustion.

(a) 1-Hexene is less stable than 2-hexene because the double bond in 1-hexene is only monosubstituted. 1-Hexene therefore has the higher heat of combustion.

(b) 2-Hexene has the higher heat of combustion because it is less stable than 2-methyl-2-pentene. In 2-hexene, the double bond is only disubstituted, whereas it is trisubstituted in its isomer.

(c) Octane has a somewhat higher heat of combustion because it is slightly less stable than 2,5-dimethylhexane; branching stabilizes an isomeric hydrocarbon.

(d) (Z)-2-pentene has a higher heat of combustion because it is slightly less stable than (E)-2-pentene because of steric interactions of the two alkyl groups in the Z-isomer, which are avoided in the E-isomer.

Ex 2.7 (a) The least stable of these three isomeric C_7H_{14} alkenes is 1-heptene, in which the double bond is only monosubstituted. More stable is 3-heptene, in which the double bond is disubstituted; and most stable is 2-methyl-2-hexene, in which the double bond is trisubstituted.

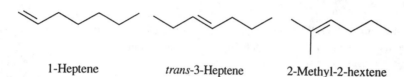

1-Heptene   *trans*-3-Heptene   2-Methyl-2-hexene

(b) Of these substituted cyclooctenes, 3-methylcyclooctene has a disubstituted double bond and is least stable. 1-Methylcyclooctene has a trisubstituted double bond and is therefore somewhat more stable than 3-methylcyclooctene, but less so than the most stable 1,2-dimethylcyclooctene, in which the double bond is tetrasubstituted.

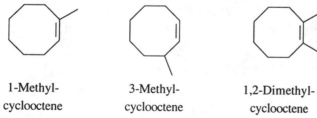

1-Methyl-   3-Methyl-   1,2-Dimethyl-
cyclooctene   cyclooctene   cyclooctene

(c) Of these three isomeric ethyloctenes, 3-ethyl-1-octene has a monosubstituted double bond, 2-ethyl-1-octene has a disubstituted double bond, and 3-ethyl-2-octene has a trisubstituted double bond, and alkene stability increases in that order.

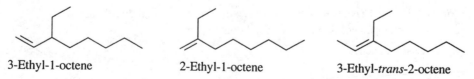

3-Ethyl-1-octene   2-Ethyl-1-octene   3-Ethyl-*trans*-2-octene

Ex 2.8 (a) For hyperconjugative stabilization of the double bond, one of the σ bonds at C-3 of 1-butene must be aligned with the p orbitals of the double bond. Three staggered rotamers (upper set) with this arrangement are shown, in which each of the substituents, H_a, H_b, and CH_3, is rotated to a position in which it can hyperconjugate with the top lobe of the p orbital at C-2. The second (lower) set exists in which this hyperconjugative interaction is maintained with the bottom lobe of C-2.

(b)   One such rotamer is shown for *trans*-2-butene in which the hydrogen atoms at C-1 and C-4 are aligned with the *p* orbitals, respectively, at C-2 and C-3.

(c)   One rotamer is shown in which the methyl groups at C-1 and C-4 are aligned for hyperconjugative interaction. A complete set of rotamers, analogous to those discussed in part *a*, is possible.

Ex 2.9   The heat of hydrogenation of an alkene corresponds to the thermodynamic relationship between the alkene and H_2 and the reduced alkane. The heat of combustion of the alkane is equal to that of the alkene plus that of hydrogen.

$$4\ CO_2 + 8\ H_2O\ +\ H_2O\quad=\quad 4\ CO_2\ +\ 9\ H_2O$$

Ex 2.10 (a)   *trans*-(or *E*)-1,3-Hexadiene. Geometric isomers exist for the internal double bond (between C-3 and C-4), but not for the terminal double bond.

(b)   *trans*-(or *E*)-1,4-Hexadiene.

(c)   1,3-Cyclohexadiene. The six-membered ring requires that all double bonds be in a *cis* relation, so it is not necessary to stipulate this as a *cis* isomer.)

(d)   1,4-Cycloheptadiene. (Too much strain would be introduced if the double bonds in this seven-membered ring compound were other than in a *cis* relation, and again stipulating this as *cis* is unnecessary.

(e)   2-Methyl-3-isopropyl-1,4-pentadiene. Neither double bond, both being at termini of the diene chain, has a geometrical isomer.

Ex 2.11 (a)   1-Hexene substituted at the 2-position by a cyclobutenyl ring whose point of attachment is C-1 of cyclobutene.

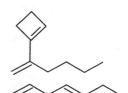

(b)   In 1,4-heptadiene, only the double bond between carbons 4 and 5 has geometric isomers. The two alkyl groups are therefore placed in a *trans* relation.

(c)   No geometrical isomers are possible in 1,2-pentadiene. (Notice that C-2 is *sp*-hybridized.)

(d)   Geometric isomers are possible only at the double bond between C-3 and C-4. At C-3, the vinyl group has priority over a methyl group, and, at C-4, a methyl group has priority over hydrogen. Thus, the vinyl and methyl groups are on opposite sides of the double bond, producing the *E*-isomer.

Ex 2.12 In each significant resonance structure, all carbon atoms are sp^2-hybridized and the maximum number of possible covalent bonds is maintained.

(a)   In tetracene, there only two resonance structures with nine π bonds distributed among 18 atoms in the four rings.

(b)   In phenanthrene, the seven π bonds are distributed among 14 atoms in four possible ways.

(c)   Seven resonance structures place positive charge on each of seven possible atoms in the cycloheptatrienyl skeleton.

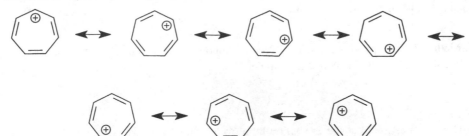

(d)   Resonance structures place positive charge at each of the five possible atoms of the cyclopentadienyl skeleton.

Ex 2.13 To calculate the number of electrons in each conjugated system, count two for each double bond, zero for each positively charged vacant p orbital, and two for each negatively charged, doubly occupied p orbital.

(a)   Cycloheptatrienyl cation has three double bonds (six electrons) and is therefore a $4n + 2$ ($n = 1$) Hückel aromatic.

(b)   Cycloheptatrienyl anion has six electrons from the three double bonds plus two electrons from the negatively charged carbon for a total of eight electrons. 8 equals $4n$ ($n = 2$). This molecule is not a Hückel aromatic, and electronic stabilization is absent.

(c)   Each of the six double bonds contributes two electrons for a total of $12 = 4n$ in which $n = 3$ in this anti-Hückel molecule which therefore lacks aromatic stabilization.

(d)   The cyclononatetraenyl cation has eight electrons, two from each of the four double bonds, forming a $4n$ ($n = 2$) system which lacks aromatic stabilization if planar.

(e)   The cyclooctatetraene dianion is a ten electron system ($4n + 2$ in which $n = 2$) and therefore exhibits aromatic stabilization.

Ex 2.14 Remember that curved arrows *always* represent the movement of electrons, and thus they start at the source of electron density and the arrowhead points to where the electrons are moving.

A double-headed arrow is used to indicate that two structures are different resonance representations for the same molecule.

Ex 2.15 (a)   3-Methyl-*n*-propylbenzene or *meta-n*-propyltoluene.

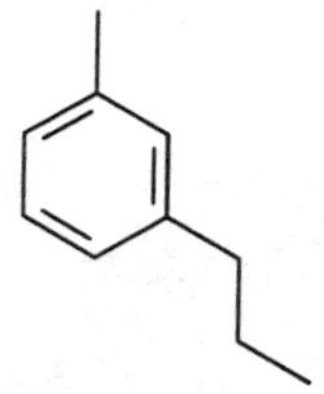

(b)   1,3-Diethylbenzene; common name: *meta*-diethylbenzene.

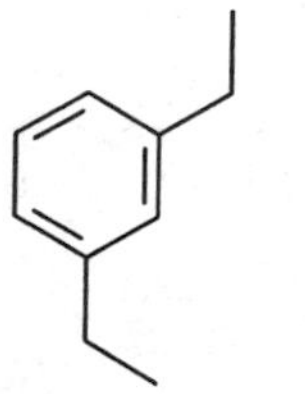

(c)   3-ethyltoluene; common name: *meta*-ethyltoluene.

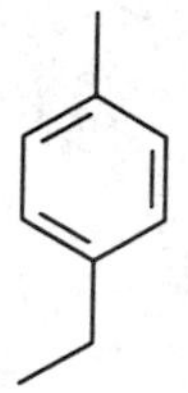

(d)   2-methylisopropylbenzene; common name: *ortho*-isopropyltoluene

Ex 2.16 The designation *ortho* refers to two substituents in a 1,2 relationship; *meta, in* a 1,3 relationship; and *para* to a 1,4 relationship of two substituents on a benzene ring.

(a)
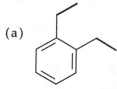
ortho-Diethylbenzene

(b)
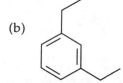
meta-Diethylbenzene

(c)
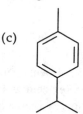
para-isopropyltoluene

Ex 2.17 This exercise is best approached as a problem in permutations. First, start by drawing a benzene ring with one of the substituents. Next, add the second substituent in each of the unique positions relative to the first. Finally, take each of these structures and add the third substituent in each of the unique positions. Notice that the symmetry of the *p*-isomer permits only two unique substitutions, unlike the situation for the *o*- and *m*-isomers.

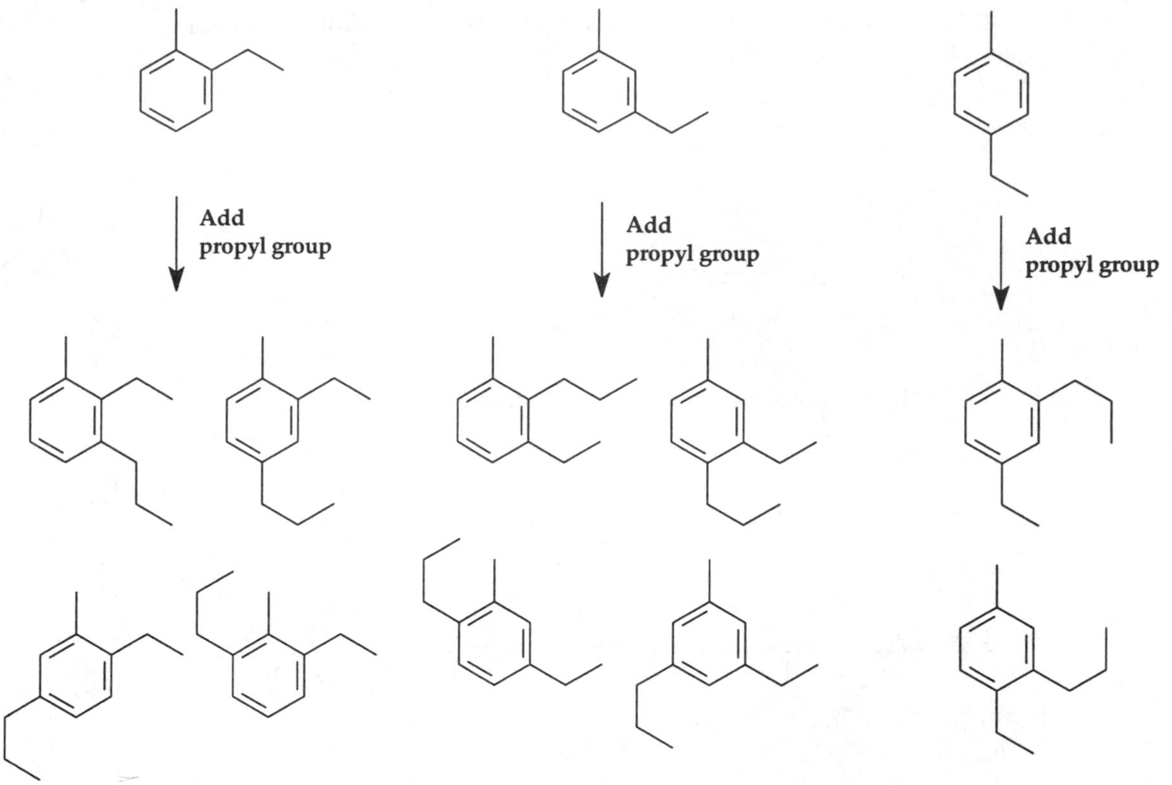

Ex 2.18 Structures can be named either as alkyl derivatives of benzene or phenyl derivatives of alkenes.

(a)   2-(1-Pentenyl)-benzene or 2-phenyl-1-pentene.

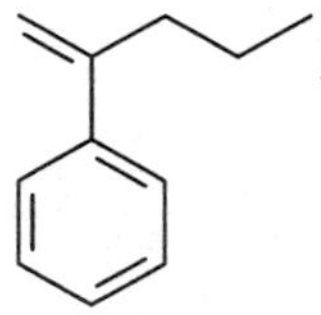

(b)   *s*-Butylbenzene or 2-phenylbutane.

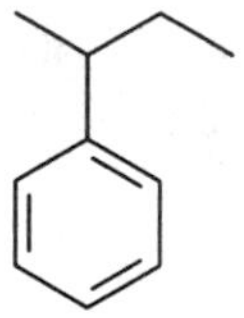

(c)   3-*n*-Pentyl-1,4-pentadiene. (Although a longer chain could be identified by including only one double bond, the correct nomenclature includes both double bond functional groups along the base hydrocarbon chain.)

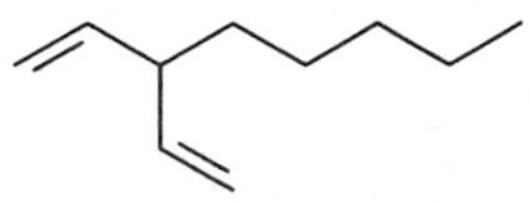

(d)   2,6-Dimethyl-4-phenyl-1-heptene. (This alkyl chain is too complex to be named as a simple derivative of benzene.)

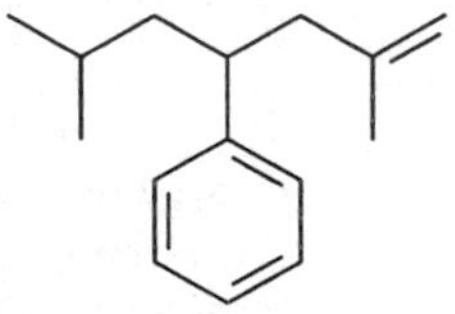

Ex 2.19 Recall that curved arrows are used to show the movement of electrons with the tail at the source and the arrowhead pointing to where the electrons wind up.

Naphthalene   Pyrene

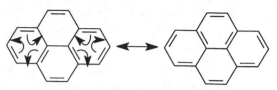

Ex 2.20 (a)   1-Hexyne   (c)   3-Phenyl-1-propyne

(b)   2-Heptyne   (d)   2,4-Hexadiyne

Ex 2.21 This exercise asks you to place a triple bond at various positions along the hydrocarbon chain.

(a)      (c)

(b)   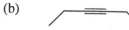   (d)

Ex 2.22 In this cumulated diene, the two double bonds are orthogonal. Hence, the hydrogen and methyl at each end of the double bond occupy orthogonal planes. There are thus two isomers possible.

Answers to Review Problems

Pr 2.1 (a)   *cis*-1,3-Pentadiene has a five-carbon chain with two double bonds. The designator *cis* refers to the geometry about the C-3–C-4 bond.

(b)   4-Methylcyclopentene has a five-membered ring with a methyl group at C-4. (The numbering starts with a carbon of the double bond and proceeds in the direction so that the other *sp²* carbon is C-2.)

(c)   3-*t*-Butyl-1-hexene has a six-carbon chain with a double bond between C-1 and C-2 and a *t*-butyl group at C-3.

(d)   *m*-Bromotoluene has a benzene ring with a bromine and a methyl substituent placed in a 1,3- relationship.

(e)   Oct-1-ene-4-yne has an eight-carbon chain with a double bond between C-1 and C-2 and a triple bond between C-4 and C-5.

Pr 2.2 This problem focuses on the different reactivity of an alkene and a cycloalkane toward catalytic hydrogenation. Alkenes are efficiently hydrogenated to the corresponding alkanes, whereas most cycloalkanes (cyclopropanes representing a possible exception) do not take up hydrogen under typical catalytic hydrogenation conditions.

(a)   Cyclohexane, lacking double bonds, does not take up hydrogen upon catalytic hydrogenation, whereas 1-hexene does.

(b)   No distinction can be made by quantitative catalytic hydrogenation, because both 1-hexene and 2-hexene take up one mole of hydrogen.

(c)   No distinction between cyclohexane and methylcyclopentane can be made by catalytic hydrogenation because both hydrocarbons, lacking double bonds, are inert to this reaction.

(d)   Cyclohexene takes up one molar equivalent of hydrogen, but cyclohexane is inert to catalytic hydrogenation.

(e)   1-Butene takes up one molar equivalent of hydrogen, whereas 1-butyne takes up two.

(f)   Both 1-butene and 1-pentene take up one molar equivalent of hydrogen, so no distinction is possible on the basis of catalytic hydrogenation.

Pr 2.3 E and Z configurations require the assignment of priorities for the two substituents at the two ends of the double bond. When the highest-priority substituents are on the same side, the configuration is Z; when on opposite sides, the configuration is E.

(a)   Priority at the left-hand carbon as drawn (C-2), methyl > H; priority at the right-hand carbon (C-3), ethyl > H; therefore, E.

(b)   No geometrical isomers because there are two identical (methyl) groups at the left-hand carbon (C-2).

(c)   Priority at the left-hand carbon (C-3), ethyl > methyl; priority at the right-hand carbon (C-2), methyl > hydrogen; therefore, Z.

(d)   Priority at the left-hand carbon (C-3), isopropyl > hydrogen; priority at the right-hand carbon (C-2), methyl > hydrogen; therefore, Z.

(e)   No geometrical isomers are possible with terminal double bonds.

(f)   Geometric isomerism is possible only at the 4-alkene. At the left-hand carbon (C-4): propenyl > H; at the right-hand carbon (C-5): methyl > H; therefore, E.

Pr 2.4   The degree of hydrogen deficiency (degree of unsaturation) is calculated by compar-ing the molecular formula with that of a saturated acyclic alkane (C_nH_{2n+2}), recognizing that the presence of either a π bond or a ring reduces by two the number of hydrogen atoms required for an acyclic alkane. Thus, a deficiency of hydrogen atoms, as pairs, indicates the number of π bonds or rings. This calculation alone makes no distinction between π bonds and rings. The following answers include a calculation of the degree of unsaturation, and a structure is then given so that you can confirm that your calculation alone agrees with the real structure. But note that, from this simple calculation, you could never have obtained (except by guessing) the unique structure shown. The number of rings is defined by the smallest number of carbon–carbon bonds that would have to be cleaved to form a completely acyclic structure.

(a)   Limonene has six fewer hydrogen atoms than does a saturated, acyclic C_{10} alkane ($C_{10}H_{22}$). Its index of unsaturation is therefore 3, indicating that the sum of the number of multiple bonds plus the number of rings is 3. In fact, limonene has two double bonds and one ring.

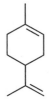

(b)   Acenaphthene has 16 fewer hydrogen atoms than does a C_{12} alkane ($C_{12}H_{26}$). This translates into an index of hydrogen deficiency of 8. In fact, acenaphthene has three rings and five formal double bonds.

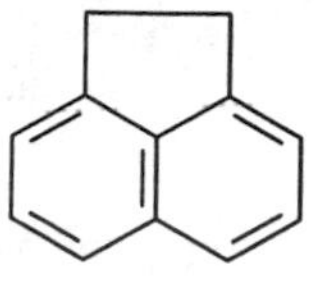

(c)   Benzo[a]pyrene has 30 fewer hydrogen atoms than does an acyclic C_{20} alkane ($C_{20}H_{42}$). This gives an index of hydrogen deficiency of 15. In fact, this carcinogen has five rings and ten formal double bonds.

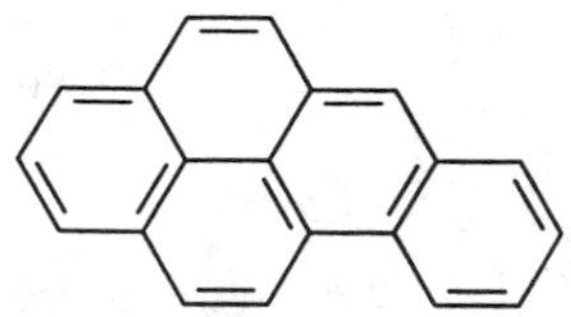

(d)   β-Pinene, with the same formula as limonene, has an index of hydrogen deficiency of 3, translated here into two rings and one double bond.

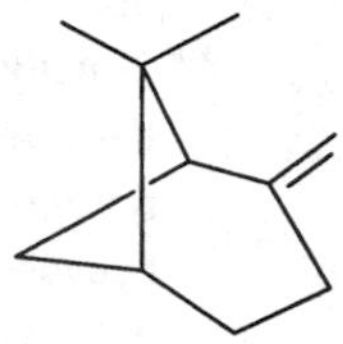

(e)   Caryophyllene has eight fewer hydrogen atoms than does an acyclic C_{15} alkane ($C_{15}H_{32}$). This gives an index of hydrogen deficiency of 4, translating into two rings and two double bonds in this case.

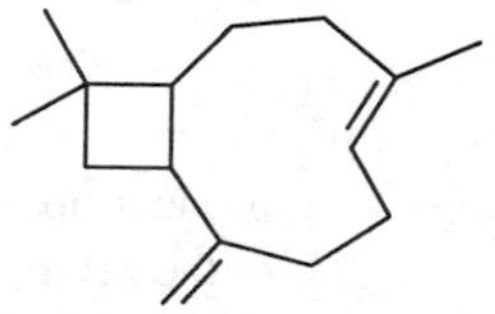

(f)   β-Cadinene has the same formula as caryophyllene and therefore an identical index of hydrogen deficiency. In fact, this compound also has two rings and two double bonds.

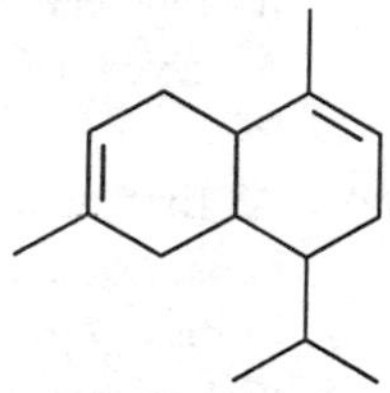

Pr 2.5   Geometric isomers are possible only when a double bond has nonidentical substituents at each end.

(a)   No geometric isomerism in 1-alkenes.

(b)   *cis-* and *trans*-2-Pentene.

(c)   No geometric isomers in terminal alkenes.

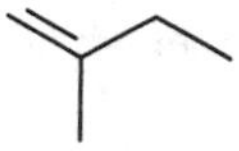

(d)   No geometric isomers in compounds bearing two identical substituents at one end of the double bond.

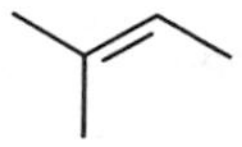

(e)   *cis-* and *trans*-isomers. Fluorine has priority over methyl, so the structure at the left is *E*. The structure at the right is *Z*.

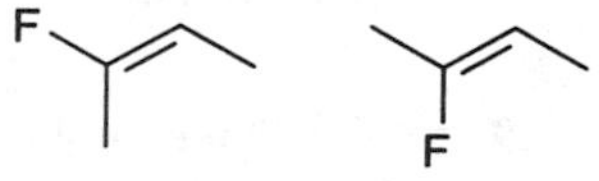

(f)   *cis-* and *trans*-Dichlorocyclohexanes because of restricted rotation in cyclic compounds.

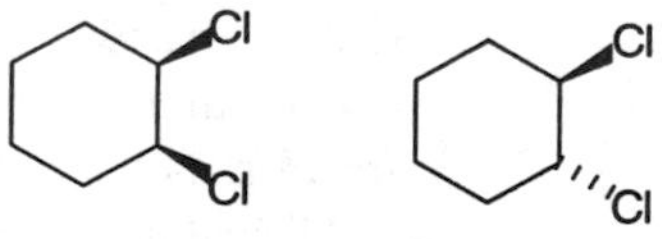

(g) No geometric isomers because *trans* double bonds in small rings are too strained.

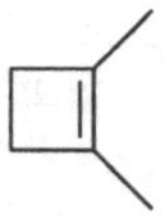

(h) *cis*- and *trans*-methyl groups. The sequence of numbering here is chosen to give the carbons of the double bond rather than the positions of the methyl groups.

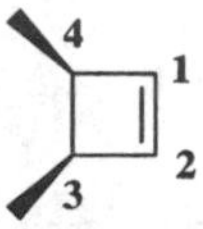

Pr 2.6 (a) These compounds are isomeric heptenes. 1-Heptene has a monosubstituted double bond; *cis*-2-heptene and *trans*-3-heptene have disubstituted double bonds; 2-methyl-2-hexene has a trisubstituted double bond; and 2,3-dimethyl-2-pentene has a tetrasubstituted double bond. The more highly substituted a double bond, the more stable is that isomer. *Trans* isomers are more stable than *cis* isomers because their geometry serves to minimize steric interactions. Thus, in order of decreasing stability: 2,3-dimethyl-2-pentene > 2-methyl-2-hexene > *trans*-3-heptene > *cis*-2-heptene > 1-heptene.

(b) Heats of hydrogenation are directly comparable for isomers that form the same product upon hydrogenation, that is, for isomers differing only in the position or the geometry of the double bond. Thus, 1-heptene, *cis*-2-heptene, and *trans*-3-heptene can be directly compared and their stabilities ordered according to their observed heats of hydrogenation, but not with their branched-chain isomers.

Pr 2.7 (a) At first, it may seem surprising that, unlike most acyclic alkenes, the *trans* isomer of a cyclic alkene is less stable than the *cis* isomer. A *trans* double bond in an eight-membered ring has appreciable bond strain, making the heat of hydrogenation of the *trans*-cycloalkene much higher than the less strained *cis*-isomer. (Use your models to construct *trans*-cyclooctene or try even to draw an eight-membered ring with a *trans* double bond.)

(b) In acyclic alkenes, steric interactions between groups on the same side of the double bond in the *cis* isomer destabilize it relative to the *trans* isomer. Thus, *cis*-2-hexene is less stable, giving it the higher heat of hydrogenation.

Pr 2.8 The double bond in 1-methylcyclohexene is trisubstituted, but that in its isomer methylidenecyclohexane is disubstituted. The more highly substituted isomer is present at higher concentrations at equilibrium because of its greater stability. The equilibrium therefore favors 1-methylcyclohexene.

1-Methyl-
cyclohexene

Methylidene-
cyclohexane

Pr 2.9 α-Phellandrene has six fewer hydrogen atoms than does a saturated acyclic C_{10} alkane, and thus an index of hydrogen deficiency of 3. After hydrogenation, the product has two fewer hydrogen atoms than does the straight-chain alkane. Two double bonds and one ring are therefore present. In fact, α-phellandrene has the structure shown at the right, but as in Problem 2.4, this structure cannot be uniquely established from the information given.

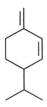

Pr 2.10   Begin by drawing the structures. Carbon atoms bearing two π bonds are *sp*-hybridized, those bearing one π bond are *sp*²-hybridized, and those bearing no π bonds are *sp*³-hybridized.

(a)   C-1: *sp*²; C-2: *sp*; C-3: *sp*²; C₄: *sp*³; C₅: *sp*³; C₆: *sp*²; C₇: *sp*².

(b)   C-1: *sp*; C-2: *sp*; C-3: *sp*³; all carbons in the ring (C-4 through C-9): *sp*².

(c)   Both carbons on vinyl group: *sp*²; three carbons in ring: *sp*³.

(d)   Ring carbons: *sp*²; methyl substituents: *sp*³.

Pr 2.11   (a)   The allyl cations represent resonance contributors because the two structures differ only in the placement of the double bonds. No atoms are at different positions.

(b)   These structures are not resonance contributors, because the position of a hydrogen atom differs in the two structures. That is, in the upper structure, C-3 bears two hydrogen atoms and C-4 two hydrogen atoms, whereas in the bottom structure, C-3 bears one hydrogen atom and C-4 three hydrogen atoms.

(c)   These are resonance structures because all atoms are at the same place, and only electrons are distributed differently.

(d)   These are not resonance structures because the allyl anion at the left bears a hydrogen at C-2, whereas the vinyl anion at the right does not. In the allyl anion, both terminal carbons bear two hydrogen atoms, whereas in the vinyl anion, one terminal carbon atom has three hydrogen substituents and the other has two.

(e)   These are not resonance structures because the placement in space of the groups in these two geometric isomers is not identical.

Pr 2.12   Isomers differ in the positions of atoms, whereas resonance structures differ *only* in the positions of electrons.

(a)   The electrons can be reorganized by moving one of the electrons of the double bond between C-1 and C-2 to a position between C-2 and C-3, thus pairing with the odd electron represented in the structure as being located on C-3 to form a double bond. One remaining electron of the original π bond is then localized on C-1.

(b)   By shifting two electrons of one of the double bonds to the adjacent bonding position, two additional resonance contributors of this pentadienyl cation are obtained.

(c)   By using the two electrons of the formal negatively charged site to form a multiple bond and by shifting the two electrons of a π bond on to a single carbon, two additional resonance structures for the pentadienyl anion are obtained.

Important New Terms

Acetylene: *see* **Ethyne** (2.4)

Aliphatic hydrocarbons: a family of compounds containing hydrogen and carbon atoms, but no aromatic rings (2.3)

Alkenes: a family of unsaturated hydrocarbons containing one or more double bonds; compounds with the empirical formula C_nH_{2n} for acyclic members with one double bond (2.1)

Alkynes: a family of hydrocarbons containing a triple bond; compounds with the empirical formula C_nH_{2n-2} for acyclic members with one triple bond (2.4)

Allene: an unsaturated hydrocarbon containing two orthogonal double bonds emanating in opposite directions from a common *sp*-hybridized carbon atom (2.4)

Allyl group: $-CH_2CH=CH_2$; an alkyl substituent in which the point of attachment is adjacent to a double bond (2.3)

Antiaromatic hydrocarbon: a planar, conjugated, cyclic, unsaturated hydrocarbon comprising sp^2-hybridized carbons, lacking the chemical stability of a Hückel aromatic; most such systems contain $4n$ π-electrons and are said to be conjugatively destabilized, although the choice of an appropriate model with which to compare them is not absolutely clear (2.3)

Antibonding molecular orbital: a molecular orbital that, when occupied by electrons, destabilizes a molecule relative to the separated atoms (2.1)

Arene: an aromatic hydrocarbon or derivative (2.3)

Aromatic hydrocarbons: a family of planar, sp^2-hybridized, conjugated, cyclic, unsaturated hydrocarbons with unusual chemical stability; according to Hückel's rule, such compounds contain $(4n + 2)$ electrons in their π systems (2.3)

Aromaticity: the special stability afforded by a planar cyclic array of *p* orbitals containing a Hückel number $(4n + 2)$ of electrons (2.3)

Aryl group: an arene fragment lacking one substituent from a ring carbon (2.3)

Biradical: a chemical species bearing two noninteracting radical centers (2.1, 2.3)

Bond alternation: a repeating sequence of short and long (single and double) bonds in an extended π system (2.3)

Bonding molecular orbital: a molecular orbital that, when occupied by electrons, stabilizes a molecule relative to the separated atoms (2.1)

Carcinogens: cancer-inducing agents (2.3)

Catalyst: a species that is not involved in the overall stoichiometry of the reaction and is recovered unchanged after a reaction, but is needed for the reaction to proceed at a reasonable rate (2.1)

Catalytic hydrogenation: the addition of one or more equivalents of H_2 in the presence of a noble metal catalyst (2.1)

cis **Isomer:** a geometric isomer in which the largest groups are on the same side of a double bond or ring (2.1)

Conjugated diene: a diene with an array of *p* orbitals on adjacent atoms; that is, a diene in which the double bonds of the π system interact directly without interruption by an intervening sp^3-hybridized atom (2.2)

Conjugation: a series of alternating single and double bonds along a carbon chain with adjacent *p* orbitals (2.2)

Constitutional isomer: relationship between two compounds of the same molecular weight but with the atoms connected in different sequences (2.1)

Cumulated diene: a diene in which the two orthogonal double bonds share a common carbon atom (2.2)

Delocalization: the spreading of π electron density over an entire π system (2.3)

Dienes: compounds containing two double bonds (2.2)

Double bond: a σ and a π bond between sp^2-hybridized atoms (2.1)

***E*-isomer**: (from the German *entgegen*, opposite): a geometric isomer in which the groups of highest priority are on opposite sides of a double bond (2.1)

Entgegen : *see* ***E*-isomer** (2.1)

Enyne: an organic compound containing a double bond and a triple bond (2.4)

Ethene (also called **ethylene**): C_2H_4; the simplest unsaturated hydrocarbon containing a double bond between sp^2-hybridized carbon atoms (2.1)

Ethyne (also called **acetylene**): $HC\equiv CH$, the simplest alkyne containing a triple bond (2.4)

Functional group: a site in a molecule at which it undergoes characteristic and selective chemical reactions (2.1)

Geometric isomers: isomers with the same connectivity along the backbone, but different spatial disposition of one or more groups around a bond with restricted rotation; *cis–trans* isomers (2.1)

Heat of hydrogenation: the heat released when one mole of an unsaturated compound is completely hydrogenated to a saturated compound (2.1, 2.3)

Hückel's rule: an empirical generalization that any planar, cyclic, conjugated system containing $4n+2$ π electrons (in which n is an integer) experiences unusual aromatic stabilization, whereas those containing $4n$ π electrons do not (2.3)

Hybrid orbitals: the orbitals formed by mixing hydrogenic atomic orbitals (2.1, 2.4)

Hydrogenation: the addition of H_2 (2.1, 2.3, 2.4)

Hyperconjugation: an orbital description of the stabilizing effect derived by interaction of an aligned σ bond with an adjacent p orbital (2.1)

Index of hydrogen deficiency: half the difference between the number of hydrogen atoms in a hydrocarbon and the number expected for a straight-chain alkane ($2n + 2$); indicative of the number of multiple bonds and/or rings present (2.1)

Isolated diene: a diene in which the double bonds do not interact directly with each other because of one or more intervening sp^3–hybridized atoms (2.2)

Kekulé structures: cyclic six-carbon structures suggested by August Kekulé which depict benzene as having localized double bonds (2.3)

***meta*-substitution**: the relationship of substituents that are 1,3 to each other on an aromatic ring (2.3)

Molecular orbitals: probability surfaces in a molecule within which an electron is likely to be found; constructed by the overlap of atomic orbitals (2.1)

Olefin: an alkene (2.1)

Orbital phasing: description of the relative wave property of electrons in orbitals that results in either favorable or unfavorable interaction; like phasing results in bonding, and unlike phasing results in antibonding interactions (2.1)

***ortho*-substitution**: the relationship of substituents that are 1,2 to each other on an aromatic ring (2.3)

***para*-substitution**: the relationship of substituents that are 1,4 to each other on an aromatic ring (2.3)

Phenyl group: a C_6H_5 fragment with one fewer hydrogen than benzene (2.3)

Pi (π) bond: a covalent bond in which electron density is symmetrically arranged above and below the axis connecting the two bonded atoms; results from the sideways overlap of p orbitals (2.1)

Polycyclic aromatic hydrocarbon: an aromatic compound containing fused rings (2.3)

Polyene: an unsaturated hydrocarbon or derivative containing more than two double bonds (2.2)

Positional isomers: isomers in which the sequence of atoms along a chain differs in the position of one or more functional groups (2.1)

Radical: a chemical species bearing a single unpaired electron on am atom; a chemical species with an odd number of electrons (2.1)

Resonance contributors: see **Resonance structures** (2.3)

Resonance hybrid: an energetically weighted composite of contributing resonance structures (2.3)

Resonance structures: valence bond representations of possible distributions of electrons in a molecule, differing only in positions of electrons and *not* in positions of atoms (2.3)

Saturation: the condition of a compound containing only sp^3-hybridized atoms; consequently, a description of a molecule lacking one or more multiple bonds (2.1)

***sp*-Hybrid orbitals**: hybrid orbitals formed by mathematically mixing one s and one p atomic orbital; the two hybrid orbitals so formed are dispersed in a linear array separated by 180° (2.4)

***sp*^{2}-Hybrid orbitals**: hybrid orbitals formed by mathematically mixing one s and two p atomic orbitals; the three hybrid orbitals so formed are dispersed in a plane and are separated by 120° (2.1)

Stereoisomers: isomers that differ only in the position of atoms in space (2.1)

trans **Isomer:** a geometric isomer in which the largest groups are on opposite sides of a double bond or ring (2.1)

Triple bond: a σ and two π bonds between adjacent *sp*-hybridized atoms (2.4)

Unsaturation: the condition of a compound containing some non-sp^3-hybridized atoms; consequently, a description of a molecule containing one or more multiple bonds (2.1)

Vinyl group: an alkene fragment lacking one substituent from the double bond (2.3)

Z-isomer: (from the German *zusammen,* together): a geometric isomer in which the groups of highest priority are on the same side of a double bond (2.1)

***Zusammen*:** *see* **Z-isomer** (2.1)

Zwitterions: neutral species that contain equal numbers of locally charged (plus and minus) centers (2.3)

Functional Groups
Containing Heteroatoms

Key Concepts

Recognizing functional groups

Polar covalent bonding

Dipole moments

Intermolecular association through hydrogen bonding and solvation

Protic and aprotic solvents

Nucleophiles and electrophiles

Homolytic cleavage

Average bond energies and bond dissociation energies

Heterolytic cleavage

Carbocation stability

Lucas test for chemical classification of primary, secondary, and tertiary alcohols

Acidity and basicity

Curved arrows to indicate electron flow

Naming functional groups

Answers to Exercises

Ex 3.1 A primary amine has one alkyl substituent bound to nitrogen; a secondary amine has two; a tertiary amine has three; and a quaternary ammonium salt has four.

(a) primary amine

(b) primary amine

(c) secondary amine

(d) primary amine

(e) quaternary ammonium salt

(f) secondary amine

(g) tertiary amine

(h) tertiary amine

Ex 3.2 As you learned in Chapter 1, formal charge calculation is accomplished by comparing the number of valence electrons for a given atom with the sum of the number of unshared electrons and half the shared electrons available to that atom. These latter quantities are obtained directly from a Lewis dot structure. We begin by drawing a valid Lewis dot structure for each compound, as we did in Exercise 1.6. We then perform the formal charge calculation by the technique learned in Exercise 1.7.

(a) H: $1 - [0 + (2 \div 2)] = 0$

C: $4 - [0 + (8 \div 2)] = 0$

N: $5 - [0 + (8 \div 2)] = +1$

O: $6 - [6 + (2 \div 2)] = -1$

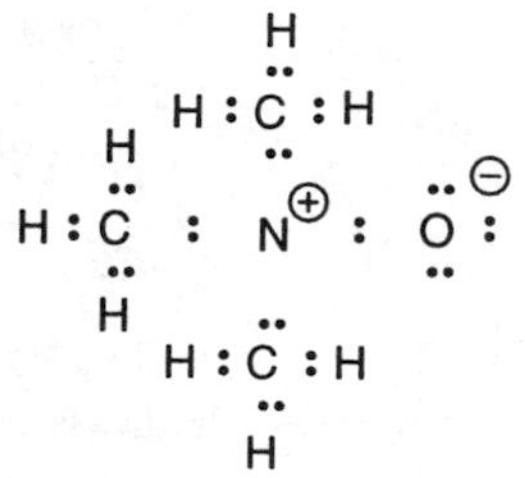

(b) H: $1 - [0 + (2 \div 2)] = 0$

Both Cs: $4 - [0 + (8 \div 2)] = 0$

N: $5 - [0 + (8 \div 2)] = +1$

O: $6 - [6 + (2 \div 2)] = -1$

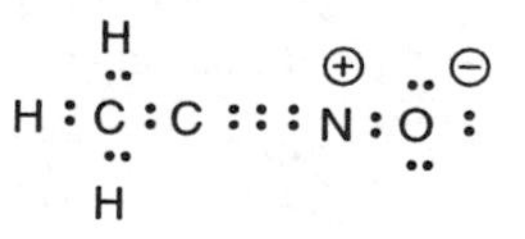

37

Ex 3.3 (a)   NH$_3$. The dipole moment of NH$_3$ results from contributions from the weakly polarized N—H bonds and the lone pair of electrons on nitrogen and is directed toward the nitrogen lone pair. In NF$_3$, the contribution to the molecular dipole is offset by the contribution of the bonds between nitrogen and the three highly electronegative fluorine atoms, whose net dipolar contribution points in the opposite direction to that of the lone pair. The resultant molecular dipole for NF$_3$ is therefore smaller than that of NH$_3$ because of the cancellation of these vectors in the former.

1.5 D

0.2 D

(b)   The structural difference between trimethylamine and 2-methylpropane is the presence in the former of a nitrogen atom with a lone pair of electrons at the position occupied by a C—H bond in the hydrocarbon. The lone pair is a site of electron density on one side of the nitrogen atom not counterbalanced by oppositely directed polar bonds. Therefore, trimethylamine has the larger dipole moment.

(c)   A larger dipole moment is observed for triphenylamine than for triphenyl-methane for the same reason as in part *b*.

Ex 3.4 (a)   In ammonia hydrogen-bonded to another ammonia molecule, one molecule acts as a hydrogen-bond donor and its partner as a hydrogen-bond acceptor. With two identical molecules, either molecule can fill either role. Because there are three polar N—H bonds and one lone pair, a single molecule can associate with up to four complementary hydrogen-bonding partners.

(b)   Both ammonia and methylamine can act as hydrogen-bond donors and acceptors. There are therefore two modes of intermolecular hydrogen bonding: ammonia as hydrogen-bond donor and methylamine as acceptor (upper) and methylamine as hydrogen-bond donor and ammonia as acceptor (lower). Ammonia can associate with as many as four molecules (three hydrogen bond donating interactions and one hydrogen bond accepting interaction), whereas methylamine can associate with only three molecules (two hydrogen-bond donating interactions and one hydrogen-bond accepting interaction).

(c)   In hydrogen bonding of methylamine with another methylamine molecule, one molecule acts as hydrogen-bond donor and the partner as hydrogen-bond acceptor.

(d)   Ammonia can hydrogen bond to another ammonia molecule as either a hydrogen-bond donor or acceptor (recall part *a*), but trimethylamine can act only as a hydrogen-bond acceptor. Hydrogen-bonding between ammonia and trimethylamine must therefore involve hydrogen-bond donation from ammonia and accepting by trimethylamine.

Ex 3.5   Intramolecular hydrogen bonding in a diamine would require one amino group to act as a hydrogen-bond donor and the other as a hydrogen-bond acceptor. The chain would therefore involve nitrogen, the sequence of carbon atoms between the amino groups, the second nitrogen, and the hydrogen atom bound to nitrogen.

(a)   (b)

Ex 3.6   (a)   Ammonia and triethylamine have similar dipole moments, but ammonia, with three NH bonds, is both a hydrogen-atom donor and a hydrogen-bond acceptor. Triethylamine can act only as a hydrogen-bond acceptor. Ammonia is therefore more soluble than triethylamine in water, a solvent that is both a hydrogen-bond donor and acceptor.

(b)   Both methylamine and *n*-octylamine are primary amines that can participate in hydrogen-bonding. The long alkyl chain of *n*-octylamine, however, is nonpolar and disrupts intermolecular association between polar water molecules. The much smaller nonpolar portion of methylamine is more readily accommodated within water. Methylamine is therefore more soluble in water.

(c)   Trimethylamine is a tertiary amine and interacts with water as a hydrogen-bond acceptor by using the lone pair on nitrogen. However, *n*-propylamine, with polar NH groups, is both a hydrogen-bond donor and an acceptor. *n*-Propylamine is therefore more soluble in water.

Ex 3.7   By definition, a Lewis base acts as an electron donor.

(a) Mg^{2+} is not a Lewis base. Rather, it is a Lewis acid, as follows from its cationic charge.

(b)   The aluminum atom in aluminum trichloride is highly electron deficient (formally Al^{3+}) and strongly attracts electrons, acting therefore as a Lewis acid rather than a Lewis base.

(c)   The boron atom in boron trifluoride has access to only six valence electrons and is surrounded by three highly electronegative fluorine atoms. Boron trifluoride is therefore a strong Lewis acid. The lone pairs of electrons of the fluorine atoms are Lewis base sites, but the overall reactivity of the molecule is dominated by the highly electron-deficient boron atom.

(d)   Triethylamine has an accessible lone pair of electrons that can coordinate effectively with partially positively charged centers, thus acting as a Lewis base.

(e)   Tin tetrachloride is characterized by a significantly electron-deficient tin atom (formally Sn^{4+}) that acts as a Lewis acid, not a Lewis base. In coordinating with electron donors (Lewis bases), the tin atom increases its coordination shell to five or sometimes six substituents.

Ex 3.8   (a)   Butane has neither polar bonds nor lone pairs of nonbonded electrons. It is, thus, neither a Lewis acid nor base. It does not interact strongly with a proton.

(b)   The OH group of 1-butanol is a proton donor and is thus a Brønsted acid. Because of the presence of two nonbonded lone pairs of electrons on oxygen, it is also a Lewis base. Reaction with a proton donor produces an oxonium ion containing a new O—H bond.

(c)   The oxygen of methyl ether has two nonbonded lone pairs of electrons. Thus, this compound is a Lewis base, protonation of which generates an oxonium ion.

(d)   Tertiary butanol, like 1-butanol in part *b*, is both a Brønsted acid (proton donor) and a Lewis base (electron donor) because of the lone pairs of electrons on oxygen. Reaction with a proton donor results in protonation on oxygen, producing an oxonium ion.

(e)   The two lone pairs of electrons on sulfur are sites of Lewis basicity. Upon protonation, a sulfonium ion is formed.

Ex 3.9 In each part of this exercise, a reduction takes place by addition of molecular hydrogen and one or more π bonds are broken. To determine the amount of energy consumed, compare the strength of the multiple bonds lost with those of the new single bonds in the product. (In each part of this exercise, the number of bonds broken is the same as the number of bonds made. This is always the case as long as all atoms start and end in their normal valence states.)

(a) This reaction is exothermic by 13 kcal/mole and is therefore thermodynamically feasible.

$$C=O \; \pi = C=O \; (\pi + \sigma) \quad -C-O \; \sigma$$
$$= \quad 179 \quad - \quad 86$$
$$= \quad 93 \text{ kcal/mole}$$

Bonds broken

C=O π   93 kcal/mole
H—H   <u>104</u>
         197 kcal/mole

Bonds made

C—H   99 kcal/mole
O—H   <u>111</u>
        210 kcal/mole

(b) This reactions is exothermic by 20 kcal/mole and is thermodynamically feasible. (Even so, this reaction is very slow without a catalyst.)

$$N≡N \; + \; 3 \; H—H \longrightarrow \; 2 \; NH_3$$

Bonds broken

2 N=N π + N–N σ   226 kcal/mole
3 H–H   <u>312</u>
           538 kcal/mole

Bonds made

6 N–H   558 kcal/mole

(c) This reaction is exothermic (by 36 kcal/mole) and is thermodynamically feasible as written.

$$H_3C—C≡N \; + \; 2 \; H—H \longrightarrow H_3C—CH_2—NH_2$$

$$2 \; C≡N \; \pi = C≡N \; 2\pi + \sigma \; - \; C-N \; \sigma$$
$$= \quad 213 \quad - \quad 73$$
$$= \; 140 \text{ kcal/mole}$$

Bonds broken

C≡N π   140 kcal/mole
H—H   <u>208</u>
        348 kcal/mole

Bonds made

C—H   198 kcal/mole
N—H   <u>186</u>
        384 kcal/mole

Ex 3.10 The formal oxidation level of a specific atom is calculated by mentally breaking all covalent bonds to the atom in question and assigning the component electrons to the more electronegative atom.

In cyclohexane, all carbon atoms are identical, so the calculation need be done only once. Each C—H bond confers two electrons on carbon because hydrogen is less electronegative than carbon; the C—C bonds confer no charge because the two carbon atoms have the same electronegativity. Thus, the oxidation level of the carbon atoms in cyclohexane is $-2 = [(2 \times (-1)) + (2 \times 0)]$.

In the same way, it can be shown that in cyclohexene, the sp^2-hybridized carbons have an oxidation level of -1 $[(1 \times (-1)) + (3 \times 0)]$ and that for the sp^3-hybridized carbon atoms is the same as for cyclohexane, -2 $[(2 \times (-1)) + (2 \times 0)]$.

Ex 3.11 Consider only those carbon atoms for which there is a change in bonding. For these, determine whether the change in substituents is from a less to a more electronegative atom.

(a)   The oxidation level of carbon in the C≡N functional group in acetonitrile is +3 and the corresponding carbon in the imine of acetaldehyde is +1 (+2 − 1). Thus, the former is at the higher oxidation level and a reduction is required to proceed from the left to the right. This order can also be recognized in the presence of three C—N bonds in CH_3CN and of two C—N bonds in $CH_3CH=NH$.

(b)   The oxidation level of the carbonyl carbon of a carboxylic acid is +3; that of an aldehyde is +1. Thus, the acid is at the higher oxidation level and a reduction is required to convert an acid to an aldehyde.

(c)   The oxidation level of the carbon bearing nitrogen in ethylamine is −1 and the carbon of the nitrile functional group in $CH_3C≡N$ is +3. Thus, an oxidation is required to convert an amine to a nitrile.

(d)   There is no change in the types of bonds to carbon in the carboxylic acid on the left and in the ester the right. Neither an oxidation nor a reduction is required.

(e)   Changing from one substituent atom to another both of which are more electronegative than carbon does not change the oxidation level. Thus, as in part *d*, neither oxidation nor reduction is required.

Ex 3.12 Using the same procedure as in Exercise 3.2, we calculate the following formal charges:

H: $1 − [0 + (2 ÷ 2)] = 0$
C (CH_3): $4 − [0 + (8 ÷ 2)] = 0$
C (isonitrile): $4 − [2 + (6 ÷ 2)] = −1$
N: $5 − [0 + (8 ÷ 2)] = +1$

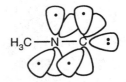

Ex 3.13 An alcohol is classified as primary if the OH group is attached to a primary carbon (that is, to one that is itself attached to only one carbon substituent); as secondary if attached to a carbon bearing two carbon substituents; and tertiary if attached to a carbon with three carbon substituents.

(a)   secondary alcohol

(b)   tertiary alcohol

(c)   primary alcohol

(d)   This is not an aliphatic alcohol, but rather a phenol, for which the designation primary, secondary, or tertiary is not used.

(e)   secondary alcohol

(f)   primary alcohol

Ex 3.14 Ethanol contains a more electronegative oxygen atom and consequently forms stronger intermolecular hydrogen bonds than those in ethylamine, which contains a less electronegative nitrogen atom. Because boiling requires disruption of this intermolecular network of hydrogen bonds, the compound with the stronger hydrogen bonds will exhibit the higher boiling point.

Ex 3.15   Ethers are named by designating the two alkyl or aryl groups attached to oxygen. If two identical groups are present, the group is mentioned only once.

(a)   Isopropyl methyl ether

(b)   Butyl ether

(c)   *t*-Butyl (2-methyl-1-butyl) ether

(d)   Allyl phenyl ether

(e) Ethyl vinyl ether

Ex 3.16 Homolytic cleavage of a single bond generates two radical species:

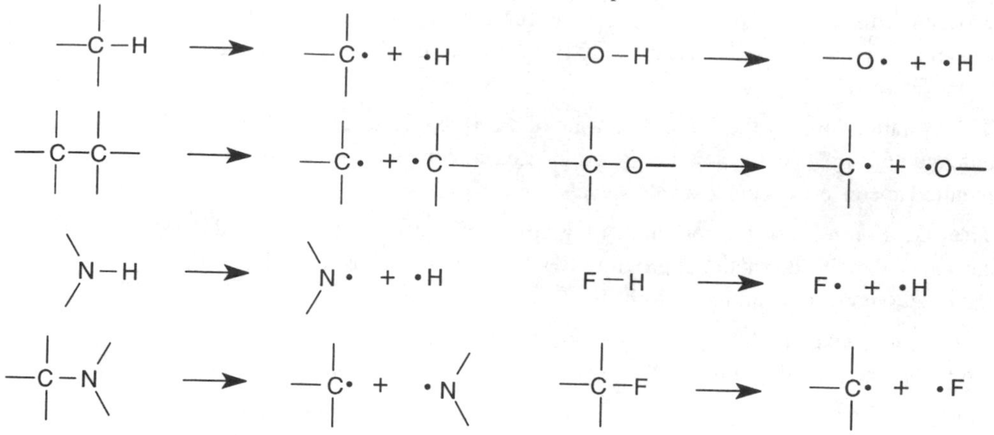

Ex 3.17 (a) Although sulfur and carbon have almost the same electronegativity, sulfur is larger and can better accommodate extra electron density. Thus, the C—S bond cleaves to produce positive charge on carbon and negative charge on sulfur.

$$H_3C\!-\!SCH_3 \longrightarrow H_3C^{\oplus} + {}^{\ominus}SCH_3$$

(b) The cleavage of a polar S—H bond places negative charge on the more electronegative sulfur atom and positive charge on hydrogen.

$$CH_3S\!-\!H \longrightarrow CH_3S^{\ominus} + H^{\oplus}$$

(c) Because of the greater electronegativity of oxygen, cleavage of an oxygen–carbon bond in an alcohol places positive charge on carbon and negative charge on oxygen.

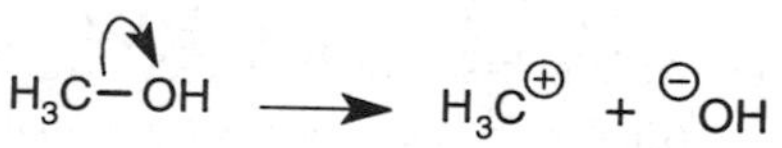

$$H_3C\!-\!OH \longrightarrow H_3C^{\oplus} + {}^{\ominus}OH$$

(d) Because of the greater electronegativity of oxygen, a C—O bond is polarized toward oxygen. Upon heterolytic cleavage of the C—O bond, oxygen takes up the bonding electrons and becomes negatively charged whereas the carbon becomes positively charge.

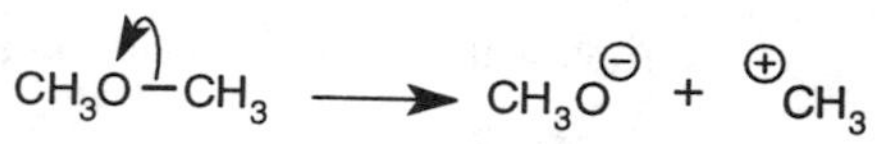

$$CH_3O\!-\!CH_3 \longrightarrow CH_3O^{\ominus} + {}^{\oplus}CH_3$$

Ex 3.18 A radical is characterized according to the classification of the carbon atom bearing the unpaired electron.

(a) secondary   (b) primary   (c) primary   (d) tertiary

Ex 3.19 The ease of movement of electrons in a heterolytic fission depends on the relative stability of the two fragment ions. In this exercise, the more electronegative atom is common to both compounds. The facility for heterolytic fission is therefore influenced significantly by the relative stability of the resulting carbocation.

(a) *t*-Butyl iodide (2-methyl-2-iodopropane). Heterolytic fission of the compound on the left produces a tertiary cation, whereas that on the right produces a primary cation, along with iodide ion in both cases.

(b) 1-Phenylethanol. Heterolytic cleavage of the structure on the left produces a primary cation, whereas that at the right produces a secondary benzylic cation and hydroxide ion. The latter is more favorable.

(c) 2,4-Dimethyl-2-iodopentane. Heterolytic cleavage of the structure on the left produces a tertiary cation, whereas that on the right produces a secondary cation, together with iodide ion. The cation produced from the left structure is more stable and is more easily formed.

(d)   *trans*-1-Bromo-2-butene. Heterolytic cleavage of the carbon–bromine bond in the compound on the left produces a primary cation, whereas that on the right produces a secondary allylic cation, along with bromide ion. The latter is more easily cleaved. The structure on the left requires less energy for heterolytic cleavage of the C—I bond.

(e)   1-Methyl-1-bromocyclohexane. The loss of bromide ion from the compound on the left produces a simple secondary cation. That from the structure on the right produces a tertiary cation. The latter cation is more stable.

Ex 3.20 The Lucas test is used to distinguish the three classes of alcohols: tertiary alcohols react immediately; secondary alcohols react slowly; and primary alcohols do not react at all.

(a)   A secondary alcohol, with a slow reaction.

(b)   A primary alcohol, with no reaction.

(c)   A tertiary alcohol, with immediate reaction.

(d)   A primary alcohol, with no reaction.

Ex 3.21 Carbocations are classified according to the number of carbon substituents bound to the positively charged carbon atom.

(a)   Secondary benzylic

(b)   Secondary

(c)   Allylic (Note that in the two possible resonance structures of this allylic cation, one places positive charge on a primary allylic site and the other at a secondary allylic site. The dominant stabilizing effect is resonance delocalization in the allylic cation rather than whether the charge density is greater at the primary or secondary site.)

(d)   Secondary benzylic

Ex 3.22 Acid-catalyzed dehydration takes place by protonation of the alcohol oxygen and subsequent loss of water from the resulting oxonium ion, to produce a carbocation. This second step is the most difficult, and tertiary alcohols react fastest because they form the most stable carbocation. A secondary alcohol reacts more slowly than a tertiary alcohol, and a primary alcohol reacts most slowly.

(a)   The structure on the left is a secondary benzylic alcohol; the center structure is a secondary alcohol; and the structure on the right is a primary alcohol. Thus, reactivity decreases from left to right.

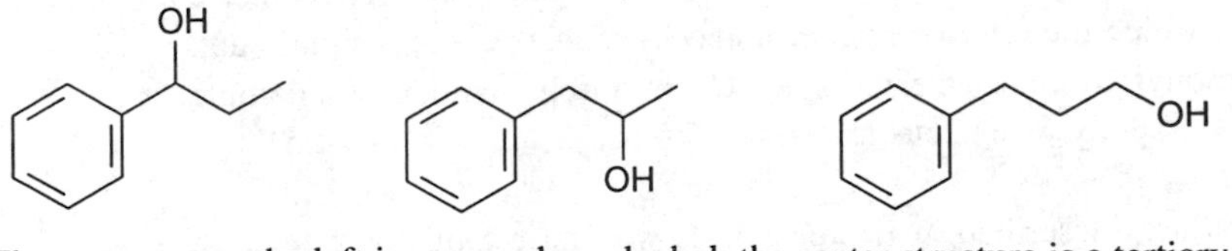

(b)   The structure on the left is a secondary alcohol; the center structure is a tertiary alcohol; the structure on the right is a primary alcohol. Dehydration takes place most rapidly from the alcohol in the center and most slowly from the structure on the right-hand side.

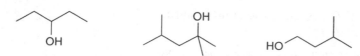

(c) The structure on the left is a tertiary alcohol; the center structure is a secondary alcohol; and the structure on the right is a primary alcohol. The tertiary alcohol on the left dehydrates most easily. The primary alcohol at the right dehydrates most slowly.

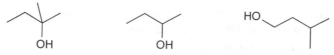

Ex 3.23 (a) Ether   (b) Aldehyde   (c) Secondary alcohol   (d) Ketone

Ex 3.24 (a) Acid chloride   (b) Aldehyde   (c) Ketone

(d) Amide (also called a carboxylic acid amide or carboxamide). Because the nitrogen atom in this compound bears one methyl group and one hydrogen in addition to the carbonyl group, it is classified as a secondary amide.

(e) Amide. This is a primary amide because the nitrogen atom is attached to only one carbon (the carbon of the carbonyl group).

(f) Anhydride   (g) Ether   (h) Ester

Ex 3.25 The carbonyl group most easily attacked by a nucleophile is that which bears the greatest fraction of partial positive charge at the carbonyl carbon and has the least resonance and hyperconjugative stabilization.

(a) Because an alkyl group is more electron-releasing and larger than hydrogen, an aldehyde (at the left) is more easily attacked by nucleophiles than is a ketone (at the right).

(b) Because the singly bound oxygen of an ester can donate a nonbonded electron pair to the carbonyl group, the carbonyl group in an ester is less electron deficient than that in an aldehyde. Thus, an aldehyde (at the left) is more easily attacked by nucleophiles than is an ester (at the right).

(c) Because the lone pair on nitrogen in a tertiary amide can be donated toward the carbonyl group in a significant resonance contributor, the carbonyl group of an amide is less electron deficient than that in a ketone. The ketone (at the left) is therefore more easily attacked by nucleophile than is the amide (at the right).

(d) In both the amide at the left and the ester at the right, the heteroatom attached to the carbonyl group can donate a lone pair of electrons toward carbon, producing a resonance contributor in which electron density has shifted onto the carbonyl oxygen. Because oxygen is more electronegative than nitrogen, the resonance contributor in which the ester oxygen bears positive charge is less significant. Thus, the carbonyl group of an ester is less electron rich than that of an amide and reacts more rapidly with nucleophiles.

Ex 3.26 By using the procedures described in Exercise 3.11, we first establish the relative oxidation level of the reactants and products. If the reactant is at a higher oxidation level than the product, a reducing agent is required; if at a lower oxidation level, an oxidizing agent is required.

(a) An aldehyde is more oxidized than an alcohol. A reducing agent is needed.

(b) An aldehyde is at a higher oxidation level than an alkene. A reducing agent is required.

(c) An aldehyde is more highly oxidized than an alkane. Again, a reducing agent is needed.

(d) An aldehyde is at the same oxidation level as a ketone. This conversion is not a redox reaction and requires neither an oxidation nor a reduction.

Ex 3.27 The greater reactivity of a thiol ester toward nucleophilic attack than with a normal (oxygen) ester results, in part, from the lower partial positive charge density on the carbonyl carbon in the ester group. To the extent that the zwitterionic structures at the right contribute electron density to the carbonyl group, the carbonyl carbon is less electron deficient and better resists nucleophilic attack. Therefore, the difference in reactivity between an ester and a thiol ester is determined by the relative stability of these two zwitterionic resonance contributors. Although oxygen is more electronegative than sulfur, there is an appreciable size mismatch between sulfur, a third-row element, and carbon. This size mismatch makes the bottom right contributor less important for a thiol ester and results in faster nucleophilic attack on carbon. Conversely, because oxygen and carbon are approximately the same size, the top zwitterionic structure, which results in reduced reactivity of the ester toward nucleophiles, is significant.

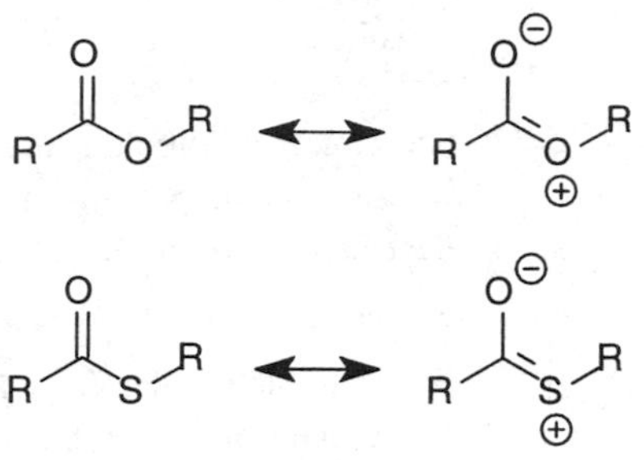

Ex 3.28 In pyridine, a six-electron aromatic delocalized system is obtained by using the three double bonds in the ring. The lone pair on nitrogen is orthogonal to these delocalized electrons and its presence or absence is not relevant to aromatic electronic stabilization. Protonation of the lone pair of electrons on nitrogen does not affect ring aromaticity, so pyridine freely acts as an effective base.

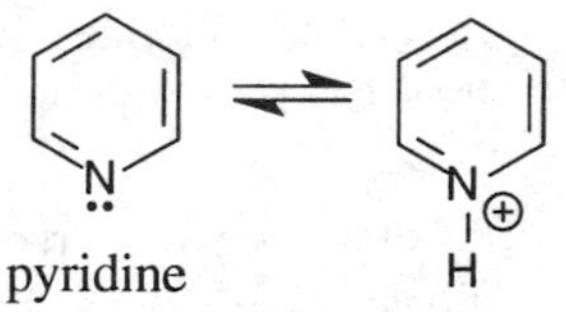

In pyrrole, a six-electron aromatic system is obtained when the ring nitrogen is sp^2-hybridized, with nitrogen's lone pair of electrons in the p-orbital, thus participating in ring delocalization. Protonation of this lone pair of electrons in an acid–base reaction renders nitrogen sp^3-hybridized and removes this atom from conjugation, thus destroying ring aromaticity. Protonation of pyrrole is therefore much more costly energetically, and pyrrole is a much weaker base than is pyridine.

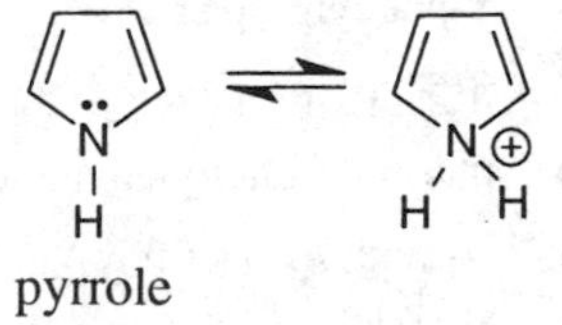

Ex 3.29 Each of the purine nitrogen atoms doubly bound to a ring carbon atom (those not bound to hydrogen) bear a lone pair of electrons orthogonal to the ring in an sp^2-hybridized orbital. Because each of these lone pairs is in the nodal plane of the ring, they do not affect ring aromaticity. In contrast, the lone pair on the remaining nitrogen atom (the one bound to hydrogen) bears its nonbonded lone pair in a p-orbital, aligned with the fused ring π system. Because the latter lone pair overlaps with the adjacent p-orbitals, its presence significantly influences π-delocalization in the ring.

Ex 3.30 When aniline functions as a base, a lone pair of electrons is localized on nitrogen, and a new N—H bond is formed. Thus, the stabilization that results from delocalization of the lone pair of electrons into the ring (shown in the four resonance contributors at the right) is lost upon protonation. As a result, protonation by acid requires more energy for aniline than for simple primary amines in which the lone pair of electrons on nitrogen does not participate in resonance delocalization.

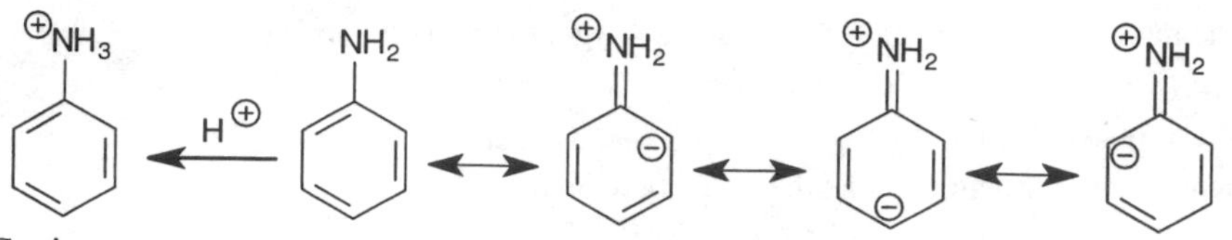

The molecular dipole is the vectorial sum, or resultant, of dipolar contributions from all polar bonds present in a molecule. (Recall from Chapter 1 that the sum of the electron density in four sp^3-hybrid orbitals is spherically symmetrical.)

Ex 3.31 The molecular dipole is the vectorial sum, or resultant, of dipolar contributions from all polar bonds present in a molecule. (Recall from Chapter 1 that the sum of the electron density in four sp^3-hybrid orbitals is spherically symmetrical.)

(a)   The four C—Cl bonds exactly cancel each other.

(b)   The three polar C—Cl σ bonds in chloroform all have dipoles that point to the same side of the molecule, resulting in a net molecular dipole moment.

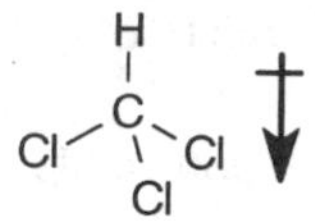

(c)   The resultant molecular dipole moment produced by summing the dipoles of the polar C—Cl bonds in dichloromethane bisects the chlorine-carbon-chlorine angle.

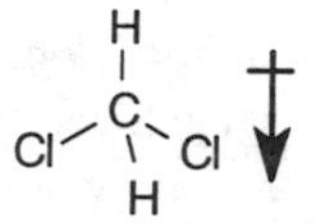

(d)   In methyl chloride, a net dipole moment results mainly from polarization of the carbon–chlorine covalent bond, with a much smaller contribution from the three C—H bonds.

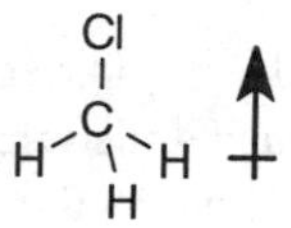

(e)   As in part *a* with carbon tetrachloride, carbon tetrabromide has no molecular dipole because the four C—Br bonds point to the four points of a tetrahedron.

Ex 3.32
(a)   1-phenylpropanone

(b)   3-phenylpropanal

(c)   methyl 2-phenylethanoate (common name: methyl phenylacetate)

(d)   *N*-methyl, *N*-ethyl, *N*-isopropyl amine

(e)   *N,N*-dimethylethanoamide (common name: *N,N*-dimethylacetamide)

(f)   2-butanone imine

(g)   3-methyl-2-butanol

(h)   benzyl allyl sulfide or benzyl allyl thioether

(i)   methyl *t*-butyl ether

Ex 3.33 The same nomenclature rules discussed in Chapters 1 and 2 apply, with a suffix specifying the identity of the functional group present.

(a)   Butanone implies the presence of the ketone group at the only possible position, C-2. (A compound with a carbonyl group at C-1 is an aldehyde; C-2 is equivalent to C-3 and C-1 is equivalent to C-4.)

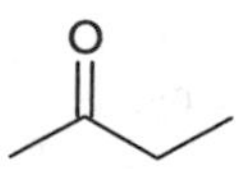

(b)   In 2-hexanone, the position of the carbonyl group must be specified by a number, unlike that in butanone.

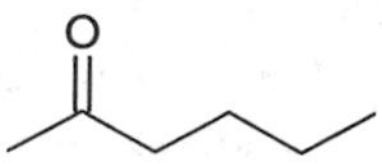

(c)   As in part *b*, the position of the ketone functional group must be specified by a number.

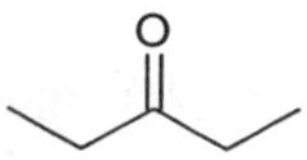

(d)   In 4-methylpentanal, as in all aldehydes, the carbonyl group is at the end of the chain, but the position of the methyl group must be specified.

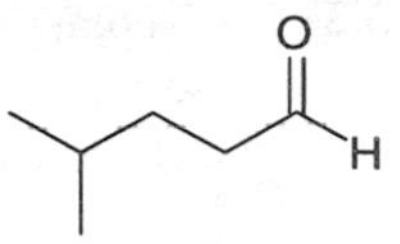

(e)   Propanoic acid defines the carboxylate group as C-1 so the chlorine is placed at C-2.

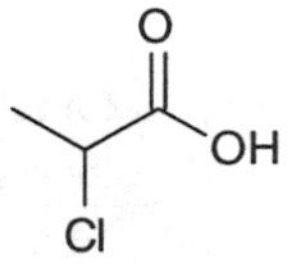

(f)   In ester nomenclature, the alkyl group mentioned first is the group attached to oxygen, with the root acid derivative named afterward.

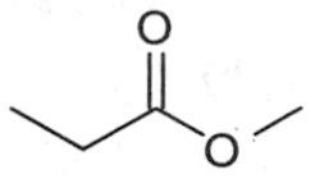

(g)   Amines are named as alkyl derivatives.

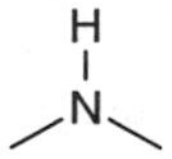

(h)   With no specified groups on nitrogen of an amide, an NH_2 group is assumed to be present.

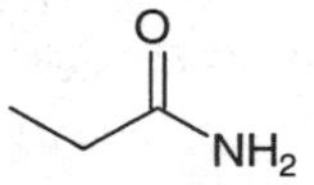

(i)   An acid chloride group is bound to a carboxylate carbon.

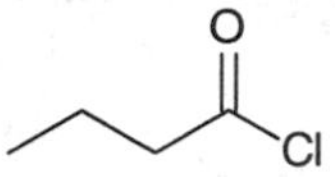

(j)   The numbering of the bromo substituent is from the carboxylate carbon as C-1.

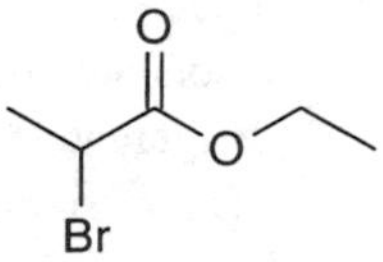

Answers to Review Problems

Pr 3.1   Ethanal imine is derived from ethanal (CH_3CHO) by replacement of the carbonyl oxygen by NH. In an imine, nitrogen has an N—H σ bond, an C—N π bond, and a lone pair in an sp^2-hybrid orbital at the end of the C=N double bond. Hydrogen can be on the same or the opposite side as the methyl group bound to the carbon terminus of the C=N bond. Because a C=N double bond is weaker than a C=C double bond, and because the imine nitrogen can rehybridize or invert, the interconversion of the geometric isomers of ethanal imine is much easier than that in 2-butene.

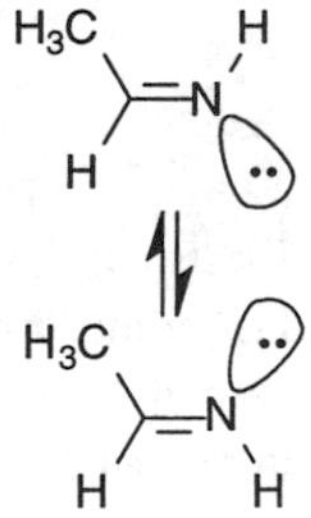

Pr 3.2   An amine is classified according to the number of alkyl substituents attached to nitrogen; an alcohol is classified according to the number of carbon substituents attached to the carbon bearing the OH group.

(a)   2-aminopropane; a primary amine
(b)   2-propanol; a secondary alcohol
(c)   N-methyl, N-isopentylamine; a secondary amine
(d)   2-methyl-1-propanol; a primary alcohol
(e)   N,N-dimethylaniline; a tertiary amine
(f)   2-phenyl-2-propanol; a tertiary alcohol

Pr 3.3   All compounds containing Lewis base sites are protonated in strong acid. The resulting oxonium and sulfonium cations formed by protonation of ethers, esters, aldehydes, and thioethers easily dissolve in a polar mineral acid.

Pr 3.4   (a)   Neutral ozone has a total of 18 electrons in a three-oxygen-atom array. To satisfy the valence requirement of each atom, the central atom is doubly bound to one oxygen atom and singly bound to the other. As a result, the central oxygen has one nonbonded lone pair and a formal positive charge, and the singly bound terminal oxygen bears formal negative charge.

   (b)   A second resonance structure can be derived from the first by shifting a pair of electrons from the negatively charged terminal oxygen toward the central oxygen atom and shifting a pair of electrons originally in a π bond onto the oxygen at the other end of the molecule. Because these resonance structures are of equal energy, they contribute equally to the hybrid, explaining why ozone has equal bond lengths (intermediate between the lengths expected for a single and a double bond) for its two oxygen-oxygen bonds.

   (c)   Because the central oxygen atom in both ozone resonance structures is sp^2-hybridized, the O—O—O angle is predicted to be approximately 120°. As a consequence, ozone is a bent molecule.

Pr 3.5   The conjugate base of benzenesulfonic acid exists as a hybrid of three equivalent resonance structures. Thus, the negative charge of the anion is delocalized to all three oxygen atoms. Because the charge density on each oxygen is only –0.33, this anion is more stable than a carboxylate ion (with a charge of –0.5 on each oxygen atom) because only two resonance structures are possible for the latter anion.

Pr 3.6   Electronegativity decreases going down a column of the periodic table. Because of the higher electronegativity of fluorine than of iodine (4.0 versus 2.5), the electrons in a carbon–fluorine bond are polarized more significantly toward the halogen atom than those in a carbon–iodine bond, producing greater charge separation and hence a higher dipole moment in fluoromethane than in iodomethane (1.85 versus 1.62 Debye). The greater polarizability of iodine is insufficient to offset this large electronegativity difference.

Pr 3.7   Although methanol has polar C—O and O—H σ bonds, charge separation in formaldehyde is enhanced by a significant contribution from a zwitterionic resonance structure in which electron density from the π-bond is shifted onto oxygen.

Pr 3.8   (a)   Each carbon in ethyne is bound to one hydrogen (and by three bonds, to another carbon). Thus, the oxidation level of these carbons is –1.

(b) The carbon of the methyl group in acetonitrile is attached to one carbon and three hydrogens; thus, its oxidation level is −3. The carbon of the nitrile has three bonds to nitrogen and one to another carbon and thus has an oxidation level of +3.

(c) The carbon of the CH_3 group in ethylamine is bound to three hydrogens and one carbon and thus has the same oxidation level as the carbon of the methyl group in acetonitrile (−3). The methylene group (CH_2) is bound to one carbon, two hydrogens, and a nitrogen and thus has an oxidation level of −1.

Pr 3.9 Ethanol, bearing an OH group, associates with itself by intermolecular hydrogen bonding, with the oxygen lone pairs acting as a hydrogen-bond acceptor and the OH group acting as a hydrogen-bond donor. Ethyl ether does not have a group that can act as a hydrogen-bond donor, although the lone pairs on oxygen do permit it to act as a hydrogen-bond acceptor. It cannot associate with itself by hydrogen bonding as can ethanol, however, and intermolecular interactions in ethyl ether are primarily through much weaker van der Waals attractions.

Pr 3.10 (a) There are four possible monochlorobutanes: two in which a hydrogen atom of *n*-butane is replaced by chlorine (1-chlorobutane and 2-chlorobutane) and two in which a similar substitution is accomplished with 2-methylpropane (1-chloro-2-methylpropane and 2-chloro-2-methylpropane).

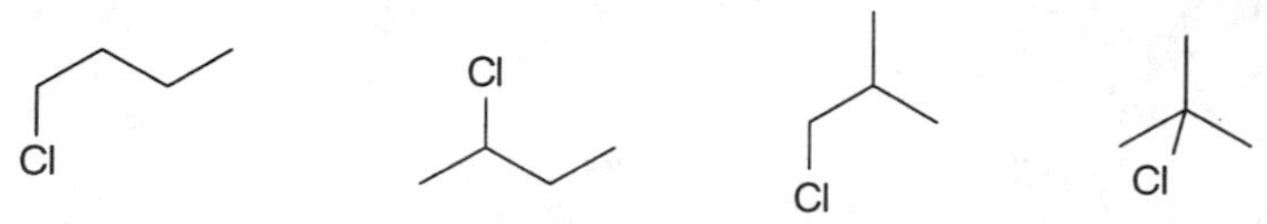

(b) There are nine dichlorobutanes: six are based on *n*-butane (1,1−, 1,2−, 1,3−, 1,4-, 2,2-, and 2,3–dichlorobutane) and three on 2-methylpropane (1,1-, 1,2-, and 1,3-dichloro-2-methylpropane).

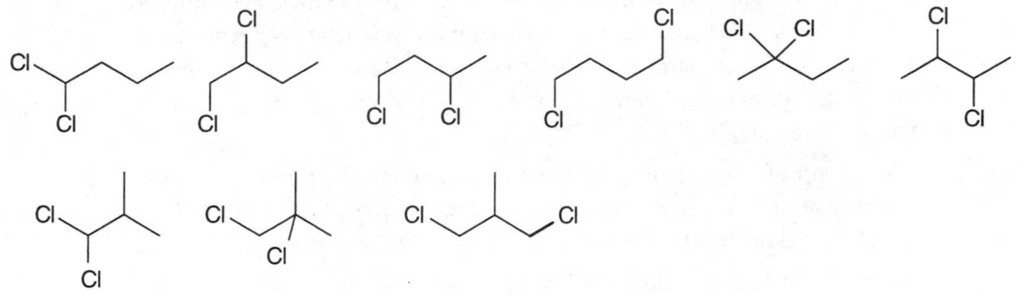

(c) There are thirteen trichlorobutanes: nine derived from *n*-butane (1,1,1−, 1,1,2−, 1,1,3−, 1,1,4−, 1,2,2−, 2,2,3−, 2,2,4−, 1,2,3−, and 1,2,4–trichlorobutane) and four derived from 2-methylpropane (1,1,1, 1,1,2, 1,1,3, and 1,2,3-trichloro-2-methylpropane and 1,3-dichloro-2-(chloromethyl)propane).

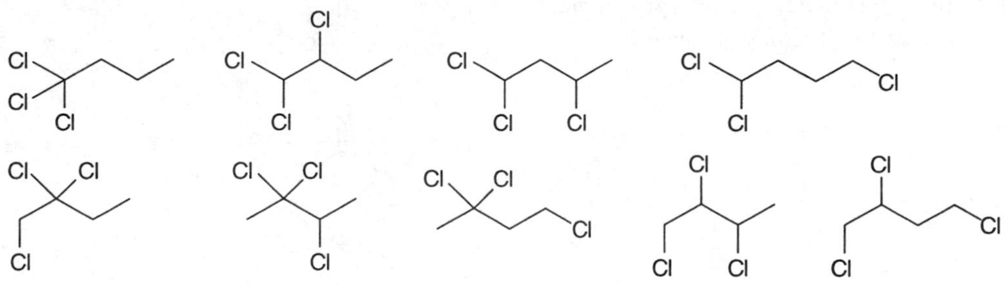

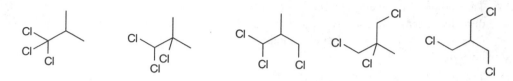

Pr 3.11 (a)   No geometric isomers are possible for 1,1,2-trichlorocyclopentane.

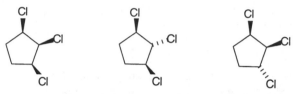

(b)   There are three 1,2,3-trichlorocyclopentanes, one with all three chlorine atoms on the same side of the ring and two with two chlorine atoms on one side and one on the other.

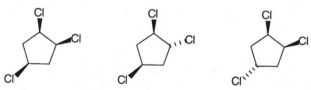

(c)   Just as in part *b*, there are three 1,2,4-trichlorocyclopentanes, one with all chlorines on the same side and two chlorine atoms with two on one side of the ring and one on the other.

Pr 3.12 (a)   Boiling point increases with molecular weight in pairs of molecules with the same type of intermolecular attractions because the kinetic energy required to disrupt van der Waals attractions between molecules containing more and heavier atoms is greater. Therefore, octane has a higher boiling point than pentane (126 veresus 36 °C).

(b)   The OH group of ethyl alcohol is both a hydrogen-bond donor and acceptor. Thus, the intermolecular hydrogen bonding in ethyl alcohol is not possible in methyl ether and the former has a higher boiling point (78 versus –25 °C).

(c)   Ethylene glycol has two sites for hydrogen bonding and, hence, a greater number of intermolecular interactions than are present with ethyl alcohol. In addition, ethylene glycol has a higher molecular weight. Thus, ethylene glycol has the higher boiling point (198 versus 78 °C)

Pr 3.13 (a)   Ethyl ethanoate, methyl propanoate, *n*-propyl methanoate, isopropyl methanoate.

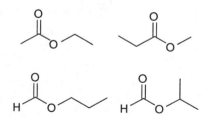

(b) Butanal, 2-methylpropanal.

(c) 2-Propanol.

(d) 3-Methyl-2-butanone, 3-pentanone, 2-pentanone.

(e) *N*-Ethyl-*N,N*-dimethylamine.

(f) 2-Bromo-2-methylpropane.

Pr 3.14 In the classification of solvents, an acidic proton is present in a protic solvent and absent in an aprotic one. An X—H bond is considered to be acidic only if X is more electronegative than carbon; that is, if X = S, O, or N. (Note that these are the same groups that participate in significant intermolecular hydrogen bonding.)

Dimethylsulfoxide (DMSO) has no acidic protons, but does have a highly polarizable S=O double bond. As shown in the resonance structure, contributions from a zwitterionic resonance contributor make this molecule highly polar, with the high negative charge density on oxygen making it favorable to associate strongly with metals and organic cations. It is therefore a dipolar aprotic solvent.

Methylene chloride has a molecular dipole by virtue of the two polar covalent C—Cl σ bonds, with the resultant electronic vector bisecting the Cl—C—Cl angle. It possesses no acidic protons, however, and lacks a highly polar zwitterionic structure like that in DMSO. It is therefore classified as a weakly polar aprotic solvent.

The zwitterionic resonance structure that causes restricted rotation about the C—N bond in amides makes dimethyl-formamide highly polar but, lacking any acidic protons, this tertiary amide is classified as a dipolar aprotic solvent.

Because of the presence of the OH group in methanol, this molecule is considered to be a protic solvent. Because of the high degree of partial positive charge on the acidic proton, methanol is also highly polar.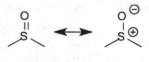

Ethyl ether has no acidic protons and no structural feature that produces high molecular polarity. Only the polar covalent C—O single bonds make ether at all more polar than a liquid alkane. It is therefore classified as a relatively non-polar solvent.

Tetrahydrofuran (THF) is a cyclic analog of ethyl ether; tying the two ends of the ethyl groups were tied together in a ring gives the structure of THF. Possessing the same functional groups as ether, THF is also a relatively nonpolar solvent, but with the alkyl groups tied back away from the ethereal oxygen lone pairs, it is a more active hydrogen-bond acceptor than is ether.

Pr 3.15 Reactive nucleophiles have regions of high electron density or negative charge; electrophiles have regions of electron deficiency or positive charge.

    (a)   The lone pair on a neutral amine makes it a good nucleophile.

    (b)   Hydroxide ion is a very active nucleophile, possessing both three free lone pairs (draw a Lewis dot structure to convince yourself) and a full negative charge. It is also a strong base.

    (c)   Lacking free electrons, positively charged metal ions behave as electrophiles.

    (d)   The two lone pairs on sulfur, together with the acidic S—H bond, places appreciable electron density on sulfur, making it an active nucleophile.

Pr 3.16 A Lewis acid, acting as an electron acceptor, has regions of electron deficiency capable of bonding with an external pair of electrons. A Lewis base has an available lone pair of electrons that can be used for bond formation to a Lewis acid. All Lewis bases can also act as Brønsted bases; that is, as proton acceptors.

    (a)   A thioether. An available lone pair of electrons on sulfur allows the thioether to act as a Lewis base.

    (b)   A sulfonic acid. An acidic proton can be easily lost from the –SO_3H group, making it a Brønsted acid. Because the proton acts as an electron acceptor, this compound is also a Lewis acid, although the Brønsted definition is usually used when an acidic proton is present.

    (c)   A primary aromatic amine (aniline). An available lone pair of electrons on nitrogen serves as the basic site in this Lewis base.

    (d)   A benzylic chloride. Three available lone pairs of electrons on chlorine make this a Lewis base.

Pr 3.17 (a)   Geometric isomers. Tthese isomers are interconverted by rotation about the C=N double bond.

    (b)   Resonance structures. All atoms are in the same positions, with the positions of electrons shifted within the π system.

    (c)   Positional isomers;. The carbonyl group is at a different position along the chain in these two structures.

    (d)   Identical. Identical substituents appear in a *meta* relationship to each other in both representations.

Pr 3.18 (a)   Donation of a lone pair of electrons from the attached atom to the ring π system produces a resonance structure in which three sites on the ring bear formal negative charge. Because of the contributions of these zwitterionic structures to the resonance hybrid, the ring is electron-rich.

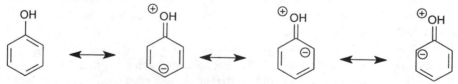

    (b)   In anisole, the oxygen atom bound to the ring bears two nonbonded pairs of electrons. Electron release from this oxygen into the aromatic π system can be represented by resonance structures analogous to those in part *a*. The effect of this contribution to the hybrid is an increase in electron density in the aromatic ring, especially at the carbons *ortho* and *para* to the oxygen substituent. The ethyl group in ethylbenzene also releases electron density into the π system through hyperconjugation. However, delocalization of σ bond electron density by hyperconjugation is less effective than that from π donation of unshared pairs of electrons. The presence of the methoxy group in methoxymethyl-benzene results in withdrawal of electron density by the highly electronegative oxygen atom from the carbon attached to the aromatic ring. (This methoxy oxygen is not in a position for conjugative interaction with the π system.) This carbon is therefore less able to donate electron density through hyperconjugation into the π system. Thus, the order of delocalization of electron density from the substituent to the aromatic ring decreases in the order:

anisole > ethylbenzene > methoxymethylbenzene.

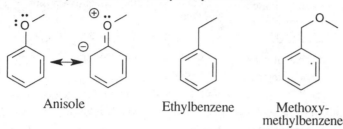

| Anisole | Ethylbenzene | Methoxy-methylbenzene |

Important New Terms

Acid chloride: RCOCl; a functional group in which a carbonyl carbon bears an alkyl or aryl group and a chlorine atom (3.9)

Adenine: $C_5H_5N_5$, a biologically important heteroaromatic base (3.11)

Alcohol: a compound bearing the OH functional group (3.5)

Aldehyde: RCHO; a functional group in which a carbonyl carbon bears a hydrogen and an alkyl or aryl group (3.8)

Alkyl halide: R—X (X = F, Cl, Br, I); a compound in which carbon is bound to a halogen atom (3.12)

Allyl cation: a resonance-stabilized carbocation in which the vacant p orbital is adjacent to a π bond (3.6)

Ambiphilicity: the tendency of an –XH group to act as both an acid and a base (3.2)

Amide: $RCONR_2$; a functional group in which a carbonyl carbon bears an alkyl or aryl group and an amino group (3.9)

Amine: an alkyl or aryl derivative of ammonia (3.1)

Amino group: an $-NH_2$ substituent (3.1)

Ammonia: NH_3; the simplest compound containing sp^3-hybridized nitrogen. (3.1)

Anhydride: RCO_2COR; a functional group in which two carbonyl carbons bearing alkyl or aryl groups are linked through an oxygen atom (3.9)

Aniline: $C_6H_5NH_2$; amino-substituted benzene (3.11)

Anion: a negatively charged ion (3.2)

Aprotic solvent: a solvent molecule lacking a polar heteroatom—H bond (3.5)

Average bond energy: the typical energy of a specific type of bond; obtained from heats of formation. (For example, the average bond energy of a C—H bond is obtained as 1/4 of the heat required to convert methane to carbon and hydrogen—that is, of $CH_4 \rightarrow C + 4H$, $\Delta H°/4 = 99$ kcal/mole. Tthe average bond energy of a C—C was obtained by measuring the heat of formation of ethane and subtracting the bond energies of six C—H bonds.) (3.3)

Benzyl cation: a resonance-stabilized carbocation in which the vacant p orbital is adjacent to an aryl ring (3.6)

Bond dissociation energy: the quantity of heat consumed when a covalent bond is homolytically cleaved (3.6)

Brønsted acid: a proton (H^+) donor (3.2)

Brønsted base: a proton (H^+) acceptor (3-2)

Carbocation: a positively charged trivalent carbon atom containing only six electrons in its outer shell (3.6)

Carbocation stability: 3°>2°>1° (3.6)

Carbonium ion: *see* **Carbocation** (3.6)

Carbonyl group: C═O; a functional group containing a carbon–oxygen double bond (3.8)

Carboxylic acid: RCO_2H; a functional group in which a carbonyl carbon bears an alkyl or aryl group and an OH group (3.9)

Carboxylic acid anhydride: see **Anhydride** (3.9)

Cation: a positively charged ion (3.5)

Conjugate acid: a product obtained by protonating a base (3.2)

Cyano group: R—C≡N; a functional group with a carbon–nitrogen triple bond. Also called a nitrile group (3.4)

Cytosine: $C_4H_5N_3O$, a biologically important hetero-aromatic base (3.11)

Dehydration: the formal loss of water, usually from an alcohol (3.5)

Dipole moment: the vector pointing from the weighted center of positive charge to the center of negative charge in a collection of atoms—typically a molecule (3.2)

Electrophile: an electron-deficient reagent that attacks centers of electron density; from the Greek *electros*, electron, and *philos*, loving (3.7)

Electrophilicity: the tendency of an atom, ion, or group of atoms to accept electron density from a carbon center (3.7)

Endothermic: describing a reaction requiring input of energy (3.6)

Ester: RCO_2R; a functional group in which a carbonyl carbon bears an OR group (3.9)

Ether: a functional group in which two alkyl or aryl groups are attached to an sp^3–hybridized oxygen atom (3.5)

Full-headed curved arrow: indicates the movement of an electron pair (3.6)

Furan: C_4H_4O; a five-atom, ring-oxygen-containing, heteroaromatic molecule (3.11)

Guanine: $C_5H_5N_5O$, a biologically important hetero-aromatic base (3.11)

Half-headed curved arrow: indicates the movement of a single electron (3.6)

Heteroaromatic molecule: an aromatic molecule containing a ring heteroatom (3.12)

Heteroatom: any atom besides carbon and hydrogen (3.1)

Heterocycle: a cyclic molecule in which the ring contains one or more heteroatoms (3.11)

Heterocyclic aromatic: see **Heteroaromatic molecule** (3.11)

Heterolysis: *see* **Heterolytic cleavage** (3.6)

Heterolytic cleavage: the cleavage of a bond in which both electrons are shifted to one atom of the bond (3.6)

Homolytic cleavage: the cleavage of a bond with one electron shifted to each of the atoms of the bond (3.6)

Hydrogen bond: the weak association of a hydrogen atom attached to one electronegative heteroatom with a non-bonded electron pair on a second electronegative atom in the same or another molecule (X—H⋯Y, where X and Y are electronegative heteroatoms) (3.2, 3.5)

Imide: RCONHCOR'; a functional group in which two carbonyl carbons bearing an alkyl or aryl group are linked through a nitrogen atom (3.9)

Imine: a functional group containing a C=N double bond (3.3)

Intermolecular hydrogen bond: a hydrogen bond connecting electronegative atoms in separate molecules (3.2)

Intramolecular hydrogen bond: a hydrogen bond connecting electronegative atoms within the same molecule (3.2)

Ketone: R_2CO; a functional group in which a carbonyl carbon bears alkyl and/or aryl groups (3.8)

Lewis acid: an electron-pair acceptor (3.2)

Lewis base: an electron-pair donor (3.2)

Lone pair: two non-bonded electrons of opposite spin accommodated in an atomic or hybrid atomic orbital (3.1)

Lucas reagent: a mixture of Brønsted and Lewis acids that induces the conversion of an alcohol to the corresponding alkyl chloride (3.6)

Lucas test: a chemical means for distinguishing tertiary, secondary, and primary alcohols by the rate of formation of the corresponding alkyl chloride from an alcohol upon treatment with the Lucas reagent (3.6)

Nitrile: RC≡N; a functional group in which an *sp*–hybridized nitrogen atom is triply bound to carbon (also called a cyano group) (3.4)

Nucleophile: an electron-rich reagent that attacks centers of positive charge; from the Greek *nucleo*, nucleus, and *philos*, loving (3.7)

Nucleophilicity: the tendency of an atom, ion, or group of atoms to release electron density to form a bond with a carbon atom (3.7)

Oxidation: a chemical transformation resulting in the loss of electrons and hydrogen atoms and/or the addition of oxygen atoms or other electronegative heteroatoms (3.3)

Oxidizing agent: an agent that effects an oxidation (3.3)

Oxonium ion: a cation produced when oxygen bears three σ bonds (3.6)

Phenol: C_6H_5OH; OH-substituted benzene (3.11)

Polar covalent bond: a chemical bond characterized by appreciable charge separation because of unequal sharing of the electrons comprising the bond between two atoms (3.1)

Polarizability: a measure of the ease with which the electron distribution in a molecule can shift in response to a change in electric field; the ability of an atom to accommodate a change in electron density (3.12).

Polar protic solvent: a solvent that has an acidic proton on a heteroatom (3.5)

Primary alcohol: RCH_2OH; an alcohol in which the O—H group is attached to a primary carbon atom (3.5)

Primary amine: RNH_2; an amine in which nitrogen is attached to one carbon substituent (3.1)

Protic solvent: a solvent molecule incorporating a polar X—H bond (3.5)

Protonated alcohol: a cationic species produced upon association of a proton with a nonbonded lone pair of the oxygen atom of an alcohol (3.6)

Proton transfer: movement of H^+ from an acidic to a basic site (3.2)

Pyramidal: description of a spatial arrangement in which a central atom and three attached groups are located at the corners of a pyramid (3.1)

Pyridine: C_5H_5N: a six-membered ring, nitrogen-containing, heteroaromatic molecule (3.11)

Pyrrole: C_4H_4NH; a five-membered ring, nitrogen-containing, heteroaromatic molecule (3.11)

Quaternary ammonium ion: a positively charged ion in which nitrogen is attached to four carbon substituents (3.1)

Radical stability: $3°>2°>1°$ (3.6)

Redox reagent: a reagent that can induce an oxidation or a reduction (3.3)

Reducing agent: an agent that effects a reduction (3.3)

Reduction: a chemical transformation induced by the addition of electrons or hydrogen atoms and/or the removal of oxygen or other electronegative atoms (3.3)

Restricted rotation: the inhibition of rotation about a σ bond (3.3)

Secondary alcohol: R_2CHOH; an alcohol in which the OH group is attached to a secondary carbon atom (3.5)

Secondary amine: R_2NH; an amine in which nitrogen is attached to two carbon substituents (3.1)

Solvation: the association of solvent molecules about a solute (3.2, 3.13)

Substitution reaction: a chemical conversion in which one group is replaced by another (3.6)

Sulfonamide: RSO_2NH_2; a functional group in which an $-SO_2NH_2$ group is attached to an alkyl or aryl group (3.10)

Sulfonic acid: RSO_3H; a functional group in which an $-SO_3H$ group is attached to an alkyl or aryl group (3.10)

Tertiary alcohol: R_3C—OH; an alcohol in which the OH group is attached to a tertiary carbon atom (3.5)

Tertiary amine: R_3N; an amine in which nitrogen is attached to three carbon substituents (3.1)

Thioether: a functional group in which two alkyl or aryl groups are attached to an sp^3-hybridized sulfur atom (3.10)

Thiol: a compound bearing the SH functional group (3.10)

Thiol ester: RCOSR; a functional group in which a carbonyl carbon bears an alkyl or aryl group and an SR group (3.10)

Thiophene: C_4H_4S; a five-membered ring, sulfur-containing, heteroaromatic molecule (3.11)

Thymine: $C_5H_6N_2O_2$, a biologically important heteroaromatic base (3.11)

Uracil: $C_4H_4N_2O_2$, a biologically important heteroaromatic base (3.11)

Water: H_2O; the simplest compound containing sp^3-hybridized oxygen (3.5)

Chromatography and Spectroscopy:

Purification and Structure Determination

4

Key Concepts

Separating mixtures by chromatography

Types of chromatography

Energy probes in different types of spectroscopy

Interpreting simple ^{1}H and ^{13}C NMR spectra: number of signals, chemical shifts, splitting patterns, integration

Interpreting simple infrared spectra: characteristic frequencies of common functional groups, use of the fingerprint region

Interpreting simple absorption spectra: shifts in absorption maxima with structure, interpretation of color

Interpreting simple mass spectra: parent ions, fragmentation patterns

Answers to Exercises

Ex 4.1   Elution from a chromatographic column is fastest for those compounds that are most poorly adsorbed onto the chromatographic support. Adsorption is enhanced by the presence of polar or polarizable functional groups and by compounds with high molecular weight (which, because of more van der Waals attractive interactions, associate more strongly with the support). On an oxide support like alumina, strong hydrogen-bonding interactions between an adsorbate that can act as a hydrogen-bond donor (for example, amines, amides, and alcohols) and a stationary phase strongly enhance adsorption and reduce the rate of elution.

   (a)   Acetone. Although the carbonyl group in acetone is polar, the alcohol group in propanol is also polar and can hydrogen-bond with alumina and other common (oxide) supports (for example, silica gel). Propanol is therefore retained longer on the column.

   (b)   Cyclohexane. Although neither of these hydrocarbons is polar, the aromatic π system in benzene is more polarizable than the σ C—H and C—C bonds present in cyclohexane. Benzene is therefore retained slightly longer than the alkane.

   (c)   Methyl acetate. The presence of an acidic OH group in a carboxylic acid results in strong hydrogen-bonding association with the support, substantially increasing its adsorption on alumina. The related ester does not interact with the support through hydrogen bonding and moves through the column more readily.

   (d)   Cyclohexyl chloride. Although polar bonds are present in both compounds, cyclohexylamine can act both as a hydrogen-bond donor and acceptor and is strongly adsorbed to the stationary support in the column.

Ex 4.2 R_f values are calculated as the ratio of distances traversed by the compound in question and the solvent front. Thus, compounds with the higher R_f value elute faster than those with a lower R_f value. Here, $R_f(A) = 6.4$ cm / 7.6 cm $= 0.84$. $R_f(B) = 5.1$ cm / 7.6 cm $= 0.67$. $R_f(C) = 2.5$ cm / 7.6 cm $= 0.33$.

Ex 4.3   Increasing gas chromatographic retention times correlate well with increasing boiling points. Therefore, the same factors that influence intermolecular association (and

7

adsorption on polar solid-liquid chromatographic supports) are important in scaling the expected gas chromatographic retention times.

(a) Butane. Because of its lower molecular weight, butane has fewer van der Waals attractions between molecules. It therefore has the lower boiling point and the shorter gas chromatographic retention time.

(b) *N,N*-Dimethylacetamide. Lacking any group capable of acting as a hydrogen-bond donor, a tertiary amide has a lower boiling point than does an amino acid, which is both a hydrogen-bond donor and acceptor and exists often as a zwitterion. *N,N*-Dimethylacetamide therefore emerges substantially earlier from a gas chromatography column than does an amino acid (if at all).

(c) 2-Butanone. Although similar in molecular weight to propanoic acid, butanone lacks a polar OH group capable of hydrogen-bond donation. Its boiling point is therefore lower, and its gas chromatography retention time shorter.

(d) Methane. The lower molecular weight of methane and the absence of the polarizable chlorines present in CCl_4 assures a lower boiling point and a shorter gc retention time.

Ex 4.4 (a) The six carbons in cyclohexane are equivalent and give rise to only one signal in its ^{13}C NMR spectrum.

(b) Cyclohexanone has a symmetry plane through C-4 and the carbonyl group. Therefore, C-2 is equivalent to C-6 and C-3 is equivalent to C-5. Four separate signals are therefore observed, for C-1, C-2 (and C-6), C-3 (and C-5), and C-4.

(c) The symmetry in the phenyl ring makes C-2 and C-6 equivalent, as are C-3 and C-5. Thus, the ^{13}C NMR spectrum has four signals for the ring, one for the attached methylene group, and one for the carboxylate carbon, for a total of six.

(d) The two ends of this molecule are the same, and C-1 is equivalent to C-4 and C-2 is equivalent to C-3. Thus, this molecule exhibits only two ^{13}C NMR signals.

(e) Because 2-butanol bears a single functional group unsymmetrically placed in the molecule, none of its carbon atoms are related by symmetry. Hence, four distinct carbon signals are observed in the ^{13}C NMR spectrum..

(f) Because of a symmetry plane through the C—Br bond, C-1 and C-5 are equivalent as are C-2 and C-4. Thus, three signals are in the ^{13}C NMR spectrum.

(g) Unlike the isomer considered in part *f*, 2-bromopentane has no symmetry element and each of the five carbons gives rise to a separate ^{13}C NMR signal.

Ex 4.5 (a) First, note that symmetry in 3-methylpentane makes C-1 equivalent to C-5, and C-2 equivalent to C-4. Thus, the four signals to be assigned are for C-1, C-2, C-3, and the C-3 methyl group. To assign each of these resonances, employ the empirical rules summarized in Table 4.1.

C-1 (and C-5) has one α (C-2, +8.0), one β (C-3, +8.0), and two γ [C-4 and the C-3 methyl group: 2 x (–2.0) = –4.0] substituents, from which to predict a chemical shift of +12.0 δ.

C-2 has two α substituents [C-1 and C-3: 2 x (+8.0) = +16.0], two β [C-4 and C-3-methyl: 2 x (+8.0) = +16.0], and one γ (C-5, –2.0) substituent, from which to predict a chemical shift of +30.0 δ.

C-3 has three α substituents [C-2, C-4, and C-3 methyl: 3 x (+8.0) = +24.0] and two β substituents [C-1 and C-5: 2 x (+8.0) = +16.0]. Because C-3 is also tertiry, we must correct for the β substituents: $2 \times (-1.5) = -3.0$. We therefore predict for C-3 a chemical shift of $+37 \delta$.

The C-3-methyl has one α substituent (C-3: +8.0), two β [C-2 and C-4: $2 \times (+8.0) = +16.0$], and two γ [C-1 and C-5: $2 \times (-2.0) = -4.0$] substituents, for a predicted shift of $+20.0 \delta$.

Thus, we expect, in order of increasing chemical shift, the following signals:

	observed	predicted
C-1 and C-5	11.6	12.0
methyl	18.9	20.0
C-2	29.5	30.0
C-3	36.6	37.0

The average error for these predictions is 0.6δ, which is excellent correspondence with the observed shifts, especially considering the range of ^{13}C chemical shifts (over 150δ).

(b)   Each of the five carbons of 2-pentanol is unique, and each is responsible for one of the five observed signals.

C-1 has one α substituent (C-2: +8.0), two β [C-3 and OH: $2 \times (+8.0) = +16.0$], and one γ carbon (−2.0) substituent, for an expected chemical shift of $+22.0 \delta$.

C-2 has three α substituents [C-1, C-3, and OH: $3 \times (+8.0) = +24.0$], and an additional +38.0 because one of the substituents is oxygen. It also has a β substituent (C-4: +8.0) and a γ carbon (C-5: −2.0), and because it is a tertiary carbon, we correct for one β substituent (−1.5). We therefore expect a chemical shift of $+66.5 \delta$.

C-3 has two α substituents [C-2 and C-4: $2 \times (+8.0) = +16$] and three β substituents [C-1, C-5, and OH: $3 \times (+8.0) = +24$], for an expected shift of about $+40.0 \delta$.

C-4 has two α substituents [C-3 and C-5: $2 \times (+8.0) = +16$], one β (C-2: +8.0), one γ oxygen (−5.0), and one γ carbon (C-1: −2.0) substituent, for an expected shift of $+17.0 \delta$.

C-5 has one α (C-4: +8.0), one β (C-3: +8.0), and one γ carbon (−2.0) substituent, for an expected shift of +14.0.

Thus, we expect, in order of increasing chemical shift:

	observed	predicted
C-5	14.0	14.0
C-4	19.1	17.0
C-1	23.3	22.0
C-3	41.6	40.0
C-2	67.0	66.5

Again, the fit is excellent, with an average error of only 1.1δ.

Ex 4.6   Allyl alcohol (isomer H) has three different types of protons: three vinyl, two aliphatic, and one hydroxyl. Vinyl protons absorb in the region 4.5 to 6.5δ, and thus the signals at 5.2 (2 H) and 6.0δ (1 H) in the spectrum of allyl alcohol are the result of these protons. A H_2C—OH group would normally appear in the region 3.3-4.0δ but in allyl alcohol the adjacent sp^2-hybridized carbon atom also withdraws electron density, resulting in a signal at 4.1δ (2 H). The position of the signal for a hydroxylic proton is quite variable (1.0-6.0 δ), appearing in this spectrum of allyl alcohol at 2.9δ.

Ex 4.7   It is often the case that the structure of an unknown compound can be determined using only a portion of the spectral data available. For example, in this case the proton spectrum exhibits a signal at 9.7 δ, which is consistent only with an aldehyde (Table 4.2). Of the possible compounds with the formula C_3H_6O (A through G), D is the only aldehyde and therefore is the correct answer.

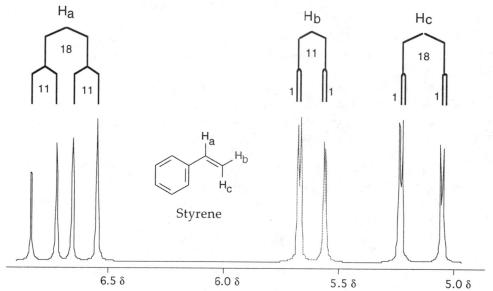

Ex 4.8   There are three equivalent methods for arriving at the same coupling pattern for a set of three interacting protons. It usually is most convenient to start with the largest coupling constant, proceeding to smaller and smaller values, as this approach avoids crossing of lines. H_a is coupled to H_c ($J = 18$ Hz) and H_b ($J = 11$ Hz); H_b is coupled to H_a ($J = 11$ Hz) and H_c ($J = 1$ Hz); and H_c is coupled to H_a (($J = 18$ Hz) and H_b ($J = 1$ Hz).

Ex 4.9   (a)   1-Propanol. These two compounds can be distinguished by symmetry. 1-Propanol lacks the symmetry of 2-propanol, and thus there are three distinct types of CH groups in 1-propanol, but only two distinct types in 2-propanol. In both isomers, the OH group appears as a broad singlet at a chemical shift that is difficult to predict, because it varies with concentration, solvent, and other physical conditions.

The C-1 protons in 1-propanol are shifted downfield by the electronegative oxygen atom of the attached OH group. Assuming that the OH rapidly exchanges, no splitting results from C-1 to OH coupling, and the two protons on C-1 appear as a triplet, split by the two protons on C-2. Signals for the two protons of C-2 are farther upfield. The coupling constants for splitting by the protons on C-1 and C-3 are not equal because these two types of protons are in different chemical environments. The C-2 protons are therefore split into a quartet by the three C-3 hydrogens and further split into an overlapping triplet by the two C-1 hydrogens. However, unless the spectrum is recorded on a high-field instrument, the resulting pattern is exceedingly complex. The C-3 protons are slightly farther upfield, split into a triplet by coupling to the C-2 protons.

This signal may overlap with the multiplet from the C-2 hydrogens. These three signals integrate in the ratio of 2:2:3. Here the broad OH signal appears at the same chemical shift as the C-2 protons, however. The integration observed is therefore 2:3:3.

In contrast with the somewhat complex spectrum for 1-propanol, the spectrum of 2-propanol is much simpler, owing to its symmetry. Thus, H-1 and H-3 are equivalent, split into a doublet by the proton on C-2. The proton on C-2 is shifted downfield by the attached OH group and split into a septet by the six equivalent protons on C-1 and C-3. These two signals integrate in a ratio of 6:1. Clearly, the distinction between these isomers is easily discerned by proton NMR spectroscopy.

(b) Benzyl alcohol. The distinction between *o*-cresol (*o*-methylphenol) and benzyl alcohol can be made on the basis of integration of the aliphatic CH singlet. For *o*-cresol, the methyl singlet and aromatic protons integrate to a ratio of 3:4 (1:1.3), whereas, for benzyl alcohol, the aliphatic signal (shifted downfield by the electronegative OH group) integrates to a ratio of 2:5 (1 to 2.5). In addition, benzyl alcohol is monosubstituted and *o*-cresol is disubstituted. These compounds give different aromatic splitting patterns.

(c) *trans*-3-Hexene. These two hexene isomers differ in symmetry, with a much simpler spectrum expected for the more symmetrical 3-hexene. For both the *cis* and the *trans* isomers (the *trans* isomer is considered here), the hydrogens on C-1 and C-6 are equivalent, as they are on C-2 and C-5 and on C-3 and C-4. The C-1 hydrogen signal appears as a triplet, split by the two C-2 hydrogens. The hydrogen on C-2 appears as a quartet because of the splitting with the C-1 hydrogens, but it is also split by the olefinic C-3 proton. The C-3 proton is split by the C-2 hydrogen neighbors but not by C-4, which is chemically equivalent. These three signals, in the aliphatic, allylic, and vinylic regions, respectively, as we move downfield, integrate to a ratio of 3:2:1.

The spectrum of 2-hexene is much more complex because the six carbons are chemically distinct. The signals from the C-5 and C-6 protons appear upfield and overlap. The allylic protons of C-1 and C-4 also probably overlap, giving a complex spectrum in the allylic region. The two vinyl hydrogens on C-2 and C-3 are not equivalent; they split each other and are split by their nearest neighbors. One might reasonably expect multiplets in the aliphatic, allylic, and vinylic regions of the spectrum, integrating in a ratio of 5:5:2.

(d) Propanoic acid. The different functional groups in these two molecules make their spectra quite distinct. The carboxylate OH has a characteristic broad signal with a chemical shift about 10 ppm downfield from TMS (tetramethylsilane), whereas the alcoholic OH group exhibits a variable shift that is not as far downfield (usually in the 1-6 ppm region). The chemical shifts of the other protons also differ in these compounds. Hydrogens adjacent to a carbonyl group appear at about 2.0–2.6 δ: five such protons are present in α-hydroxyacetone, but only two α protons are present in propanoic acid.

In α-hydroxyacetone, the unsubstituted methyl group appears as a singlet at about 2.0 δ, and the methylene carbon bearing the OH group is shifted farther downfield, appearing as a second singlet. The hydroxyl group appears as a broad singlet at a chemical shift dependent on the conditions employed for the measurement of the spectrum. These signals respectively integrate to a ratio of 3:2:1.

Propanoic acid has a two-proton signal in the region near 2.5 δ that appears as a quartet by virtue of splitting by the three equivalent C-3 hydrogens. The C-3 methyl group is upfield, in the aliphatic region, split into a triplet by the two equivalent C-2 methylene protons. Although the integration for these signals is the same as for α-hydroxyacetone (3:2:1), the differences in splitting patterns and in chemical shift make the spectra of these constitutional isomers easily distinguishable.

Ex 4.10 To answer this question, classify each unique proton in the molecule according to its functional group, as listed in Table 4.2. Then count the number of nearest neighbor protons (those on adjacent carbons) from which to predict splitting patterns. The number of each unique type of hydrogen provides the expected integration (d = doublet, t = triplet, q = quartet).

(a) H-1: 3.3–4.0, t, 2 H; H-2: 1.2–1.4, t of t, 2 H; H-3: 1.2–1.4, t of q, 2 H; H-4: 1.2–1.4, t, 3 H; OH: 2.5–7.0, broad singlet.

(b) H-1: 9.6–10.0, broad singlet, 1 H; H-2: 2.1–2.4, t, 2 H; H-3: 1.2–1.4, t of q, 2 H; H-4: 1.2–1.4, t, 3 H.

(c) H-1: 1.2–1.4, d, 3 H; H-2: 3.3–4.0, t of q, 1 H; H-3: 1.2–1.4, d of q, 2 H; H-4: 1.2–1.4, t, 3 H; OH: 2.5–7.0, broad singlet.

(d) H-1: 2.1–2.4, singlet, 3 H; H-3: 2.1–2.4, q, 2 H; H-4: 1.2–1.4, t, 3 H.

Ex 4.11 Nodes result when two (or more) signals of different frequencies are mixed, because at various points in time, the intensities of the components will be equal in strength but opposite in sign. At each of these points the sum will be zero, resulting in a node.

Ex 4.12 The absorption at 1725 cm^{-1} is in the region characterisitc for carbonyl groups. Of the isomers with formula C_3H_6O, only acetone (C) and propanal (D) have carbonyl groups.

These isomers could be differentiated easily by either ^{1}H or ^{13}C NMR spectroscopy. Acetone has six equivalent protons that would therefore appear as a singlet, whereas propanal has three different types of protons with the characteristic aldehyde proton appearing in the range 9.5-10.0 δ. The ^{13}C NMR spectrum of acetone would exhibit only two different signals because the two methyl groups are identical, whereas each carbon atom of propanal is unique and would give rise to its own signal.

Ex 4.13 The O—H stretching absorption in the infrared spectrum of alcohols changes appearance with concentration because of hydrogen bonding. In very dilute solution, each molecule of an alcohol is surrounded by the solvent (in this case, CH_2Cl_2), resulting in a sharp absorption for this single species. As the concentration increases, hydrogen bonding between pairs of alcohol molecules increases, and in very concentrated solutions, there are trimers and even tetramers. The stretching frequency of the O—H bond varies with the degree of hydrogen bonding. Thus, at higher concentrations, there is a broad signal that is the composite resulting from dimer, trimer, and tetramer.

Ex 4.14 The strength of an infrared absorption is directly dependent on the polarity of the bond. Nonpolar bonds do not absorb at all, whereas quite polar bonds absorb strongly. The presence of the hydroxyl group in allyl alcohol significantly increases the polarity of the C=C bond, resulting in a stronger signal for allyl alcohol than for propene.

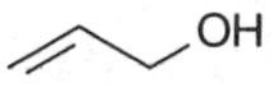

Allyl alcohol   Propene

Ex 4.15 The ^{13}C NMR spectrum for each of these differ in the number of signals for the sp^3-hybridized carbon atoms and the sp^2-hybridized carbon atoms of the benzene ring.

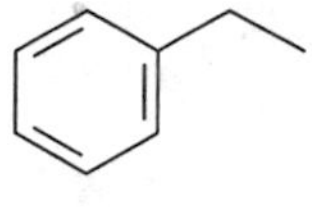

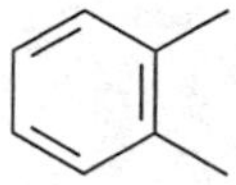

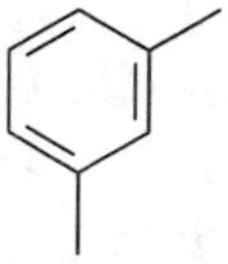

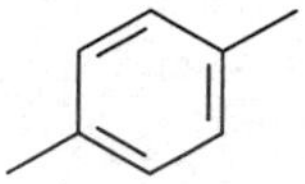

	A	B	C	D
sp^3	2	1	1	1
sp^2	4	3	4	2

The ^{1}H NMR spectrum of A will differ from those of B, C, and D, each of which will exhibit a single signal for the six identical protons of the two methyl groups. In theory, the number of aromatic signals will also differ but it is not possible to rule out coincidental overlap. Indeed, ethylbenzene (A) exhibits a broad singlet for the three different aromatic protons.

Ex 4.16 This problem is best solved by reference to Table 4.3, which lists typical infrared bands for characteristic bond stretches in common functional groups.

(a) 2-Butanone has a strong carbonyl stretch (1640–1780 cm^{-1}), whereas the vinyl ether does not.

(b) 2-Butanone has a strong carbonyl stretch (in the region 1640–1780 cm^{-1}), which the allyl alcohol lacks. Conversely, the alcohol exhibits a broad band for the O—H stretch at 3400–3650 cm^{-1}, which is missing from the IR spectrum of the ketone.

(c) The characteristic nitrile stretch at 2210–2260 cm^{-1} is absent from the amine. The nitrile, however, does not exhibit the characteristic N—H stretch of a primary amine at 3360–3500 cm^{-1}.

(d) Benzene shows aromatic C—H stretches above 3000 cm^{-1} (at about 3020–3100 cm^{-1}), whereas cyclohexane shows only aliphatic C—H stretches below 3000 cm^{-1} (at about 2850–2960 cm^{-1}). Benzene also shows a pair of peaks at about 1500 and 1600 cm^{-1}, indicative of ring distortions; these peaks are absent in the IR spectrum of cyclohexane.

Ex 4.17 Absorption maxima are shifted to longer wavelength (lower energy) with increasing conjugation.

(a) The conjugated styrene derivative at the left absorbs at longer wavelengths than does the allyl isomer at the right in which the double bond is isolated from the aromatic ring.

(b) In the structure at the right, a conjugated π system connects the keto and ester carbonyl groups electronically. This conjugation is absent in the β-ketoester at the left and thus the conjugated ketoester absorbs at longer wavelengths.

(c) Because the central ring of the structure at the right has two sp^3-hybridized carbons, the two aromatic rings are not conjugated. Thus, anthracene, the fused aromatic hydrocarbon at the left, absorbs at longer wavelengths because it is a fully conjugated system.

(d)   The two π bonds in 1,3-cyclohexadiene at the right are conjugated, but those in 1,5-hexadiene at the left are not. Because conjugation shifts absorption maxima to lower energies, cyclohexadiene absorbs at longer wavelengths.

Ex 4.18   The parent peak in a mass spectrum is the cation that corresponds to the molecular weight of the molecule, produced by loss of an electron without fragmentation. The mass-to-charge ratio (m/z) can thus be obtained by calculating the molecular weight. Fragment ions are obtained when weak bonds are cleaved after ionization, often at or adjacent to functional groups.

(a) $C_4H_{10}O$: m/z 74. Base peak: cleavage adjacent to the carbon bearing the –OH group: CH_2OH^+: m/z, 31.

(b) C_4H_8O: m/z 72. Base peak: cleavage adjacent to the carbonyl group: CH_3CO^+: m/z, 41, and $CH_3CH_2CO^+$: m/z, 57.

(c) C_3H_6O: m/z 58. Base peak: cleavage adjacent to the carbonyl group: HCO^+: m/z, 29.

(d) C_3H_7Cl: m/z 78 and 80 (there are two chlorine isotopes that occur in a 2-to-1 ratio). Base peak: loss of a chlorine atom: $CH_3C^+HCH_3$: m/z, 43.

Answers to Review Problems

Pr 4.1   Longer chromatographic retention times result from stronger adsorption on the stationary support. Because adsorption strength correlates with higher polarity and higher molecular weight, the correct prediction rests on choosing the more polar of a pair of closely related molecules or the heavier molecule when the functional groups are the same.

(a) 2-butanol   (b) octanoic acid   (c) cyclohexanol   (d) guanine

Pr 4.2   Gas chromatography is used for volatile, low-molecular-weight compounds. High-pressure liquid chromatography is used primarily for compounds of low-to-medium molecular weight that exist as neutral species. Gel electrophoresis is used for polymeric compounds and for highly polar, ionic, or zwitterionic low-molecular-weight species. Gas chromatography is most frequently used as an analytical technique, although preparative separations by gas chromatography also can be conducted. The other two methods are used routinely for preparative purposes, as well as for analysis.

(a)   Gas chromatography. These compounds are sufficiently volatile to pass through a gas chromatographic column readily. Their low molecular weights make them less appropriate for liquid chromatographic methods because they are not retained well on a normal solid column. Because neither compound is easily ionized, gel electrophoresis is not applicable.

(b)   Reverse-phase, high-pressure liquid chromatography. A carboxylic acid is relatively nonvolatile and passes through a gas chromatographic column only with difficulty. Because of its high polarity, a carboxylic acid also binds tightly to most liquid chromatographic columns. Because reverse-phase columns are hydrocarbon-like, they do not bind excessively to a carboxylic acid.

(c)   Gas chromatography. Both compounds are relatively volatile: the analysis described in part *a* can be used.

(d)   High-pressure liquid chromatography or gel electrophoresis. The presence of heteroatoms in the aromatic ring appreciably raises the boiling point, making gas chromatographic separation unsatisfactory for guanine. The presence of basic nitrogen atoms allows for converting both compounds into ions by protonation, producing species that migrate under the influence of an electrical field. These compounds are too polar and too basic for the use of high-pressure liquid chromatography, except with reverse-phase columns.

Pr 4.3

(a) 1,1,2-Trichloroethane. The downfield proton, attached to the carbon that also bears two chlorines, is split into a triplet by the two protons on the adjacent carbon. The protons at 3.95 δ, those on the monochlorinated carbon, are split into a doublet by the single proton on the adjacent carbon. The infrared spectrum contains absorptions for only aliphatic C—H stretches at 2950 cm^{-1} and C—Cl stretches below 1000 cm^{-1}.

(b) Acetaldehyde. The aldehydic proton appears as a broad singlet far downfield, whereas the protons on the α methyl group at 2.20 δ are split into a doublet by the aldehydic proton. The infrared band at 1730 cm^{-1} is indicative of the presence of a carbonyl group, here an aldehyde.

(c) Acetic acid. The carboxylic acid proton appears far downfield as a broad singlet. The α methyl group at 2.10 δ is also a singlet because the adjacent carbon bears no hydrogens. The infrared spectrum shows both a strong carbonyl stretch and a broad band at 3600 cm^{-1} for the stretching of the O—H group.

(d) Methyl formate. The formate proton appears far downfield in the aldehyde region as a broad singlet, whereas the methoxy group appears as a broad singlet at 3.77 δ because of the electron withdrawal by the attached oxygen atom. The carbonyl band in the infrared spectrum is shifted to 1745 cm^{-1}, indicative of an ester, and is accompanied by a C—O stretch at 1250 cm^{-1}.

(e) Ethanol. The broad singlet at δ 2.58 is the OH group. (However, the position of this absorption can change with solvent and concentration differences.) The triplet at 1.22 δ is assigned to hydrogens on C-2, split by the two equivalent protons on the carbon bearing the OH group. The methylene group is shifted downfield by the electronegative oxygen atom and is split into a quartet by the three methyl hydrogens on the adjacent atom. The characteristic infrared band indicates the presence of an alcohol group.

(f) α-Chloropropanoic acid. The broad signal far downfield is attributable to the carboxylate proton. The α carbon bears both a proton and a chlorine, causing a further downfield shift from the normal chemical shift of protons on a carboxylate α carbon. The signal is split into a quartet by the three protons on the adjacent methyl group. The proton on the α carbon, in turn, splits the signal for the methyl group into a doublet, which appears in the normal region for an aliphatic carbon. The infrared spectrum shows bands assigned to an O—H stretch (3600 cm^{-1}), a carboxylate carbonyl (1710 cm^{-1}), and a C—Cl stretch (below 850 cm^{-1}).

(g) Methoxyacetonitrile. The methoxy group appears as a singlet at δ 3.47 because of electron withdrawal by the adjacent oxygen. The methylene group is further downfield, being not only adjacent to an oxygen atom but also α to the nitrile group. The presence of the nitrile functionality is indicated by the presence of a band at 2250 cm^{-1} in the infrared spectrum, as well as the absence of N—H stretches. The band at 1100 cm^{-1} is assigned as the C—O stretch in the methoxy group.

(h) Trimethylene oxide. The two equivalent methylene groups attached to the ring oxygen are split into a triplet by the common adjacent methylene group. The CH$_2$ group, in turn, is split by the four equivalent hydrogens into a quintet. The band at 1120 cm^{-1} is indicative of a C—O stretch, and the absence of a C=O carbonyl or OH stretch indicates that oxygen is present as an ether.

(i)   Allyl alcohol. The three protons on the double bond appear as multiplets in the vinylic region, splitting each other as well as being split by the allylic methylene group. The CH_2 group also appears as a multiplet because of long-range splitting by the vinylic hydrogens. The alcohol proton at δ 3.58 appears as a broad singlet in the proton NMR spectrum. The O—H stretch is indicated by a broad band at 3600 cm^{-1} in the infrared. The vinylic protons appear in the IR spectrum as C—H stretches above 3000 cm^{-1}, whereas the methylene hydrogens are at a frequency just below 3000 cm^{-1}. The remaining IR band is attributable to a C—O stretch.

(j)   *p*-Chloroaniline. The presence of an aromatic ring is indicated by the pair of doublets in the aromatic region, with the *para*-substitution pattern indicated by the symmetry of the two doublets. (The *para* substitution makes H-2 and H-6 equivalent, as well as H-3 and H-5.) The chemical shift of the two-proton singlet is indicative of the proximity of an electron-withdrawing atom, as is encountered in the NH_2 group. The primary amino group is also indicated by the pair of bands at 3520 and 3400 cm^{-1}. Aromatic protons are indicated by the C—H stretch above 3000 cm^{-1} and by the pair of bands assigned to ring deformations at 1490 and 1590 cm^{-1}. The presence of chlorine is inferred from the C—Cl stretch at 910 cm^{-1}.

Pr 4.4   The carbon NMR spectrum shows the presence of ten unique carbons from which it can be concluded that there is no symmetry element in the molecule. The likely presence of an aromatic ring derives from the index of unsaturation, as well as the complex pattern in the aromatic region of the proton NMR spectrum. The presence of a methoxy group is indicated by the singlet in the proton NMR in the region indicative of attachment to an electron-withdrawing atom. The phenolic proton appears as a broad singlet in the proton NMR and as a broad band in the O—H stretch region of the infrared spectrum. The assignment of the placement of these functional groups within the molecule would be ambiguous based on spectral data alone because of the complexity of the ^1H NMR spectrum caused by overlapping absorptions. In most cases, a practicing chemist would have additional chemical information such as the structure of the starting material and the identity of the reagents from which the material was obtained.

Pr 4.5   To predict features of a proton NMR spectrum, begin by drawing a structure to determine the various types of carbons and protons present and the number of each type. Then use Table 4.2 to estimate chemical shifts for each type of proton. From the structure, count the number of nearest neighbors to predict splitting patterns. Characteristic infrared bands for many functional groups can be predicted from the list in Table 4.3. A more detailed analysis, especially in the fingerprint region, is difficult, however. The absence of characteristic bands also can be helpful in choosing between possible isomeric structures.

(a)   There are two types of protons in ethyl bromide: those on the carbon bearing the bromine atom and those in the methyl group. The two H-1 protons are shifted downfield by the electronegative bromine and appear as a quartet because of splitting by the three hydrogens of the C-2 methyl group. The three H-2 protons appear in the normal aliphatic region as a triplet because of splitting by the two H-1 protons. These two signals integrate, respectively, as 2:3. Two ^{13}C NMR signals are expected, one shifted downfield by bromine. There are only aliphatic C—H stretches in the IR spectrum (below 3000 cm^{-1}) and C—Br stretches in the fingerprint region below 1000 cm^{-1}.

(b) There are three nonequivalent carbons in propyne and hence three ^{13}C NMR signals. There are two types of hydrogens: the acetylenic hydrogen on C-1 and the methyl hydrogens on C-3, with the acetylenic hydrogen appearing downfield from the aliphatic proton signal. Both signals appear as singlets, because neither carbon has nearest-neighbor hydrogens, with an integration ratio of 1:3. This unsymmetrical alkyne shows an infrared band at about 2200 cm^{-1}, in addition to the aliphatic C—H stretches just below 3000 cm^{-1}.

$HC \equiv C - CH_3$

(c) There are three types of carbon in 2-propyne-1-ol, with the hydroxyl-substituted carbon appearing downfield from the acetylenic carbons. There are also three types of protons: the OH proton; the H-1 proton on the carbon bearing the OH group; and H–3, the acetylenic proton. All three signals are singlets, lacking nearest neighbors for observable coupling. The chemical shift of the OH group is difficult to predict because it depends on solvent and concentration, but H-1 appears downfield from a typical aliphatic carbon because of the deshielding effect of both the triple bond and the hydroxyl group. H–3 appears in the typical region for sp-hybridized carbons. Integration of these three signals appears in a ratio of 1:2:1. The infrared spectrum shows not only an unsymmetrical triple bond, as in part b, but also a broad absorption band at about 3600 cm^{-1}, as is characteristic of an alcohol.

$HC \equiv C - CH_2OH$

(d) There are three types of protons and carbons in allyl bromide and, hence, three ^{13}C signals, one appearing downfield from the allylic region (by virtue of deshielding by the electronegative bromine atom) and two in the vinylic region. The two-proton signal from H-1 on the carbon bearing bromine is split into a doublet by the vinylic hydrogen on C-2. The hydrogens on C-2 and C-3 appear downfield in the vinylic region, with H-2 appearing as a triplet of triplets, being split not only by H-1, but also by H-3. H-3 appears as a doublet because of coupling to H-2. These signals integrate in a ratio of 2:1:2, respectively. The infrared spectrum of allyl bromide exhibits aliphatic C—H stretches at about 2850 cm^{-1}, as well as vinylic C—H stretches and C—Br stretches in the fingerprint region.

(e) Because of symmetry in 2-nitropropane, C-1 and C-3 are equivalent. There are two signals in the ^{13}C NMR spectrum, appearing, respectively, in the normal aliphatic region and farther downfield because of strong electron withdrawal by the nitro group. Similarly, the proton spectrum has a six-proton doublet in the aliphatic region, in which the observed splitting is due to coupling with the single proton on C-2. The H-2 signal appears as a septet shifted appreciably downfield from the aliphatic region by the nitro group. The infrared spectrum shows aliphatic C—H stretches at about 2850 cm^{-1}, as well as bands at 1560 and 1350 cm^{-1} for N—O stretches in the nitro group.

(f) Two types of carbon are present in N,N-dimethylformamide: the formyl carboxamide carbon and the methyl groups on the amide nitrogen. The carboxamide carbon appears significantly downfield from the methyl groups, which are themselves shifted downfield somewhat from the typical aliphatic region by the electronegative nitrogen atom. Analogous shift effects are seen in the proton spectrum, in which these hydrogens appear as singlets, integrating 1:6. The infrared spectrum is dominated by the strong carboxamide carbonyl stretch at about 1750 cm^{-1}. Although amides often show broad stretches at about 3400 cm^{-1}, such bands are absent here because the N—H amide bonds to which they are assigned have been replaced by N-methyl groups.

(g) The three types of carbon in methyl ethyl sulfide appear as separate ^{13}C NMR signals, with the methyl and methylene groups bound to sulfur shifted downfield from the remaining aliphatic carbon atoms. The methyl group attached to sulfur appears in the proton spectrum as a singlet shifted downfield, possibly overlapping with the quartet from the methylene protons of the ethyl group. The remaining ethyl carbon appears as a triplet in the normal aliphatic region. These three proton signals integrate, respectively, in a ratio of 3:2:3. The infrared spectrum shows aliphatic C—H stretches, as well as broad C—S stretches.

(h)   Like 2-nitropropane, 2-propanol is symmetrical and exhibits only two carbon signals: one in the aliphatic region and the other shifted downfield by the OH group. The proton spectrum is also similar to that described in part *e*, except that the magnitude of the chemical shift of H-2 is smaller and an additional broad singlet for the OH group is evident. As before, the downfield C-2 signal appears as a septet and the upfield signal as a doublet, integrating in a ratio of 1:6. The infrared spectrum shows a characteristic OH stretch in the region about 3600 cm^{-1}, with aliphatic C—H stretches also present.

(i)   Four types of carbon are present in vinyl acetate: farthest upfield is the acetate methyl group, followed farther downfield by the two vinyl carbons and the carboxylate ester carbon. The three protons on methyl appear at about 2.0 δ as a singlet, and those on the vinyl group appear, respectively, as a one-proton doublet of doublets for the hydrogen on the carbon attached to the ester oxygen and as two one-proton doublets of doublets for the two protons at C-2 of the vinyl group (on the same and opposite sides of the double bond as the ester oxygen). The infrared spectrum shows a strong carbonyl stretch for the ester group (1750 cm^{-1}), as well as alkyl and alkenyl C—H stretches.

(j)   Each of the four carbons in this α-bromoacid appears as a separate signal in the ^{13}C NMR spectrum, with the carboxylate carbon farthest downfield, followed by the C-2 carbon, which is shifted by the electronegative halogen, and C-3 and C-4 in the normal aliphatic region. The carboxylic acid proton appears far downfield as a singlet, with H-2 appearing as a triplet, shifted downfield from the aliphatic H-3 and H-4 signals. The last two signals are likely to overlap; and, because H-3 is expected to appear as a triplet of quartets and H-4 as a triplet, they are likely to appear together as a complex multiplet integrating to indicate the presence of five protons. The infrared spectrum shows both a strong carboxylate carbonyl stretch at about 1700 cm^{-1} and a broad O—H stretch at about 3600 cm^{-1}.

(k)   The four carbons of 2-butanone are unique, giving rise to four ^{13}C signals. The carbonyl carbon C-2 appears farthest downfield, with the two α carbons C-1 and C-3 being shifted downfield from the aliphatic region, where C-4 appears. The H-1 and H-3 protons appear at about δ 2.0 ppm, respectively, as a three-proton singlet and a two-proton quartet, possibly overlapping. H-4 appears upfield as a three-proton triplet. The characteristic carbonyl stretch of the ketone at 1700 cm^{-1} is the strongest band in the infrared spectrum.

(l)   Like butanone, butanal has four unique carbons, with the carbonyl carbon appearing farthest downfield and the α carbon shifted downfield from the aliphatic region, where C-3 and C-4 signals are evident. The aldehydic proton appears as a singlet at about δ 9.0 ppm, and the H-2 signal appears as a triplet at about 2.0 δ ppm. H-3 and H-4 are likely to overlap, appearing as a five-proton multiplet in the aliphatic region of the spectrum. The integration of these four signals therefore is 1:2:5. As with butanone, the infrared spectrum is dominated by the carbonyl stretch at 1750 cm^{-1}, with aliphatic C—H stretches also being evident.

(m) There are five unique carbons in this compound: four along the main hydrocarbon chain and one in the methoxy group. C-1 and C-3 are shifted downfield, together with the methoxy group, because of attachment to the electronegative oxygen atom. C-2 and C-4 appear in the normal aliphatic region. The proton spectrum exhibits a singlet for the methoxy group at about δ 3.6 ppm, in the same region in which signals from C-1 and C-3 protons appear. The C-1 protons appear as a triplet and the C-3 proton as a triplet of quartets. As a result, the region where these three protons appear (together with the singlet for the methyl of the methoxy group) is extremely complex, unless a high-field instrument is used. This region therefore exhibits a multiplet integrating to show the presence of six protons. The protons on C-2 and C-4 also appear together, as a doublet of doublets and a triplet, respectively, and integrating for five protons. Aliphatic C—H and C—O stretches are present in the infrared spectrum, as well as a broad O—H stretch at about 3600 cm^{-1}.

(n) A mirror plane of symmetry in monosubstituted benzenes makes C-2 in the ring equivalent to C-6 and C-3 to C-5, with C-1 and C-4 being unique. With the attached methyl group, there are, therefore, five ^{13}C signals. The proton spectrum shows a five-proton signal in the aliphatic region and a three-proton singlet in the benzylic region at about δ 2.5 ppm. The infrared spectrum shows both aliphatic and aromatic C—H stretches below and above 3000 cm^{-1}, respectively, as well as the characteristic pair of bands at 1600 and 1500 cm^{-1} caused by ring deformations.

Pr 4.6 First focus on the feature that differentiates the two compounds, and then devise a spectroscopic probe that is unique to each.

(a) An *n*-butyl group can be most easily distinguished from a *t*-butyl group by proton or carbon NMR. The four carbons in the *n*-butyl substituent are unique and give rise to a complex aliphatic proton spectrum; those in *t*-butyl are equivalent and appear as a nine-proton singlet. Four signals appear in the ^{13}C NMR spectrum of *n*-butylamine, and two in that of *t*-butylamine.

(b) An analysis similar to that for part *a* applies to *s*-butylamine, which has four unique types of hydrogen compared with the two unique types in *t*-butylamine.

(c) These two compounds differ in molecular weight, with 1-methylcyclohexene possessing one more carbon and two more hydrogens than methylenecyclopentane. They can therefore be differentiated by comparing the parent ions in the mass spectrum. They also differ in symmetry, with methylenecyclopentane having only four unique carbons by virtue of the symmetry plane passing through the long axis of the olefinic group and the C-3–C-4 σ bond. 1-Methylcyclohexene lacks this symmetry element and shows seven unique carbons in its ^{13}C NMR spectrum.

(d) The methyl group of *m*-cresol appears as a singlet in the proton NMR spectrum, which integrates in the ratio 3:4 with the aromatic protons. The benzylic proton singlet in benzyl alcohol integrates to a ratio of 2:5 (1:2.5) with the aromatic protons. Furthermore, there is no symmetry in *m*-cresol, so that all seven carbons appear as unique signals, whereas the symmetry plane through C-1 and C-4 causes benzyl alcohol to exhibit only five ^{13}C NMR signals.

(e)   The methyl singlet in the proton spectrum of *p*-cresol integrates in a ratio of 3:4 to the aromatic protons, whereas the methoxy singlet in anisole integrates 3:5. The chemical shifts of these two singlets also differ, with the methoxy singlet appearing farther downfield than the benzylic methyl group. Again, the symmetry in both aromatic compounds reduces the number of observable carbon absorbances to four in *p*-cresol and five in anisole. The infrared spectrum of a phenol shows a strong O—H stretch at about 3600 cm^{-1}, which is absent from the IR spectrum of anisole.

(f)   The presence of a carbonyl stretch in the infrared spectrum of a ketone is absent in that of the epoxide. The proton NMR spectrum of acetophenone shows a singlet at about 2.0 δ for the acetyl group, whereas the three nonaromatic hydrogens of styrene oxide appear as a one-proton triplet and a two-proton doublet.

(g)   The molecular weights of acetophenone and *p*-anisaldehyde differ by one oxygen atom; their mass spectral parent ions therefore differ by 16 mass units. Furthermore, the chemical shift of a methyl group attached to an oxygen atom differs from that attached α to a carbonyl group. The aldehyde also exhibits a characteristic downfield shift for the aldehydic proton that is absent in the ketone.

(h)   *m*- and *p*-Xylene differ in symmetry. In *p*-xylene, all four nonalkylated ring carbons are equivalent, as are the two alkylated carbons and the appended methyl groups. This isomer therefore exhibits three absorbances in its ^{13}C NMR spectrum. In *m*-xylene, a symmetry plane passes through C-2 and C-5, causing five types of carbon to be observable in the ^{13}C NMR spectrum. Analogous considerations apply to the analysis of the complexity of the splitting pattern observed in the proton NMR spectrum.

Pr 4.7   A hydrocarbon with a parent mass peak at 86 is most likely to have C_6H_{14} as its formula. (A structure containing fewer than six carbons cannot accommodate enough hydrogens to reach the observed parent mass. A C-7 hydrocarbon of mass 86 would have to be C_7H_2, which would be a highly unsaturated species.) It is therefore a saturated C_6 alkane. From the ^{13}C NMR spectrum, we can tell that the molecule is symmetrical with only two types of carbons present. Of all the possible C_6H_{14} isomers (recall Exercise 1.10, part *b*), only 2,3-dimethylbutane has only two types of carbons; all other isomeric hexanes have more. The presence of four methyl groups also accounts for the major fragment in the mass spectrum corresponding to the loss of a methyl group.

Pr 4.8   The index of hydrogen deficiency is consistent with the presence of either one double bond or one ring in these compounds. The presence of two signals far downfield in the carbon NMR is consistent with the presence of a double bond. The ^{13}C NMR further indicates that the molecule lacks symmetry; that is, that all six carbons are different. This specifically eliminates all structures in which the placement of substituents is symmetrical.

No symmetry:
all six carbons unique

Symmetry:
two or more identical carbons

Is it possible to distinguish the unsymmetrical isomers from each other from the ^{13}C NMR spectrum alone? First, by referring to Table 4.1, we can predict the chemical shifts for the sp^2-hybridized carbons of each of these isomers.

a	b	c	d	e	f	g	isoA	isoB
113.0	121.0	113.0	121.0	105.0	113.0	113.0	124.9	123.9
137.0	129.0	145.0	129.0	145.0	137.0	137.0	131.6	130.7

We can conclude that isomers A and B are b and d because the predictions for the other isomers would have much larger errors. We could predict that isomer B is the *cis* isomer because its absorptions for the sp^2-hybridized carbons are approximately 1 δ unit upfield from that of isomer A, but these differences are not sufficiently large to allow for an unambiguous assignment of which is which. At this point, we could calculate the shifts for all of the other carbons of b and d, but let us see if we can shorten the task somewhat. Looking at Table 4.1, we find that the effect of an α *trans* double bond is opposite that of an α *cis* double bond on an sp^3-hybridized carbon. Thus, we predict that C-1 in b is at 18.5, whereas C-1 in d is at 13.5 (both have one α and one β substituent). C-6 in each of these geometric isomers also has one α and one β substituent, as well as one γ carbon, resulting in a prediction of 14.0. Because all other carbons have more α/β substituents (or are sp^2-hybridized), C-1 and C-6 in b and d should be the most upfield absorptions. The highest field shifts for isomer B of 12.7 and 13.7 thus fit better with a prediction of 13.5 and 14.0 than they do for 14.0 and 18.5. In fact, isomer B is *cis*-2-hexene, structure d, and isomer A is *trans*-2-hexene, structure b.

Pr 4.9 Compounds containing only σ bonds absorb only at wavelengths below 200 nm; a single π bond, at about 200 nm; a conjugated diene, at about 230 nm; and a triene, at about 250 nm. Thus, cyclohexane and tetrahydrofuran do not absorb significantly in the ultraviolet region. Toluene has a relatively weak band at 278 nm, and acetone has a weak band at 315 nm. The extended conjugation in the aromatic system of naphthalene produces a much stronger absorption at 286 nm and a weak absorption at 312 nm. Both cyclohexane and tetrahydrofuran are therefore suitable solvents for measuring the spectrum of naphthalene. In contrast, although the absorptions of toluene and acetone are weak, these compounds would absorb essentially all the light near their absorption maxima and thus could not be used as solvents in which to observe the absorption of naphthalene at 286 nm.

Pr 4.10 The parent ion has an *m/z* ratio that is the same as the molecular weight calculated using the mass of the most prevalent isotope of each element present.

(a) 86.073. Aliphatic aldehydes readily cleave at the C—C bond to the carbonyl carbon, leaving the charge with the oxygen fragment. This ion, CHO^+ (*m/z* 29), is thus produced from all such aldehydes, and its presence is diagnostic for this functional group.

(b) 120.058. Fragmentation of aryl ketones takes place readily to form an acylium ion (*m/z* 105) stabilized by the aromatic ring. This ion then loses CO to form $C_6H_5^+$ (*m/z* 77).

(c) 46.041. The major fragmentation pathway for aliphatic alcohols is by cleavage of a C—C bond to the carbinol carbon, with positive charge remaining with the oxygen fragment.

(d) 74.073. Ethers are cleaved in two principal ways: at the C–O bond, with positive charge remaining on carbon; and at the C–C bond one removed from the oxygen with the charge residing on the residue containing the oxygen. For diethyl ether, these cleavages produce ions at 29 and 59 *m/z*.

Important New Terms

Absorption spectroscopy: the measurement of the dependence of the intensity of absorbed light on wavelength, for light in the visible and ultraviolet regions (4.3)

Adsorption: association with a solid surface, often reversible (4.2)

Applied field (H_{app}): the external magnetic field applied to a sample in a nuclear magnetic resonance spectrometer (4.3)

Baseline separation: an efficient separation of two compounds in which the peaks detected as representative of elution of the component molecules do not overlap; that is, the detector response returns to the base line between peaks (4.2)

Base peak: the most intense peak in a mass spectrum (4.3)

Chemical shift: the magnitude of the change of the observed resonance energy for a given nucleus relative to that observed for a standard (usually, tetramethylsilane); the position on an NMR spectrum at which a given nucleus absorbs (4.3)

Chromatogram: a plot of a detector response as a function either of the volume of effluent flowing through the column or of time (4.2)

Chromatographic resolution: the degree of separation of a mixture of compounds (4.2)

Chromatographic separation: the isolation of individual components of a mixture through a chromatographic technique (4.2)

Chromatography: the technique by which components of a mixture are partitioned between two different phases, attaining separation because of a difference in solubility of the component molecules in each phase (4.1)

Column chromatography: liquid chromatography conducted with an open chromatography column through which the eluent flows in response to gravity (4.2)

Coupling: the interaction of the magnetic spin of a nucleus with one or more neighboring nuclei in nuclear magnetic resonance (NMR) spectroscopy, causing a signal to be split into a characteristic pattern reflecting the number of magnetically active neighboring nuclei (4.3)

Coupling constant: the magnitude of splitting of an NMR signal by one or more magnetically active neighboring nuclei (4.3)

Detector: a device that produces a signal in response to the presence of a compound of interest (4.2)

Doublet: a two-line multiplet (4.3)

Downfield: the chemical shift of a nucleus that resonates at a higher δ value than a reference nucleus; that is, one shifted to a lower frequency; deshielded; left-hand portion of an NMR chart (4.3)

Effective field (H_{eff}): the net magnetic field felt at a nucleus of interest in an NMR scan; differs from the applied field by the tiny local magnetic field (H_{loc}) induced by the electron cloud surrounding the nucleus (4.3)

Electromagnetic radiation: a particle (called a photon) or a wave traveling at the speed of light; includes infrared, visible, ultraviolet, and x-ray ranges. When regarded as a wave, light is described by its wavelength (λ) or its frequency (ν) (4.3)

Electrophoresis: the migration of a charged molecule under the influence of an electric field; used to separate charged organic species, often proteins, nucleic acids, and other polyelectrolytes (4.2)

Eluent: the mobile phase in liquid chromatography (4.2)

Elution: the motion of solute and solvent through the stationary phase in a chromatography column (4.2)

Elution time: the time required for a given compound to pass through a chromatography column (4.2)

Excited state: an electronic configuration with a higher energy content than the ground state; often produced by absorption of a photon, promoting an electron from a bonding or nonbonding molecular orbital to an antibonding molecular orbital (4.3)

Extraction: the selective partitioning of a compound between two immiscible liquids, often a nonpolar organic phase and an aqueous or alcoholic phase (4.2)

Fingerprint region: the region in the infrared (400 cm^{-1} to about 1100 cm^{-1}) that usually exhibits a series of complex, low-energy bands that are characteristic of a specific molecule (rather than a functional group) (4.3)

Flame ionization detector: a gas chromatography detector that senses the presence of ions that are generated as the effluent from the column is burned in a hydrogen flame (4.2)

Fragmentation pattern: a molecule-specific set of fragment ions obtained by bombarding a neutral molecule with high-energy electrons in a mass spectrometer (4.3)

Gas chromatography: a chromatographic technique in which a vaporized sample is carried by a gaseous mobile phase over a stationary phase (usually either a solid or a solid coated with a nonvolatile liquid) (4.2)

Gel electrophoresis: a separation technique that uses an electric field to induce movement of polyelectrolytes through a gel. (*see also* **Electrophoresis**) (4.2)

Ground state: the most stable, lowest-energy electronic configuration (4.3)

High-performance liquid chromatography: *see* **High-pressure liquid chromatography** (4.2)

High-pressure liquid chromatography (HPLC): liquid chromatography in which the mobile phase is driven through a sealed chromatography column by a mechanical pump (4.2)

HOMO: highest occupied molecular orbital (4.3)

Infrared spectroscopy: a technique that measures the absorption light of energies of about 4000 to 400 cm^{-1} (4.3)

Integration: the measurement of the relative area under each peak of a spectrum (4.2)

Integration curve: a measure of the area under each peak of a spectrum or chromatogram (4.3)

Liquid chromatography: a chromatographic technique in which a solid or liquid sample is carried by a liquid mobile phase over a stationary phase (usually a solid composed of small particles around which the liquid phase can flow) (4.2)

LUMO: lowest unoccupied molecular orbital (4.3)

Magnetic resonance imaging (MRI): a three-dimensional map of water concentration in an object; often used in medical applications for visualizing organs or anomalous growths (4.3)

Mass spectroscopy: a technique that determines the mass of ions formed when molecules are bombarded with high-energy electrons (4.3)

Mobile phase: the flowing medium used in chromatography to carry a mixture through the stationary phase; flow can be induced by gravity, pressure, or capillary action (4.2)

Mobility: a measure of the ease with which a given compound can move (for example, through a chromatography column) (4.2)

Molecular ion: in mass spectrometry, an unfragmented (parent) ion formed by loss of an electron from a molecule; has the same mass as the sample being analyzed (4.3)

Multiplet: a pattern obtained by splitting the signal for a magnetically active nucleus into several lines (4.3)

Multiplicity: the number of peaks into which a signal is split (4.3)

Normal phase chromatography : a liquid chromatographic technique in which less–polar compounds elute first through a polar stationary phase, often unmodified silica gel or alumina (4.2)

n,π* transition: an electronic transition of an electron from one of the nonbonded, lone pairs of electrons to a π* (antibonding) orbital (4.3)

Nuclear magnetic resonance (NMR) spectroscopy: a spectroscopic technique for measuring the amount of energy needed to bring a nucleus (most commonly ^{1}H or ^{13}C in organic molecules) into resonance when a molecule is placed in a strong magnetic field and is irradiated with radio-frequency waves (4.3)

Paper chromatography: a chromatographic technique in which a mixture of compounds is separated by elution by the liquid phase passing by capillary action through a sheet of chromatographic paper (4.2)

Photoexcitation: the process by which a photon ($h\nu$) is absorbed by a molecule, causing the promotion of one of the electrons from a bonding to an antibonding orbital (4.3)

π,π* transition: an electronic transition taking place through the promotion of an electron in a π (bonding) orbital to a π* (antibonding) orbital (4.3)

Polyelectrolyte: a high-molecular-weight molecule that readily ionizes to form a multiply charged species when dissolved in water or other polar solvents (4.2)

Proton decoupling: the simplification of a nuclear magnetic resonance (NMR) spectrum by irradiation of the sample with radio-frequencies either at a specific region or over the entire chemical shift range at which protons absorb; results in saturation the populations in the high spin state and loss of coupling to the irradiated nuclei; a technique used routinely to simplify ^{13}C NMR spectra (4.3)

Quartet: a four-line multiplet (4.3)

Refractive index: the ratio of the speed of light in a vacuum to the speed of light in a material. The path of light is bent upon passing from one medium to another of different refractive index (4.2)

Refractive index detector: device that produces an electrical signal in response to the difference in refractive index of a solvent with and without a solute; often used in conjunction with high-pressure liquid chromatography (HPLC) (4.2)

Resonance (in nuclear magnetic resonance): condition in which the applied radio-frequency energy matches the energy difference between the parallel and anti-parallel spin states of the nucleus, so that the energy is absorbed, causing its spin to "flip" from the lower energy parallel state to the higher anti-parallel energy state (4.3)

Retention time: the interval required for a molecule to elute from a chromatography column; influenced by the magnitude of noncovalent interactions between the compounds being separated and the stationary phase (4.2)

Reverse-phase chromatography: a liquid chromatographic technique in which more polar compounds elute first through a nonpolar stationary phase, often silica gel coated with a long-chain alkylsilane (4.2)

R_f value: the ratio of the distance migrated by a substance compared with the solvent front (4.2)

Shielding: the shift of a nuclear magnetic resonance (NMR) signal from that expected from the applied field caused by donation of electron density to the observed nucleus (4.3)

Solvent front: the furthest point reached by the solvent in chromatography (4.2)

Spectroscopy: a set of techniques that measure the response of a molecule to the input of energy (4.1)

Spectrum: a display of peak intensity detected for a given spectroscopic method as a function of incident energy (4.3)

Stationary phase: an immobile medium (usually a solid or highly viscous liquid) through which a mixture passes in chromatography (4.2)

Thermal conductivity detector: a gas chromatography detector that measures the difference in thermal conductivity (heat capacity) between the carrier gas alone and that observed as a sample elutes from the column (4.2)

Thin layer chromatography (tlc): chromatographic technique in which a mixture of compounds is separated by elution by the liquid phase by capillary action through a flat solid support such as a sheet of glass, plastic, or aluminum foil coated with a thin layer of silica gel or alumina (4 2)

Triplet: a three-line multiplet (4.3)

Ultraviolet spectroscopy: a technique that measures a molecule's tendency to absorb light of wavelengths of 200 to 400 nm (a region of energy just higher than that detectable by the human eye) (4.3)

Upfield: the chemical shift of a nucleus that resonates at a lower δ value than a reference nucleus; that is, for most uncharged molecules, at a higher frequency and thus closer to that of tetramethylsilane; right-hand portion of an NMR spectrum (4.3)

Visible spectroscopy: technique that measures a molecule's tendency to absorb light of wavelengths of about 400 to 800 nm (the region of energy detectable by the human eye) (4.3)

Wavelength: distance from peak to peak of a wave; describes the energy of electromagnetic radiation (4.3)

Stereochemistry

Key Concepts

Geometric isomerization

Torsional strain: staggered and eclipsed conformers

Conformational analysis of butane: energy diagrams

Barriers to rotation

Gauche and *anti* conformers

Newman projections

Torsional strain in cycloalkanes

Chair and boat cyclohexanes

Ring-flipping

Axial and equatorial substituents

Chirality

Measuring optical activity

Racemic mixtures

Specifying absolute configuration

Relationships between stereoisomers:

      enantiomers, diastereomers, *meso* compounds

Fischer projections

Resolution

Answers to Exercises

Ex 5.1 The interconversion of geometrical isomers requires rotation about a bond at which rotation is restricted. The cases discussed in this chapter deal with carbon–carbon double bonds. In this exercise, you are asked to compare the barrier to free rotation for the indicated compound with that in a simple alkene.

(a) These two isomers of 1,3-butadiene differ with respect to the orientation of the two π bonds, which are either on opposite sides of the C-2—C-3 single bond (at the left, called the *s-trans* isomer) or on the same side of the C-2—C-3 single bond (at the right, called the *s-cis* isomer).

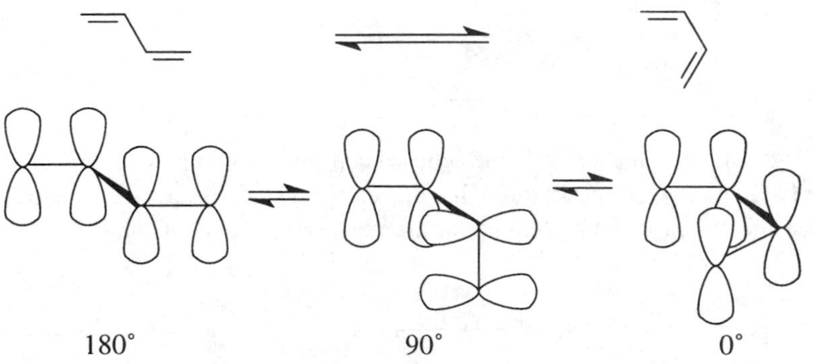

    180°   90°   0°

Interconversion of these structures requires rotation about the C-2—C-3 σ bond, moving the dihedral angle between the planes containing the 1,2 and the 3,4 π bonds from 0 to 180°. At 90°, the π bonds are perpendicular and cannot interact. At this geometry, conjugative stabilization is completely lost, although the two π bonds remain intact. The energy barrier for this rotation thus measures the conjugation energy. Because this energy is lower than that of a π bond itself, this interconversion is much easier than in a simple alkene.

(b) These two geometric isomers have methyl groups on the opposite sides (left) and on the same side (right) of the imine double bond. Their interconversion requires a rotation through an orthogonal geometry in which the C=N π bond is broken. Because a π bond containing a heteroatom is weaker than one between carbons, the rotational barrier is lower than that in a C=C double bond.

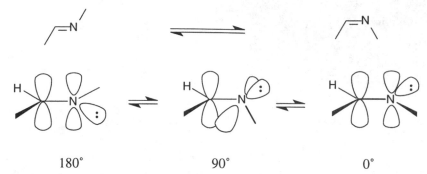

(c) As in part *b*, these two geometric isomers have methyl groups on opposite sides (left) and on the same side (right) of the nitrogen–nitrogen double bond. Their interconversion again requires a rotation through an orthogonal geometry in which the N=N π bond is broken. Because this π bond contains two heteroatoms, it is even weaker than that in the imine, and the rotational barrier is still lower than that for the C=N double bond in part *b*.

Ex 5.2 (a) In the eclipsed conformation of 2,2,3,3-tetramethylbutane, each of the three methyl groups on C-2 aligns with one of the three methyl groups on C-3.

(b) To obtain the staggered conformation from the eclipsed conformation in part *a*, the front (or the back) carbon is rotated through a dihedral angle of 60°. Thus, each of the three methyl groups on C-2 bisects the positions of two of the methyls on C-3.

(c) In the staggered conformation of propane, each of the hydrogens on C-1 bisects the positions of two substituents on C-2.

(d)   By rotation of the front carbon of the staggered conformation in part *c* through a 60° dihedral angle, each bond on C-1 aligns with a bond on C-2.

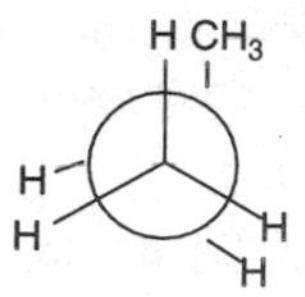

Ex 5.3   Begin with the most stable, *anti*, staggered conformation shown at the left in the profile below, assigning it an arbitrary conformational energy of zero. Rotation of the back carbon through a 60° dihedral angle increases torsional energy to a maximum at the eclipsed conformation in which ethyl eclipses hydrogen. A second 60° dihedral angle rotation restores the staggered arrangement, minimizing torsional strain but introducing a *gauche* interaction between the ethyl groups on C-3 and C-4. Because an ethyl group is larger than the methyl group discussed in the text, destabilization caused by steric interaction in the *gauche* conformer is greater here. The next 60° dihedral angle rotation again leads to a torsionally strained eclipsed conformer, this time with ethyl eclipsing ethyl, thus producing additional destabilization compared with the eclipsed conformer in which ethyl eclipses hydrogen. The fourth 60° rotation leads to a *gauche* conformer identical in energy with that attained at 120°. The next 60° rotation produces an eclipsed conformation of energy identical with that at 60°. Finally, the last rotation restores the substituents to their original *anti* conformation at the zero of energy.

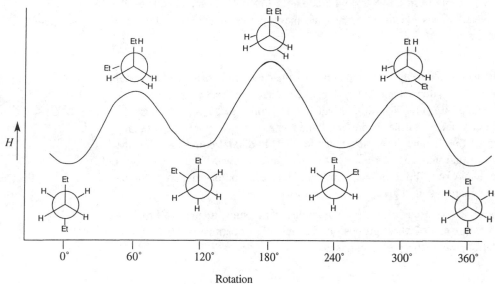

Ex 5.4   From Table 5.2, we estimate the energy differences necessary to establish the given distribution:

(a)   less than 0.5 kcal/mole

(b)   about 0.6 kcal/mole

(c)   about 5.5 kcal/mole

Ex 5.5   It is exceedingly difficult to illustrate these stereochemical relationships without physical manipulation of models. The effort at constructing the models and rotating around all possible bonds will illustrate for you how molecules are constantly sampling the infinite number of possible spatial relationships.

Ex 5.6   Your models might pop open if you allow too much stress at a given site while you are manipulating the models as suggested. Don't worry if this happens: this will illustrate which conformations a real molecule would avoid as it explores

C25

conformational interconversions. Since different conformations are produced during
this twisting, the bond angles must indeed change.

Ex 5.7   Try to feel the strain induced when your molecular model flips through a
conformation in which at least three ring carbons are planar and contrast that stress
with that induced by the twist-boat conformation you constructed in Exercise 5.6.

Ex 5.8   With all the alkyl substituents except *t*-butyl, it is possible to rotate about the ring
carbon—axial alkyl substituent bond so that a hydrogen atom (rather than an alkyl
group) is directed over the ring (illustrated below with axial isopropylcyclohexane).
With a *t*-butyl substituent, a highly unfavorable 1,3-diaxial interaction between a
methyl group and the axial hydrogen atoms at C-3 and C-5 must take place.

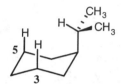

Ex 5.9   (a)   The *t*-butyl group acts as a conformation lock; it is found exclusively in
an equatorial position. A methyl group at C-2 is axial if *cis* and equatorial
if *trans*. Because the diequatorial conformer is preferred, the *trans* isomer is
more stable.

(b)   The isopropyl group is large and prefers the equatorial position to a greater
extent than does a methyl group. With one substituent equatorial, the
second group at the 4-position is equatorial if *trans*, axial if *cis*. Because the
diequatorial conformer is preferred, the *trans* isomer is more stable. A ring
flip of this isomer forms the much less stable diaxial conformer, but the
conformational equilibrium is strongly shifted toward the diequatorial
isomer. A ring-flip in the *cis* isomer interconverts conformers, so that axial
and equatorial alkyl groups switch positions.

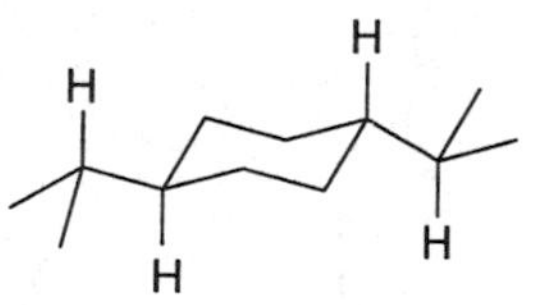

(c)   The analysis used in part *b* also applies if the substituents are bromine
atoms instead of alkyl groups, but because the substituents are in a 1,3-
relation, the *cis* isomer is more stable.

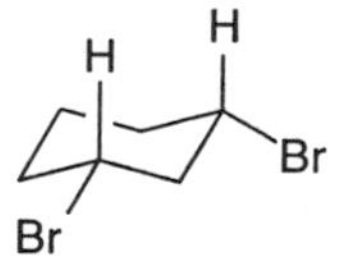

(d)   The *t*-butyl group again acts as a conformational lock in the equatorial
position. The *cis* isomer has the ethyl group equatorial for 1,3-substitution,
whereas the *trans* isomer bears an axial ethyl group. The diequatorial isomer
is preferred, so the *cis* isomer is more stable.

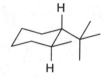

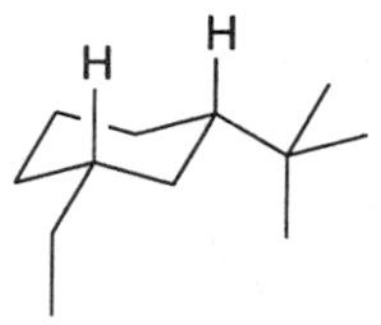

Ex 5.10   In a *cis*-decalin, one of the bonds at the ring fusion has an axial orientation and the
other is equatorial. This relationship is also present in the ring-flipped isomer shown
at the right. There are two stereoisomers for 2-methyldecalin and two for 3-
methyldecalin, with the methyl group *cis* and *trans* to the hydrogen at C-1. In all
cases, the structures in which the methyl is in an equatorial position are the more
stable isomers.

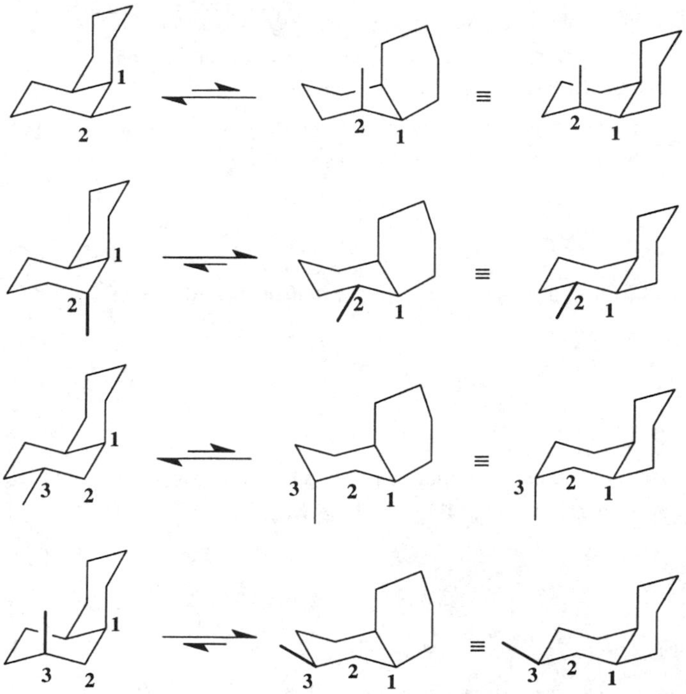

Ex 5.11 (a)   There are three mirror planes, each at right angles to the others.

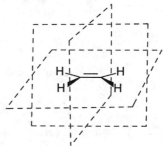

(b)   Benzene has even higher symmetry than does ethylene. In addition to the plane containing all 12 atoms, there are six additional planes at right angles to the first, three going through carbons at opposite positions on the ring and three going between carbons. Two of these are shown at the right, along with the plane containing all the atoms.

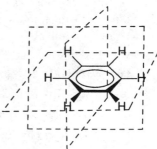

(c)   There are only one plane of symmetry for *anti*-butane.

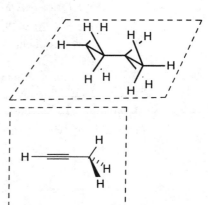

(d)   There are three planes of symmetry for propyne; one is shown at the right. Each is defined by C-2, C-3, and one of the three hydrogen atoms of the methyl group.

Ex 5.12 A center of chirality is indicated by the absence of a mirror plane of symmetry at that atom. Usually such centers are at atoms bearing four different substituents. The centers of chirality are indicated in the following structures by asterisks.

(a)   Because the ring carbons at the right and left of the carbon bearing bromine are identical, this molecule has no center of chirality and does not have a configurational isomer.

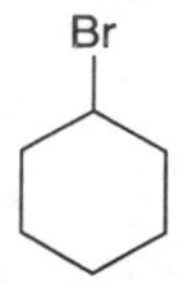

(b)   There are two centers of chirality, at the carbons substituted by bromine and chlorine, respectively.

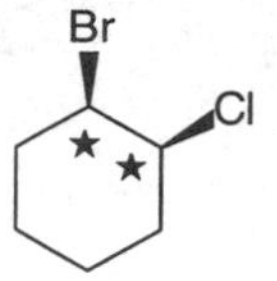

(c)   There is no center of chirality in this molecule. It possesses a mirror plane through C-1 and C-4, as well as one through the bonds between C-2 and C-3 and between C-5 and C-6.

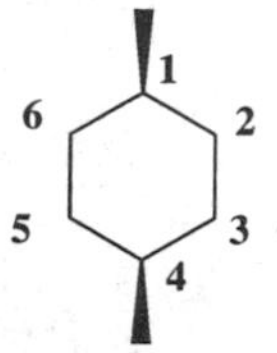

(d)   The single center of chirality is at the carbon bearing both a chlorine and a bromine substituent.

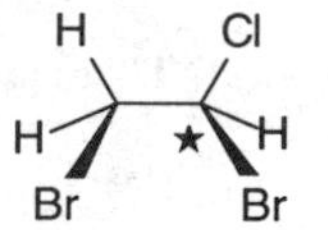

(e)   A center of chirality exists at C-2, the carbon to which the chlorine is bound.

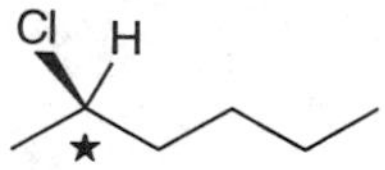

Ex 5.13 To assign absolute configuration, position yourself so that the group of lowest priority is farthest away. That is, visualize a Newman projection obtained by viewing down the carbon-to-lowest-priority-group bond. When the progression from highest to lowest priority of the other three groups attached to carbon is clockwise, the center is R; if counterclockwise, S.

(a) R (priority: Br > Et > Me > H)

(b) R (priority: Br > Cl > Et > Me)

(c) S (priority: I > Br > Et > Me)

(d)   C-1: R (priority: Br > CH_2 CH(OMe)R > CH_2CH_2R > H)

C-3: S (priority: OMe > CH_2 CHBrR > CH_2CH_2R > H)

(e) S (priority: Br > Cl > R > Me) Note that the carbon bearing two chlorines is not a center of chirality.

Ex 5.14 To draw absolute configurations of a given configuration, orient the group of lowest priority away from the observer and place groups of decreasing priority in clockwise fashion for the R enantiomer and in counterclockwise fashion for the S enantiomer.

(a)   Priority: Br > n-Pr > Me > H

(b)   Priority: Br > Cl > n-Pr > Et

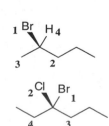

(c)   Priority: Cl > F > Et > Me

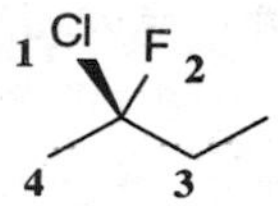

(d)   Priority at C-1: Br > CHFR > CH_2 CH_2 R > H; at C-2: F > CHBrR > CH_2 CH_2 R > H

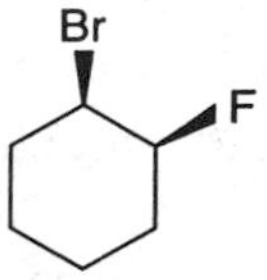

Ex 5.15 In a mixture of enantiomers, the R and S enantiomers have identical magnitudes of specific rotation but opposite signs. If we let x = mole fraction of one of the enantiomer (A) and y = mole fraction of the other enantiomer (B), then $x + y = 1$, and $y = 1 - x$. The observed rotation is the sum of the contributions of the two enantiomers. Let us assume that enantiomer A has $+100°$ specific rotation. Then, observed rotation = [(specific rotation A) $\times x$] + [(specific rotation B) $\times x$].

(a) $10° = [100° \times x] + [(-100°) \times (1 - x)]$

$x = 0.55$

$y = 045$

Optical purity = (0.55 – 0.45) x 100% = 10%

(Notice that this same value is obtained by taking the ratio of the observed rotation to the specific rotation.)

(b)   Similarly,

$50° = [200° \times x] + [(-200°) \times (1 - x)]$

$x = 0.625$

$y = 0.375$

Optical purity = (0.625 – 0.375) × 100% = 25%

(Notice again that this same value is obtained by taking the ratio of the observed rotation to the specific rotation.)

(c)   Enantiomeric excess   = % major enatiomer – % minor enatiomer
                                 = (3.5÷ 4.5) – (1.0 ÷ 4.5) = 78% – 22% = 55%

Ex 5.16 To determine the number of possible stereoisomers, first obtain the number of centers of chirality, n. The number of possible stereoisomers is then obtained as 2^n – the number of *meso* compounds.

(a)   Because there is no center of chirality, no isomers are possible.

(b)   With two centers of chirality and no possibility of *meso* compounds, there are $2^2 = 4$ stereoisomers possible.

(c)   Because there is no center of chirality, no isomers are possible.

(d)   With a single center of chirality, there exists a pair of enantiomers; that is, two stereoisomers.

(e)   With a single center of chirality, there exists a pair of enantiomers; that is, two stereoisomers.

Ex 5.17 A necessary and sufficient condition for the existence of a *meso* compound is the presence of a mirror plane of symmetry that relates each center of asymmetry in the molecule.

(a)   The existence of a mirror plane through C-2 and C-5 makes this a *meso* compound.

(b)   Although there is a symmetry element (an axis of symmetry) in this molecule, there is no mirror plane and hence this molecule is optically active.

(c)   Because a mirror plane bisects the bonds between C-1 and C-2 and between C-4 and C-5, this is an optically inactive *meso* compound.

(d)   Lacking a mirror plane, this molecule is optically active.

(e)   Rotation by 180° about the C-2–C-3 σ bond results in the conformation shown at the right (as a sawhorse representation) in which there is a mirror plane of symmetry bisecting the C-2–C-3 bond. This compound is therefore a *meso* compound.

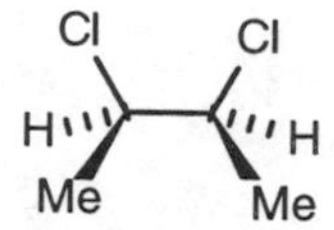

(f)   A mirror plane of symmetry bisects the C-2–C-3 bond. As in part *e*, this compound is therefore a *meso* compound and optically inactive.

(g)   Rotation by 120° about the C-3–C-4 σ bond results in eclipsing of the ethyl groups, as shown at the right. This conformation does not have a mirror plane of symmetry, and because it is the only arrangement in which the ethyl groups are placed symmetrically, no conformation has mirror symmetry. Therefore, this compound is not a *meso* compound and is optically active.

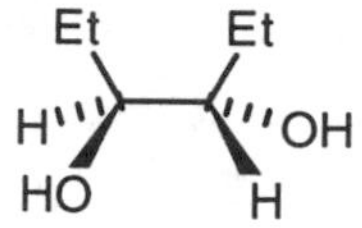

Ex 5.18   This exercise can be approached in several ways. One of the easiest is to convert the structures in the problem into the eclipsed sawhorse representations as was done in Exercise 5.17, parts *e-g*. Then, by visualizing from the first center of asymmetry to the second, one obtains the following representations.

(e)   (f) 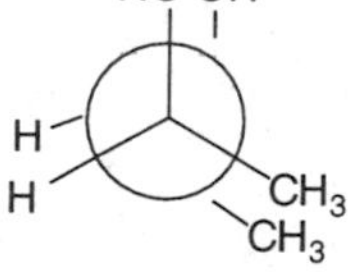   (g)

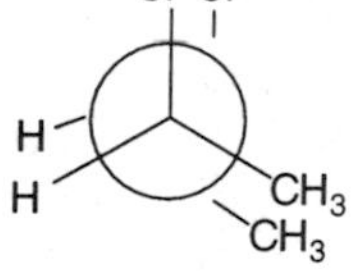

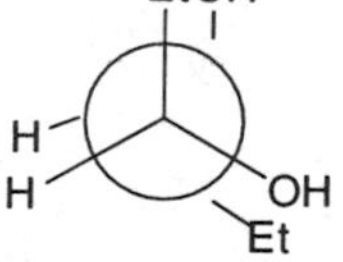

Ex 5.19   To solve this problem, orient the molecules in such a way as to look for a mirror-image relation between the pair. Enantiomers are nonsuperimposable mirror images. Diastereomers are nonmirror image stereoisomers. Constitutional isomers have a different sequence of bonds in two molecules with the same net chemical formula. Identical compounds are exactly superimposable. (If it proves difficult to see the stereochemical relation between structures, make models and physically attempt the indicated superimposition.)

(a)   Counterclockwise rotation by 120° of the C-4 center of asymmetry in the structure at the left produces an eclipsed structure that is identical with that at the right. These are therefore two representations of the same molecule.

(b)   The structure at the left has a mirror plane of symmetry between C-1 and C-2, which the structure at the right lacks. These two structures are therefore not mirror images and are diastereomers.

(c)   The sequence of attachment of atoms is different in 1,4-dimethylcyclohexane at the left and in 1,3-dimethylcyclohexane at the right, irrespective of any stereochemical considerations. These two compounds are therefore constitutional isomers.

(d)   Fischer projections represent eclipsed representations of molecules bearing stereocenters. The molecule at the left clearly has a plane of symmetry between C-2 and C-3 that is absent from the molecule at the right. The molecule at the left is therefore a *meso* compound and is diastereomeric to the isomer on the right.

(e)   Inversion of the molecule at the right makes it exactly superimposable on the molecule at the left. These are therefore two representations of the same molecule.

Ex 5.20 By convention, the groups attached to the horizontal lines in a Fischer projection come toward the observer, and those attached to the vertical lines go away from the observer. The carbon skeleton in a Fischer projection is usually drawn vertically so that any substituents present on centers of asymmetry in the molecule are oriented toward the observer. Thus, the Newman projections from Exercise 5.17 can be conveniently converted into Fischer projections by viewing them from the direction indicated.

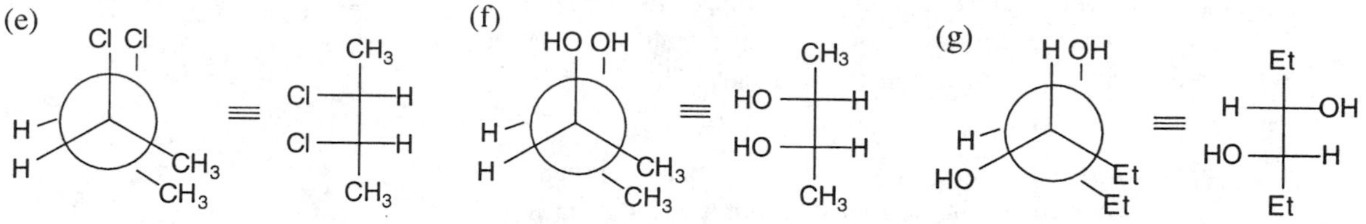

(e) (f) (g)

Answers to Review Problems

Pr 5.1 In the *E* isomer, the groups of highest priority on each terminus of the double bond are directed to opposite sides. Here, both the methyl and acid groups have precedence over hydrogen.

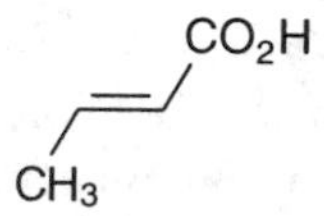

Pr 5.2 (a) The most stable conformer is the *anti* staggered form in which the large methyl groups are directed as far as possible from each other. The least stable form is that in which the methyl groups are eclipsed.
(b) From the values in Table 5.1, we estimate that these conformers differ by the energetic cost of two methyl–methyl eclipsing interactions (2 x 2.7 kcal/mole) + the cost of one hydrogen–hydrogen eclipsing interaction (0.9 kcal/mole) for a total of 6.3 kcal/mole conformational destabilization. The *anti* staggered isomer will predominate by more than 99.9%.

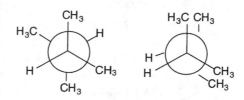

Pr 5.3 The conformational energy diagram for rotation about the C-2–C-3 bond of 2,2,3,3-tetramethylbutane is similar to that for ethane, differing only in the magnitude of the energy difference between the staggered and eclipsed conformations. In the diagram at the right, one methyl group on the front carbon and one methyl group on the back carbon are shown in bold face so that we can see the result of rotation first by 60° to the eclipsed conformation and then by another 60° to the staggered conformation. Because this molecule has three-fold symmetry (when viewed along the C-2–C-3 bond about which we are considering conformational isomerism), there is only one unique staggered and one unique eclipsed conformation. This process can be repeated three times, resulting in a total rotation of 360°.

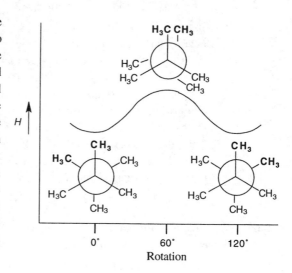

Pr 5.4 In its more stable conformer, the methyl group in methylcyclohexane occupies an equatorial position. Upon ring-flipping, this position becomes axial, making the ring-flipped conformer less stable. In the equatorial position, the methyl group is *anti* to the ring carbons, as is shown in the Newman projection. When methyl is axial, however, it is *gauche* to the ring carbons, which explains the destabilization of the ring-flipped form. There are *gauche* interacttions between the axial methyl group and the axial hydrogen atoms on C-3 and C-5.

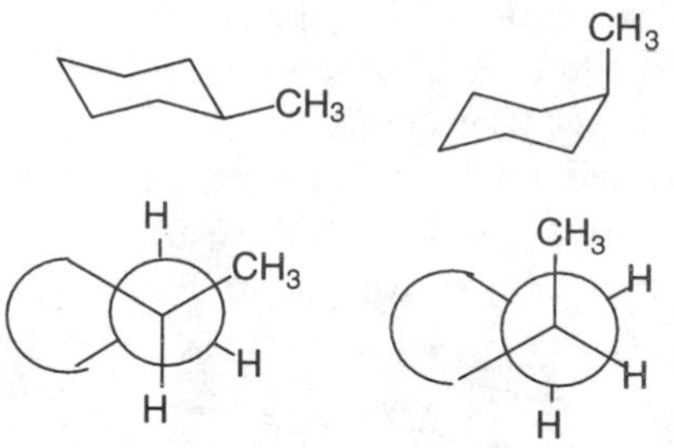

Pr 5.5 (a) In *trans*-1,2-dimethylcyclohexane, both alkyl groups occupy equatorial positions. (There is a ring-flipped conformer in which both alkyl groups are axial, but this is less stable than the diequatorial conformer and does not dominate the conformational equilibrium.) In *cis*-1,2-dimethylcyclohexane, one methyl group must be axial and the other equatorial. This situation persists in the ring-flipped conformer as well. Thus, the diequatorially substituted, *trans* isomer is the more stable conformer.

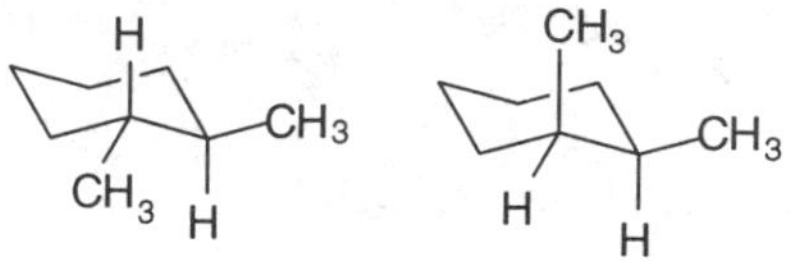

(b) In *cis*-1,3-dimethylcyclohexane, both alkyl groups occupy equatorial positions. (There is also a ring-flipped conformer in which both alkyl groups are axial, but this is less stable than the diequatorial conformer and does not dominate the conformational equilibrium.) In *trans*-1,3-dimethylcyclohexane, one methyl group must be axial and the other equatorial. This situation persists in the ring-flipped conformer as well. Thus, the diequatorially substituted, *cis* isomer is the more stable conformer.

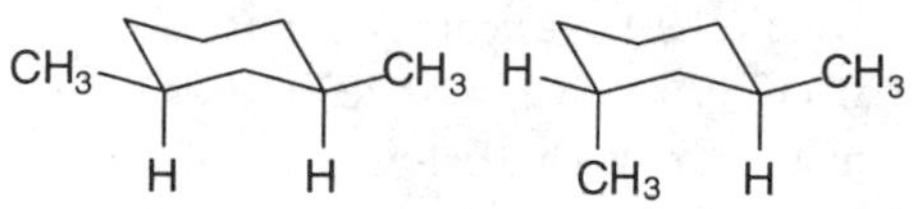

(c) In *trans*-1,4-dimethylcyclohexane, both alkyl groups occupy equatorial positions. (There is also a ring-flipped conformer in which both alkyl groups are axial, but this is less stable than the diequatorial conformer and does not dominate the conformational equilibrium.) In *cis*-1,4-dimethylcyclohexane, one methyl group must be axial and the other equatorial. This situation persists in the ring-flipped conformer as well. Thus, the diequatorially substituted *trans* isomer is the more stable conformer.

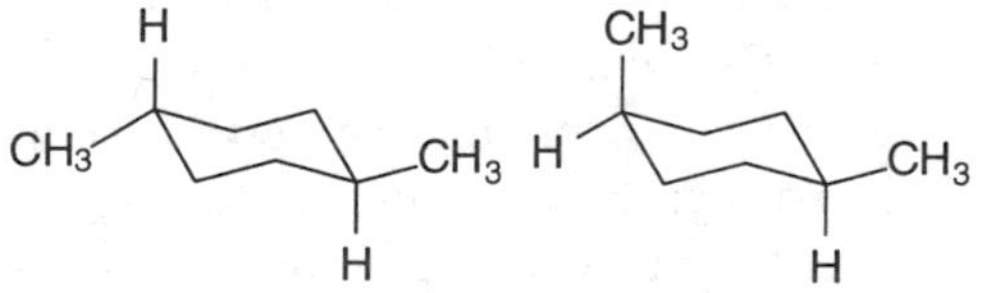

Pr 5.6 To evaluate conformational stability, draw the ring-flipped isomers and analyze the energetic factors influencing each.

(a) A ring-flip takes the axial methyl group to an equatorial position, thus forming the more stable stereoisomer, shown here at the right.

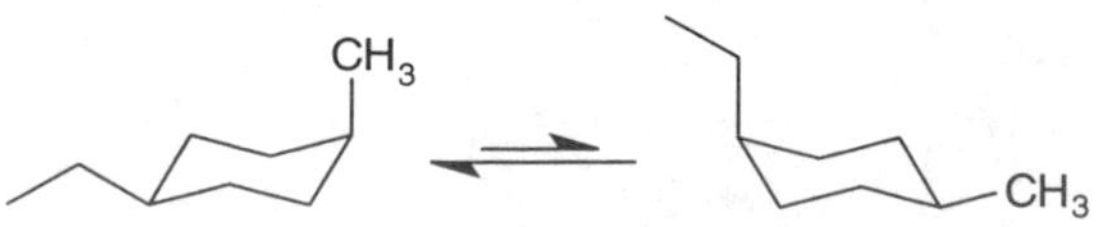

(b) A ring-flip takes the axial methyl to an equatorial position and the equatorial ethyl to an axial position. Because the *gauche* interaction between the ethyl group and the ring carbons is slightly greater than that for the smaller methyl group, the conformer in which the ethyl group is equatorial is the more stable.

(c)   The axial bromine in the conformer at the right is converted by a ring-flip into the more stable equatorial conformer at the left.

(d)   A ring-flip takes the conformer at the left (with two equatorial bromine atoms) to the less stable conformer at the right in which the bromine atoms are both at axial positions.

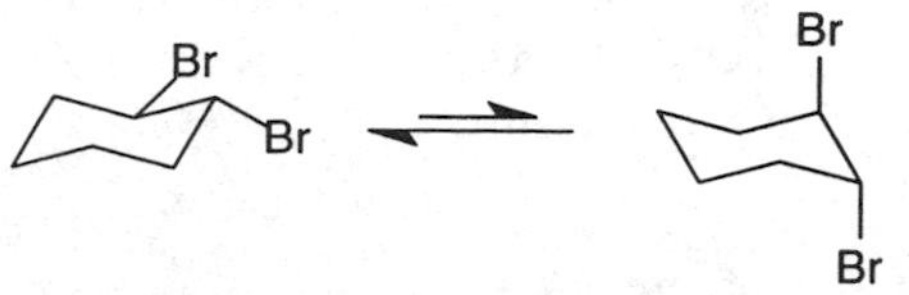

(e)   As in part *b*, the larger isopropyl group prefers an equatorial position, forcing the smaller methyl group into an axial position in the more stable isomer shown at the right.

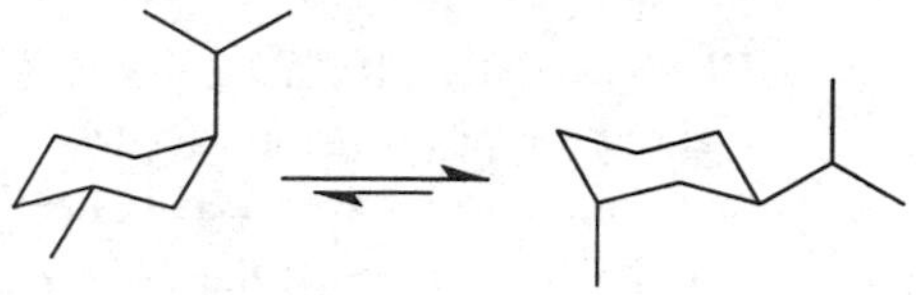

Pr 5.7   The difference in energy between the axial and equatorial isomers is influenced significantly by electrostatic interactions. With the OH in an axial position, the hydroxyl oxygen at the negative end of the C—O and O—H polarized bonds is closer to the ring carbon (*), which is the positive end of the C—O bond indicated. This electrostatic effect is sufficient to outweigh the 1,3-diaxial interactions with the hydroxyl group. (Remember, however, that a hydroxyl group is quite small.)

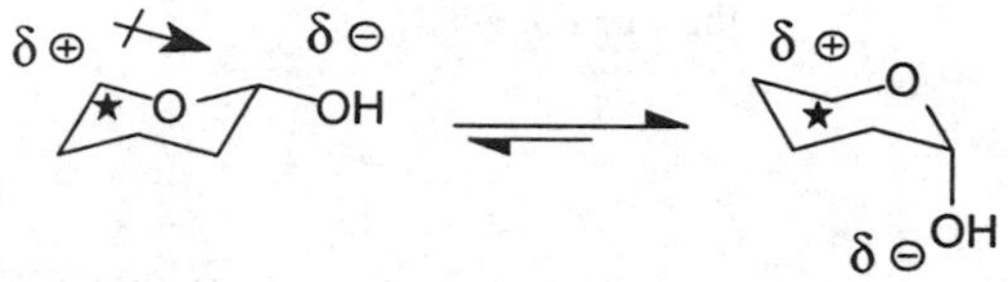

Pr 5.8   In addressing this problem, first identify any centers of asymmetry and assign priorities to the four attached groups. Then look at each center from a point of view that places the group of lowest priority furthest from you. Then decide whether the progression from groups of highest to lowest priority is clockwise (*R*) or counterclockwise (*S*).

(a)   At the ring center of chirality, the priorities are O > C(OH)= > C(OH)H > H: therefore *R*;. At the center of chirality on the chain, the priorities are OH > C(O)C > C(OH)H$_2$ > H: therefore *R*.

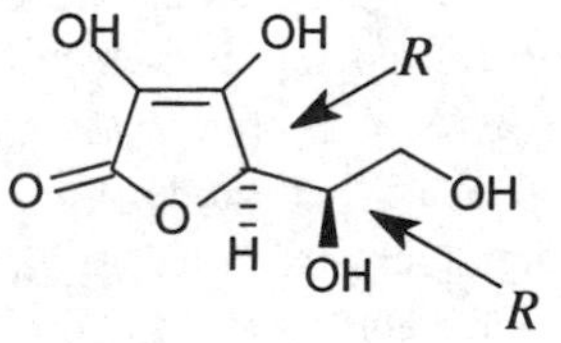

(b)   There are five centers of asymmetry in α-D-glucose. Starting with the center at the right of the structure as drawn, the priorities are OR > OH > C(OH) > H: therefore *S*. Moving clockwise around the ring, the priorities at the next center are OH > C(OH)OR > C(OH)C > H: therefore *R*. At the next carbon, the priorities are OH > CH(OH)COR > CH(OH)COH > H: therefore *S*. At the next carbon, the priorities are OH > CH(OR)CH$_2$(OH) > CH(OH)CH(OH) > H: therefore *R*. At the last carbon the priorities are OR > CH(OH) > CH$_2$OH > H: therefore *R*.

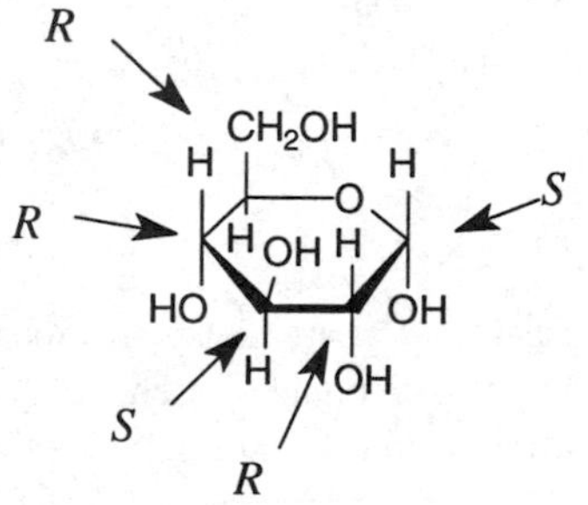

(c)   There is a single center of asymmetry with priority NH$_2$ > CO$_2$H > CH$_2$CO$_2$H > H: therefore *S*.

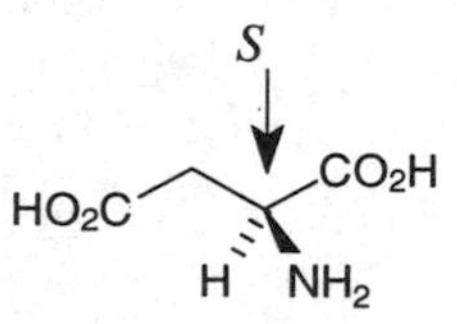

88   Chapter 5

(d)   There are four centers of asymmetry in cocaine. At the two bridgehead positions, the priorities are NCH_3 > CH_2CHO > CH_2CH_2 > H: therefore one center is R and the other S. At the carbon bearing the CO_2CH_3 group, the priorities are CO_2CH_3 > CH(OR) > CH(NR) > H: therefore R. At the carbon bearing the OCOPh group, the priorities are OCOPh > $CHCO_2CH_3$ > CH_2 > H: therefore S.

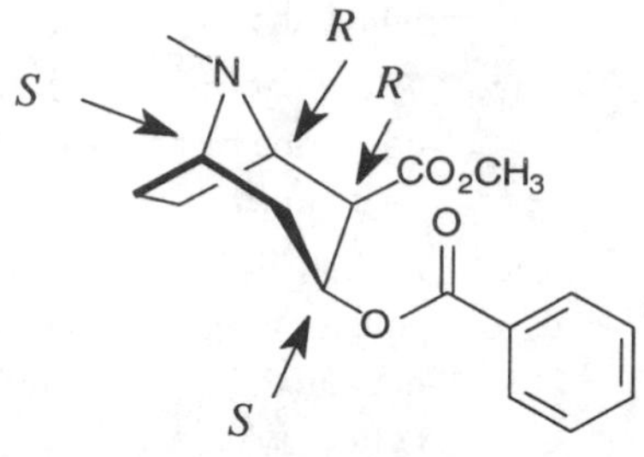

(e)   There are four centers of asymmetry in quinine. At the carbon bearing the vinyl group on the bicyclic ring, the priorities are CH_2N > CHCRCR > CHC= > H: therefore S. At the bridgehead carbon, the priorities are CHCC > CH_2CHR > CH_2CH_2 > H: therefore S. At the carbon α to nitrogen, the priorities are NR_2 > CH(OH)R > CH_2> H: therefore S. At the center of chirality on the side chain, the priorities are OH > CHN > CH= > H: therefore R..

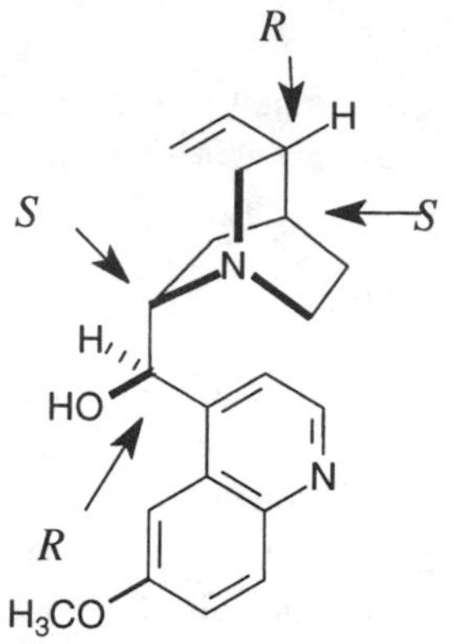

(f)   There are three centers of chirality in xylose. At C-2 (α to the aldehyde group), the priorities are: OH > CHO >CH(OH) > H: therefore R. At C-3, the priorities are OH> CH(OH)CHO > CH(OH)CH(OH) > H: therefore S. At C-4, the priorities are OH> CH(OH)R > CH_2OH > H: therefore R.

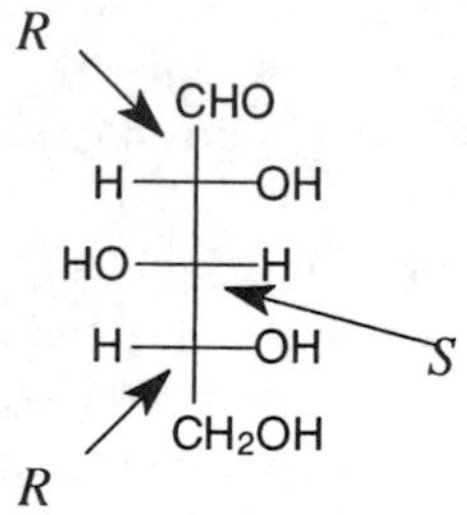

Pr 5.9 In a Fischer projection, horizontal substituents represent groups projected toward the observer, and vertical substituents represent those projected away from the observer. To solve this problem, assign priorities and arrange the four substituents so as to derive a clockwise order for substituents in decreasing priority at each R center and a counterclockwise order for each S center. There are four possible Fischer projections for each center of asymmetry. Only one is shown here.

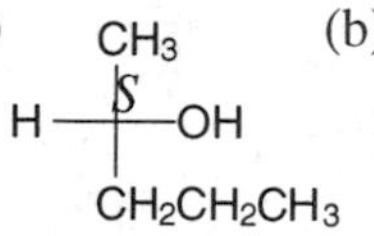

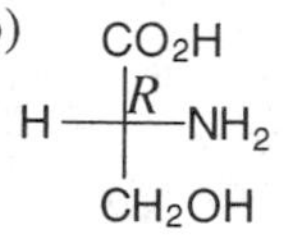

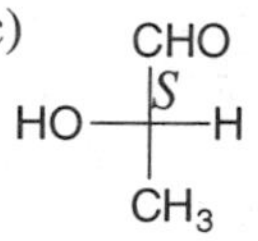

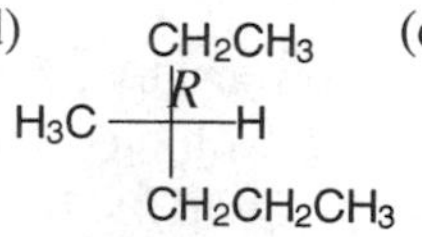

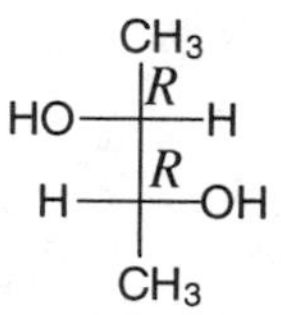

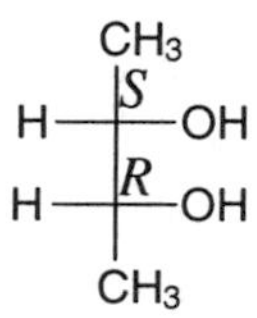

Pr 5.10 Enantiomers are nonsuperimposable mirror images. Diastereomers are stereoisomeric nonmirror images. Identical compounds are completely superimposable, possibly after a conformational interconversion, but without the making or breaking of covalent bonds.

Make models of both compounds and compare in three dimensions, if you have trouble seeing the relationship. Alternatively, you can assign R or S designations to any centers of asymmetry. In enantiomers, each S center in one compound is reflected as an R center in the other. In diastereomers, this R-to-S relationship is not retained.

(a)   enantiomers
(b)   identical: no center of chirality
(c)   diastereomers: geometric isomers
(i)   enantiomers
(j)   identical: no center of chirality
(k)   diastereomers

(d) identical

(e) enantiomers

(f) positional (not stereo) isomers

(g) diastereomers: geometric isomers

(h) diastereomers: geometric isomers

(l) identical

(m) identical

(n) identical

(o) identical

Pr 5.11 There are nine centers of asymmetry in cholesterol and $2^9 = 512$ possible stereoisomers. Because of the lack of symmetry in the molecule, there can be no *meso* isomers. Only one stereoisomer occurs naturally.

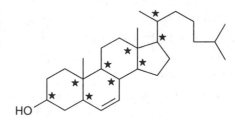

Important New Terms

(+): dextrorotatory; *see d* (5.8)

(–): levorotatory; *see l* (5.8)

Absolute configuration: the three-dimensional structure of a molecule that has one or more centers of chirality (5.6)

Absolute stereochemistry: the unambiguous specification of all spatial positions about a center of chirality (5.6)

Achiral: descriptor of a molecule in which at least one conformation has a mirror plane of symmetry; lacking handedness (5.5)

Activation energy ($\Delta H^{\ddagger}$ or E_{act}): the energy difference between a ground state reactant and the transition state (5.1)

Anti **conformer**: a conformational isomer in which two large groups on adjacent atoms are separated by a 180° dihedral angle (5.2)

Arrhenius equation: $k = Ae^{\frac{-\Delta H^{\ddagger}}{RT}}$; mathematical correlation of the rate of a reaction with its activation energy (5.2)

Axial: descriptor of a group pointing roughly orthogonally from the pseudoplane of a chair conformation (5.4)

Boat conformation: the eclipsed conformation of cyclohexane or an analogous six-atom cyclic compound in which the spatial placement of C-1 and C-4 roughly resembles the bow and stern of a boat (5.4)

Bridgehead atom: an atom that is common to both rings in a bicyclic (or multicyclic) compound (5.4)

Cahn–Ingold–Prelog rules: used in specifying absolute stereochemistry (5.6)

Center of chirality: a tetrahedral atom (usually carbon) bearing four different groups (5.5)

Chair conformation: the staggered conformation of cyclohexane or an analogous six-atom cyclic compound roughly resembling the back, seat, and footrest of a chair (5.4)

Chiral: descriptor of the property of handedness; when applied to molecules, lacking a mirror plane through any conformation (5.5)

Chiral center: *see* **Center of chirality** (5.5)

Chirality: handedness; the property of an object (in this context, a molecule) whereby the object is not superimposable on its mirror image (5.5)

Chiral molecule: a molecule lacking an internal plane of symmetry; a molecule that is not superimposable on its mirror image; the most common indicator of chirality is the presence of a carbon atom bonded to four different groups (5.5)

Configurational isomers: stereoisomers that can be interconverted only by the breaking and reforming of a covalent bond (5.1)

Conformational analysis: energetic description of conformational interconversion; relates the relative atomic positions to the changes in potential energy during rotation about a σ bond (5.2)

Conformational anchor: a substituent (usually large) that so strongly prefers the equatorial position that it blocks conformational flipping of the six-membered ring to which it is attached (5.4)

Conformational isomers: stereoisomers that can be interconverted by rotation about a σ bond (5.2)

Conformational lock: *see* **Conformational anchor** (5.4)

Conformer: a conformational isomer (5.2)

Constitutional isomers: isomers having the same molecular formula but with the atoms attached in different sequences (5.1) *d*: relative stereochemical designator of a molecule with a positive (dextrorotatory) rotation: from the Greek for "right-rotating" (5.8)

d,l: indicator of an optically inactive racemic modification (5.8)

D: absolute stereochemical descriptor that relates substituent disposition at a given center of chirality to that in natural *D*-glyceraldehyde (5.8)

Decalin: bicyclo[4.4.0]decane; two fused six-membered rings (5.4)

Dextrorotatory: *see d* (5.8)

1,3-Diaxial interaction: the steric interaction between axial substituents bound to carbon atoms in a six-membered ring, resulting in a steric destabilization (5.4)

Diastereomers: nonmirror-image stereoisomers (5.8)

Dihedral angle: the angle formed by two intersecting planes (5.1)

Eclipsed conformation: a spatial arrangement in which each σ bond at one carbon atom is coplanar with a σ bond on an adjacent atom (dihedral angle = 0°); when viewed in a Newman projection, the conformation has aligned bonds on adjacent atoms (5-2)

Enantiomeric excess: the predominance of one enantiomer over the other (5.7)

Enantiomers: stereoisomers related to each other as nonsuperimposable mirror images; stereoisomers with opposite configuration at each center of chirality (5.5)

Energy barrier: the amount of energy required to reach the most unfavorable point along the path followed in the conversion of one species to another (5.1)

Energy of activation: *see* **Activation energy** (5.1)

Enthalpy change: ($\Delta H°$); heat of reaction; the difference between the bond energies of the reactants and products (5.1)

Entropy: disorder; free motion (5.1)

Entropy change: ($\Delta S°$); the difference in disorder between reactants and products (5.1)

Equatorial: descriptor of a group pointing roughly parallel with the pseudoplane of a chair conformation (5.4)

Equilibrium: the state in which the forward rate of an ideally reversible reaction is equal to the reverse rate (5.2)

Flagpole hydrogens: the two hydrogens located in a 1,4-relationship in a boat cyclohexane that point at each other (5.4)

Fischer projection: a stick notation used to indicate absolute configuration in which the intersection of two lines indicates the position of a chiral carbon, with horizontal lines indicating substituents directed toward

the observer and vertical lines indicating substituents directed away from the observer (5.8)

Gauche **conformer**: a conformational isomer in which two large groups on adjacent atoms are separated by a 60° dihedral angle (5.2)

Geometric isomer: an isomer in which restricted rotation in a ring or at a multiple bond determines the relative spatial arrangement of atoms (5.1)

Half-chair: a high-energy conformation that represents the transition state obtained upon converting a chair to a boat conformation; has all but one atom of the ring in the same plane (5.4)

Heat of reaction: the energy difference between a reactant and a product (5.1)

l : relative stereochemical designator of a molecule with a negative (levorotatory) specific rotation: from the Greek for "left-rotating" (5.8)

L: absolute stereochemical descriptor that relates substituent disposition at a given center of chirality to that in natural *L*-glyceraldehyde (5.8)

Levorotatory: *see l* (5.8)

Meso **compound**: an optically inactive molecule that contains a mirror plane or center of symmetry interrelating centers of chirality in the molecule (5.8)

Mirror image: a reflected projection of an object (5.5)

Mirror plane: a plane through which each part of an object on one side of the plane is reflected to an identical part on the opposite side (5.5)

Newman projection: a representation used to indicate stereochemical relationships between groups bound to adjacent carbon atoms; conformational descriptor in which a triad juncture inscribed within a circle represents dihedral angles between σ bonds on one carbon and those attached to the adjacent atom (5.2)

Nitrogen inversion: the rapid redisposition of the non-bonding lone electron pair of an amine to the opposite side of the molecule, converting the starting amine to its mirror image (5.10)

Optical isomers: isomers that differ in the three-dimensional relationship of substituents about one or more atoms (5.1)

Optically active: descriptor of a sample that rotates the plane of polarized light; a sample containing an excess of one enantiomer of a chiral molecule (5.7)

Optically inactive: descriptor of a sample that does not rotate a plane of polarized light (5.7)

Optical purity: the degree of excess of one enantiomer over the other in a mixture as determined by comparison of the optical rotation of the sample with that of a sample presumed to be a single enantiomer (5.7)

Plane of symmetry: a symmetry element that bisects a molecule such that half the molecule is the mirror image of the other half (5.5)

Plane-polarized light: light that has the electric vectors of all photons aligned in a single plane; obtained by passing ordinary light though a polarizer (5.7)

Polarimeter: an instrument used in quantitatively measuring optical rotation (5.7)

Polarized light: *see* **Plane-polarized light** (5.7)

Potential energy well: an energy minimum along a potential energy diagram representing a molecule or intermediate with a real-time existence (5.2)

Puckered: descriptor of a nonplanar cycloalkane that has fewer eclipsing C—H interactions and lower torsional strain than its planar analog (5.3)

R: absolute stereochemical designator employed in the Cahn–Ingold–Prelog rules; used to describe the stereoisomer in which a clockwise rotation is required to move from the highest to the next-to-lowest priority groups attached to a chiral tetravalent atom when the substituent of lowest priority is directed away from the observer (5.8)

Racemate: *see* **Racemic mixture** (5.7)

Racemic mixture: an optically inactive mixture composed of equal amounts of enantiomers (5.7)

Racemic modification: *see* **Racemic mixture** (5.7)

Relative stereochemistry: the specification of the stereochemical relationship between two molecules (5.6)

Resolution: the separation of a racemic mixture into two pure enantiomers; often accomplished by forming and then separating diastereomers, followed by regeneration of the original reactant (5.8)

Ring-flip: conformational interconversion of one ring conformation to another of the same type; often used to describe chair-to-chair or boat-to-boat interconversions (5.4)

S: absolute stereochemical designator employed in the Cahn–Ingold–Prelog rules; used to describe the stereoisomer in which a counterclockwise rotation is required to move from the highest to the next-to-lowest priority of groups attached to a chiral tetravalent atom when the substituent of lowest priority is directed away from the observer (5.8)

Specific rotation ($[\alpha]$): the extent to which a given molecule (on a weight basis) rotates a plane of polarized light. The observed rotation is the product of the specific rotation, the concentration in the sample compartment, and the pathlength of the sample cell; $\alpha = [\alpha] \times l \times c$, where l = pathlength (in dm) and c = concentration (in g/ml) (5.7)

Staggered conformation: a spatial arrangement in which each σ bond on one carbon atom is fixed at a 60° dihedral angle from a σ bond on an adjacent atom; when viewed end-on in a Newman projection, the conformation in which the bonds on one atom exactly bisect those on the adjacent atom (5.2)

Stereoisomers: isomers that differ only in the way that atoms are arranged in space (5.1)

Steric effect: *see* **Steric strain** (5.2)

Steric strain : the destabilization resulting from van der Waals repulsion between groups that are too close to each other (5.2)

Superimposable: descriptor of the relationship of two molecules for which a conformation exists so that each of the four substituents at a center of chirality can be placed upon each other and are thus oriented in exactly the same direction in space

Syn **eclipsed conformer**: a conformational isomer in which the largest groups on adjacent carbons are disposed with a dihedral angle of 0° (5.2)

Torsional strain: destabilization of an eclipsed conformation relative to a staggered conformation (5.2)

Transition state: highest energy arrangement of atoms along a reaction pathway (5.1)

Twist-boat conformation: a distorted boat conformation in which the steric interaction of the flagpole hydrogens has been relieved by conformational twisting (5.4)

Understanding Organic Reactions

Key Concepts

Reaction thermicity from an energy diagram

Characterization of transition states

Enthalpy and entropy changes

Activation energy barriers

Hammond postulate

Intermediates in energy diagrams

Early and late transition states

Identifying the rate-determining step in multistep reactions

Concerted reactions

Diels-Alder reaction

Reactive intermediates: carbocations, radicals, carbanions, carbenes, radical ions

Acid-base equilibria

pK_a

Factors affecting acidity

Thermodynamic vs. kinetic control

Microscopic reversibility

Keto-enol tautomerism

Kinetics of unimolecular and bimolecular reactions

Collision theory

Answers to Exercises

Ex 6.1   (a)   In an exothermic reaction, the potential energy of the product, P, is lower than that of the reactant, R. A small activation energy barrier implies that the transition state differs in energy only slightly from the reactant. (From the Hammond postulate, a reaction whose transition state differs in structure only slightly from the reactant is said to proceed through an early transition state.)

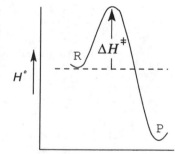

Reaction coordinate

  (b)   As in part *a*, the energy of the product is lower than that of the reactant, but the high activation barrier implies a transition state at a much higher energy than the reactant. Nonetheless, because the reaction is exothermic, the transition state still more closely resembles the reactant, both in energy and structure, than the product.

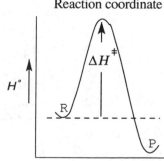

Reaction coordinate

(c) In an endothermic reaction, the product lies at a higher energy than the reactant. With a small activation energy barrier, the transition state energy is only slightly higher than that of the product. From the Hammond postulate, the transition state most closely resembles the product, therefore, both in energy and in structure. This is a typical profile of a reaction proceeding through a late transition state.

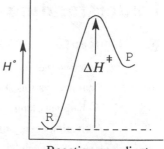

Reaction coordinate

(d) In a thermoneutral reaction, the energies of the reactant and product are identical. A high activation energy barrier implies a high-energy transition state and, hence, kinetic difficulty with the reaction. The transition state resembles the reactant and product equally; that is, its geometry is distorted halfway between the reactant and product.

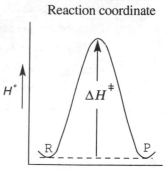

Reaction coordinate

Ex 6.2   To determine the thermicity of these reactions, estimate the bond energies of all bonds broken and formed, having identified each as a primary, secondary, or tertiary bond of the relevant type. (Use the bond dissociation energy table, 3.2, on page 102 of the textbook.)

(a) The C—Cl and H—Br bonds that are formed produce 168 kcal/mole, and the bonds broken, C—Br and H—Cl, consume 171 kcal/mole. The reaction is therefore endothermic in the direction written.

![reaction (a): isopropyl bromide + HCl → isopropyl chloride + HBr]

(b) A C—O and an H—Br bond are formed (181 kcal/mole), and an O—H and C—Br bond are broken (181 kcal/mole). The reaction is thus thermoneutral.

![reaction (b): isopropyl alcohol + CH₃Br → isopropyl methyl ether + HBr]

(c) The C—H and Br—Br bonds that are broken consume 142 kcal/mole, and the bonds formed, C—Br and H—Br, release 155 kcal/mole. The reaction is therefore exothermic, in the direction written, by 13 kcal/mole.

![reaction (c): cyclohexane + Br₂ → bromocyclohexane + HBr]

Ex 6.3   In parts *a-d* of this exercise, a carbonyl group, the tautomer of an enol is given. Enols are provided in parts *e* and *f* ; the tautomer is the corresponding carbonyl compound.

(a)

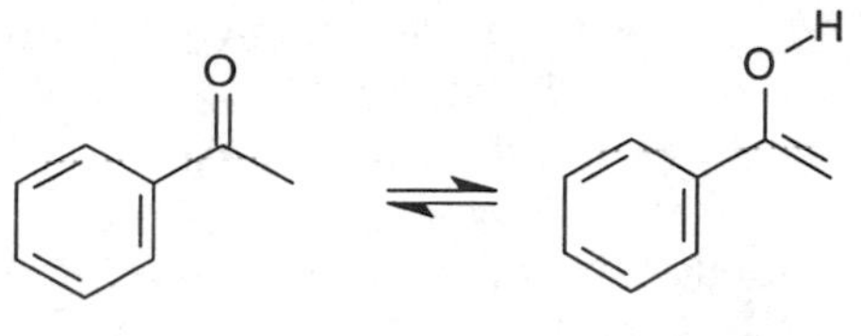

(d)

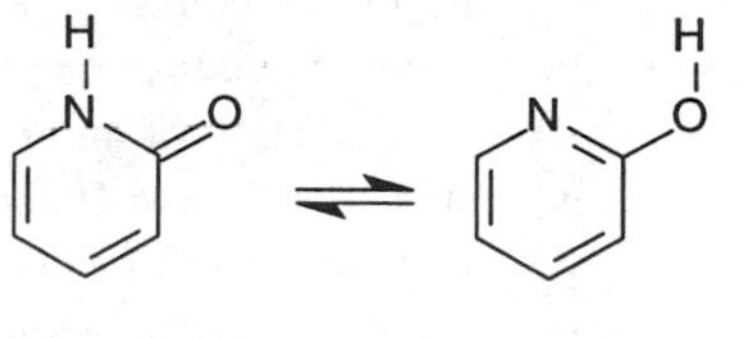

(b)

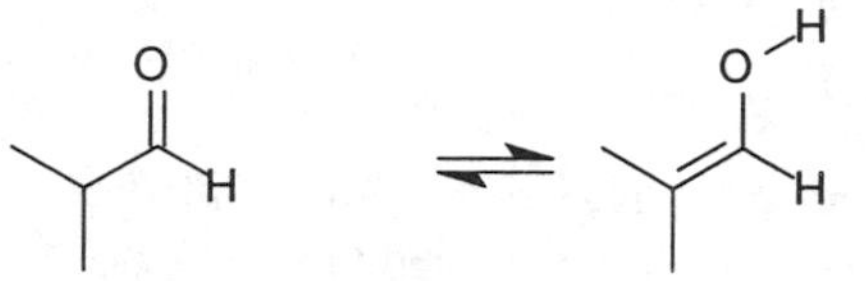

(e)

(c)

(f)

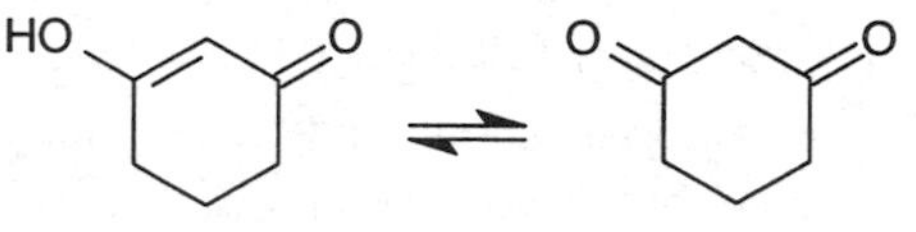

Ex 6.4   A reversible reaction is driven in the direction dictated by favorable release of free energy. Because enthalpy and entropy contributions to free energy enter the equation $\Delta G° = \Delta H° - T\Delta S°$ with opposite signs, the position of an equilibrium can be shifted when the $T\Delta S°$ term is equal to or greater than the $\Delta H°$ term. This problem asks us to solve $\Delta H° = T\Delta S°$ for $\Delta S°$, where $\Delta H°$ is given and T is approximately 300 K. This value is then the minimum entropy change needed to reverse the equilibrium at about room temperature.

(a) $\Delta S° = \dfrac{\Delta H°}{T} = \dfrac{1{,}000\ \frac{cal}{mole}}{300\ K} = 3.3\ \dfrac{cal}{mole \cdot K}$

(b) $\Delta S° = \dfrac{\Delta H°}{T} = \dfrac{-3{,}000\ \frac{cal}{mole}}{300\ K} = -10\ \dfrac{cal}{mole \cdot K}$

(c) $\Delta S° = \dfrac{\Delta H°}{T} = \dfrac{-10{,}000\ \frac{cal}{mole}}{300\ K} = -33\ \dfrac{cal}{mole \cdot K}$

Ex 6.5   In the conversion of an alcohol to an ethyl halide, a primary C—OH bond is broken and a primary C—X is formed. In addition, the O—H bond of water is formed and the H—X bond of the halogen acid is broken. When the reaction proceeds through a single step, the Hammond postulate asserts that a highly exothermic reaction proceeds through an early transition state (resembling the reactant), and an endothermic conversion takes place through a late transition state (resembling the product).

We find from the table of bond dissociation energies at the back of the book that the relevant bond dissociation energies are:

C—Cl	81	H—Cl	103
C—Br	68	H—Br	87
C—I	51	H—I	71
O—H	111	C—O	86

For the reaction of an alcohol with HCl, the energy consumed in breaking the C—O and H—Cl bonds is 197 kcal/mole, whereas that released in forming the C—Cl and O—H bonds is 199 kcal/mole. Thus, the reaction is endothermic by 2 kcal/mole. For a concerted reaction, the transition state is slightly more like the product than the reactant.

For the reaction of an alcohol with HBr, the energy consumed in breaking the C—O and H—Br bonds is 181 kcal/mole, whereas that released in forming the

C—Br and O—H bonds is 187 kcal/mole. Thus, the reaction is exothermic by 6 kcal/mole, with the transition state resembling the reactant more than the product.

For the reaction of an alcohol with HI, the energy consumed in breaking the C—O and H—I bonds is 165 kcal/mole, whereas that released in forming the C—I and O—H bonds is 172 kcal/mole. This reaction is exothermic (7 kcal/mole), as is the reaction with HBr, and would have an early transition state, therefore resembling the starting materials, if concerted.

Ex 6.6   Because the acidic proton is shown in the exercise, you do not need to decide which one is removed by base. Therefore, begin with base attacking the indicated proton, forming a B—H bond and releasing the pair of electrons that formed the original C—H covalent bond. This sequence generates a carbanion. In each case, the anionic site is adjacent to a π system into which charge can be delocalized, producing the resonance structures shown.

(a)

(b)

(c)

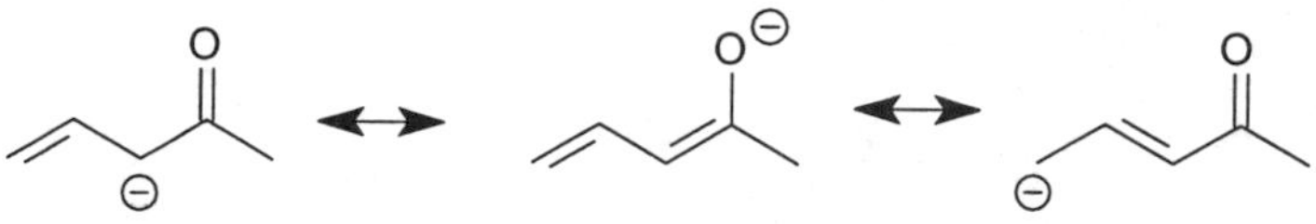

(d)

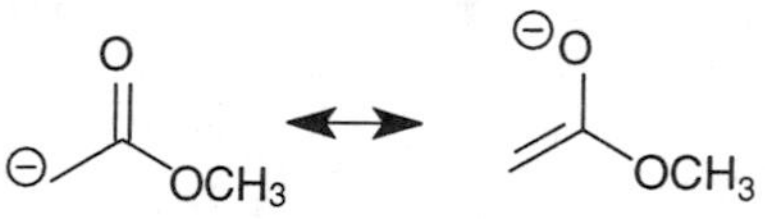

Ex 6.7   Carbocationic carbon atoms bearing three σ bonds are always sp^2-hybridized, with a vacant p orbital. Carbanionic carbon atoms are sp^3-hybridized, unless the lone pair of electrons interacts by resonance with an adjacent π bond. In the latter case, the carbanionic carbon atom rehybridizes to sp^2, with the electron pair being placed into a p orbital so that negative charge can be spread over the conjugated system.

(a)   In the allyl cation, the p orbitals on all three carbon atoms are aligned, with significant resonance contributors placing charge at each terminal carbon.

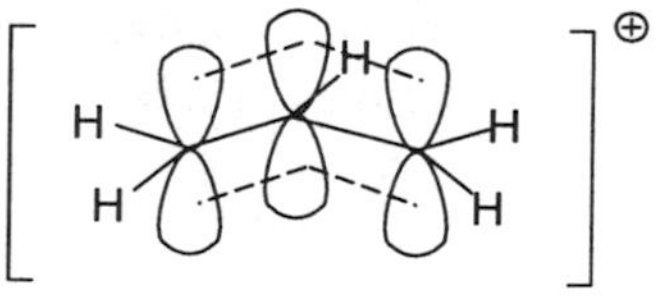

(b)   In the cyclopropyl anion, sp^3-hybridization at the carbanionic carbon makes that center tetrahedral.

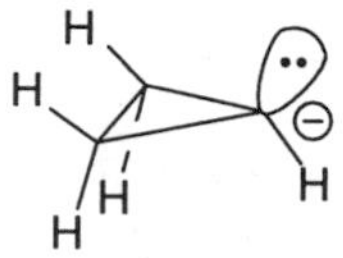

(c)   In the cyclopropyl cation, sp^2-hybridization at the cationic carbon makes that center planar.

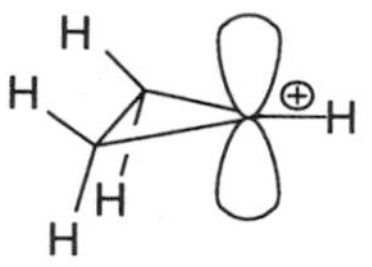

(d)   In the allyl anion, the carbanionic center is sp^2-hybridized in order to interact by resonance with the adjacent π bond, producing a structure similar to that observed in the allyl cation but with two more electrons. Negative charge is at the two terminal carbons, and the structure can be described by two equally significant resonance structures with negative charge at a terminal carbon atom. (Recall Exercise 6.6, part *a*.)

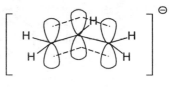

Ex 6.8   Formal charge at any atom is calculated as the number of valence electrons less the number of unshared electrons less half the number of shared electrons.

(a)   carbon in methylene (:CH$_2$): $4 - 2 - [4 \div 2] = 0$

(b)   carbon in dichlorocarbene: $4 - 2 - [4 \div 2] = 0$

Ex 6.9   (a)   Removal of one electron from the π system of benzene to form a cation radical leaves five electrons in the π system. One of the formal π bonds in a Kekulé structure therefore has only one electron. This can be indicated by placing formal positive charge at one carbon and a single electron on the adjacent carbon. Resonance structures can then be written so as to move the singly occupied *p* orbital around the ring (recall resonance theory from Chapter 2). With the positive charge fixed at C-1, there are three such structures. Then move the positive charge to C-2, for which three additional resonance contributors can be written. By successively moving the positive charge to each position about the ring, a total of 18 resonance contributors are obtained.

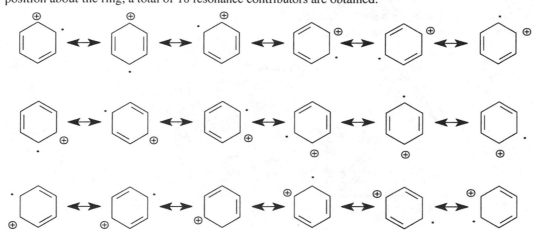

(b)   In the formation of the naphthalene anion radical from naphthalene, one extra electron is added to an antibonding π orbital. Thus, one of the formal π bonds in a Kekulé representation has three electrons, which can be indicated by placing formal negative charge on one carbon (the result of the presence of two electrons in a *p* orbital) and a single electron in the *p* orbital of the adjacent atom. Resonance structures can then be written so as to move the singly occupied *p* orbital around the ring, keeping the position of the doubly occupied *p* orbital fixed. Five such contributors are shown here for the structures in which negative charge is at C-1. Five analogous structures can be drawn for contributors in which negative charge is located at each of the ten carbons in the naphthalene ring (not drawn), for a total of 50 resonance contributors.

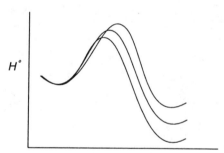

(c) Adding an electron to a carbonyl π bond populates the π antibonding orbital and, as in the naphthalene anion radical, causes formal disruption of the π bond, while placing a pair of electrons on one atom and a single electron on the other. Thus, two resonance structures can be drawn, with carbon as a radical and oxygen as an anion, or vice versa. Because oxygen is more electronegative than carbon, these structures are *not* of equal energy: that in which oxygen is negatively charged is the more stable.

Ex 6.10 The Hammond postulate asserts that the transition state will resemble most closely that species (starting material or product) to which it is closest in energy. In an exothermic reaction, the transition state resembles more the starting material, and the more exothermic the reaction, the less influence the energy of the product has on the transition state. Nonetheless, for similar reactions, the most exothermic will have the transition state of lowest energy, and therefore the fastest rate of reaction. The difference in energy among the three transition states will be less than the difference in energy among the three products.

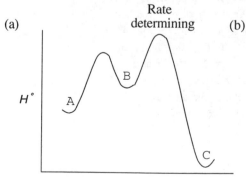

Reaction coordinate

Ex 6.11 The rate-determining step is that in which the transition state is of highest energy. In reaction profile (a), the second step (the conversion of intermediate B into product C) is therefore rate-determining. In reaction profile (b), the first step (the conversion of the starting material D to intermediate E) is rate-determining.

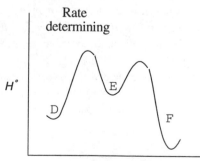

Ex 6.12 Protonation takes place most rapidly at the site of highest electron density. Because oxygen is more electronegative than carbon, it bears a greater fraction of negative charge in the enolate anion and is the primary site for hydrogen bonding with an attacking proton donor.

Ex 6.13 The key features that let us decide if a reaction is under kinetic or thermodynamic control are the relative rates of the forward and backward reactions. When these rates are comparable, under reasonable reaction conditions for the forward process, the backward reaction also takes place. When two different products are formed, the backward reactions to starting material provide pathways for interconversion of these two products, and their ratio is an indicator of their relative stabilities, not of their initial rates of formation. On the other hand, when the activation energy for the backward reaction is much larger than that for the forward reaction (a highly exothermic reaction), interconversion of various products through reversal of the reaction is quite slow, and product ratios are indicators of their rates of formation (kinetic control).

(a) In this example, the energies of the starting material and the two products, P_1 and P_2, are not very different. Therefore, the activation energies are similar for the forward and backward reactions. Thus, P_1 and P_2 are formed under thermodynamic control and, because P_2 is slightly more stable, it predominates. (Notice that the activation energy for the formation of P_2 is larger than that for P_1.)

(b) The situation here is similar to that in part *a* except that now the starting material is much less stable than either product. The rates of reversal of the reactions that form P_1 and P_2 are slower than the forward reactions, and the ratio of the two products indicates their relative rates of formation ($P_1 > P_2$).

(c) Here P_1 and P_2 are in equilibrium (through the starting material), just as in part *a*. However, this situation is different in that the more stable product, P_2, is favored under both thermodynamic and kinetic control.

Ex 6.14 An equilibrium constant can be calculated as either the ratio of the rates of the forward and reverse reactions or the ratio of the concentrations of reactants and products at equilibrium.

(a) $K = \dfrac{k_1}{k_2} = \dfrac{10^{10} \text{ M}^{-1}\text{s}^{-1}}{10^8 \text{ M}^{-1}\text{s}^{-1}} = 100$

(b) $K = \dfrac{[C][D]}{[A][B]}$

From the stoichiometry indicated in the equation, the concentration of C = the concentration of D. Thus, from the data given, $[C] = [D] = 1/2 [A_o]$. From the stoichiometry and from the data stating that the initial concentration of B equals that of A, we know that $[A] = [B] = 1/2 [A_o]$.

Therefore, $K = \dfrac{(1/2 [A_o])(1/2 [A_o])}{(1/2 [A_o])(1/2 [A_o])} = 1$

(c) $\Delta G° = -RT \ln K$.

Therefore:

$\ln K = -\dfrac{\Delta G°}{RT} = -\dfrac{-1,000 \text{ cal/mole}}{\dfrac{2 \text{ cal}}{\text{K·mole}} \, 300 \text{ K}} = +1.67 \qquad K = 5.3$

(d) $\ln K = -\dfrac{\Delta G°}{RT} = -\dfrac{-10,000 \text{ cal/mole}}{\dfrac{2 \text{ cal}}{\text{K·mole}} \, 300 \text{ K}} = +16.7 \qquad K = 1.8 \times 10^7$

(e) $\ln K = -\dfrac{\Delta G^\circ}{RT} = -\dfrac{-30{,}000 \text{ cal/mole}}{\dfrac{2 \text{ cal}}{\text{K·mole}}\, 300 \text{ K}} = +50$ $\qquad K = 5.2 \times 10^{21}$

Ex 6.15 pK_a is defined as the negative logarithm of K_a, and K_a is the equilibrium constant, K, divided by the concentration of water (55.6 molar).

(a) $pK_a = -\log\left(\dfrac{4 \times 10^{-6}}{55.6}\right) \quad = \quad 7.14$

(b) $pK_a = -\log\left(\dfrac{3 \times 10^{-40}}{55.6}\right) \quad = \quad 41.3$

(c) $pK_a = -\log\left(\dfrac{1250}{55.6}\right) \quad = \quad -1.35$

(d) $pK_a = -\log\left(\dfrac{1}{55.6}\right) \quad = \quad 1.75$

(e) $pK_a = -\log\left(\dfrac{5}{55.6}\right) \quad = \quad 1.05$

Ex 6.16 Begin by identifying the most acidic proton in both compounds, and consult Table 6.1 or the table of pK_as at the back of the book for a representative pK_a.

(a)   An N—H proton in a primary amide is more acidic than a proton α to the carbonyl group of a tertiary amide, and therefore has the lower pK_a (approximately 17 versus 30).

(b)   The O—H proton of the carboxylic acid is much more acidic than the α-proton in acetone (approximately 5 versus 19).

(c)   The N—H proton in methylamine is a stronger acid than the C—H protons of trimethylamine (approximately 40 versus 45).

(d)   The O—H proton of benzyl alcohol is much more acidic than the C—H ring protons of benzene (approximately 16 versus 43).

Ex 6.17 (a)   Sulfur is to the right of phosphorus in the third row of the periodic table, making it more electronegative. This, in turn, polarizes the S—H bond more than the P—H bond and makes sulfur better able to accommodate negative charge as the anionic conjugate base. H_2S is therefore more acidic.

(b)   The halogens are to the right of the group VI elements in the periodic table, and chlorine appears in the third row, making it more polarizable than second-row elements. A proton attached to chlorine is more acidic than one bound to oxygen. HCl is more acidic than H_2O.

(c)   Sulfur and oxygen are in the same column in the periodic table, but sulfur, being a third-row element, is more polarizable than oxygen, and can better accommodate negative charge. Therefore, H_2S is the stronger acid.

(d)   As in part *c*, carbon and silicon are in the same group (IV) of the periodic table. The larger silicon atom is more polarizable and better able to take on negative charge as the anionic conjugate base is formed. Although SiH_4 is a stronger acid than methane, both are very weak.

(e)   Because oxygen is more electronegative than nitrogen, as it is to the right of nitrogen in the periodic table, methanol is more acidic than methylamine.

(f)   Iodine is below and to the right of sulfur in the periodic table, making HI the stronger acid.

Ex 6.18 This exercise requires you to identify the most acidic proton and then to determine how its acidity is affected by other structural changes in the molecule.

(a)   Both compounds are α-substituted ethanoic acids, so that the acidic proton is the carboxylate OH. In both cases, acidity is enhanced by inductive withdrawal by a halogen atom. The more electronegative halogen withdraws electron density from the adjacent σ bond system to a greater extent, giving the fluorine-substituted acid the lower pK_a (pK_a 2.59 versus 2.86). This difference is much smaller than that between acetic acid and, for example, chloroacetic acid because fluorine and chlorine are both highly electronegative (4.0 versus 3.0). Indeed, all four α-haloacetic acids have very similar pK_a values:

| pK_a | 2.59 | 2.86 | 2.90 | 3.18 |

(b)   The two compounds are isomeric chlorobutanoic acids in which the distance between the electronegative chlorine atom and the acidic proton differs. The primary effect of the chlorine atom is inductive electron withdrawal from the carboxylate: the fewer intervening bonds, the greater is its effect. Thus, the α-chloro isomer is more acidic than the β isomer (pK_a 2.86 versus 4.05).

(c)   Inductive withdrawal of electron density by fluorine is effective in stabilizing the negative charge placed on oxygen upon deprotonation of an alcohol. Thus, α,α,α-trifluoroethanol is more acidic than ethanol (pK_a 12.4 versus 15.9).

Ex 6.19 Potassium hydroxide and other compounds that ionize upon solution in water generate cations and anions. In general, when these ions are relatively small, the solvent interacts with both, with the lone pairs of electrons on oxygen interacting with the cations, and the hydrogen atoms of water participating in hydrogen bonds with the anions. The presence of the methyl group in methanol reduces the number of hydrogen atoms to one, decreasing hydrogen bonding. Further, the methyl group interferes by steric hindrance with the interaction of the cations with the lone pair of electrons on oxygen. As the size of the alkyl group is increased in the progression from methanol to ethanol to t-butanol, the steric interaction increases, decreasing solvation of the cation.

Ex 6.20 In the compounds in parts a-d of this exercise, the α hydrogen is most acidic and can be removed by base. For the diketone in part e, the most acidic hydrogen is the one that is α to the two carbonyl groups. For the enone in part f, the γ-hydrogen is the most acidic because of conjugative delocalization of the negative charge over three atoms.

(b)  H₃C—C(=O)—CH₂⁻ ↔ H₃C—C(—O⁻)=CH₂

(c) ⁻H₂C—C(=O)—H ↔ H₂C=C(—O⁻)—H

(d) 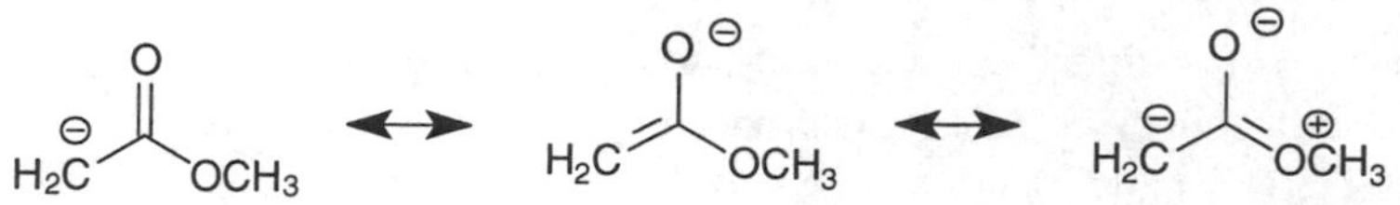 ⁻H₂C—C(=O)—OCH₃ ↔ H₂C=C(—O⁻)—OCH₃ ↔ ⁻H₂C—C(=O)—O⁺CH₃

(e) 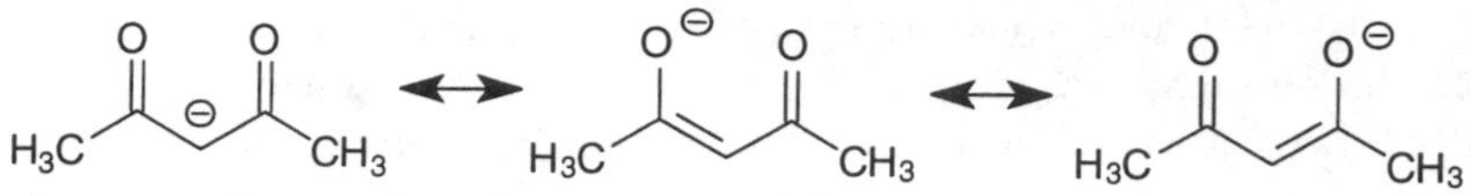 H₃C—C(=O)—CH⁻—C(=O)—CH₃ ↔ H₃C—C(—O⁻)=CH—C(=O)—CH₃ ↔ H₃C—C(=O)—CH=C(—O⁻)—CH₃

(f) 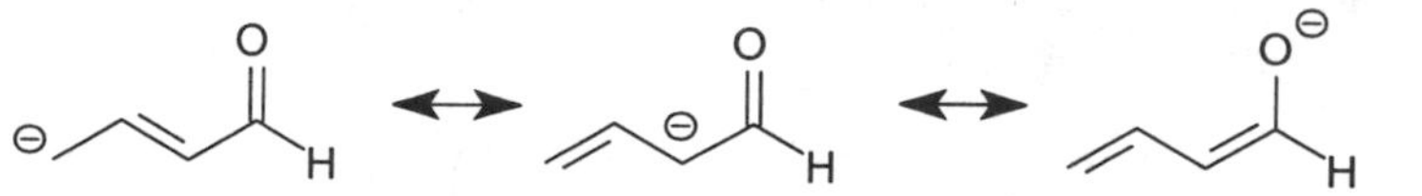 ⁻CH₂—CH=CH—C(=O)—H ↔ CH₂=CH—CH⁻—C(=O)—H ↔ CH₂=CH—CH=C(—O⁻)—H

Ex 6.21 Consulting Table 6.1 (and our chemical intuition), identify the following hydrogen atoms as most acidic:

(a) The protons α to the carbonyl group are more acidic than those on the β position because the anion derived by deprotonation of the former is resonance-stabilized with negative charge spread over carbon and oxygen. In the anion resulting from β-deprotonation, charge is highly localized on carbon.

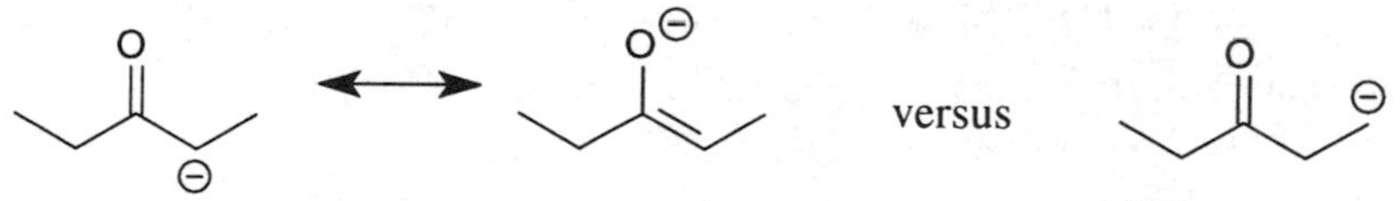

versus

(b) The carboxylic acid is a stronger acid than either the alcohol O—H group or any of the C—H bonds in the molecule.

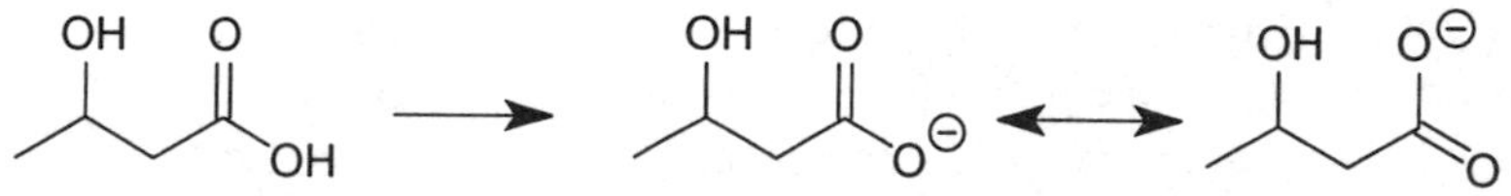

(c) The α-C–H bond at C-3 is most acidic because of the existence of resonance structures that delocalize charge both to oxygen and to the γ position in the dienolate anion.

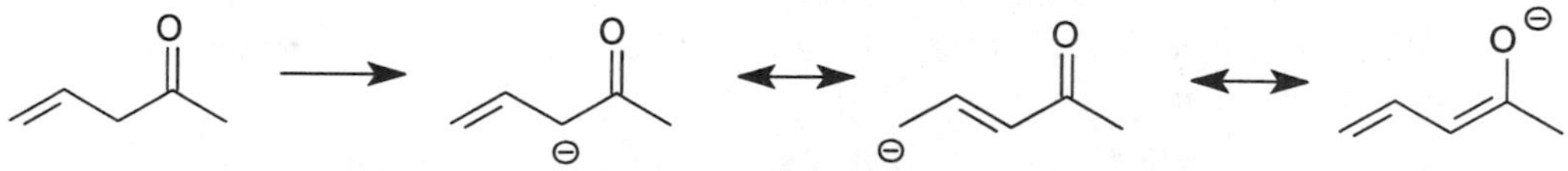

(d The α-C–H bond in this ester is more acidic than the C–H bonds on the alkyl group because of resonance stabilization of the ester enolate anion.

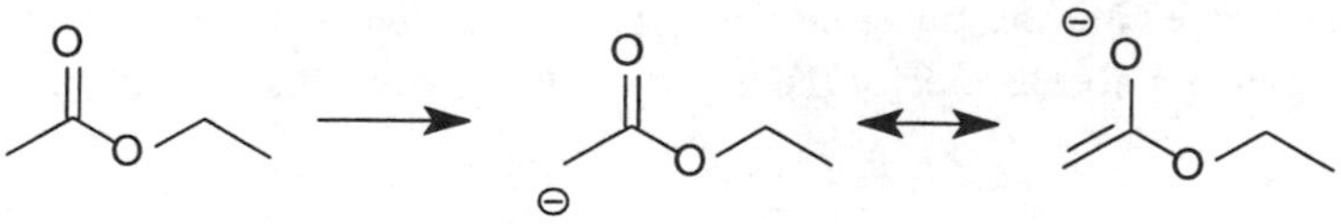

Ex 6.22 Resonance structures differ from each other only in the position of the electrons. For any given resonance structure of the cycloheptatrienyl anion, the negative charge is represented as localized on a single carbon. However, there are a total of seven resonance structures, and each of the seven carbon atoms bears negative charge on one of the structures.

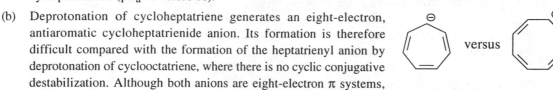

Ex 6.23 A planar, conjugated anion containing $(4n + 2)$ electrons exhibits aromatic stabilization, facilitating deprotonation of a C—H bond to produce such an anion. An anion containing $4n$ electrons, however, is expected to be antiaromatic and hence destabilized. Deprotonation to generate a $4n$ electron anion is therefore difficult.

(a) Deprotonation of cyclopentadiene generates a stable, aromatic, six-electron cyclopentadienide anion. Deprotonation of benzene does not influence the ring aromaticity, but the resulting phenyl anion is highly localized in an sp^2-hybridized orbital. The phenyl anion lacks appreciable stabilization; therefore, benzene is less acidic than cyclopentadiene (pK_a 43 versus 16).

(b) Deprotonation of cycloheptatriene generates an eight-electron, antiaromatic cycloheptatrienide anion. Its formation is therefore difficult compared with the formation of the heptatrienyl anion by deprotonation of cyclooctatriene, where there is no cyclic conjugative destabilization. Although both anions are eight-electron π systems, the orbitals of the cyclooctatriene anion do not form a cyclic system; thus, this anion is not antiaromatic.

(c) In contrast with the aromaticity of the cyclopentadienide anion, the anion formed by deprotonation of cyclopropene is antiaromatic because its cyclic conjugated system contains four electrons. The acidity of the C—H bond in cyclopropene is therefore much lower than that in cyclopentadiene.

(d) In both the parent (unsubstituted) cyclopropene and 3-bromo-cyclopropene, the acidity of the sp^3-hybridized carbon of the C–H bond is low because of formation of an antiaromatic cyclopropenide anion. Although both anions are unstable, the electronegative bromine atom further withdraws electron density from the ring carbon, enhancing the acidity of the C—H bond of 3-bromo-cyclopropene over that of the parent. (In fact, the instability of the anion formed upon deprotonation makes bromocyclopropenide anion so unstable that it suffers rapid loss of bromide ion, forming a carbene.)

Ex 6.24 Because the Arrhenius equation tells us that:

$$k = A e^{\frac{-\Delta H^{\ddagger}}{RT}}, \qquad \text{then:} \qquad \frac{k_1}{k_2} = \frac{A_1 e^{\frac{-\Delta H_1^{\ddagger}}{RT}}}{A_2 e^{\frac{-\Delta H_2^{\ddagger}}{RT}}}$$

If the pre-exponential factors A_1 and A_2 are approximately equal, as is reasonable for similar chemical reactions, then:

$$\frac{k_1}{k_2} = \frac{e^{\frac{-\Delta H_1^{\ddagger}}{RT}}}{e^{\frac{-\Delta H_2^{\ddagger}}{RT}}} = e^{\frac{-(\Delta H_1^{\ddagger} - \Delta H_2^{\ddagger})}{RT}}$$

(a) $\quad \dfrac{k_1}{k_2} = e^{-0} = 1$

(b) $\quad \dfrac{k_1}{k_2} = e^{-\left(\frac{-1000 \text{ cal/mole}}{(2 \text{ cal/mole K})(300 \text{ K})}\right)} = e^{1.67} = 5.3$

(c) $\quad \dfrac{k_1}{k_2} = e^{-\left(\frac{-2000 \text{ cal/mole}}{(2 \text{ cal/mole K})(300 \text{ K})}\right)} = e^{3.33} = 28$

(d) $\quad \dfrac{k_1}{k_2} = e^{-\left(\frac{-5000 \text{ cal/mole}}{(2 \text{ cal/mole K})(300 \text{ K})}\right)} = e^{8.33} = 4.16 \times 10^3$

Notice that because there is an exponential relation between the relative rates and the difference in activation energy, comparatively small energy differences lead to large differences in rate. A simple rule is that each 1.4 kcal/mole difference in activation energy produces an order of magnitude difference in rate. Thus, with 5 kcal/mole of energy difference, the difference in rate is approximately 3.5 orders of magnitude.

Ex 6.25 The mean energy is given by the equation:

$$\text{mean energy (cal/mole)} = \frac{8RT}{\pi}$$

Thus, at $T = 100$ K, the mean energy is 506 cal/mole, and at $T = 200$ K, it is 1012 cal/mole, (Recall that the relationship is linear.)

Ex 6.26 The relative rate of two reactions as a function of changes in either activation energy or temperature can be evaluated by the equation:

$$\text{relative rate} = \frac{e^{-\frac{\Delta H^{\ddagger}}{RT_1}}}{e^{-\frac{\Delta H^{\ddagger}}{RT_2}}}$$

(a) The relative rates of a reaction with an activation energy of 30 kcal/mole run at 0 °C and 100 °C is thus 2.7×10^6.

(b) The change in rate upon a 10 °C change in temperature is less than that for a 100 °C change: relative rate = 7.1.

Ex 6.27 (a) A Boltzmann curve is shifted toward higher energies as the temperature is increased; that is, to the right in a plot of population against potential energy.

(b) A Boltzmann curve is shifted toward lower energies as the temperature is decreased; that is, to the left in a plot of population against potential energy.

(c) A change in activation energy does not change the shape of a Boltzmann curve, although the specific energy cut-off above which a molecule has sufficient energy to reach the transition state is shifted to the right in a plot of population against potential energy. Thus, a smaller fraction of molecules have sufficient energy to react.

Answers to Review Problems

Pr 6.1 In an exothermic reaction, the bond energy content of the product is lower than that of the reactant; in an endothermic reaction, the opposite is true. A concerted reaction is one that proceeds smoothly from reactant to product, encountering no intermediates along the reaction surface. Intermediates are indicated by an energy minimum. The rate-determining step is the step with the highest activation barrier and, thus, the slowest step.

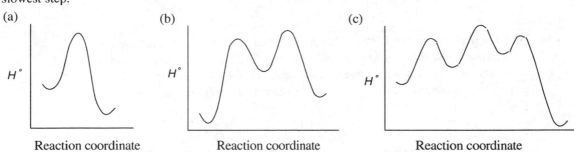

(a) Reaction coordinate   (b) Reaction coordinate   (c) Reaction coordinate

Pr 6.2 (a) Because alkyl groups can hyperconjugate to stabilize adjacent positive centers, carbocation stability follows the order: 3° > 2° > 1° > methyl. Thus:

(b) Because radicals are electron deficient (lacking one electron from that needed for a complete valence shell), they follow the same order of stability as carbocations: 3° > 2° > 1° > methyl. Thus:

(c) A benzylic anion is stabilized by resonance and is the most stable of this group. Because alkyl groups behave in solution as if they release electrons to anionic sites, increasing the number of alkyl groups at the carbanionic carbon destabilizes the anion slightly. As we saw in the acidity of alcohols, alkyl groups also interfere with solvation of the negatively charged site and with ion-pairing. Thus:

(d) Carbenes are electron-deficient species that can be stabilized by either phenyl substitution or alkylation. Hence:

(e) Because oxygen is more electronegative than carbon, ketyl radical anions are more stable than hydrocarbon radical anions. Conjugative interaction with an adjacent aromatic ring stabilizes a phenyl-substituted ion radical compared with an unsubstituted analog. Therefore:

(f) An aromatic cation radical is stabilized by electron-donating or hypercon-
jugating substituents and is destabilized by electron-withdrawing substituents.
Hence:

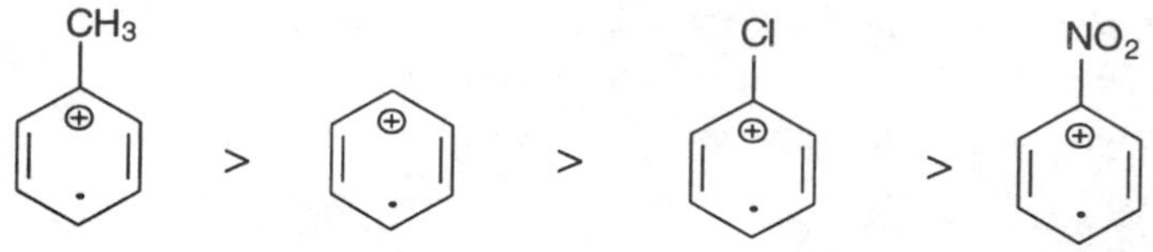

Pr 6.3 (a) The naphthalene cation radical has positive charge and radical character at each
carbon about the ring, as indicated by the possible resonance structures. Here,
only one set (that with positive charge at C-1) is shown. (Compare with the
exactly analogous structures written for the radical anion in Exercise 6.9, part
b.)

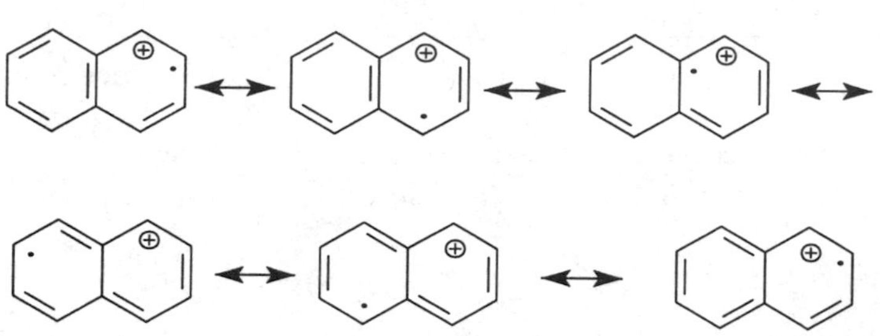

(b) The benzophenone anion radical has several significant resonance structures in
which charge or spin density is located not only on the carbonyl (as in the
acetone ketyl anion radical, Exercise 6.9, part *c*), but also on adjacent aromatic
rings. In the set of structures shown below, the radical center on carbon is
delocalized into one of the adjacent rings. An analogous set exists in which the
radical center is on oxygen, with negative charge being delocalized into the
attached rings. Each structure also has another Kekulé structure in which the
double bonds of the ring are shifted.

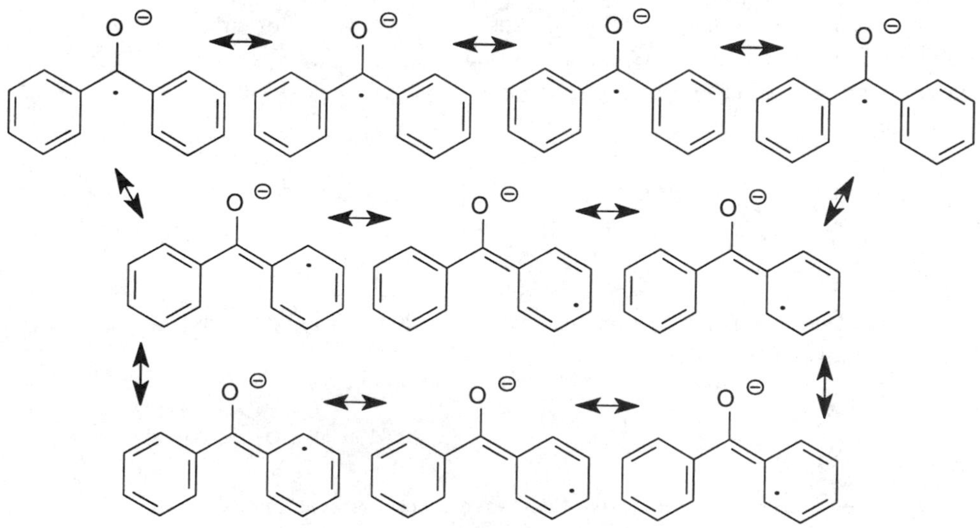

Pr 6.4 Deprotonation at C-1 or C-2 of the cyclopropane produces a benzyl anion, deuteration of which gives the observed product. That the C—D bond in the product is not formed only on the same side as the C—H bond in the reactant implies either that the anionic center is fully planar, as would be expected if it were sp^2-hybridized, or that the rate of inversion of the tetrahedral carbanion is faster than deuteration. Because the data provided does not differentiate between these possibilites, it can not be determined if the carbanionic carbon is chiral (pyramidal) or achiral (planar).

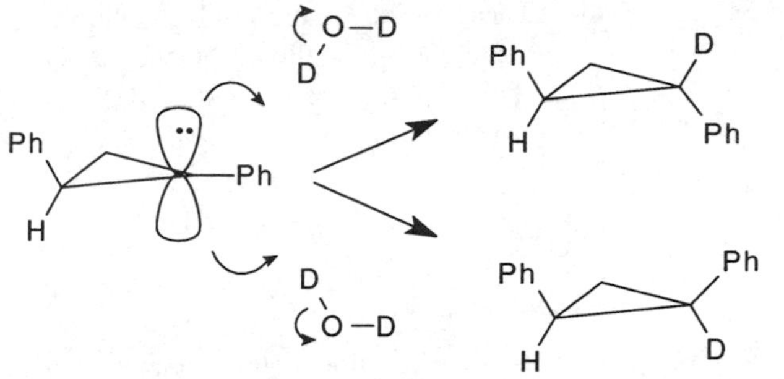

Pr 6.5 Entropy favors molecular disorder, and hence the breakup of a large molecule into several small ones.

   (a) Because a more disordered system is found in the product, the entropy change is favorable (positive).

   (b) The formation of a complex molecule from two or more simpler fragments is unfavorable entropically (negative).

   (c) Because the number of molecules present as products (acid and alcohol) and as reactants (ester and water) is the same, and because solution association (for example, the extent of aggregate hydrogen bonding) is similar in the reactant and product, there is only a negligibly small entropy change in this reaction.

Pr 6.6 A Diels-Alder reaction forms a six-membered ring from a diene and an alkene (dienophile). The diene part can be identified by the presence of a double bond in the product.

   (a) This product is a cyclohexadiene and could have been formed, in principle, from either of two combinations: 1,3-butadiene and dimethyl acetylenedi-carboxylate, or 2,3-bis-carbomethoxy-1,3-butadiene and acetylene. The former set of reagents is preferred because Diels-Alder reactions work best if the dienophile is electron-deficient.

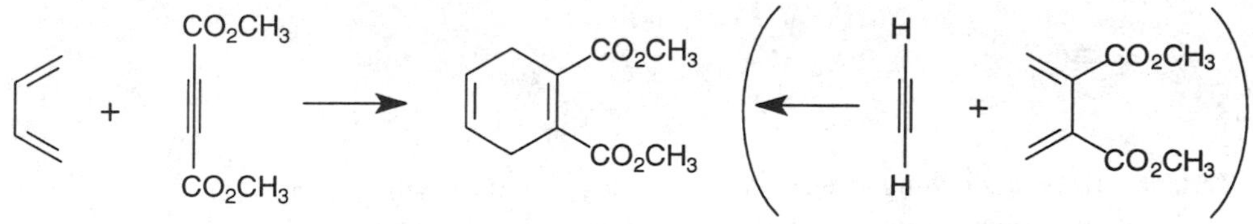

   (b) This product is a bicyclic compound produced from the reaction of a cyclic diene (cyclopentadiene) with maleic anhydride.

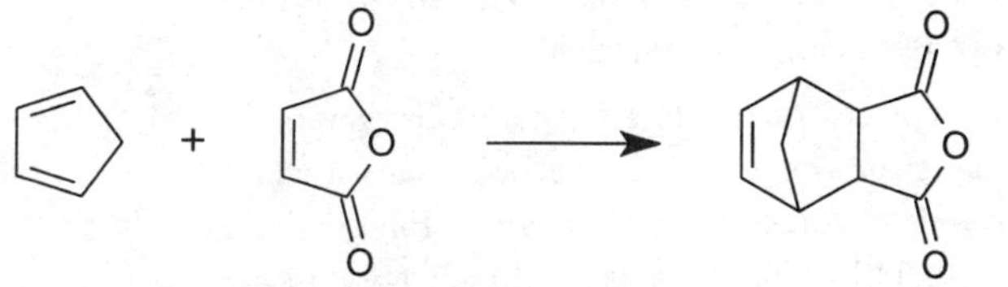

   (c) With only a single double bond in the cyclohexenyl ring, this product must have been formed from 1,3-pentadiene and methyl acrylate.

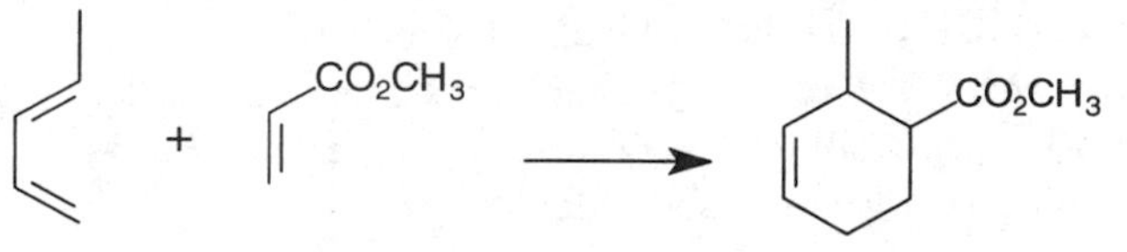

(d) The position of the double bond in the cyclohexene product suggests 1,3-butadiene and methyl 2-butenoate as reactants. (Because no stereochemistry is shown for the product, either the *cis* or the *trans* isomer of the unsaturated ester can be used.

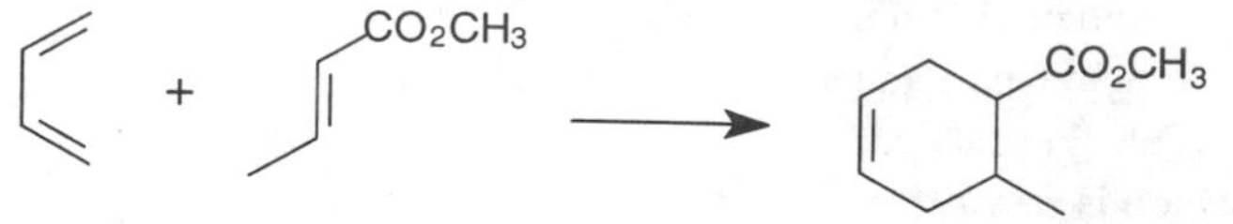

Pr 6.7 A Diels-Alder reaction requires an *s-cis* diene so that the ends of the diene system can interact directly with the dienophile to form a six-membered ring. This is not possible in a fused bicyclic molecule because rotation about the C-2–C-3 σ bond is restricted by the ring.

Pr 6.8 Increasing pK_a values imply decreasing acidity through a series.

(a) A carboxylic acid is a stronger acid than a phenol than an alcohol. This follows because negative charge on oxygen is delocalized to the carbonyl group in a carboxylate anion and to three ring carbons in phenoxide, but is highly localized in an alkoxide.

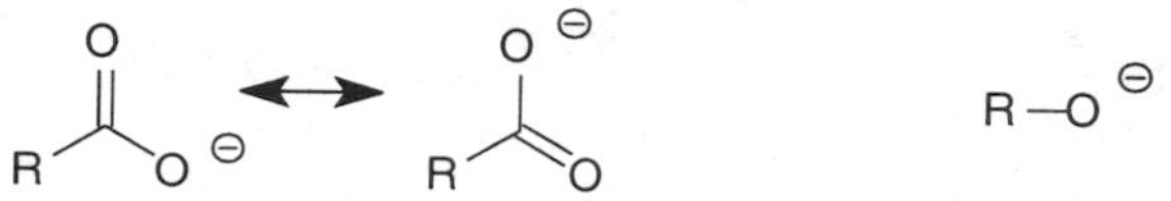

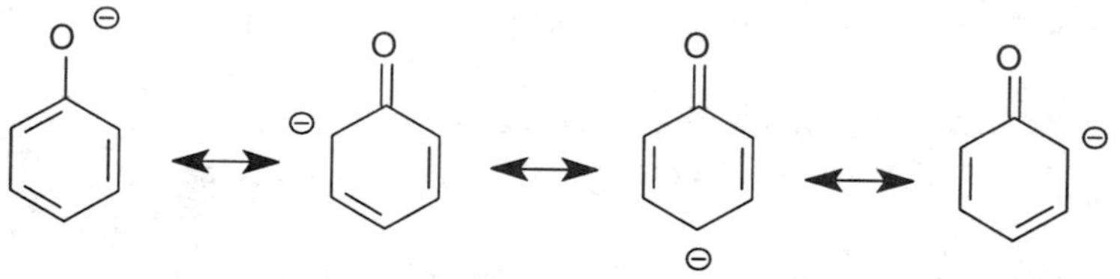

(b) Deprotonation at a C—H bond takes place most readily at a site with a greater fraction of *s* character; that is, an anion is more stable at an *sp*- than at an sp^2- than at an sp^3-hybridized atom. Hence, 1-butyne is more acidic than 1-butene, which is, in turn, more acidic than butane.

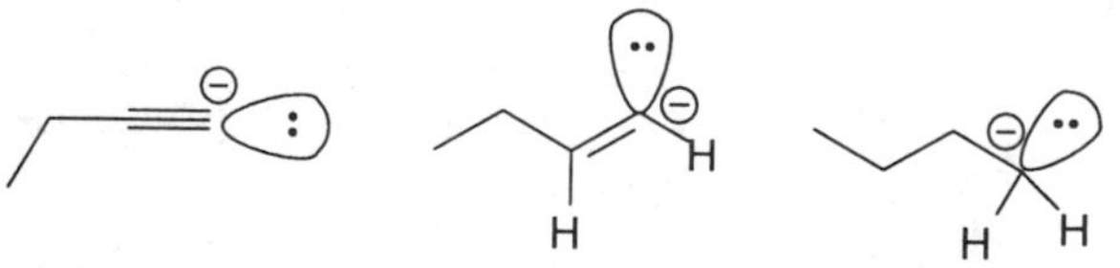

(c) An electron-withdrawing group enhances acidity, and the smaller the number of σ bonds that must be polarized, the stronger is the effect. The nitro group, in which nitrogen bears formal positive charge, is a stronger electron withdrawer than an electronegative bromine atom. Thus, acidity decreases in the order: 2-nitropropanoic acid, 3-bromopropanoic acid, propanoic acid.

(d) An anion in which negative charge is placed on an electronegative oxygen atom is more stable than an analogous structure in which carbon bears formal negative charge. The benzylic protons of toluene are more acidic than the CH protons of benzene because of the possibility for resonance delocalization in the benzyl anion: phenol > toluene > benzene.

(e) Ethanol, bearing a highly polar O—H bond, is most acidic. The α-C—H proton of methyl acetate ($CH_3CO_2CH_3$) can be removed by strong base because of resonance stabilization of the resulting ester enolate anion. Deprotonation of dimethyl ether takes place only with very strong base because the resulting carbanion has no significant structural stabilization.

(f) An NH group is more acidic than a CH group. In this series, the negative charge in the anion can be delocalized into the ring or into the adjacent carbonyl group. Thus, acidity decreases in the order: hexanoamide, aniline, hexylamine.

(g) A chlorine atom acts as an electron-withdrawing group, enhancing the acidity of a benzoic acid. The larger the number of electron-withdrawing groups, the more stable is the anionic conjugate base: 2,4,6-trichlorobenzoic acid, *p*-chlorobenzoic acid, and benzoic acid.

(h) Because the polarity of a carbonyl group places partial positive charge on carbon, it acts as an electron-withdrawing group to an adjacent carboxylic acid. Because this effect is inductive, it is greater the closer the electron-withdrawing group is to the acidic group of interest. Thus, oxalic acid (HO_2CCO_2H) is more acidic than malonic acid ($HO_2CCH_2CO_2H$), which is more acidic than acetic acid (H_3CCO_2H).

(i) The conjugate acids of these compounds are species in which the nitrogen lone pair has been protonated. In both pyrrole and *N*-methylpyrrole, deprotonation of the ammonium salt produces a resonance-stabilized, six-electron, aromatic neutral molecule. Because alkyl groups are electron releasing, *N*-methylpyrrole is slightly less acidic than pyrrole because of the methyl group on nitrogen. The aromaticity of the pyridine is less significantly affected by the protonation of the nitrogen lone pair which is located perpendicular to the π system of the ring. There is no aromaticity gain upon deprotonation of pyridinium ion; hence, it is the weakest acid (making pyridine the strongest base).

Pr 6.9 The γ-hydrogen atoms of crotonaldehyde are conjugated to the carbonyl group, affording a conjugated enolate anion upon deprotonation, in which negative charge is distributed over a five-atom system. There are three significant resonance contributors, and negative charge in the anion is found on three different atoms (two carbons and one oxygen).

Pr 6.10 The insolubility of octylamine in water implies that the energy gained by hydrogen bonding at the NH_2 group is insufficient to compensate for the disruption of solvent (water) hydrogen bonding accompanying the solvation of the nonpolar octyl group. Its contrasting solubility in aqueous sulfuric acid derives from the protonation of the basic lone pair on nitrogen to form a cationic ammonium salt, which (with a full positive charge) is water-soluble. Octanoamide similarly resists dissolving in water because of the difficulty in solvating the nonpolar alkyl chain. The existence of resonance stabilization in the amide group spreads the electron density from the amide nitrogen lone pair over three atoms, reducing the electron density at one site where acid can conveniently approach.

A base, however, can abstract an acidic amide (N—H) proton, generating a resonance-stabilized anion that is water-soluble.

Pr 6.11 To answer this problem, order the functional groups according to their acidities. (See the table of pK_a values at the back of the book.)

    (a)   The proton on the *sp*-hybridized acetylenic carbon is more acidic than either the benzylic or the ring protons.

    (b)   The phenolic OH group is more acidic than either the α carbonyl protons or those on the ring.

    (c)   The amide NH proton is more acidic than either the α carbonyl protons or those on the ring.

    (d)   The sulfonic acid OH group is more acidic than the relatively acidic phenol OH group and the protons on the α carbon of the ketone.

Pr 6.12 (a)   The greater stability of the product in this exothermic reaction is indicated by the lower energy minimum in the product than in the reactant. An intermediate appears as an energy minimum along the reaction coordinate.

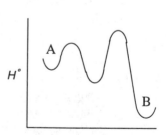

Reaction coordinate

    (b)   The shape of the curve does not change, but the fraction of reactant molecules able to overcome the reaction barrier increases, and, hence, the rate is faster.

    (c)   The rate is proportional to the concentration of A.

Pr 6.13 The most acidic proton(s) is shown boldfaced for part of this question. In each case, removal of the indicated proton(s) results in an anion that is resonance stabilized.

    (a)   (b)   (c)   (d)

Pr 6.14 This reaction is bimolecular so the concentrations of both reagents appear in the reaction-rate expression.

   (a) Doubling the concentration of C doubles the rate of the reaction.

   (b) Doubling the concentration of D doubles the rate of the reaction, rather than halving it.

   (c) Doubling the concentration of both C and D quadruples the rate of the reaction.

   (d) Increasing the temperature of any ground-state reaction increases the rate, unless the temperature increase is so large as to open the possibility for competing reactions that were impossibly slow at the lower temperature.

Pr 6.15 This energy diagram corresponds to a two-step reaction mechanism in which the second step has the higher activation energy.

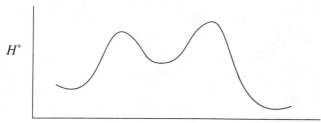

Reaction coordinaate

Pr 6.16 (a) Because the first step in this two-step reaction is endothermic, the transition state resembles the intermediate more than the starting material.

   (b) The second step is exothermic therefore resembles the intermediate more than the product.

   (c) The rate-determining step in a multistep mechanism is also the step with the transition state of highest energy. Thus, the second step is rate limiting.

Pr 6.17 There are three significant resonance structures of methylvinylketone.

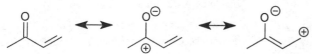

Important New Terms

Activation energy barrier: *see* **Activation energy** (6.1)

Bimolecular reaction: a reaction that requires a collision between two reactants in the rate-determining step (6.9)

Carbanion: a negatively charged, trivalent carbon bearing an unshared electron pair (6.4)

Carbene: a neutral reactive intermediate in which a carbon atom bears two σ bonds and two unshared electrons; contains only six electrons in its outer shell (6.4)

Concerted reaction: a reaction that proceeds directly from reactant to product through a single transition state and without intermediates (6.1)

Conjugate acid: a species obtained by the addition of a proton to a Brønsted base (6.8)

Conjugate base: a species obtained by the removal of a proton from a Brønsted acid (6.8)

Diels-Alder reaction: the concerted cyclization of a conjugated diene and an alkene (called a dienophile) to produce a cyclohexene; the most frequently encountered (4 + 2) cycloaddition (6.6)

Dienophile: in a Diels–Alder reaction, the alkene component that reacts with a diene (6.6)

Digonal: a carbon atom with only two bonds (6.4)

Early transition state: a reactant-like transition state (6.3)

Effective collision: a collision between two reactants with the correct orientation and with sufficient energy to overcome the activation energy barrier (6.9)

Endergonic reaction: a chemical transformation in which free energy input is needed; a reaction in which the free energy content of the products is higher than that of the reactants (*see also* **Endothermic reaction**) (6.1)

Endothermic reaction: a conversion with a positive enthalpy change (*see also* **Endergonic reaction**) (6.1)

Energy diagram: a graphic representation of the change in free energy (or enthalpy) encountered during the course of a reaction (6.1)

Enol: a functional group in which a hydroxyl group is attached to an alkenyl carbon (6.2)

Enolate anion: a resonance-stabilized anionic intermediate obtained by removal of a proton from the α position of a carbonyl compound or the OH group of an enol (6.2)

Equilibrium constant: (K = [C][D]/[A][B]; a measure of the equilibrium position of the reaction A + B = C + D; the ratio of the forward and reverse rate constants of a reversible reaction at equilibrium (6.7)

Exergonic reaction: a reaction in which free energy is released; a reaction in which the total free energy content of the products is lower than that of the reactants (6.1)

Exothermic reaction: a conversion with a negative enthalpy change (6.1)

Free energy: a state property of a system; has contributions from both enthalpy ($H°$) and entropy ($S°$); measure of the potential energy of a molecule or group of molecules (6.1)

Free energy change: a measure ($\Delta G°$) of the potential energy change during a chemical reaction; includes enthalpy ($\Delta H°$) and entropy ($\Delta S°$) components; $\Delta G° = \Delta H° - T\Delta S°$; related to the equilibrium constant as $\Delta G° = - RT\ln K$ (6.1, 6.6)

Hammond postulate: an assertion that a transition state most closely resembles the stable species that lies closest to it in energy (6.3)

Hybridization effect: the influence of mixing of s and p orbitals; the greater the fraction of s character (50% in an sp hybrid; 33% in an sp^2-hybrid; 25% in an sp^3-hybrid) of the hybrid orbital, the more electronegative is the atom (6.7)

Inductive effect: the charge polarization through a series of σ bonds, causing a shift of electron density from or to a charged or polar site (6.8)

Intermediate: *see* **Reactive intermediate** (6.1, 6.4)

Ion pairing: the electrostatic association between oppositely charged ions (6.4)

Irreversible reaction: an exothermic reaction in which the activation energy for the reverse reaction is sufficiently large that the reaction proceeds only in the forward direction under practical conditions (6.1)

Isoelectronic: describing two atoms with the same electronic configuration (6.4)

K_a: the acid-dissociation equilibrium constant;

$$K_a = K[H_2O] = \frac{[A^-][H_3O^+]}{[HA]} \ (6.7)$$

Keto-enol tautomerization: the process by which a proton is shifted from the α carbon of a ketone to the carbonyl oxygen, or from the OH group of an enol to the remote alkenyl carbon; a 1,3 shift of a proton in an aldehyde or ketone (6.6)

Ketyl: a radical anion obtained when an electron is added to the carbonyl group of a ketone (6.4)

Kinetic control: descriptor of a chemical reaction in which the reverse reaction takes place slowly or not at all, so that the relative concentration of products directly correlates with the relative rates of their formation rather than their relative stabilities (6.6)

Kinetics: a description of factors influencing the rate at which a reaction proceeds (6.5)

Late transition state: a transition state that is product-like (6.3)

Microscopic reversibility: a requirement that the same transition state is encountered in the forward and backward directions in any reversible chemical reaction (6.5)

pK_a: the negative logarithm of K_a; a larger positive value indicates a weaker acid (6.6)

Potential energy surface: a plot of the changes in potential energy taking place as a reaction proceeds (6.1)

Radical anion: a reactive intermediate with one more electron than needed for the electron configuration of a stable neutral molecule (6.4)

Radical cation: a reactive intermediate lacking one electron from the complement needed for a stable neutral molecule (6.4)

Rate-determining step: the step in a multistep sequence whose transition state lies at highest energy (6.3)

Rate-limiting step: *see* **Rate-determining step** (6.3)

Reaction coordinate: the variation of a specific structural feature (e.g., bond length or angle) that measures how far a reaction has proceeded (6.1)

Reaction profile: *see* **Energy diagram** (6.1)

Reactive intermediate: a metastable species with a high energy relative to a reactant and product; encountered at an energy minimum (in a potential energy well) along a reaction coordinate (6.1, 6.4)

Resonance effect: the stabilization by delocalization of π electrons; the donation or withdrawal of electron density by overlap with a neighboring π system (6.8)

Retro-Diels–Alder reaction: the concerted fragmentation of cyclohexene (or a derivative) to butadiene (or a derivative) and an alkene (or a derivative); the reverse of a Diels-Alder reaction (6.6)

Reversible reaction: a reaction that can proceed backward or forward with similar ease many times (6.1)

Singlet: a molecule in which all electrons are paired, generally with two electrons of opposite spin being paired in each molecular orbital (6.4)

Tautomerization: a change from one structure to another in which the only changes are the position of attachment of a hydrogen atom and the position of π bond(s); typically, a 1,3 (or 1,5) shift of a proton to or from a heteroatom in a three-atom system containing a double bond; catalyzed by acid or base (6.2)

Tautomers: constitutional isomers that differ only in the position of an acidic hydrogen along a three-atom segment containing a heteroatom and a double bond (6.2)

Termolecular reaction: a reaction that requires a collision between three reactants in the rate-determining step; termolecular reactions are rare (6.9)

Thermodynamics: a description of the relative energies of the reactants and products and the equilibrium established between them (6.5)

Thermoneutral reaction: a conversion in which the reactants and products have the same energy content (6.1)

Through bond: transmission of some effect through the electron density connecting atoms in covalent bonds (6.7)

Thermodynamic control: descriptor of a chemical reaction in which the reverse reaction takes place at a rate not substantially different from the forward reaction, establishing equilibrium (6.6).

Transition state theory: a theory that asserts that the rate of a reaction varies exponentially with the energy required to reach the transition state (6.9)

Trigonal: a carbon atom with three σ bonds (6.4)

Triplet: a molecule in which not all electrons are spin-paired, with two electrons of the same spin being accommodated in two different orbitals (6.4)

Unimolecular reaction: a reaction involving only a single species in the rate-determining step (6.9)

Mechanisms of Organic Reactions

Key Concepts

Reaction classification: addition, elimination, substitution, condensation, rearrangement, isomerization, redox

Calculating $\Delta H°$

Using curved arrows to represent reaction mechanisms

Four ways of studying reactions:

1) Predicting product, given reactant, reagents, and conditions

2) Suggesting reagents and conditions needed to convert a reactant to a product

3) Defining the reactant needed to produce a given product under specified conditions

4) Describing the reaction mechanism, including any important intermediates

Concerted, bimolecular, back-side attack, with inversion of configuration, in an S_N2 (substitution) reaction

Rate-determining formation of a carbocation in an electrophilic addition

Unimolecular, heterolytic fragmentation to a carbocation, with racemization, in an S_N1 (substitution) reaction

Steps in a radical chain reaction: initiation, propagation, termination

Radical formation and reaction in the propagation steps of free-radical halogenation

Factors affecting selectivity in a radical reaction

Answers to Exercises

Ex 7.1 To classify a reaction type, begin by comparing the number and type of atoms present in the reactant and product. If more atoms are present in the product, ask if any atoms or group of atoms from the reactant are missing. If there are fewer atoms in the product, determine which atoms have been replaced to establish whether a more complex group has been replaced by a simpler one, or whether one or more groups have been lost. If the number and type of atoms in the reactant and product are unchanged, a rearrangement or isomerization is likely.

(a) Addition. The formula for the reactant is C_6H_{12} and for the product, $C_6H_{12}Br_2$. Thus, two bromine atoms have been added.

(b) Substitution. The formula for the reactant is C_6H_{12} and for the product, $C_6H_{11}Br$. Thus, a bromine atom has replaced a hydrogen atom originally present in the reactant.

(c) Substitution. Methanol (CH_3OH) is converted to methyl chloride (CH_3Cl). An OH group has been replaced by Cl.

(d) Addition. The formula of the reactant is C_5H_{10} and that of the product is C_5H_{12}. Thus, two hydrogen atoms have been added to the reactant.

(e) Condensation. Two simpler organic molecules ($C_2H_4O_2$ and CH_4O) yield a product ($C_3H_6O_2$) that differs from the sum of the formulas of the reactants by one water molecule.

(f) Substitution. The product (C_3H_5OI) differs from the reactant (C_3H_6O) by having one iodine atom in place of one hydrogen atom.

(g)   Isomerization (positional and skeletal). The formulas of the reactant ($C_5H_{11}Br$) and product are the same, but carbon skeleton is different as is the position of bromine in the product is different.

(h)   Addition. The formula of the product (C_6H_{10}) is the sum of the formulas of the reactants (C_2H_4 and C_4H_6).

Ex 7.2   For each reaction, identify the kind of bond(s) made and broken.

(a)   A primary C—H bond (100 kcal/mole) is broken as H—Br (87 kcal/mole) is formed: $\Delta H° = +13$ kcal /mole.

(b)   A secondary C—H bond (96 kcal/mole) is broken as H—Br (87 kcal/mole) is formed: $\Delta H° = +9$ kcal /mole.

(c)   A tertiary C—H bond (93 kcal/mole) is broken as H—Br (87 kcal/mole) is formed: $\Delta H° = +6$ kcal /mole.

(d)   A primary C—H bond (100 kcal/mole) is broken as H—Br (87 kcal/mole) is formed: $\Delta H° = +13$ kcal/mole.

Ex 7.3   To calculate the bond energy changes accompanying a given chemical reaction, we must define which bonds are broken and which are formed. By assigning a positive sign to those broken and a negative sign to those formed, we can simply add the energies to determine overall bond energy changes.

(a)   Here a tertiary C—Br bond is broken and a tertiary C—Cl bond is formed, as H—Cl is broken and H—Br is formed: $68 - 82 + 103 - 87 = 2$ kcal/mole.

(b)   Here a methyl C—Br bond is broken and a methyl C—Cl bond is formed, as H—Cl is broken and H—Br is formed: $70 - 85 + 103 - 87 = + 1$ kcal/mole.

(c)   Here a methyl C—I bond is broken and a methyl C—Cl bond is formed, as H—Cl is broken and H—I is formed: $57 - 85 + 103 - 71 = + 6$ kcal/mole.

Ex 7.4   Where possible, ask how the atomic formula of the product differs from that of the reactant. Then classify each reaction type, and scan a library of known reactions for a sequence of reagents that can accomplish the transformation. If the product is not given, ask what the expected reaction with the given reagent might be, and predict which part of the molecule is affected by the indicated chemical treatment. (You may find it helpful to compile a series of index cards that list various reaction types as they are introduced in the text: they will be used repeatedly throughout the course.)

(a)   The product (C_4H_{10}) differs from the reactant (C_4H_8) in that two additional hydrogens are present. This is therefore an addition, and the original double bond is no longer present. The addition of two hydrogens across a C=C double bond is accomplished by catalytic hydrogenation (Chapter 2).

(b)   The product is not given, so ask yourself what reaction type is accomplished with the indicated reagents. The reagents shown here are those typically used for catalytic hydrogenation (Chapter 2). Hydrogen adds across a double bond under these conditions, and one double bond is present in the reactant. Therefore, the expected product is cyclohexane.

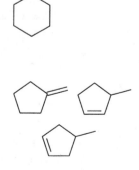

(c)   The reagents are those required for catalytic hydrogenation, and the product is a saturated cyclic hydrocarbon consistent with an addition reaction. Therefore, in this reaction, one or more equivalents of hydrogen gas have been added across C=C double bond(s). Although it is possible to add hydrogen to more than one such double bond, the stated stoichiometry requires that only one C=C was present in the starting material. Nonetheless, some ambiguity in the answer still persists, because the product gives us no indication of the specific position of the double bond in the reactant. Therefore, we can say only that our reactant must have been one of the monoalkenyl isomers of C_6H_{12} consisting of a single five-carbon ring and a one-carbon substituent. All of the structures shown here to the right would produce methylcyclopentane upon reduction with H_2 in the presence of a metal catalyst.

(d) The conversion of an alcohol to an alkene requires the formal loss of water. This can be effected by treatment of the alcohol with acid.

(e) A comparison of the formulas of the reactant and product indicates that the starting material ($C_4H_{10}O$) differs from the product (C_4H_8) by H_2O. We can therefore classify this reaction as an elimination of water; that is, a dehydration. As we have seen in this Chapter, such reactions are accomplished by heating in acid. (For now, do not worry about how the dehydration gives *trans*-product rather than *cis* or how it leads to 2-butene instead of 1-butene: this level of question will be treated in Chapter 9.)

(f) The product is an alkene attached to a phenyl ring. The conditions given are those for acid-catalyzed dehydration like that discussed in part *e*. Therefore, this product could have been formed by loss of water from a reactant differing from the product by additional H and OH groups. There are two ways in which these groups could have been attached to the skeleton, as shown. Again, we cannot distinguish these two reactants without more information.

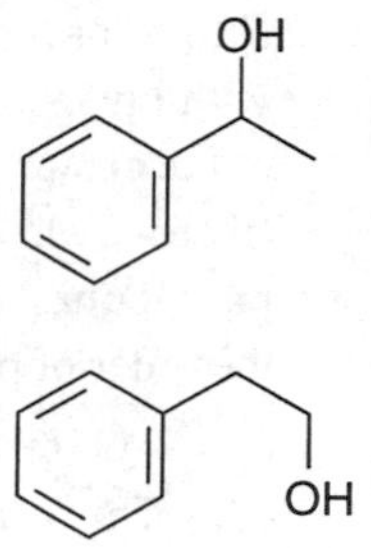

Ex 7.5 First, determine whether the indicated cleavages are homolyses or heterolyses. If the former, the two electrons of the cleaving σ bond are split so that one goes to each fragment. If the latter, the two electrons go to one fragment, leaving the other fragment positively charged. In that case, establish which of the two fragments can better accommodate positive or negative charge. Recall that a half-headed arrow indicates the movement of one electron; a full-headed arrow indicates that two electrons move as a pair.

(a) Homolysis, producing two radicals.

(b) Heterolysis; the two electrons of the π bond form a σ bond to the attacking electrophile.

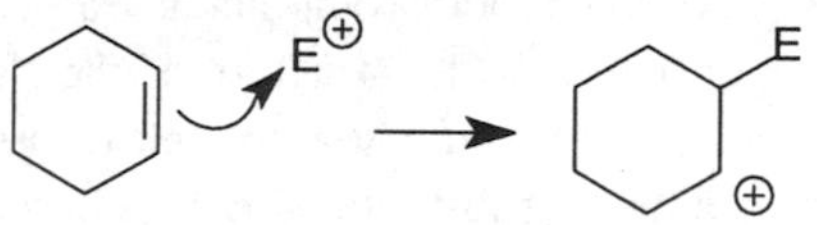

(c) Heterolysis; the two electrons go to the more electronegative bromine atom, forming a tertiary carbocation.

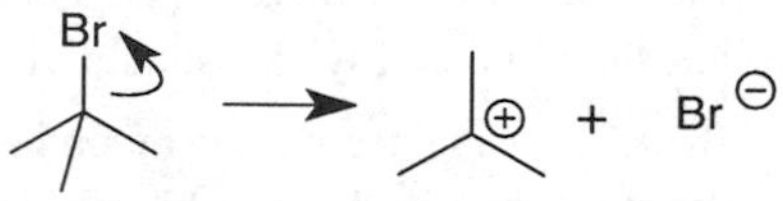

(d) Heterolysis: a proton is removed by base to produce water and a resonance-stabilized enolate anion.

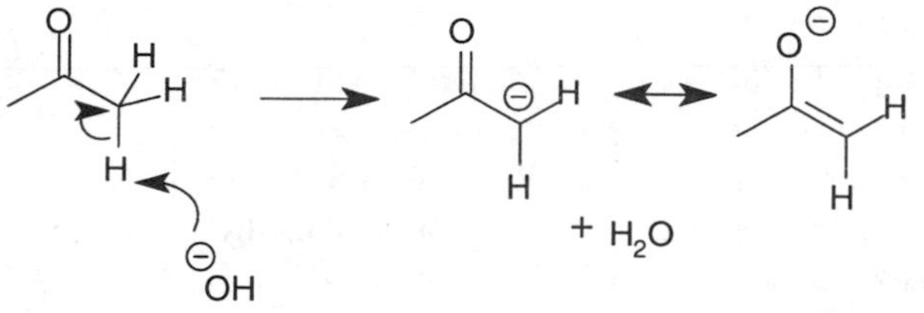

Ex 7.6 Although reaction 6 describes an S_N2 reaction that takes place with inversion of configuration, the self-exchange indicated will produce a racemic modification that must be optically inactive. As the reaction proceeds, more and more of the inverted (S) configuration is produced, until all identity of which enantiomer is product and which is reactant is lost. An alternative analysis that leads directly to the same answer is to consider the pentavalent transition state for this reaction: here the incoming bromide ion and the bromide ion leaving group are equally bound. In this geometry, a mirror plane exists (that which contains carbon, its two alkyl substituents, and the C—H bond). The existence of a mirror plane permits no chirality, so that all optical activity must be lost for any molecule proceeding through this geometry.

Ex 7.7 Inversion of configuration takes place in all S_N2 reactions. Because the Cahn-Ingold-Prelog rules for assigning absolute configuration are artificial constructs, inversion may or may not result in a change in the assignment of absolute configuration from R to S (or S to R). Whether the notation of the absolute configuration also changes with inversion of configuration depends on whether the priorities for the substituents at the center of chirality in the product follow the same order as those in the reactant (that is, whether the nucleophile that becomes the new substituent at carbon in the product has the same priority order as the leaving group did). In the examples here, the order of priorities is maintained and the absolute notation changes.

(a) The reactant has an R center of chirality at C-2 (priorities: I > Et > Me > H). This center is S in the product (priorities: Br > Et > Me > H).

(b) The S center of chirality at C-1 (priorities: Br > Ph > Me > H) in 1-phenyl-1-bromoethane is inverted by the reaction. In the product, group priorities at the chiral center (SH > Ph > Me > H) make this center R.

(c) The R center of chirality at C-2 (priorities: CHClR > CH$_2$R > Me > H) in *cis*-1-chloro-2-methylcyclohexanol is unaffected by the reaction. The center of chirality at C-1 is S (priorities: Cl > CH(Me)R > CH$_2$R > H) in the reactant and R (priorities: I > CH(Me)R > CH$_2$R > H) in the product.

Ex 7.8 Nucleophilic substitution take place most readily at sp^3-hybridized sites that can undergo back-side attack with relatively little steric interference.

(a) Although both compounds are secondary alkyl bromides, the carbon bearing bromine in the structure at the right (2-bromo-3,3-dimethylhexane) is more hindered than that at the left (2-bromo-6-methylheptane) because of the bulky dimethyl substitution on the adjacent carbon.

(b) Back-side displacement is much easier in the sp^3-hybridized allyl bromide at the left than in the vinyl bromide at the right.

(c) Back-side displacement is much easier in the sp^3-hybridized benzyl bromide at the right than in the aryl bromide at the left. Indeed, there are no examples of aryl halides undergoing an S_N2 reaction. (The nucleophile would have to approach the carbon undergoing substitution from within the aromatic ring.)

Ex 7.9 Because of its sensitivity to steric bulk in the back-side displacement, S_N2 nucleophilic substitution is more readily accomplished at the site with easiest steric access to the nucleophile: Me > 1° (allyl, benzyl) > 2° ≫ 3°. Reading from left to right in the exercise, we find a tertiary bromide, a primary bromide bearing a methyl group on the adjacent carbon, a secondary bromide, and a primary bromide in which the adjacent carbon is a CH$_2$ group. Thus, the predicted reaction order is:

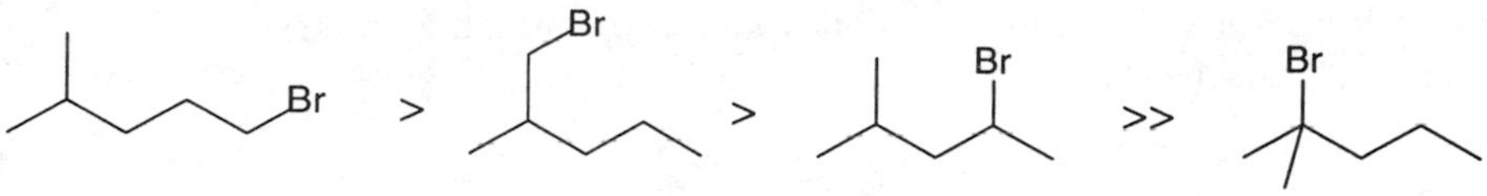

Ex 7.10 S_N2 reactions occur most easily at unhindered sp^3-hybridized carbon atoms.

(a) Benzyl bromide: easy substitution.

(b) Tertiary chloride: no S_N2 substitution.

(c) Tertiary bromide: no S_N2 substitution.

(d) Secondary bromide: easy substitution.

(e) Allyl chloride: easy substitution.

Ex 7.11 This question asks us to identify the nucleophile required for each of these S_N2 reactions.

(a) $HOCH_2CH(CH_3)_2$ or $^-OCH_2CH(CH_3)_2$

(b) NH_3 or NH_2^-

(c) $HSCH_2CH_2CH_3$ or $^-SCH_2CH_2CH_3$

(d) N_3^-

(e) Cl^-

Ex 7.12 The more active nucleophile is that with the greater electron density on the attacking atom, which is therefore better able to release electron density for the formation of a new bond to the electron-deficient site in the electrophilic reactant.

(a) A negatively charged amide ($^-NH_2$) ion is a more reactive nucleophile than a neutral amine (lone pair).

(b) NH_3. Because oxygen is more electronegative than nitrogen, the lone pairs of electrons on oxygen in water are held more tightly than those on nitrogen in ammonia.

(c) I^-. Iodide ion is less electronegative and more polarizable than chloride ion.

(d) HS^-. The sulfur in thiolate anion is less electronegative (and more polarizable) than the oxygen in hydroxide ion.

Ex 7.13 By extension of Markovnikov's rule, protonation is expected at the site that leads to the more stable carbocation.

(a) Protonation takes place at either end of the diene system, forming a secondary allylic carbocation. A less stable, simple secondary cation is formed by protonation at C-2.

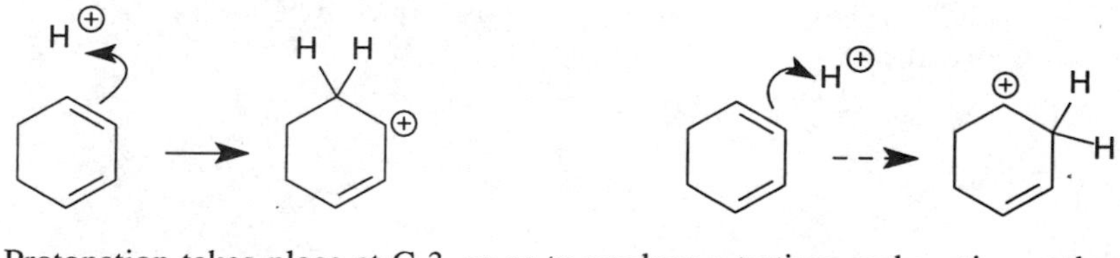

(b) Protonation takes place at C-3, so as to produce a tertiary carbocation, rather than at C-2, which would produce a secondary cation.

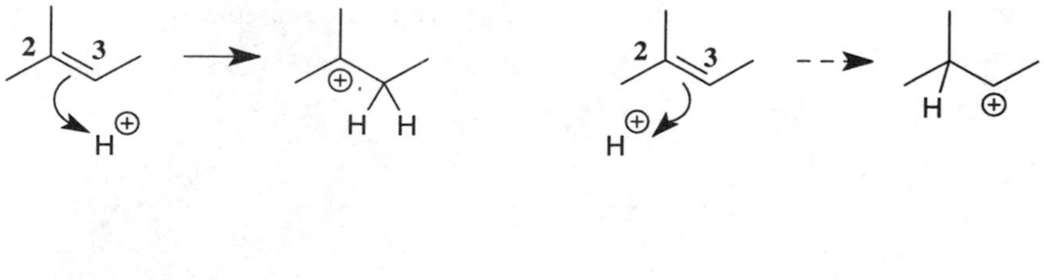

(c)   Protonation takes place at C-1, so as to produce a tertiary carbocation, rather than at C-2, which would yield a primary cation.

(d)   Protonation takes place at C-3, so as to produce a tertiary benzylic carbocation, rather than at C-2, which would yield a secondary cation.

Ex 7.14 The rate-determining step in an S_N1 reaction consists of heterolytic cleavage to form a carbocation. The Hammond postulate allows us to infer that the transition state for this cleavage closely resembles the cation. Therefore, any factor that stabilizes the cation also stabilizes the transition state leading to it. The order of carbocation stabilities is: benzyl ≈ tertiary > allylic ≈ secondary > primary > methyl.

(a)   2-Chloro-2-methylbutane. Heterolytic cleavage of the C—Cl bond produces a secondary cation from the reactant at the left and a tertiary cation from the reactant at the right.

(b)   Benzyl bromide. Cleavage of the C—Br bond in the structure at the left produces a benzylic cation; in the structure at the right, a primary cation.

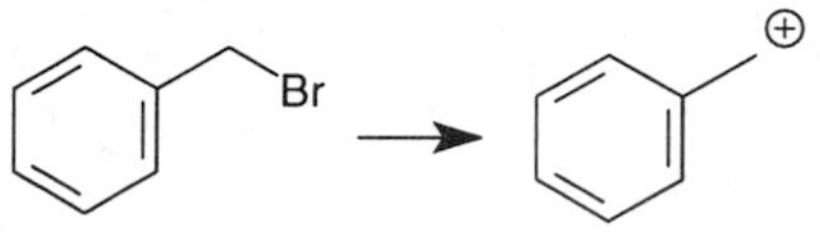

(c)   *t*-Butyl bromide. A tertiary cation is produced as the C—Br bond in the structure at the left is cleaved, but a primary cation is produced by the analogous pathway in the compound at the right.

(d)   1-Methyl-1-chlorocyclohexane. A tertiary cation is formed by C—Cl cleavage in the structure at the right, whereas a secondary cation is formed by C—Cl cleavage in the structure at the left.

Ex 7.15 The conversion of cyclohexanol to bromocyclohexane takes place in several steps: protonation to an oxonium ion, loss of water to produce a carbocation, and capture of the carbocation by bromide ion.

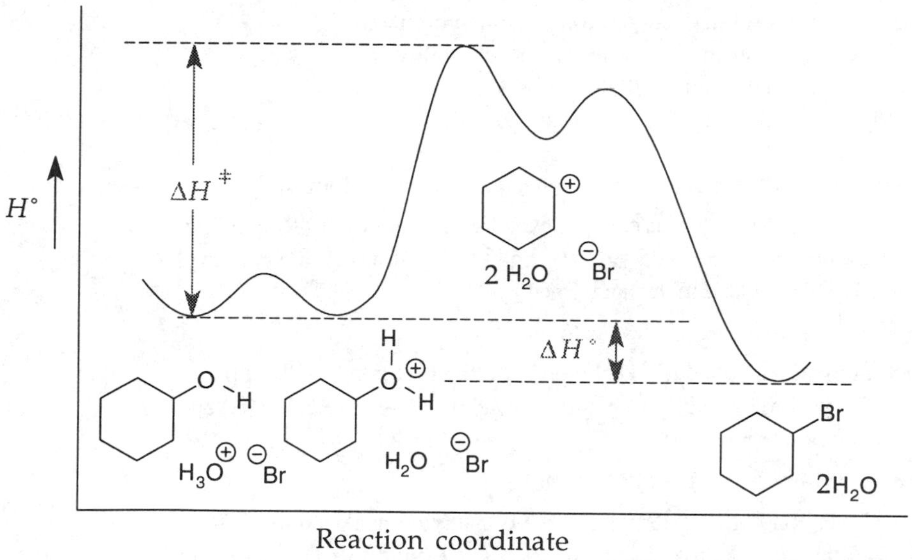

Reaction coordinate

Ex 7.16 Hydrolysis through an S_N1 mechanism takes place more rapidly when a more stable carbocation is produced as the intermediate in the rate-determining step.

(a) Cyclohexyl bromide. The secondary carbocation produced is much more stable than the phenyl cation that would be formed by C—Br cleavage in bromobenzene. (*Note*: The proposed hydrolysis of bromobenzene would produce a phenyl (C_6H_5), not a benzylic ($C_6H_5CH_2$) cation

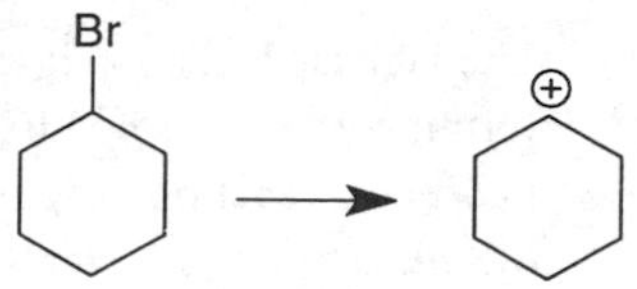

(b) 2-Bromo-2-methylbutane. Cleavage of the C—Br bond in the structure at the right produces a tertiary cation; in the structure at left, a secondary cation.

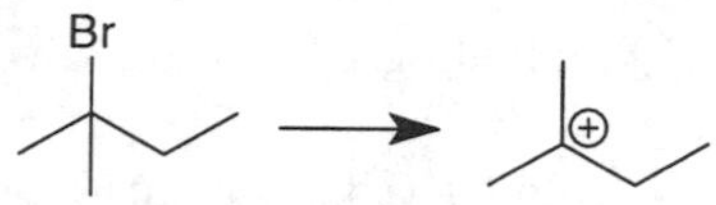

(c) 1-Bromo-1-phenylethane. Cleavage of the C—Br bond produces a secondary benzylic cation from the structure at the left, but a primary cation from the isomer at the right.

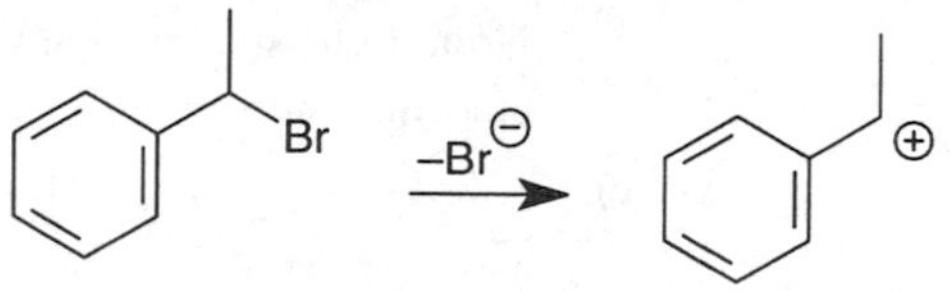

Ex 7.17 (a) 1-Bromo-1-methylcyclohexane. The cation produced by loss of water from the protonated alcohol is secondary, and a rearrangement to a more stable tertiary can take place rapidly when the hydrogen atom in the tertiary C—H bond shifts with its electrons to the secondary site.

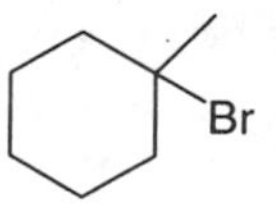

(b) 3-Bromo-3-ethylhexane. One of the ethyl groups bound to the adjacent carbon shifts toward the incipient carbocation expected when water is lost from the protonated alcohol. When the shift takes place at the same time that water is lost, a tertiary cation is formed directly, avoiding the intermediacy of a highly unstable carbocation. Capture of this carbocation by bromide ion completes the reaction.

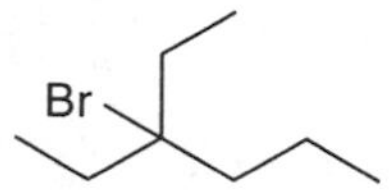

(c) 2-Bromobutane. A hydrogen atom shifts from C-2 to C-1 as water is lost from the protonated alcohol.

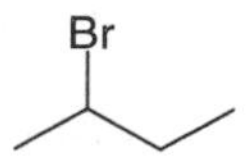

(d) 1-Bromo-2-methyldecalin. Dehydration of the protonated alcohol produces a secondary carbocation, whose stability can be enhanced if the adjacent methyl groups shifts with its electrons to that site, thus producing a tertiary carbocation.

Ex 7.18 We must begin by identifying the character of each bond cleaved and formed.

(a) A secondary C—H bond (96 kcal/mole) in cyclohexane and the Br–Br bond (46 kcal/mole) are broken. A secondary C—Br bond (68 kcal/mole) in cyclohexyl bromide and H—Br (87 kcal/mole) are formed.

$\Delta H° = (96 + 46) - (68 + 87) = -13$ kcal/mole.

(b) A primary allylic C—H bond (86 kcal/mole) and the I—I bond (36 kcal/mole) are broken. They are replaced in the products by allylic C—I (41 kcal/mole) and H—I (71 kcal/mole) bonds.

$\Delta H° = (86 + 36) - (41 + 71) = +10$ kcal/mole.

(c) A primary C—H bond (100 kcal/mole) in 2,2-dimethylpropane and the Cl—Cl bond (59 kcal/mole) are broken. A primary C—Cl bond (80 kcal/mole) in 1-chloro-2,2-dimethylpropane and H—Cl (103 kcal/mole) are formed.

$\Delta H° = (100 + 59) - (80 + 103) = -25$ kcal/mole.

Ex 7.19 (a) The indicated reaction involves breaking a secondary C—H bond of propane and forming an H—Br bond: $\Delta H° = 96 - 87 = +9$ kcal/mole. This endothermic step makes the reaction very selective. That the reaction proceeds at all is related to the high exothermicity of the second propagation step, in which the alkyl radical attacks Br$_2$.

(b) Hydrogen is abstracted by chlorine at a primary C—H bond: $\Delta H° = 100 - 103 = -3$ kcal/mole. The radical formed in this abstraction is less stable than those produced by alternative abstractions from secondary or benzylic positions. Nonetheless, the reaction is exothermic. For this reason, free-radical chlorination is much less selective than bromination.

(c) A benzylic C—H bond is broken, replaced in the products by an H—Cl bond. Thus, $\Delta H° = 88 - 103 = -15$ kcal/mole. The reaction is exothermic.

Ex 7.20 This exercise asks us to examine symmetry in each molecule to determine where unique substitution products are obtained.

(a)
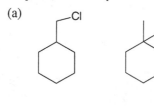
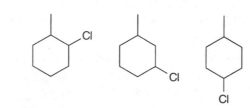

(b)

(c)
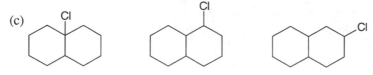

(d)

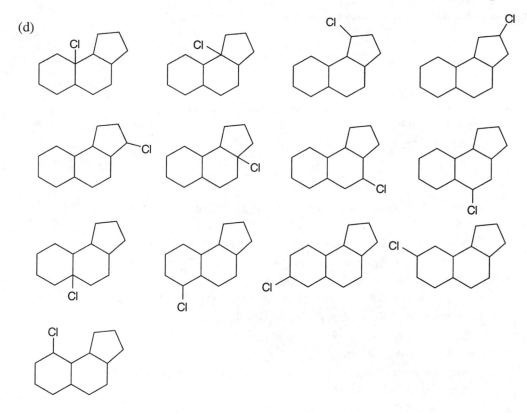

Ex 7.21 In propane, there are two secondary and six primary hydrogen atoms. The expected ratio of reactivity toward a chlorine atom for each secondary to each primary hydrogen is therefore $(56/2) \div (44/6) = 3.8$ to 1. In butane, there are four secondary hydrogens and six primary hydrogens. Thus, we expect $(x/4) \div ([1-x]/6) = 3.8$, and $x = 0.74$.

Ex 7.22 In 2-methylpropane, there is one tertiary and nine primary hydrogens. Using the same logic as in Exercise 7.21, we expect $(x/1) \div ([1-x]/9) = 5$, and $x = 0.36$.

Ex 7.23 Free-radical bromination is selective and results in the cleavage of the weakest C—H bond, which leads to the more stable radical. Radical stability follows the order $3° > 2° > 1°$.

(a) All of the hydrogen atoms are identical, and thus the product is the same regardless of which one is replaced by bromine.

(b) The secondary position is attacked rather than the primary one.

(c) The tertiary position is attacked rather than the primary or secondary sites.

(d) The benzylic position is attacked rathar than of either the primary site or the ring hydrogens.

(e)   The tertiary benzylic site is attacked rather than the primary benzylic site or the ring hydrogens.

(f)   A hydrogen atom is preferentially removed from the secondary allylic site rather than the vinylic, primary, or simple secondary sites. The resulting allylic radical abstracts bromine at both C-1 and C-3 to form a mixture of two allyl bromides.

Ex 7.24 In this exercise, decide if halogenation is selective. If so, determine which site is selectively attacked.

(a)   All of the hydrogen atoms in cyclohexane are the same, and even with chlorine atom (an unselective halogenating reagent), there is only one possible monochlorination product.

(b)   Only a single type of hydrogen is present in this highly symmetrical molecule. Therefore, the indicated bromination product is formed.

(c)   Bromination is a selective free-radical halogenation, producing 2-bromobutane instead of the 1-bromobutane shown.

(d)   Free-radical iodination is endothermic. Thus, the indicated reaction cannot be carried out by standard free-radical halogenation.

Answers to Review Problems

Pr 7.1   To solve this problem, look for changes in molecular formula to determine the reaction type. In an addition reaction, the product formula consists of the sum of the formulas of the reactants. In an elimination reaction, the product formula differs from the reactant's by the loss of two atoms or groups of atoms. In a substitution reaction, an atom or group of atoms is replaced by another. In a condensation reaction, a complex organic product is formed from two simpler organic reactants, accompanied by loss of a small stable molecule (for example, water). In a molecular rearrangement, the bond connectivity of the skeleton is altered. In a geometric isomerization, the connectivity of second-row atoms remains constant, but the spatial disposition about a bond with restricted rotation is switched. In a redox reaction, the formal oxidation level of one or more carbon atoms is changed: often such a reaction can be recognized by changes in the numbers of bonds to heteroatoms.

(a)   In this molecular rearrangement, the position of a substituent has been shifted two carbons down the chain. Because the formula of the product is identical to that of the reactant, this is also a positional isomerization (but not a geometric one).

(b)   This is an aldol condensation in which two molecules of aldehyde become bound to form a more complex product, with loss of water.

(c)   An azide group replaces bromine in this substitution reaction.

(d) The formula of this addition product differs from the reactant by the presence of two additional hydrogens and one oxygen. This reaction is therefore an addition (or, more specifically, a hydration). Because a hydrogen atom at C-3 in the reactant is at a different position in the product, the reaction is also a rearrangement.

(e) The elements of toluenesulfonic acid (tosylate anion and a H^+) are lost from the reactant in this elimination reaction.

(f) In this geometric isomerization, the *trans*-double bond (C-3–C-4) of the reactant is transformed into a *cis*-double bond in the product.

(g) In this geometric isomerization, the relative positions of two groups on a cyclohexane ring have switched from *cis* to *trans*.

Pr 7.2 (a) In a catalytic hydrogenation, two (or more) additional hydrogen atoms appear in the product as σ bonds to carbon in place of a carbon–carbon double bond. We have seen in Chapter 3 that a multiple bond is equivalent in formal oxidation level to an additional bond to one heteroatom. The replacement of a π bond by two σ bonds makes additional electrons available to carbon. For example, in ethene (C_2H_4), the formal oxidation level of carbon is –2: $–(4 \times +1)/2$, whereas in ethane (C_2H_6), it is –3: $–(6 \times +1)/2$. Thus, this reaction can be classified as both an addition and a reduction.

(b) The formula of 2-propanol (C_3H_8O) differs from that of 2-propanone , also called acetone, (C_3H_6O) by two hydrogen atoms. In the oxidation of 2-propanol to acetone, two hydrogen atoms are removed, one from oxygen and one from the adjacent carbon. Because the product has two fewer atoms, this reaction is an elimination. The three carbons in acetone must compensate for a formal charge of –4: –[–{6 hydrogens (+1) + 1 oxygen (–2)], whereas in 2-propanol the three carbons must compensate for a formal charge of –6: [8 hydrogens (+1) + 1 oxygen (–2)]. Fewer electrons are available to one or more of the carbons in acetone, so the reaction is also classified as an oxidation.

Pr 7.3 To solve this problem, identify the bond to be made, and determine which of the fragments is more stable as an anion or active nucleophile.

(a) A C—N bond between *n*-BuBr and $^-N_3$ (azide ion).

(b) A C—C bond between *n*-BuBr and cyanide (^-CN). Notice that this product has five carbons, whereas the analogous reaction with azide gave a product with four carbons.

(c) A C—O bond between methyl bromide and methoxide ion (CH_3O^-)

(d) For a C—O bond to close a ring, the nucleophile must be linked through a chain of atoms to the electrophilic site: $Br(CH_2)_4OH$. The reaction can be activated by treatment with base to convert the less reactive ROH group to the more reactive RO^- group.

(e)   A C—S bond between *n*-BuBr and HS⁻.

(f)   A C—S bond between *n*-PrBr and EtS⁻ or between EtBr and *n*-PrS⁻.

(g)   A C—O bond between methyl bromide and ⁻OSO₂Ph.

(h)   A C—P bond between ethyl bromide and triphenylphosphine.

Pr 7.4   This reaction includes basic deprotonation that converts the alcoholic OH group to the more nucleophilic alkoxide ion. The negatively charged oxygen atom then rapidly attacks the secondary bromide from the back side, leading to ring closure with inversion of configuration.

Pr 7.5   The ease of an S_N2 reaction is controlled by how easily a pentavalent transition state is reached. As a result, the S_N2 reaction is very sensitive to size of groups attached to and near the carbon undergoing substitution. The large *t*-butyl group adjacent to the primary carbon in neopentyl bromide is more sterically blocking than the two smaller groups (methyl and *n*-propyl) at C-2 of 2-bromopentane.

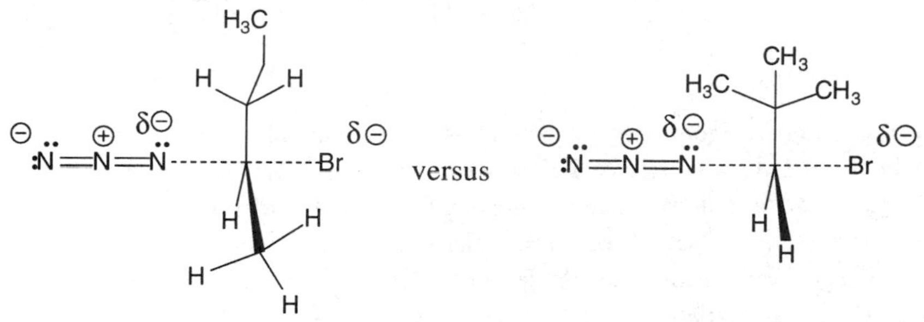

Pr 7.6   (a)   An S_N2 reaction proceeds with back-side attack, leading to a product with inverted stereochemistry at the carbon undergoing substitution (in this case, C-2).

(b)   An S_N2 reaction is bimolecular, and its rate is therefore proportional to the concentrations of both the alkyl halide and the nucleophile. Doubling the concentration of the alkyl halide doubles the rate of reaction.

(c)   As in part *b*, a reaction run with a doubled nucleophile concentration takes place at twice the initial rate. This would not be true if the reaction proceeded through an S_N1 mechanism, in which the rate of reaction is independent of the concentration of the nucleophile.

Pr 7.7   The rate-determining step of acid-catalyzed hydration is protonation to form the more stable carbocation. The choice between compounds is dictated by the relative stabilities of the more stable carbocation accessible from each reactant. Carbocation stabilities follow the order: benzyl ≈ tertiary > primary allylic ≈ secondary > primary > methyl.

(a)   Terminal protonation of the alkene at the left forms a secondary carbocation. Protonation at C-2 of the structure at the right provides a more stable benzylic carbocation. Capture of the cation by water affords the benzylic alcohol shown.

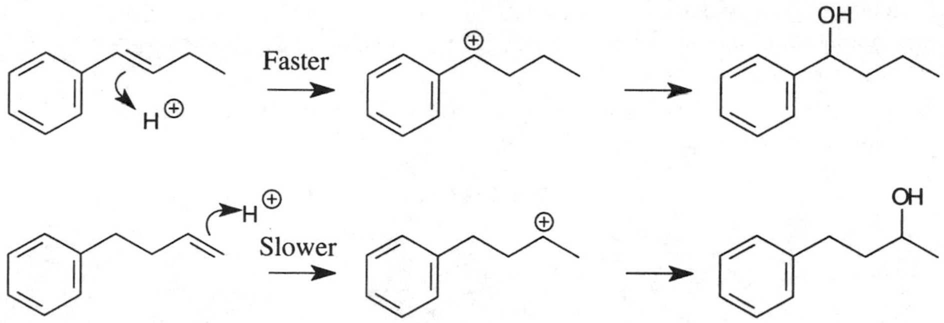

(b) Protonation at C-3 of the structure at the left produces a tertiary carbocation; protonation at C-2 or C-3 of the structure at the right, a secondary one. The cation produced from the reactant at the left is therefore preferred. Trapping by water, followed by loss of a proton, produces the alcohol bearing an OH group at the tertiary site.

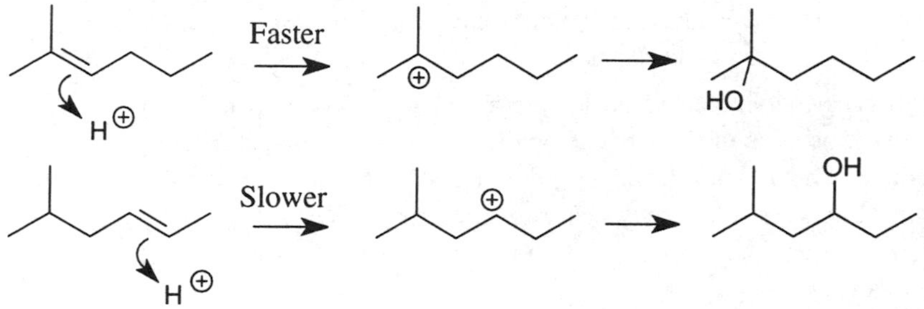

(c) Protonation of 2-butene forms a secondary cation, and protonation of 1,3-butadiene at C-1 forms a secondary allylic ion that is more stable. Because the positive charge in an allylic cation is distributed over two carbons, water adds at two sites and two regioisomeric alcohols are produced.

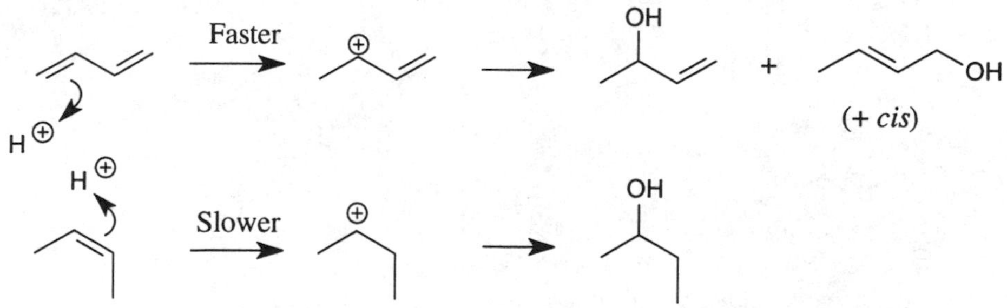

Pr 7.8 Treatment with acid results in protonation of one of the lone pairs of the oxygen of (R)-2-butanol, converting the OH group to a better leaving group (water). Water is then lost to form a planar achiral carbocation. Capture of this cation by water forms an oxonium ion, deprotonation of which regenerates 2-butanol. Because the intermediate carbocation is planar, the reaction with water takes place with equal facility on either face, leading to equal amounts of the R and the S isomer of the alcohol product. Racemic mixtures are always optically inactive.

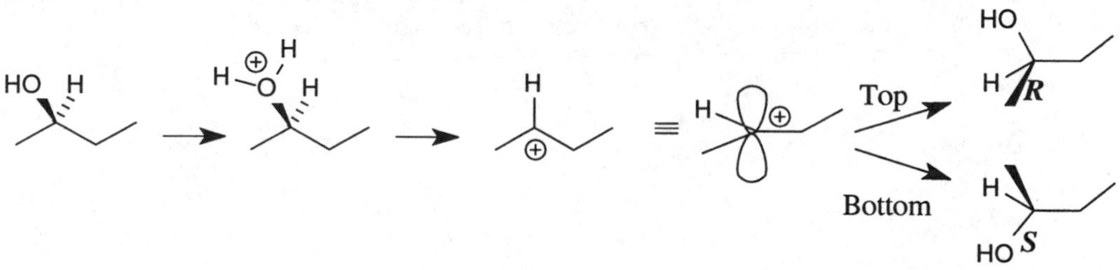

Pr 7.9 This reaction begins by deuteration with D_3O^+ to form a secondary carbocation. Because a more stable tertiary cation is formed if a hydrogen shifts with the electrons in the C—H σ bond from C-3 to C-2, this shift is rapid. The resulting tertiary cation is trapped by D_2O to yield a deuterated oxonium ion, dedeuteration of which produces 4-D-2-methyl-2-butanol–OD.

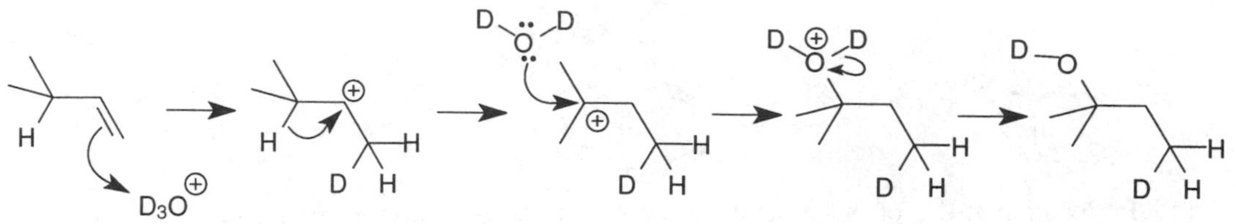

Pr 7.10 The rate-determining step of an acid-catalyzed dehydration consists of the loss of water to form a carbocation. The more stable the cation, the easier is the elimination.

(a) *t*-Butyl alcohol > *s*-butyl alcohol > *n*-butyl alcohol. Respectively, these alcohols form a tertiary, secondary, and primary carbocation upon loss of water from the oxonium ion (protonated alcohol).

(b) *p*-Methoxybenzyl alcohol > benzyl alcohol > *p*-nitrobenzyl alcohol. A methoxy group in the *para* position of the benzene ring is electron-releasing, stabilizing the intermediate benzyl cation (formed by loss of water from the protonated alcohol) by making accessible the resonance structure shown. The *p*-nitro group is destabilizing toward the analogous benzylic cation because it is strongly electron-withdrawing.

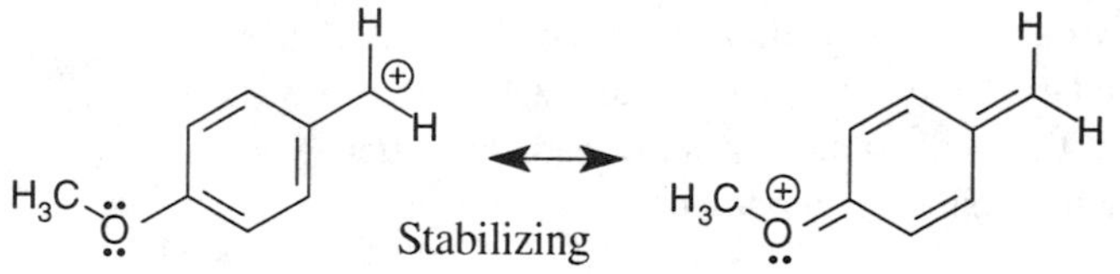

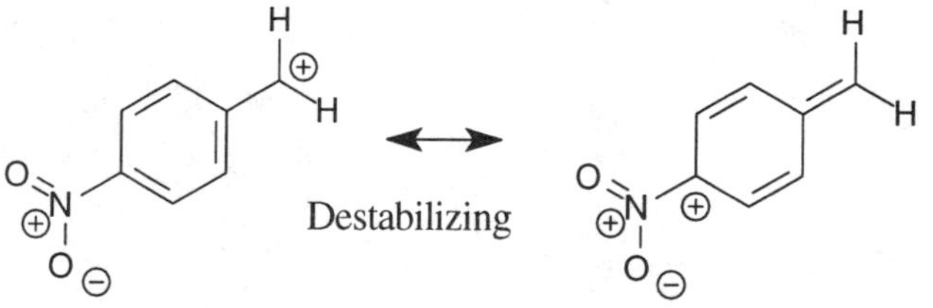

(c) α,α-Dimethylbenzyl alcohol > benzyl alcohol >> *p*-methylphenol. A tertiary benzylic cation is formed from α,α-dimethylbenzyl alcohol, whereas an unsubstituted benzyl cation is formed from benzyl alcohol itself. Loss of water from a phenol, even with an electron-releasing ring substituent, would produce an extremely unstable phenyl cation.

Pr 7.11 (a) This reaction begins by protonation of one of the lone pairs of the ethereal oxygen, forming an oxonium ion. Cleavage of the C—O bonds to that oxygen is made difficult because of the instability of the resulting primary carbocation. Simultaneous deprotonation of the β carbon avoids the formation of this cation.

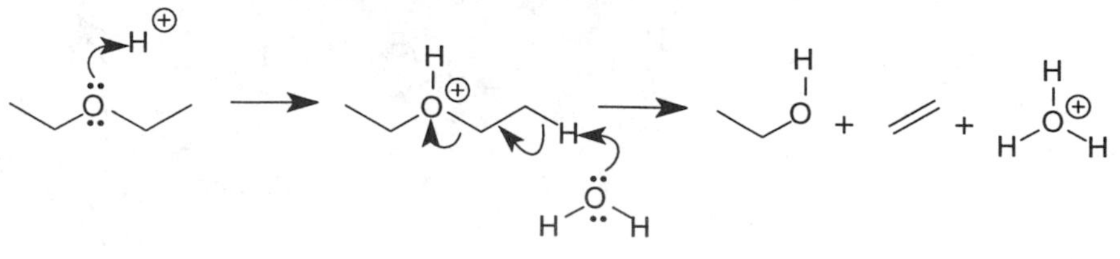

(b)   Parallel reactions with tetrahydrofuran and dioxane give the alkenols shown.

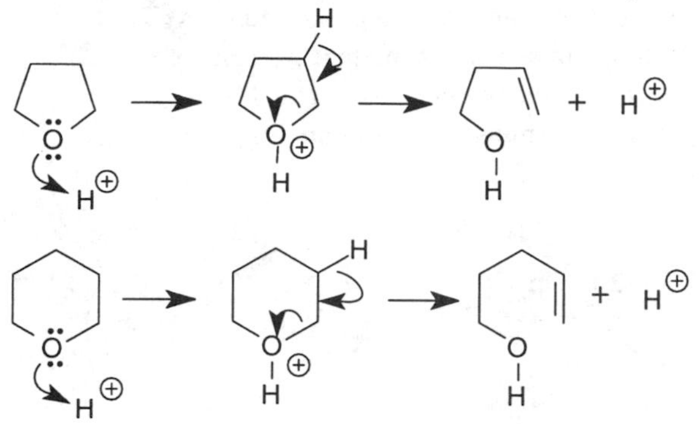

Pr 7.12 As with acid-catalyzed dehydration, the course of the reaction is driven by the formation of the most stable carbocation possible. Like the acid-catalyzed dehydration of alcohols, the pinacol rearrangement begins by protonation of one of the lone pairs of the OH group. This converts OH to a better leaving group (H_2O), producing a carbocation after C—O bond cleavage. Although this cation is tertiary, an even more stable cation can be formed when one of the adjacent methyl groups migrates to the carbon from which water was lost, because this places positive charge on a carbon to which an atom (oxygen) bearing two available lone pairs of electrons is bound. The resulting stabilized carbocation then loses a proton, providing the carbonyl product observed in this reaction.

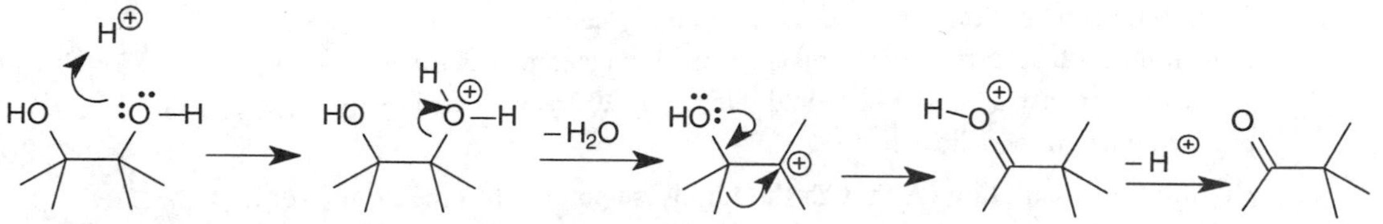

Pr 7.13 As in the hydrolytic reaction discussed in Problem 7.10, the rate of solvolysis of alkyl halides is controlled by the stability of the cation formed upon heterolytic cleavage of the C—X bond.

(a)   A tertiary bromide forms the most stable cation, followed by the secondary and primary alkyl bromides. The presence of methyl groups at C-3 and C-4 does not significantly influence the cleavage of the C—Br bond at C-1.

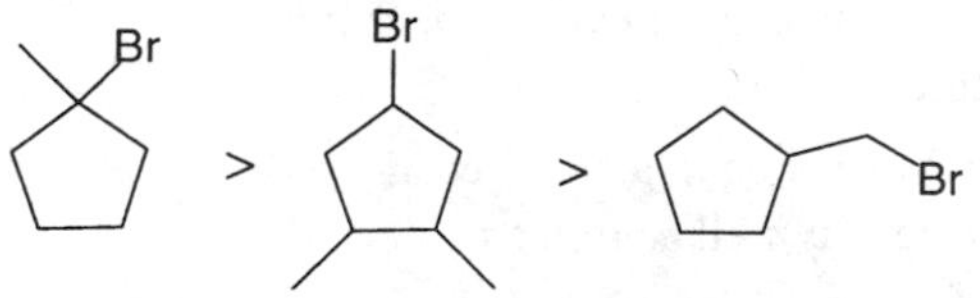

(b)   A benzyl cation is more stable than a secondary cation, which in turn is more stable than a primary cation.

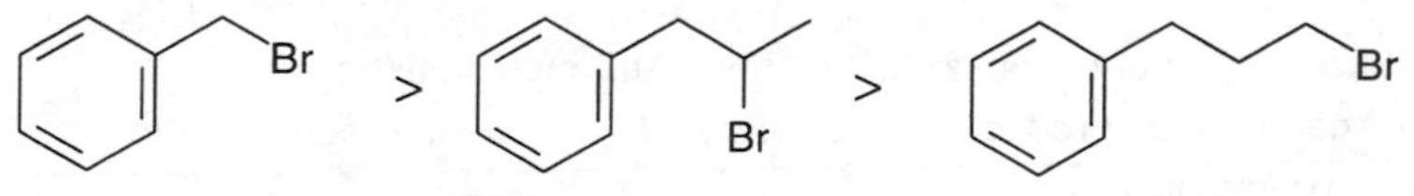

(c)   A tertiary cation is more stable than a secondary allylic cation , which is more stable than a simple secondary cation.

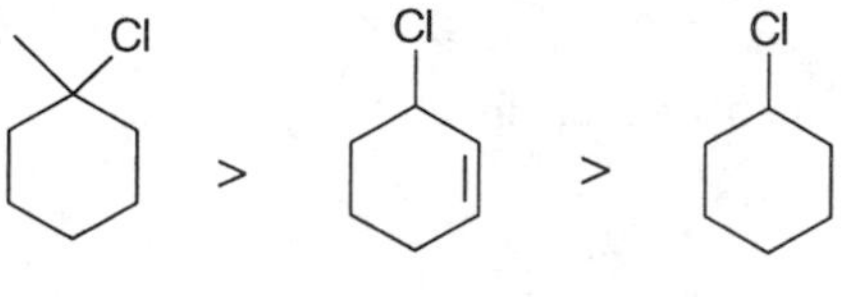

Pr 7.14 Heterolytic cleavage of the C—I bond would generate a primary carbocation. Formation of this unstable intermediate is avoided if the cleavage is assisted by Ag^+ acting as a Lewis acid (to form AgI, which precipitates). A simultaneous shift of a C—C bond produces a secondary, expanded-ring carbocation. The resulting cyclohexyl cation is then trapped by water, forming an oxonium ion that is deprotonated to produce the alcohol product.

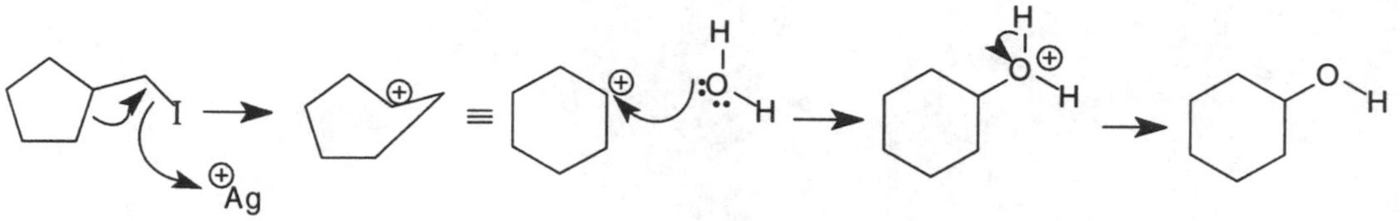

Pr 7.15 Begin by classifying each reaction to compare the observed product with that expected of each class.

(a) An acid-catalyzed hydration (electrophilic addition). The product is wrong because Markovnikov's rule tells us that protonation would take place so as to form the more stable cation; that is, at the more highly substituted carbon. That carbon would then be trapped by water, appearing in the product as alcohol. The correct product is shown at the right.

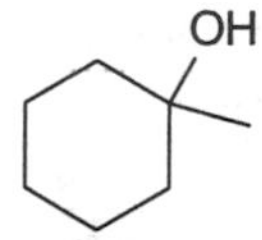

(b) Nucleophilic substitution under S_N1 conditions. The cation formed by cleavage of the C—Cl bond rearranges to the more stable tertiary benzylic cation, which is trapped by water.

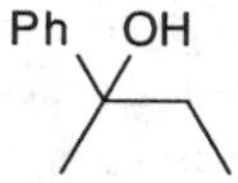

(c) Homolytic substitution. Bromine atom is highly selective, and the secondary radical that is a required intermediate for the indicated product is not the most stable radical. Instead, bromine would selectively abstract the tertiary hydrogen, producing a tertiary radical.

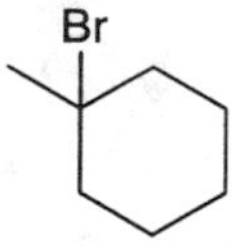

(d) Homolytic substitution. Although the highly selective bromine atom would attack the indicated carbon, the product is achiral and there is only one structure possible.

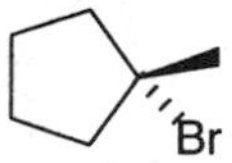

(e) This reaction does proceed as written. Nucleophilic substitution under classic S_N2 conditions. The reaction proceeds with inversion of configuration at the center of asymmetry, giving the indicated product. This structure is correct.

Pr 7.16 This problem focuses on the greater selectivity of bromine over chlorine in free radical substitution. Where selectivity is necessary, bromination is needed. If not, chlorination leads to the analogous chlorinated alkane.

(a) The symmetry of cyclohexane makes all hydrogens equivalent. Because no selectivity is needed, cyclohexyl chloride could equally well be formed.

(b) Bromine selects for the tertiary hydrogen, despite a 9:1 excess of primary hydrogens. Chlorine would not be so selective, and a mixture would result. Better stick to bromine.

(c) Bromine selects for substitution at the benzylic position. Although statistics do not so heavily favor the "wrong" product here as in part *b*, a cleaner product mixture is obtained by bromination.

(d) All 12 hydrogens are equivalent, and no selectivity is required. Chlorination is equally useful.

Pr 7.17 To form the indicated product, chlorine must bond to carbon, homolytically cleaving a C—C bond to form a primary radical, as indicated.

(a)

(b) In the first step, a C—C bond is broken and a C—Cl bond is formed: $\Delta H° =$ 84 kcal/mole – 81 kcal/mole = +3 kcal/mole. In the second step, a Cl—Cl bond is broken and a C—Cl bond is formed: $\Delta H° =$ 58 kcal/mole – 81 kcal/mole = –23 kcal/mole. The net reaction is therefore exothermic by 20 kcal/mole, but because the first step is endothermic, this route is very slow compared with the exothermic abstraction of hydrogen. (See the energetics of this reaction in the text.)

Pr 7.18 This problem looks at the bond dissociation energies of various kinds of bonds. All homolytic cleavages are endothermic, and those that require the least energy input (that is, those in which the weakest bond is broken) are, thus, most favorable. From the bond energies listed, we expect cracking to effect C—C cleavage.

$$CH_3CH_2CH_3 \rightarrow CH_3CH_2\cdot + CH_3\cdot \qquad \Delta H° = +86 \text{ kcal/mole}$$

$$CH_3CH_2CH_3 \rightarrow CH_3CH_2CH_2\cdot + H\cdot \qquad \Delta H° = +100 \text{ kcal/mole}$$

$$CH_3CH_2CH_3 \rightarrow CH_3\cdot CHCH_3 + H\cdot \qquad \Delta H° = +96 \text{ kcal/mole}$$

Important New Terms

Addition reaction: a chemical conversion in which two reactant molecules combine to form a product containing all the atoms of both reactants (7.1)

Aldol condensation: the production of a more complex α,β-unsaturated aldehyde (or ketone), with the elimination of water, upon treatment of two equivalents of an aldehyde (or ketone) with acid or base (7.1)

Arrow notation: the use of single- or full-headed curved arrows to indicate electron motion in a chemical reaction mechanism (7.2)

Arrow pushing: the use of curved arrows to describe the movement of electrons as a reaction proceeds (7.2)

Back-side attack: the approach of a nucleophilic reagent from the side opposite that from which the leaving group is displaced (7.4)

Beckmann rearrangement: the acid-catalyzed reaction through which the oxime of a ketone is converted to an amide in which one of the carbon substituents originally on the carbonyl carbon has migrated to nitrogen (7.1)

Catalyst: a reagent that facilitates a reaction without itself ultimately forming chemical bonds in the product or appearing in the stoichiometric equation describing the reaction (7.1)

Catalytic hydrogenation: a reaction catalyzed by a heterogeneous catalyst (usually a noble metal) in which hydrogen is added across one or more multiple bonds (7.1)

Chain reaction: a chemical conversion in which one of the products is a reactive species that initiates another cycle of the reaction; a reaction which, after initiation, repeats a cycle of propagation steps until one of the reactants is consumed (7.6)

Condensation reaction: a chemical conversion in which two molecules combine to form a more complex product, with the loss of a small molecule, usually water or an alcohol (7.1)

Dehydration: the formal loss of water from an alcohol (7.1)

Dehydrohalogenation: the formal loss of HX from an alkyl halide (7.1)

Electrophilic addition: an addition reaction initiated by attack by an electron-deficient reagent (an electrophile), often a proton from a hydronium ion (7.5)

Elimination reaction: the inverse of an addition reaction in which a single complex molecule splits into two simpler products (7.1)

Free-radical halogenation: a homolytic substitution of halogen for hydrogen, often in an alkane (7.6)

Geometric isomerization: a chemical conversion in which the relative positions of groups bound to a functional group with restricted rotation are reversed (7.1)

Homolysis: the cleavage of a bond in which the electrons are shifted, one to each of the atoms of the bond; synonymous with homolytic cleavage (7.2)

Homolytic substitution: *see* **Free-radical halogenation** (7.6)

Hydration: the addition of water to a multiple bond (7.1)

Hydrolysis: a reaction in which water displaces a leaving group or is added across a multiple bond (7.1, 7.5)

Initiation step: the first step of a radical reaction in which the number of radicals produced is greater than the number of radicals present in the reactants (7.6)

Initiator: a substance with an easily broken covalent bond that fragments to radicals that can induce a radical chain reaction (7.6)

Inversion of configuration: the reversal of configuration at a center of chirality attained by forming a new bond on the opposite face from the site where a bond is broken (7.4)

Isomerization: a chemical conversion in which compounds with the same molecular formula, but different structures, are interconverted (7.1)

Leaving group: a group displaced from a reactant in a substitution or elimination reaction (7.5)

Le Chatelier's principle: the observation that the position of an equilibrium A + B = C + D can be shifted to the right by either increasing the concentrationa of A and/or B or decreasing the concentrations of C and/or D (7.1)

Markovnikov's rule: an empirical prediction that the regiochemistry of the addition of HX to an unsymmetrical alkene takes place so as to locate the proton on the less substituted carbon atom of the multiple bond (7.5)

Mechanism: *see* **Reaction mechanism** (7.1, 7.3)

Mechanistic organic chemistry: the subarea of organic chemistry that focuses on the study of how reactions take place (7.3)

Microscopic reversibility: the principle dictating that the pathway followed in the forward and reverse directions of a given reaction must be the same (7.4)

Nucleophilic substitution: a chemical conversion in which a leaving group is displaced by an electron-rich (nucleophilic) reagent (7.4, 7.5)

Organic synthesis: a subarea of organic chemistry that focuses on the construction of interesting new molecules or complex existing molecules (for example, natural products) (7.3)

Physical organic chemistry: a subarea of organic chemistry that relates structure to reactivity in explaining reaction mechanisms (7.3)

Pinacol rearrangement: the acid-catalyzed conversion of a 1,2-diol to a ketone with migration of a carbon–carbon bond (7.1)

Positional isomerization: a chemical conversion in which the position of a functional group is altered (7.1)

Propagation steps: the principal product-forming sequence in a free-radical chain reaction in which a reactant radical is converted to product and a different radical; the number of product radicals in a propagation step is equal to the number of reactant radicals; a step in a free-radical chain that carries on the chain (7.6)

Protonation: the covalent attachment of a proton (H^+) to an atom bearing either a nonbonded lone pair of electrons or a π bond (7.5)

Racemization: the loss of optical activity when one enantiomer is converted to a 50:50 mixture of enantiomers (7.5)

Radical chain reaction: a chain reaction in which a free radical is produced in the initiation and propagation steps and consumed in the termination steps (7.7)

Rate-determining step: the slowest step in a multi-step reaction (7.5)

Reaction mechanism: the sequence of bond-breaking and bond-making by which a reactant is converted to a product; a detailed description of the electron flow, including the identity of any intermediate(s) formed, that takes place during a chemical reaction (7.1, 7.3)

Rearrangement reaction; a chemical conversion in which the molecular skeleton is altered so that the sequence in which atoms are attached is changed (7.1)

Redox reaction: a reaction involving an oxidation or reduction (7.1)

Regiochemistry: the orientation of a chemical reaction on an unsymmetrical substrate (7.6)

Regiocontrol: the formation of one regioisomer to a greater extent than others in a chemical reaction (7.6)

Regioselective: descriptor of a reaction in which there is a clear preference for one of two or more possible regioisomers (7.6)

Retrosynthetic analysis: a procedure for planning a chemical synthesis in which a route to a target product is chosen by selecting a precursor for the ultimate product, which in turn has a logical precursor, and so forth (7.3)

Selectivity: the preference for reaction with one reagent over another at one site rather than another (7.6)

Self-exchange: a substitution reaction in which the incoming and leaving groups are identical (7.4)

S$_N$1 reaction: a stepwise, unimolecular, nucleophilic substitution that proceeds through an intermediate carbocation (7.5)

S$_N$2 reaction: a concerted bimolecular, nucleophilic, substitution that takes place by back-side attack of a nucleophile and leads to a substitution product with inverted configuration at the substituted carbon (7.4)

Solvolysis: a reaction in which the solvent displaces a leaving group or is added across a multiple bond (7.5)

Substitution reaction: a chemical conversion in which one atom or group of atoms in a molecule is replaced by another (7.1)

Termination reaction: a reaction that stops a chain reaction by consuming a reactive intermediate without producing another or by converting two reactive intermediates to one stable product (7.6)

Substitution by Nucleophiles at *sp*³-Hybridized Carbon

Key Concepts

Mechanism of bimolecular nucleophilic substitution (S_N2)

Mechanism of unimolecular nucleophilic substitution (S_N1)

Conversion of alcohols to alkyl halides and to tosylate esters

Synthesis of ethers and thioethers by the Williamson ether synthesis

Synthesis of primary amines from alkyl halides and sulfonates by the Gabriel synthesis

and by reaction with excess ammonia

Synthesis of phosphonium salts and ylides

Methods for preparing nucleophilic carbon reagents: insertion, transmetallation, deprotonation

Bond polarization in organometallic reagents

Introduction of deuterium by quenching of organometallics with D_2O

Synthesis of nitriles by displacement by ¯CN ion

Synthesis of alkynes by displacement by acetylide anions

Synthesis of carbon–carbon bonds by displacement by organometallics

Answers to Exercises

Ex 8.1 In all S_N2 reactions, the leaving group is an electronegative atom, often a halogen or the oxygen atom of a sulfonate ester. In each case, the carbon atom that bears the leaving group is that substituted with a halogen or oxygen. You are asked to determine here the classification of the atom to which this electronegative atom or group is attached, where (as described in Chapters 1 and 3) this classification counts the number of carbon substituents attached to this same atom.

(a) primary bromide

(b) tertiary bromide

(c) secondary chloride

(d) tertiary allylic bromide

(e) tertiary chloride

(f) benzylic iodide

(g) secondary bromide

(h) primary methanesulfonate

Ex 8.2 First, classify the type of substitution reaction as S_N2 in each conversion in order to determine the predicted stereochemistry.

(a) (*R*)-2-Iodoheptane. This is a classic example of an S_N2 reaction; therefore, inversion of configuration is expected. The absolute configuration of the starting material is *S*, and because the Cahn-Ingold-Prelog priorities of the leaving group and new substituent are the same in relation to the other unchanged substituents, C-2 in the product is an *R* center.

(b) (1*S*,3*S*)-3-methylcyclohexanol. This S_N2 reaction inverts configuration at the site of the substitution but leaves the absolute configuration of the other center of chirality (C-3) unaltered. The absolute configuration of the reactant was (1*R*,3*S*)-1-chloro-3-methylcyclohexane. Because the hydroxide ion can act as both a nucleophile and a base, some elimination, to produce *(3S)-3-methylcyclohexene* and *(4S)-4-methylcyclohexene*, is likely to take place along with the indicated substitution.

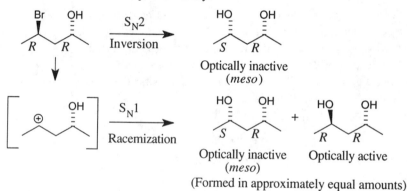

(c) *(3R)*-hexanol. An S_N2 reaction inverts the configuration at the chiral center of *(3S)*-iodohexane. As in part *b*, some elimination (to 2-hexene and 3-hexene) is likely to compete with the desired substitution.

Ex 8.3 (a,b) The S_N2 reaction occurs with inversion of configuration, and the 2(R) center in the reactant bromide is converted to a 2(S) center in the product. This product has a mirror plane of symmetry through C-3, and the two chiral centers are of opposite configuration; it is therefore a *meso* compound and optically inactive. In contrast, the S_N1 reaction proceeds through a planar cation, which is nearly equally attacked on each face. Although the presence of the unaffected center of chirality at C-4 causes a small preference for attack on one face, the (2R,4R) (optically active) and (2S,4R) (optically inactive *meso* compound) isomers are formed in nearly equal amounts. The mixture of products from the S_N1 reaction therefore exhibits some optical activity.

(c) The use of (2S,4R)-2-bromo-4-pentanol leads to the same result under S_N1 reaction conditions, because the same carbocation intermediate is formed regardless of the configuration of the carbon bearing bromine in the starting material. On the other hand, conversion of this bromide to the alcohol under S_N2 conditions results in the formation of the optically active (2R,4R) diol.

Ex 8.4 The cations produced in the indicated cleavages are the benzyl and allyl cations. The two resonace structures for the allyl cation (are shown here as well as in Figure 3.19) are identical in energy. The resonance contributors for the benzyl cation, shown in Figure 3.20, are also reproduced here. The two at the right are identical in energy.

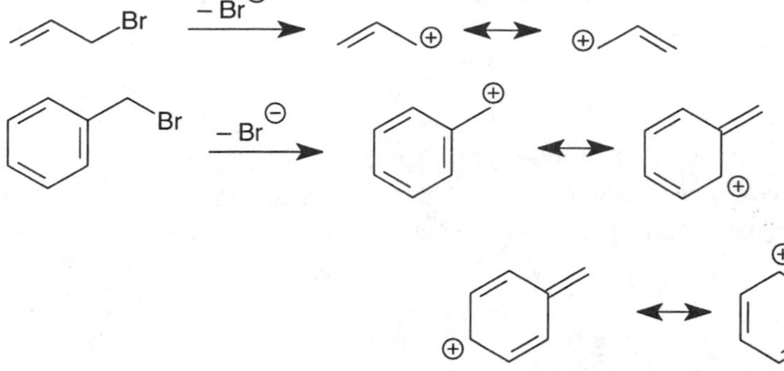

Ex 8.5 An S_N1 mechanism is expected for tertiary, benzylic, and allylic halides and tosylates. An S_N2 mechanism is expected for primary halides and tosylates. Either mechanism can be followed for secondary reactants. The path followed is often inferred from the identity of the isolated products. An S_N2 pathway is indicated by an inversion of configuration at a center of chirality, whereas an S_N1 reaction would take place via racemization. S_N2 reactions are also favored in polar aprotic solvents as compared with polar protic solvents.

(a) Inversion of configuration at a secondary bromide: S_N2

(b) S_N2 at a primary bromide

(c) Although there is no center of chirality in this secondary chloride, the solvent favors an S_N2 reaction.

(d) S_N1 reaction at a benzylic chloride

(e) S_N1 reaction at a tertiary chloride

(f) Inversion of configuration at a secondary mesylate

Ex 8.6 The Williamson ether synthesis is an S_N2 reaction in which an alkoxide anion displaces a halide or tosylate. Because the order of reactivity in an S_N2 reaction is Me > 1° > 2° >> 3°, the alkyl halide or tosylate must be sp^3-hybridized and must be methyl, primary, or secondary. Here, the substrates are represented as bromides, but chlorides or tosylates also can be used.

(a)

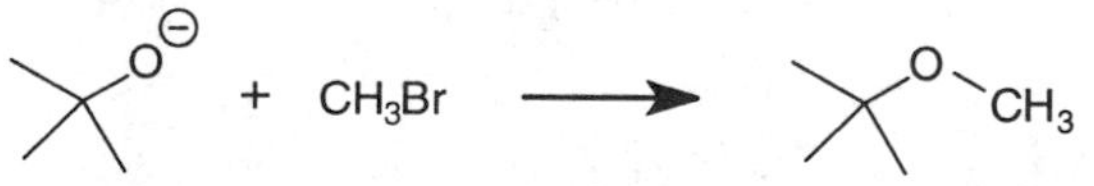

(b)

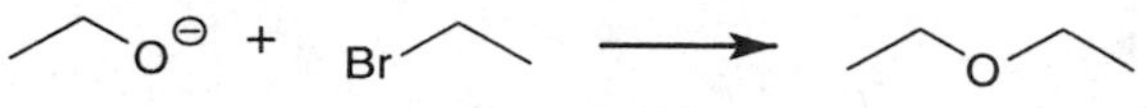

(c) Because both carbon atoms attached to oxygen in this ether are secondary, there are two possible combinations of reagents.

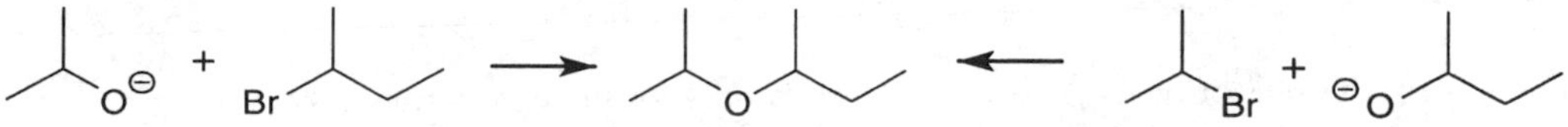

(d)

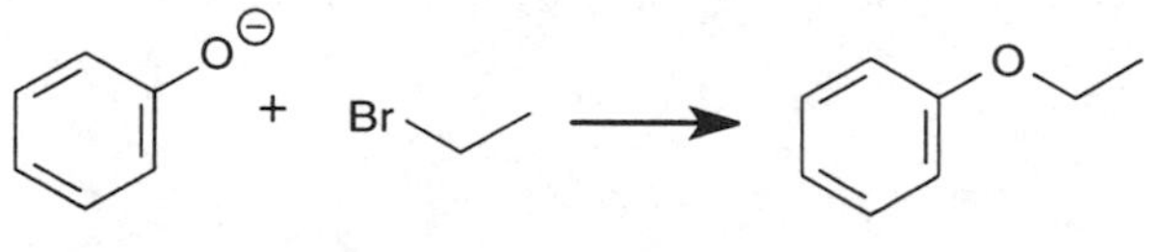

Ex 8.7 In the anion of a sulfonic acid, three resonance contributors of equal energy exist in which charge is distributed equally to the three oxygen atoms.

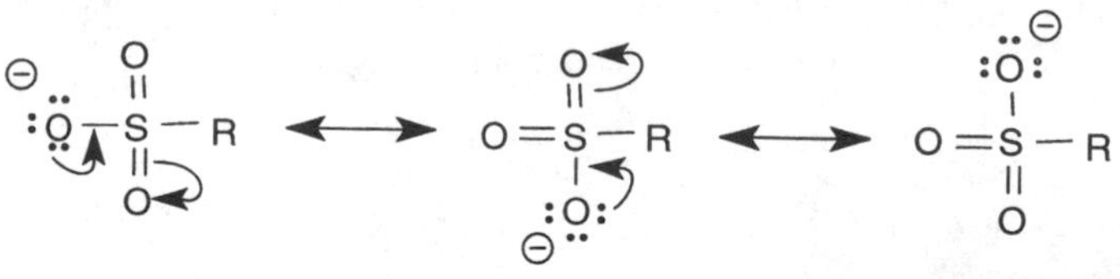

Ex 8.8 The S_N1 reaction takes place in two steps: first, cleavage of the carbon–oxygen bond of the chlorosulfite ester, producing a carbocation, SO_2, and chloride ion; and then capture of the carbocation by the highly nucleophilic chloride ion.

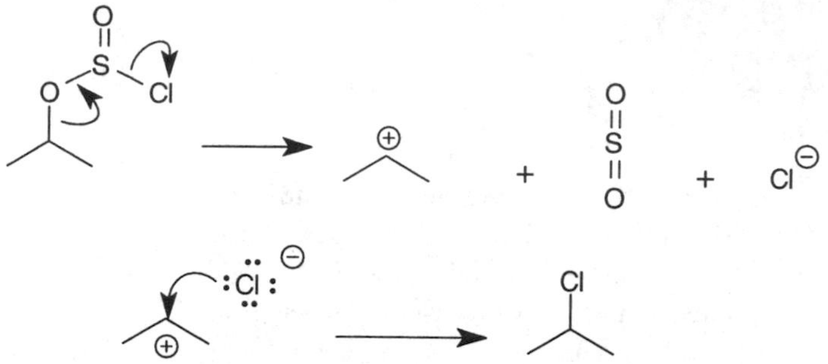

The S_N2 reaction occurs in one step, with chloride ion effecting a backside displacement of the leaving group (SO_2 and Cl^-).

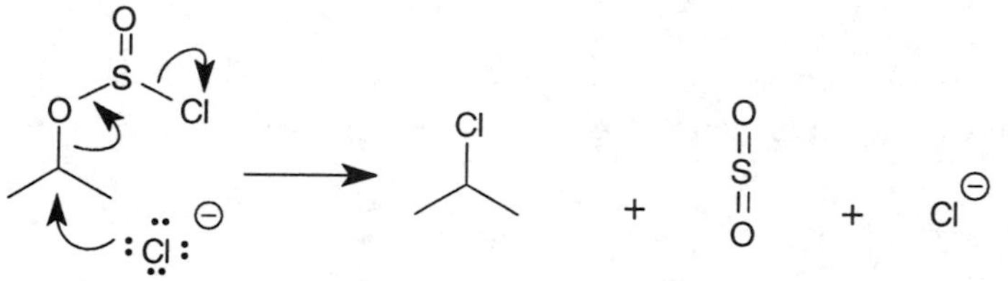

Ex 8.9 As in all S_N2 reactions, the rate of the S_N2 reaction of a chlorosulfite ester depends on the concentration of both the substrate and the nucleophile. As we saw in Exercise 8.8, the chloride ion is the nucleophile. It is produced in an acid–base equilibrium with HCl formed along with the chlorosulfite ester as the alcohol is treated with thionyl chloride. In the presence of a base like pyridine, the concentration of free chloride ion is increased, and with it, the rate of the S_N2 reaction.

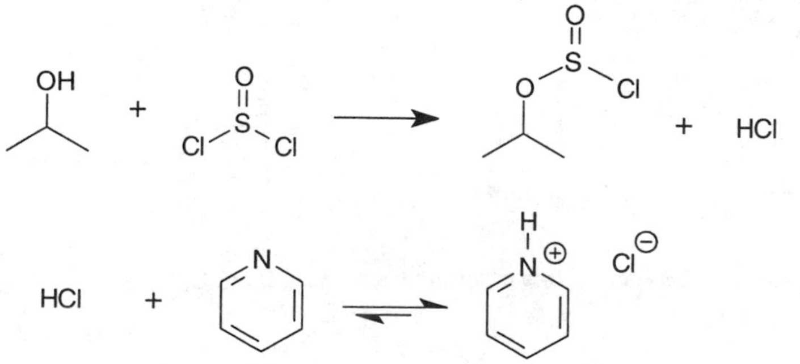

In contrast, the rate of the S_N1 reaction depends only on the concentration of the substrate. An increased chloride ion concentration induced by the presence of base will not affect the rate determining step.

Ex 8.10 Applying LeChatelier's principle to the indicated equilibrium, we find that an increased concentration of NaSH will shift the equilibrium to the right, producing a higher concentration of NaSCH₃.

$$H_3C{-}SH \;+\; \overset{\oplus}{Na}\;\overset{\ominus}{SH} \;\rightleftharpoons\; H_3C{-}\overset{\ominus}{S}\;\overset{\oplus}{Na} \;+\; H_2S$$

Ex 8.11 Like the Williamson ether synthesis, these thioethers are produced in highest yield through an S$_N$2 reaction path involving a primary or secondary alkyl halide or sulfonate ester. Allylic and benzylic halides are also useful reagents. Alkyl bromides are shown here, but chlorides, iodides, mesylates, or tosylates could also be used .

(a) Either the anion of methanethiol with benzyl bromide, or the anion benzylthiol of with methyl bromide.

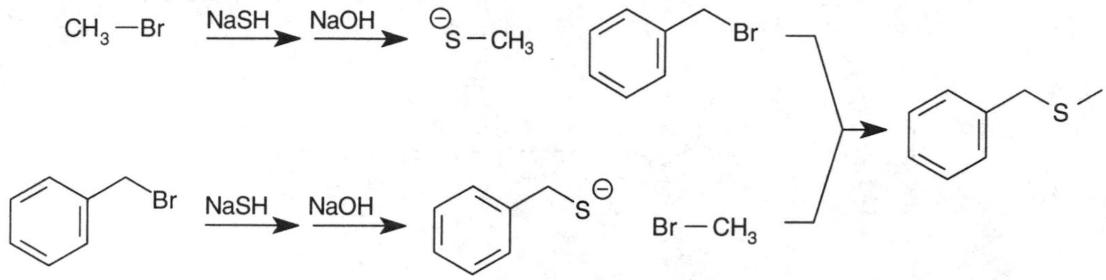

(b) Cyclohexyl bromide with NaSH.

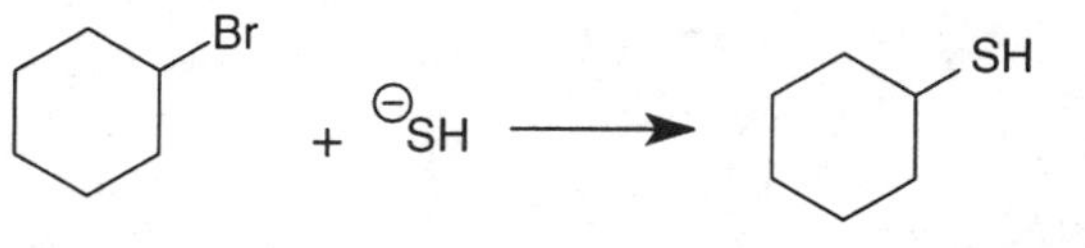

(c) Either the anion of 2-propanethiol with ethyl bromide, or the anion of ethanethiol with 2-bromopropane.

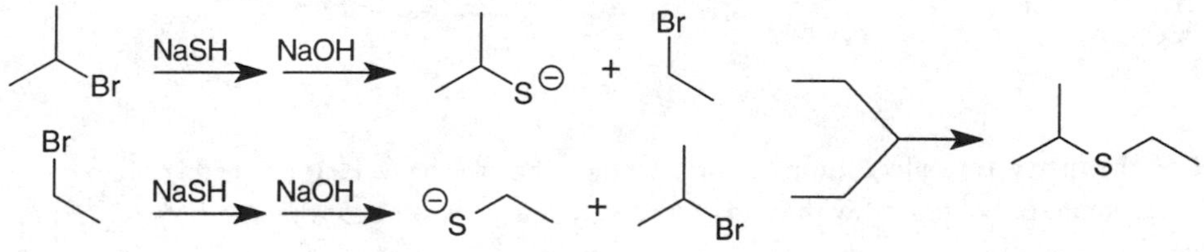

(d) 1,5-Dibromopentane with NaSH.

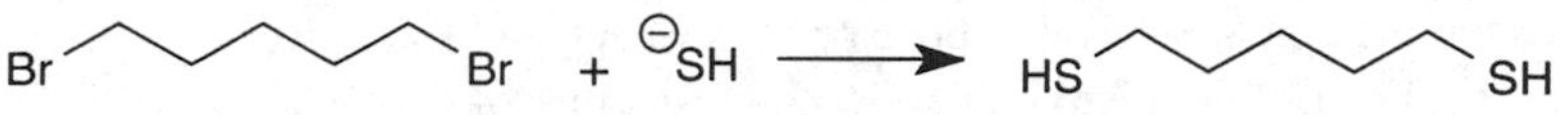

Ex 8.12 (a) The anion of phthalimide is stabilized by delocalization of the negative charge onto both of the adjacent carbonyl oxygen atoms. Its nucleophilicity is still high because of the high electron density in this anion.

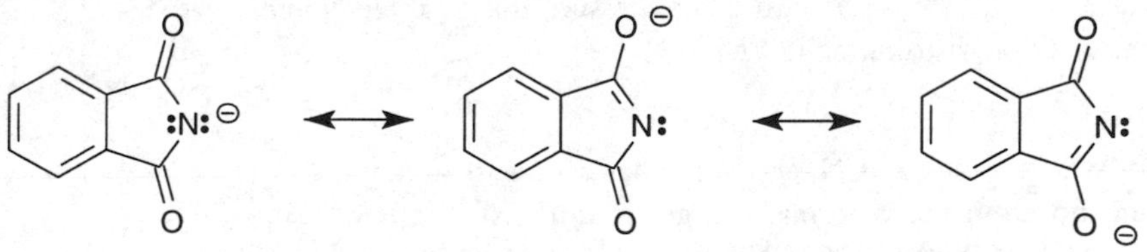

(b) Resonance in neutral phthalimide forms a zwitterion as the lone pair of electrons on nitrogen is shifted toward one of the two carbonyl oxygen atoms. To the extent that such zwitterionic forms contribute to the electronic structure of phthalimide, nucleophilic reactivity at nitrogen is reduced. Such resonance contributors dramatically reduce the local electron density on nitrogen compared with that in a free amine like ammonia and thereby make it a less reactive nucleophile than the phthalimide anion in part *a*.

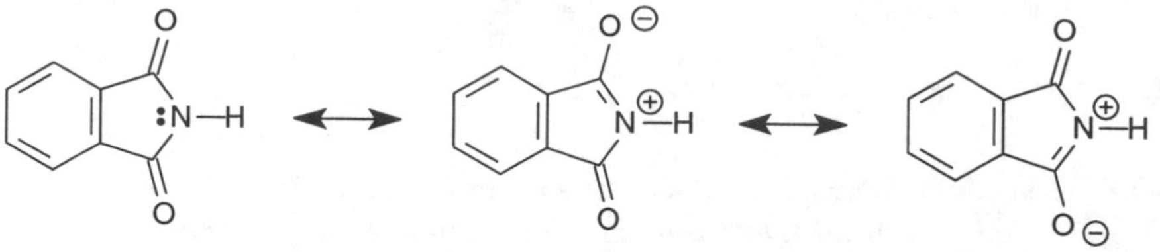

Ex 8.13 Although each of these primary amines could be prepared by treating the corresponding halide or sulfonate ester with a large excess of ammonia, this Exercise asks that we avoid a large excess of any reagent. If we are to comply, and to avoid a distribution of primary, secondary, and tertiary amines along with quaternary ammonium salts, we will employ the Gabriel synthesis.

(a)

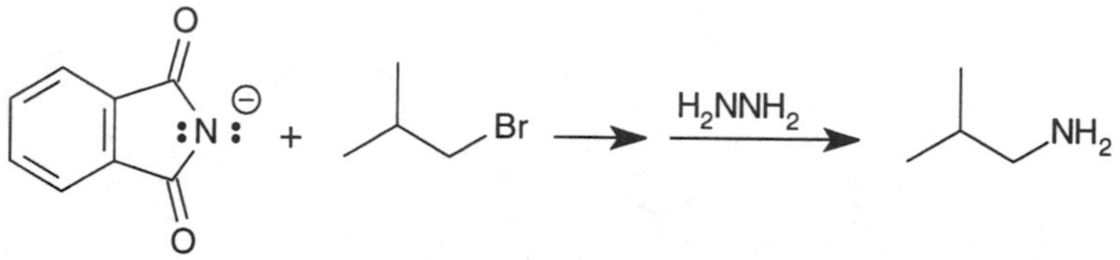

(b)

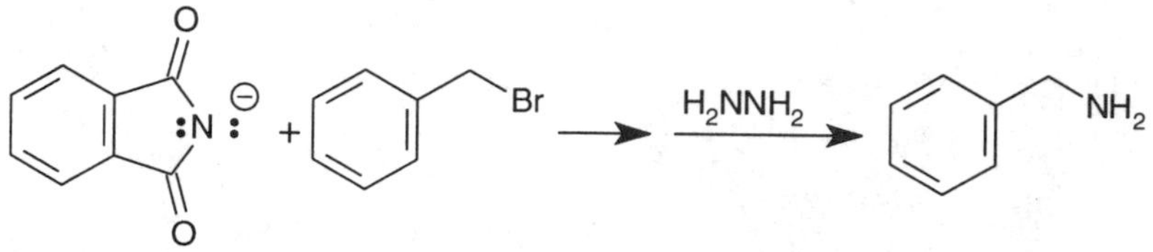

(c)

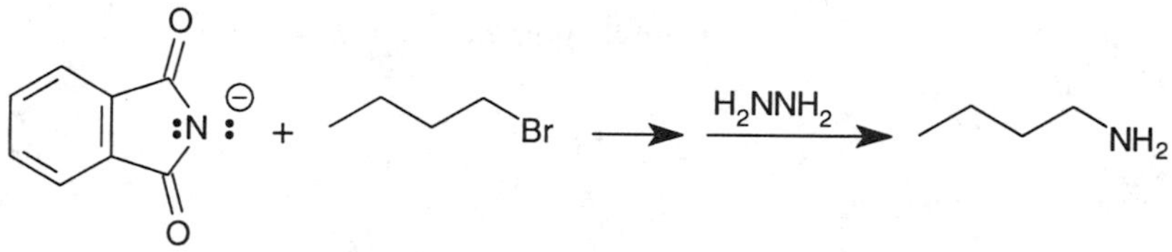

(d) The chemistry for substitution to form amines that we have learned in this Chapter can only be used with primary and secondary alkyl halides (and other leaving groups).

Ex 8.14 Using the rules for determining oxidation level as developed in Chapters 2 and 3, we calculate the oxidation level of C-1 in *n*-butyl chloride as –1 and that of C-1 in *n*-butyllithium as –3. The insertion of lithium into a C—Cl bond is therefore a reduction.

Ex 8.15 In the resonance contributor in which negative charge is localized on nitrogen, carbon lacks access to an octet of electrons. As a result, this is a very minor contributor to the electron distribution in CN⁻.

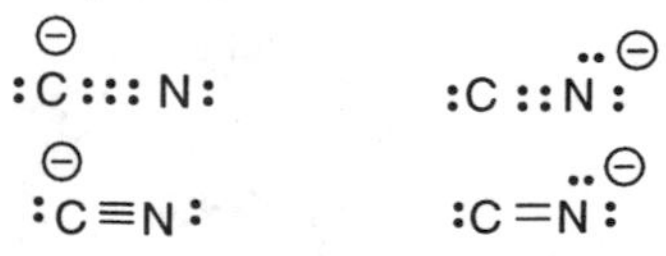

Ex 8.16 In the Lewis dot structure of CH_3CN, each atom has access to an octet of electrons, and no atom bears formal charge. In contrast, the two important resonance contributors to CH_3NC exist either as a zwitterion or in a structure in which carbon lacks access to an octet of electrons.

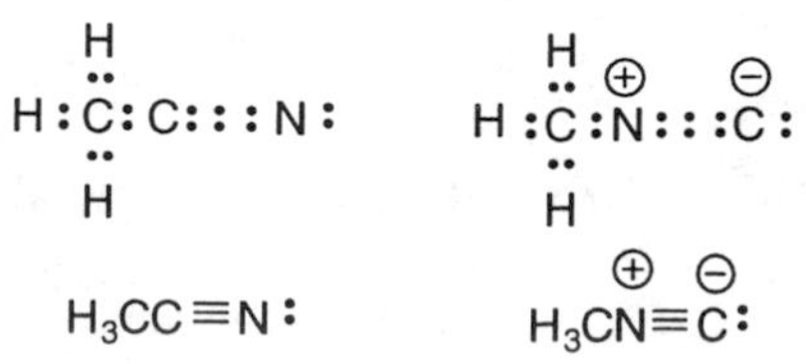

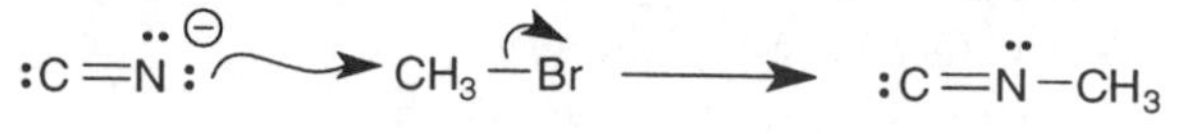

Ex 8.17 In this exercise, either cyanide ion or an acetylide anion is used as a nucleophile for an S_N2 displacement. The former can be supplied as NaCN, but the latter must be generated from the corresponding terminal alkyne by treatment with $NaNH_2$. Although the reaction partner shown here is the alkyl bromide, other halides or sulfonate esters would be equally useful.

(a)   2-(Cyclohexyl)-1-bromoethane + NaCN

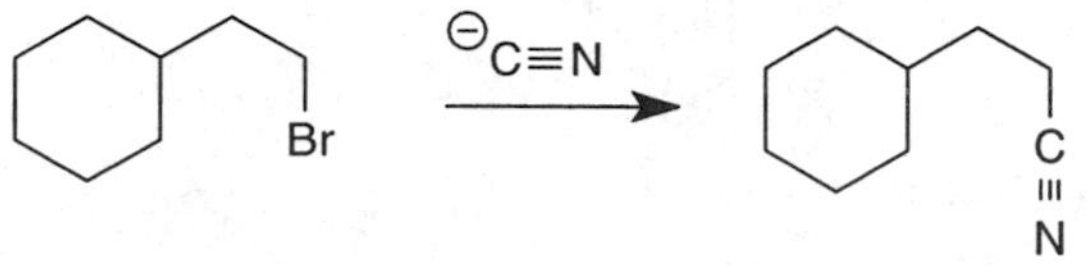

(b)   1-Bromopentane + NaCN

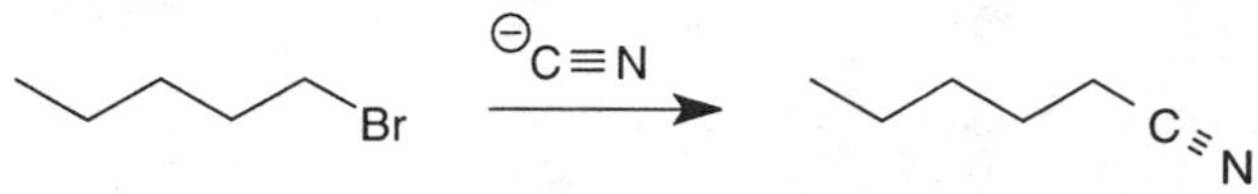

(c)   Benzyl bromide + anion of propyne (propyne + $NaNH_2$), or

Methyl bromide + anion of 3-phenylpropyne (3-phenylpropyne + $NaNH_2$)

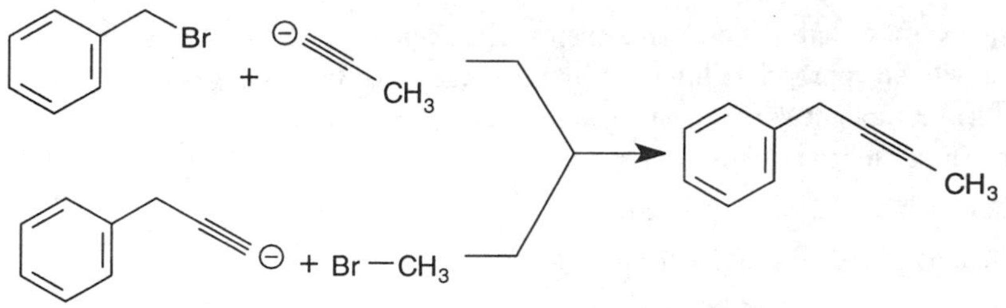

(d)   1-Bromopropane + anion of pentyne (pentyne + $NaNH_2$)

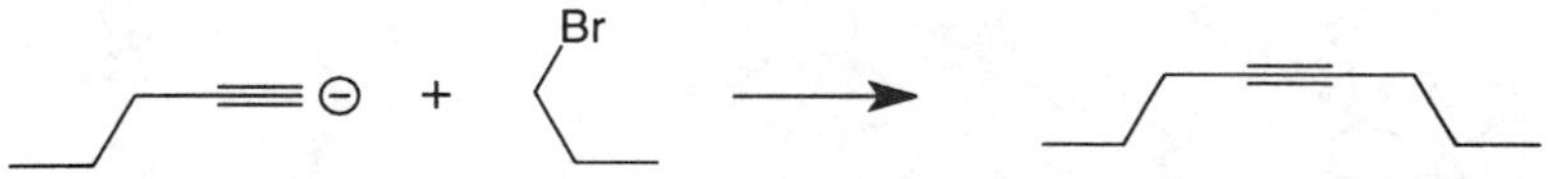

Ex 8.18 S_N2 reactions take place readily at primary and secondary halides and tosylates. Thus, displacements will take place readily with the primary and secondary chlorides represented in parts *a* and *b*. Elimination is the more likely consequence with the tertiary halide in part *c*.

(a)

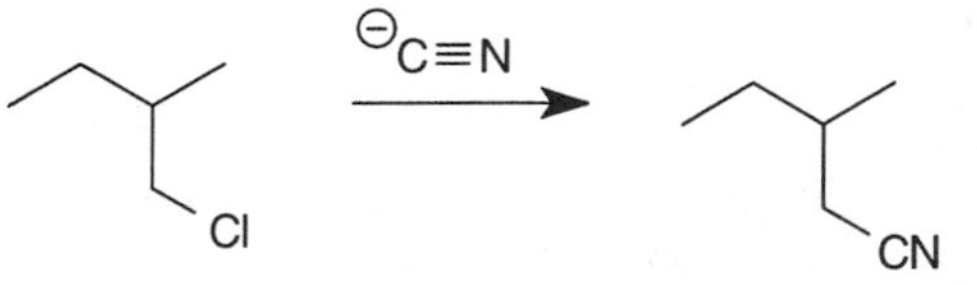

(b)

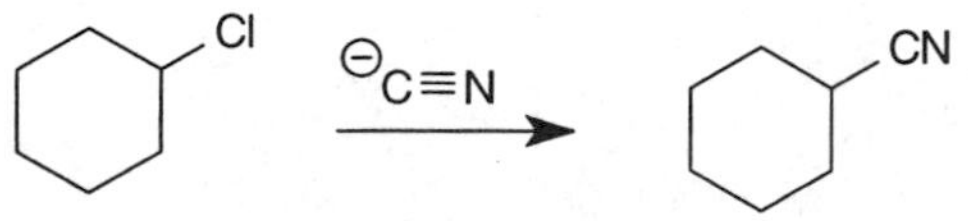

(c)

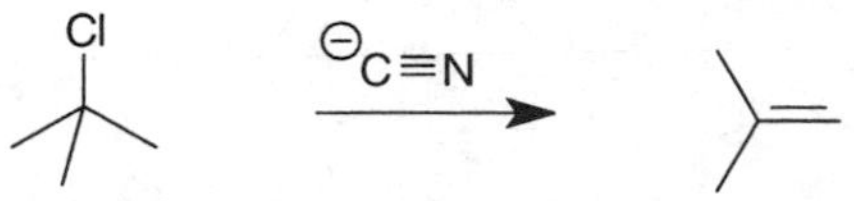

Ex 8.19 The reaction of a Grignard reagent with ethylene oxide adds a $-CH_2CH_2OH$ unit to the carbon skeleton of the alkyl halide.

(a)   Bromobenzene to phenylmagnesium bromide

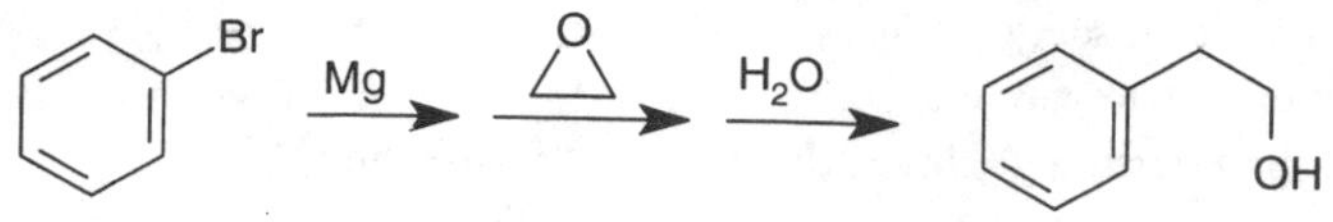

(b)   Cyclohexylmethyl bromide to cyclohexylmethylmagnesium bromide

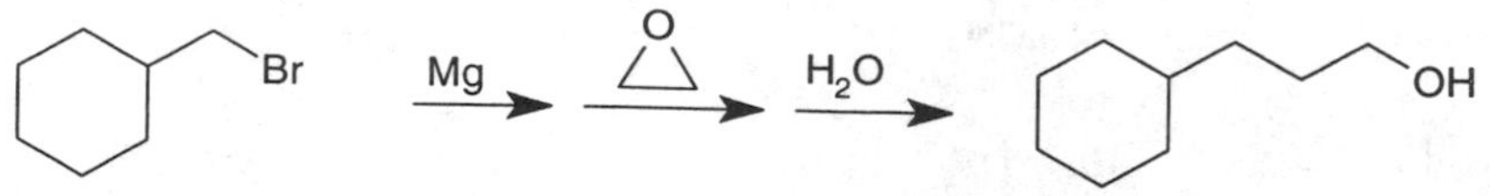

(c)   2-Propyl bromide to 2-propylmagnesium bromide

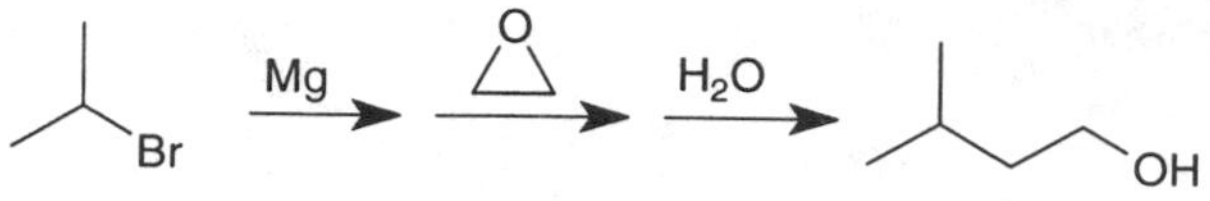

(d)   *n*-Butyl bromide to 1-butylmagnesium bromide

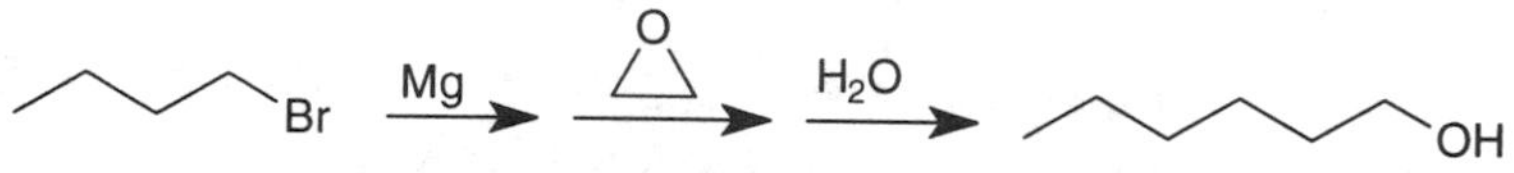

Ex 8.20 There are five unique carbon–carbon bonds in 2-methylhexane, any of which can be formed by coupling with an appropriate lithium dialkylcuprate. To obtain the proper combinations, perform a thought experiment in which each of the unique bonds is broken in sequence. There are ten possible routes:

1)   Lithium dimethylcopper + 2-bromohexane, or
2)   Lithium di(2-hexyl)copper + bromomethane

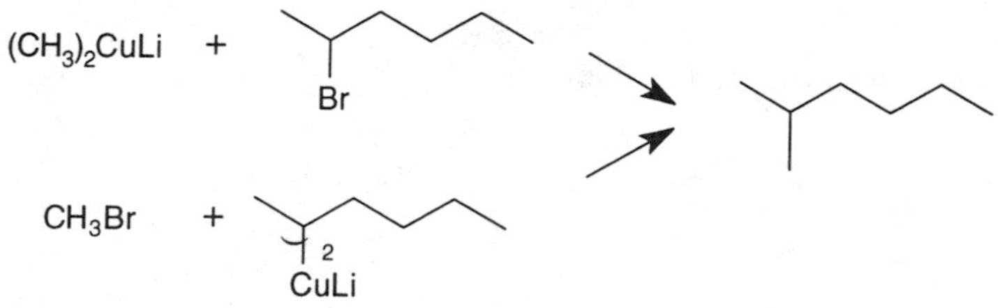

3)   Lithium di-*n*-butylcopper + 2-bromopropane, or
4)   Lithium diisopropylcopper + 1-bromobutane

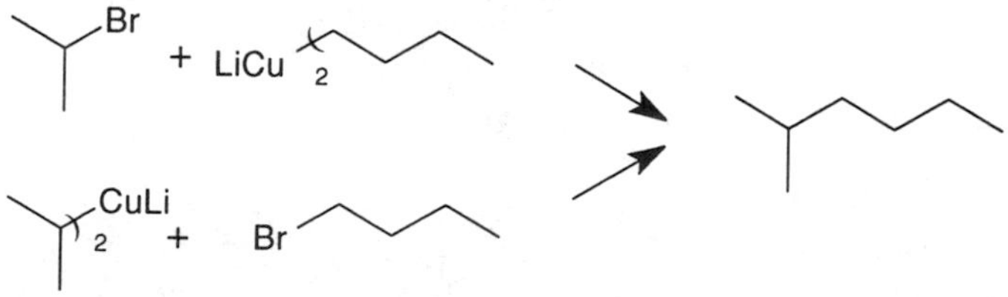

5)   Lithium dipropylcopper + 1-bromo-2-methylpropane, or
6)   Lithium diisobutylcopper + 1-bromopropane

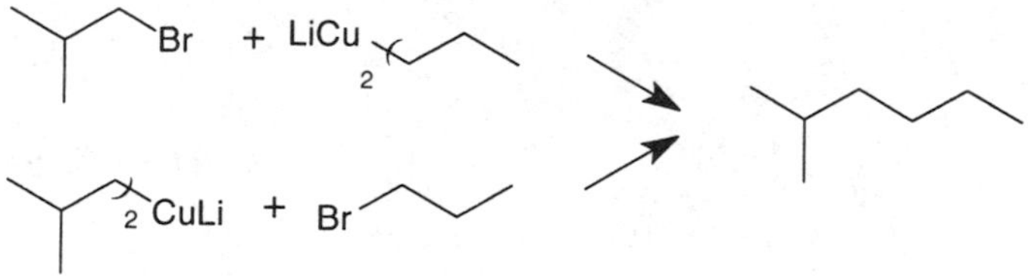

7)   Lithium diethylcopper + 1-bromo-3-methylbutane, or
8)   Lithium di(3-methyl-1-butyl)copper + bromoethane

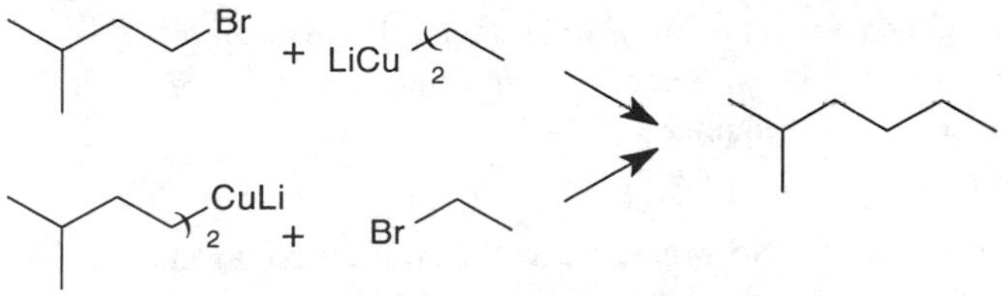

9)   Lithium dimethylcopper + 4-methyl-1-bromopentane, or

10)   Lithium di(4-methyl-1-pentyl)copper + bromomethane

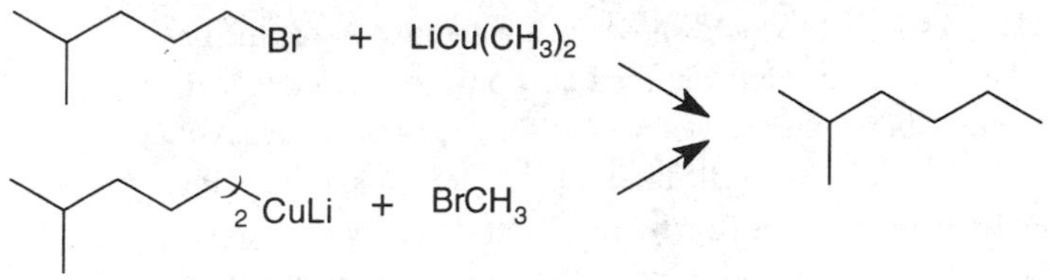

Ex 8.21 Magnesium inserts into most carbon–halogen bonds. For example, stable Grignard reagents can be made from aryl, vinyl, and alkyl halides. However, if an acidic functionality is also present in the molecule, rapid proton transfer to the highly basic C—Mg bond takes place. Thus, the C—X bond is converted to a C—H bond through the intermediacy of an unstable organometallic reagent. Therefore, this exercise asks whether there is an acidic functional group in the starting material.

(a)   2-Bromo-1-butanol has an acidic OH group, making the formation of a stable Grignard reagent impossible.

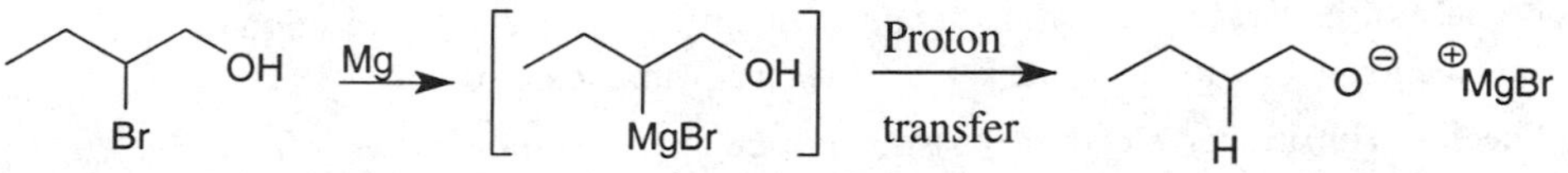

(b)   2-Chloro-1-phenylpropane has no protons sufficiently acidic to protonate a C—Mg bond, so magnesium insertion results in a stable Grignard reagent.

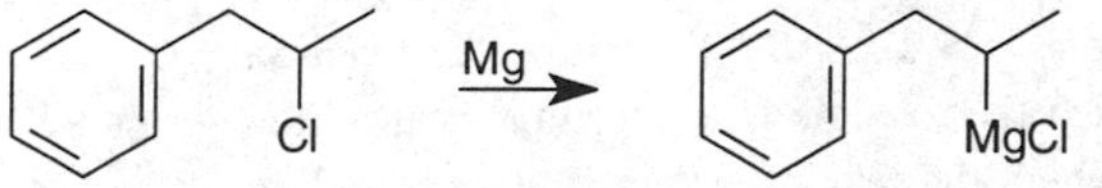

(c)   *p*-Bromotoluene reacts readily with magnesium. Because there are no acidic hydrogens in this molecule, a stable Grignard reagent is formed.

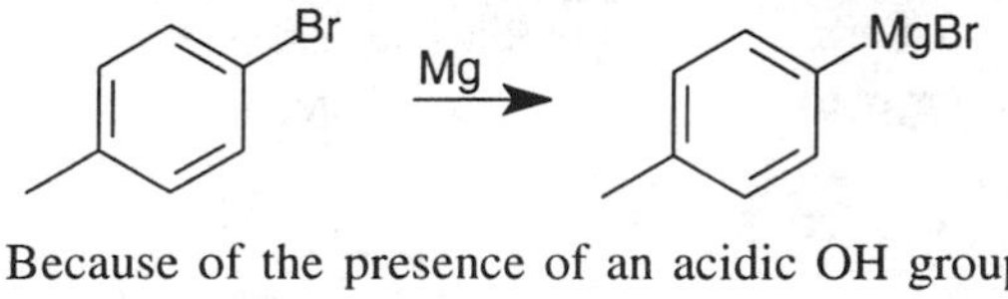

(d)   Because of the presence of an acidic OH group in *p*-bromophenol, it is not possible to form a stable Grignard reagent from this molecule.

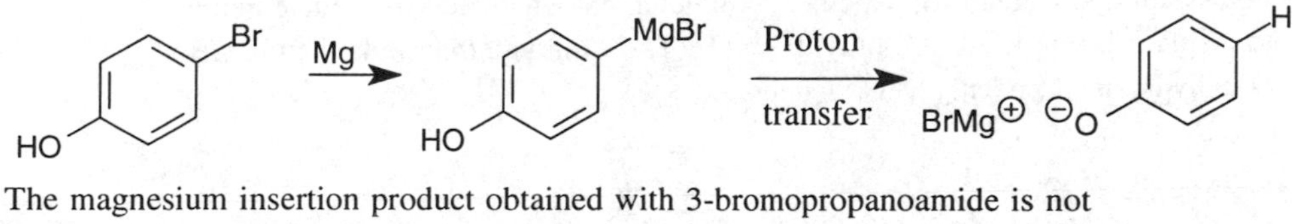

(e)   The magnesium insertion product obtained with 3-bromopropanoamide is not stable because of the acidic NH protons.

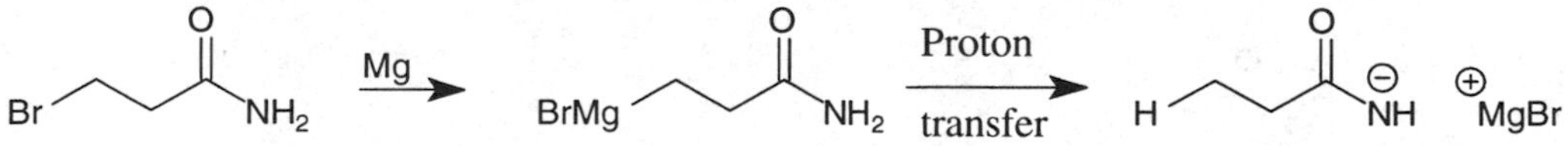

Ex 8.22 The quantity of water (with two acidic protons) required to protonate 1 mmole of *n*-butyllithium is 0.5 mmole. With a molecular weight of 18 g/mole and a density of 1g/mL, we can calculate the volume of water required:

$$5 \times 10^{-4} \text{ mole } (18 \text{ g/mole}) (1 \text{ mL/g}) = 9 \times 10^{-3} \text{ mL}$$

Compared with 10 mL THF, this is < 0.1%. No wonder special care must be taken to use very dry ethers as solvents for the preparation of Grignard reagents!

Ex 8.23 Many spectral techniques would make it possible to follow the course of this reaction. For example, the infrared spectrum would show the appearance of the nitrile band at 2210–2260 cm^{-1} in the desired product. But the question asks not merely to follow the course of the reaction, but rather to determine the fraction of product appearing as the alkene by-product. Several techniques are possible. Gas chromatography, with the products being identified by retention times by co-elution with an authentic sample of each product, would provide this information, as would integration of the ^{1}H NMR spectrum to compare the relative intensity of the olefinic region (2H from the alkene) with the downfield pentet (1H adjacent to the nitrile group).

Answers to Review Problems

Pr 8.1 The rate of an S$_N$2 reaction is affected by the leaving group, the nucleophile, and the degree of substitution (steric hindrance) of the carbon atom undergoing substitution.

(a) Iodine ion is better able to accommodate negative charge than is chloride ion because of its larger size and higher polarizability. As a result, the reaction of cyanide ion with *n*-iodoheptane is faster than that with *n*-chloroheptane.

(b) Species with higher electron densities are better nucleophiles (when other factors are equal). Thus, the reaction of *n*-butyl bromide with ethoxide ion is faster than that with ethanol.

(c) Reactivity in S$_N$2 reactions is affected by steric access of the nucleophile to the carbon bearing the leaving group: Me > 1° > 2° >> 3°. Therefore, the reaction of azide with *n*-butyl tosylate, in which the tosylate is at a primary carbon, is faster than with *s*-butyl tosylate, in which the leaving group is at a secondary site.

(d) The two options here are with an alkoxide nucleophile at a primary site (the reaction of isopropoxide with ethyl bromide) or at a secondary site (the reaction of ethoxide with 2-bromopropane). The former is easier and leads to a faster reaction rate.

Pr 8.2 This problem probes the types of product that can be formed by S$_N$2 reactions.

(a) Potassium hydroxide forms the phthalimide anion, which acts as a nucleophile to displace bromide from *n*-butyl bromide. The *N*-alkylphthalimide then reacts with hydrazine to release *n*-butylamine.

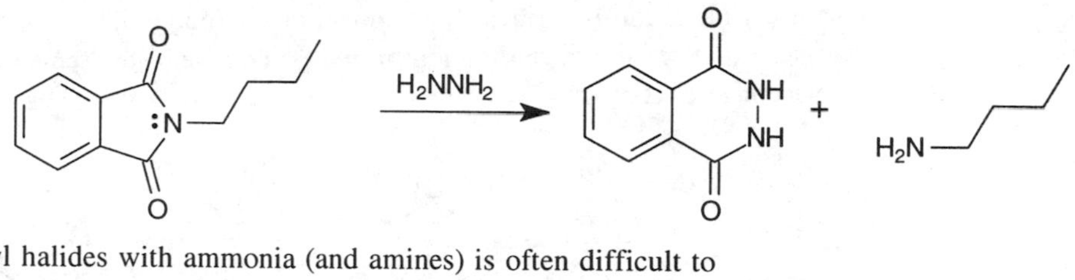

(b) The reaction of alkyl halides with ammonia (and amines) is often difficult to control, so that only one alkyl group is added to nitrogen except when a large excess of ammonia is used. One would therefore expect formation of mono-, di-, and trialkylamines, as well as some tetraalkylammonium bromide if stoichiometric quantities of ammonia are used. With a large excess of ammonia, reaction of the alkyl halide with ammonia is much more likely than with the product primary amine. Thus, under these conditions, the reaction effectively stops at the monoalkylation stage.

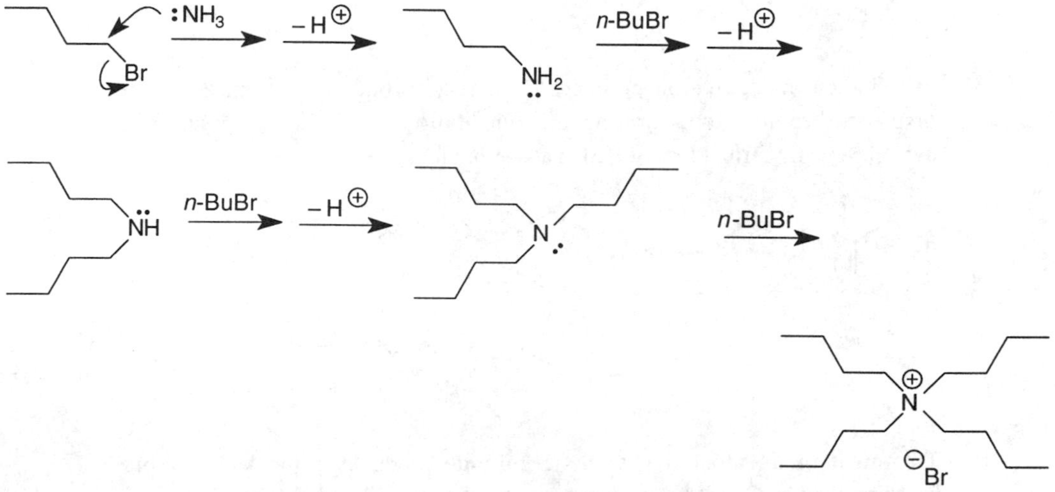

(c) Phenol is sufficiently acidic that treatment with sodium carbonate forms the phenoxide anion, which in turn acts as a nucleophile to attack the chiral bromide in a Williamson ether synthesis. This S_N2 reaction proceeds with inversion of configuration.

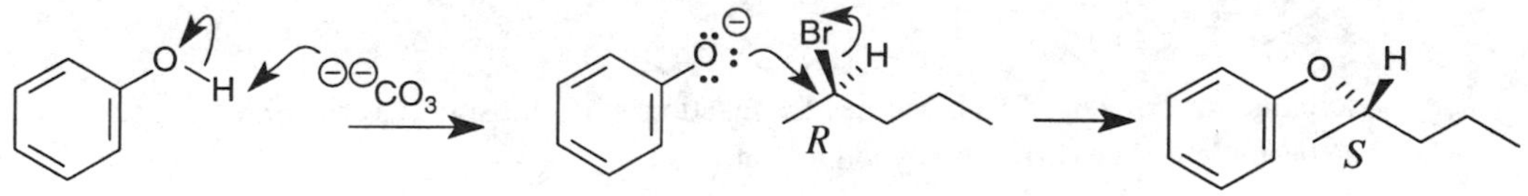

(d) Alcohols are converted to chlorides by treatment with thionyl chloride. This chloride is then displaced by triphenylphosphine to form a phosphonium salt.

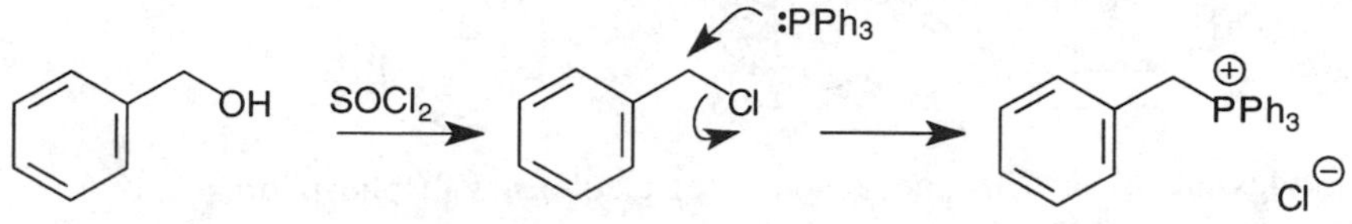

(e) Cyanide ion is a reactive nucleophile that displaces tosylate ion from this allylic tosylate.

(f) An acetylide anion is formed by treatment of a terminal alkyne with sodium amide. Addition of a deuterium ion from D_2O completes the sequence for the replacement of H by D.

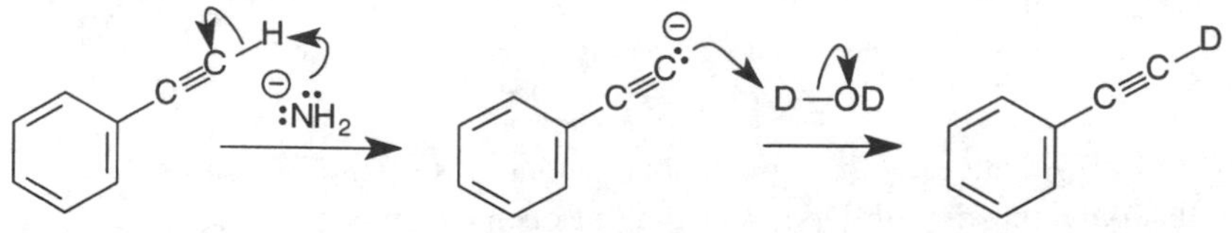

(g) Cyanide ion effects S_N2 displacement of iodide ion with inversion of configuration. This produces a product with two centers of asymmetry with the same chirality (R). Thus, the product is chiral and optically active.

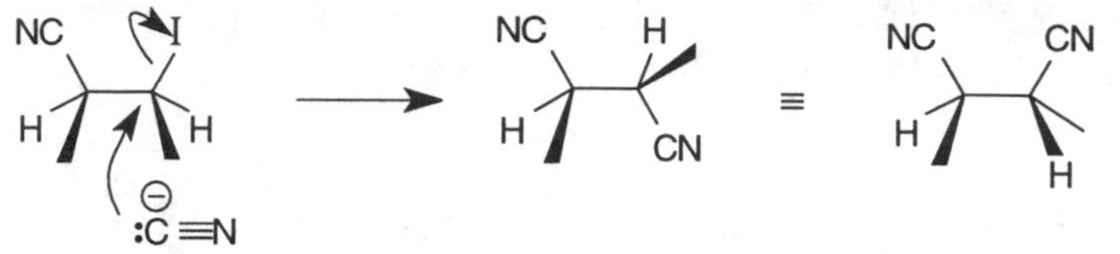

(h) Free radical halogenation produces cyclohexyl bromide. Sodium acetylide displaces bromide ion, forming the substitution product. (Elimination to cyclohexene by formal loss of HBr is a side-product.)

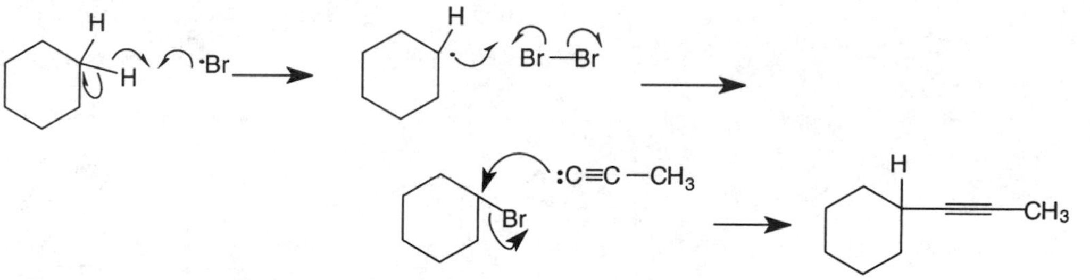

(i) Treatment of an alcohol with tosyl chloride produces a tosylate without breaking the C—O bond. As a result, the tosylate has the same configuration as the alcohol. S_N2 displacement by cyanide ion results in inversion of configuration at that site.

(j) Alcohols are converted to bromides by reaction with phosphorus tribromide. The bromide is then displaced by iodide ion.

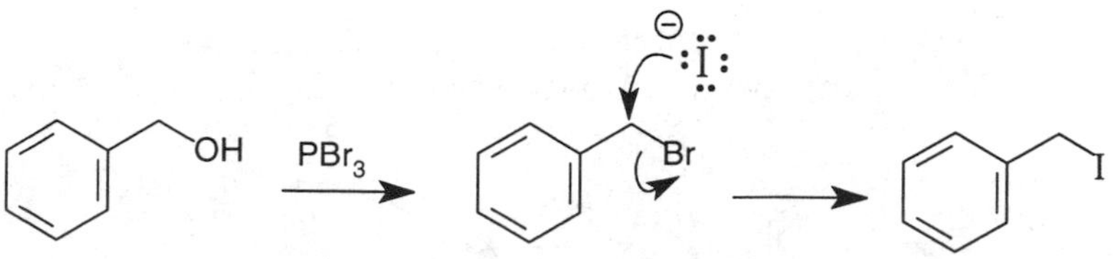

(k) Alkyl chlorides are formed by the treatment of alcohols with phosphorus trichloride. Because the C—O bond is broken in an S_N2 process, the alkyl chloride has the absolute configuration opposite that of the starting alcohol. Reaction of this chloride with ethoxide ion results in a second S_N2 reaction that takes place with inversion of configuration (a Williamson ether synthesis). The product therefore has the same absolute configuration as the reactant.

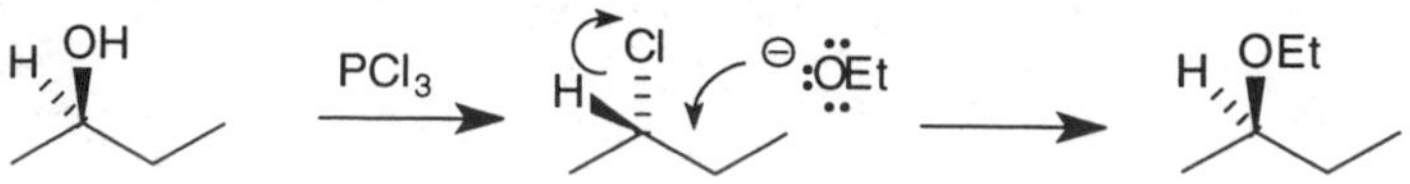

(l)   Analogous to the formation of alkoxide ions from alcohols, sodium thiolates are formed by reaction of thiols with sodium metal. These anions are reactive nucleophiles, undergoing the same types of reactions as alkoxides, in this case forming a thioether.

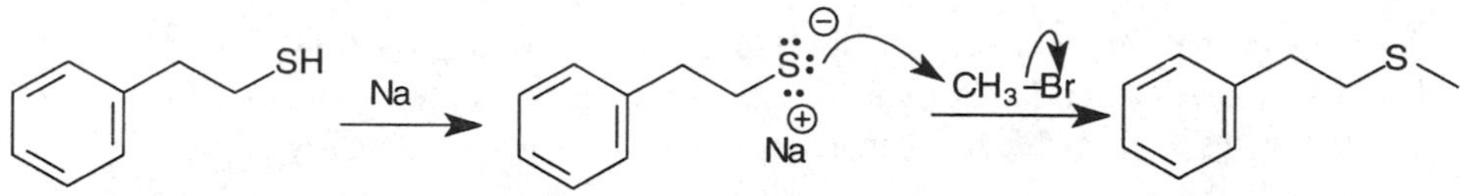

(m)   Thiolates can also be formed by treatment of thiols with base. The higher activity of the thiolate than of the thiol permits direct displacement on a tosylate.

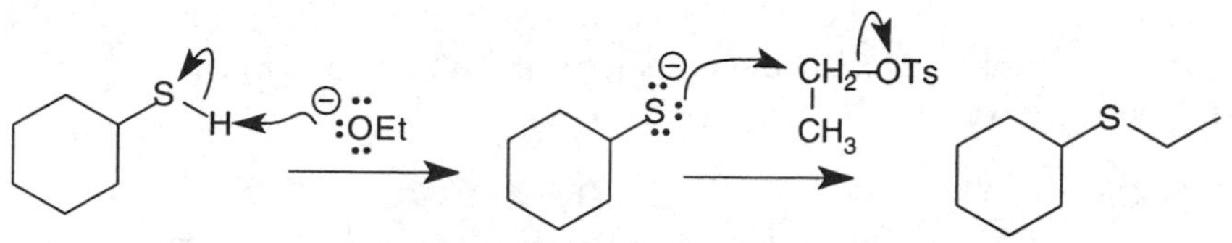

(n)   An ylide is formed by deprotonation of the phosphonium salt produced by the reaction of triphenylphosphine with an alkyl tosylate.

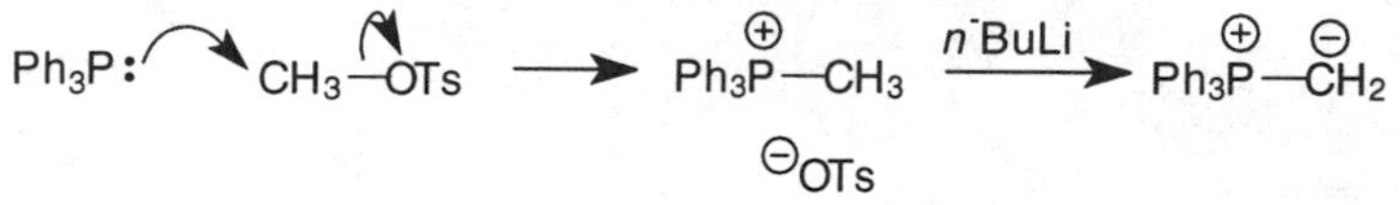

Pr 8.3   Alkylations of alkynyl anions take place by S_N2 displacement. Several answers are possible: one is given here for each part of this problem.

(a)   The anion of 1-propyne can displace a leaving group from an ethyl halide or tosylate. (Reaction with ethyl bromide is shown here.)

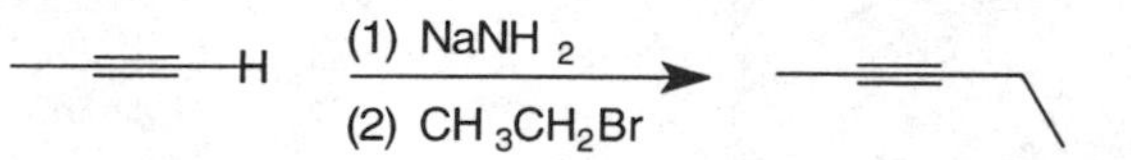

(b)   Treatment of the anion formed by deprotonation of 1-propyne with methyl bromide gives 2-butyne, catalytic hydrogenation (Chapter 2) of which produces *cis*-2-butene.

(c)   Alkylation of the anion of propyne (as in part *a*) with *n*-propyl bromide produces a six-carbon alkyne. Reduction of both π bonds of the triple bond affords *n*-hexane.

(d)   2-Butanol can be prepared from *cis*-2-butene, prepared as in part *b*, by acid-catalyzed hydration. This alcohol can be converted to a 2-bromobutane by treatment with PBr₃ and transformed into *s*-butylamine using the Gabriel synthesis or by using a large excess of ammonia.

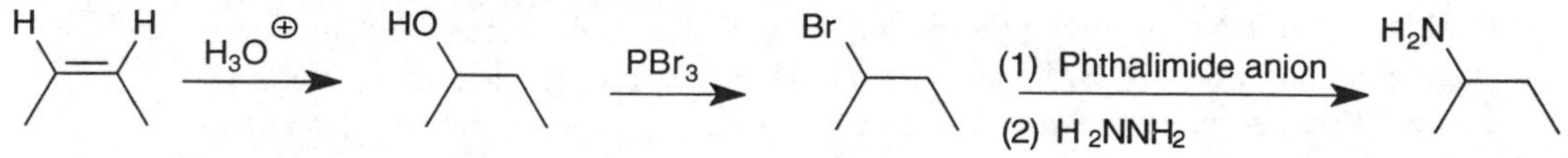

Pr 8.4   Organometallic reagents are usually prepared by metallation of the alkyl halide, deprotonation with a metal salt, or transmetallation.

(a)   Treatment of benzyl chloride with lithium metal produces benzyllithium.

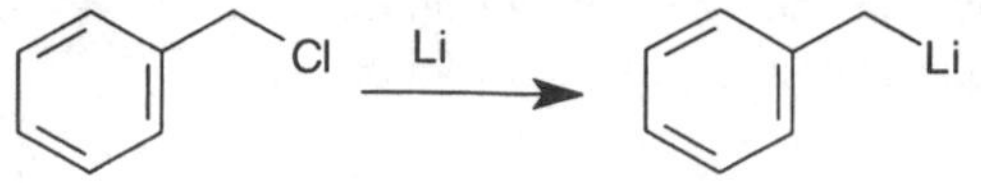

(b)   To prepare the Grignard reagent, introduce bromine through a free-radical halogenation (Chapter 7). The resulting benzyl bromide is then converted to benzylmagnesium bromide by treatment with metallic magnesium in dry ether.

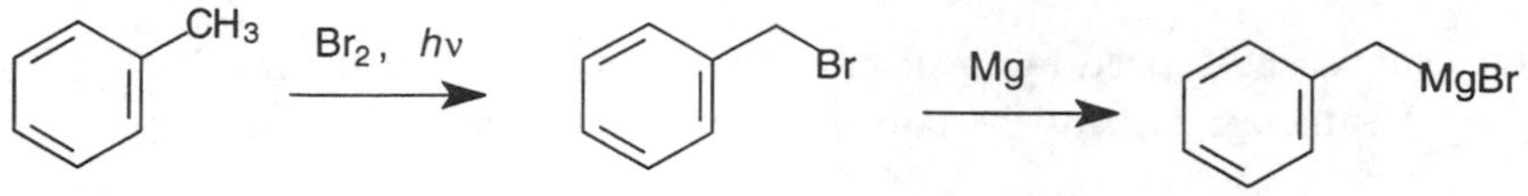

(c)   Reaction of the alcohol with thionyl chloride affords benzyl chloride, which is then converted into benzyllithium in a process parallel to that in part *a*, and then into lithium dibenzylcuprate by treatment with CuI.

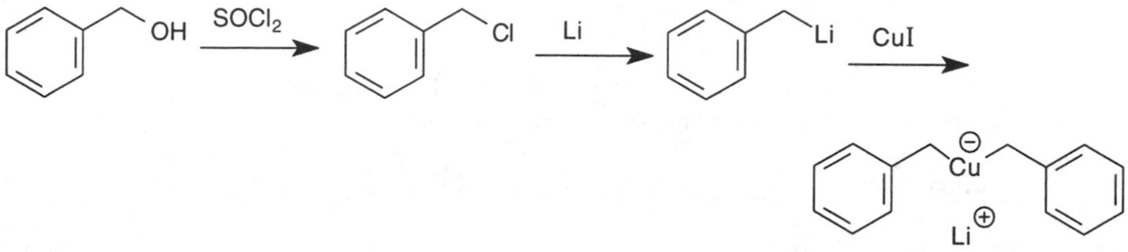

Pr 8.5   The reaction of a Grignard reagent with ethylene oxide extends the carbon chain by two atoms.

(a)   The precursor to the required Grignard reagent is bromobenzene.

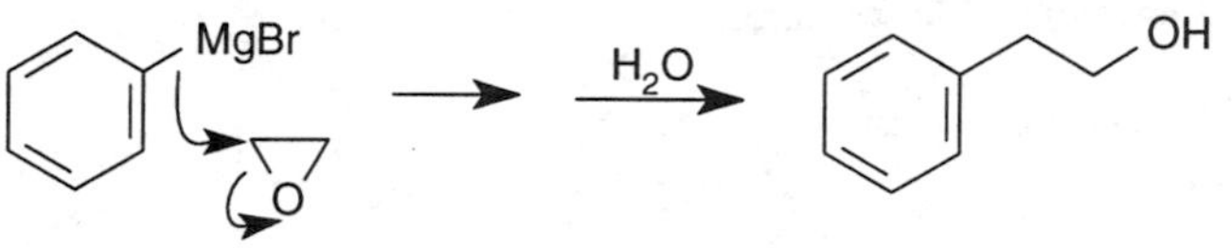

(b)   *n*-Butylmagnesium bromide is prepared from *n*-butyl bromide.

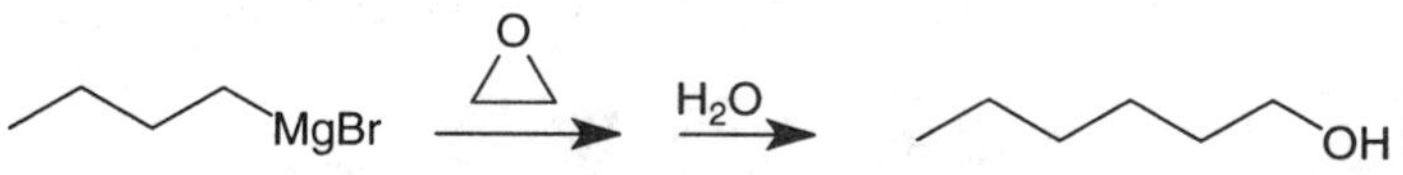

(c)   *trans*-4-Methyl-1-bromocyclohexane is converted to the required Grignard reagent by reaction with magnesium metal.

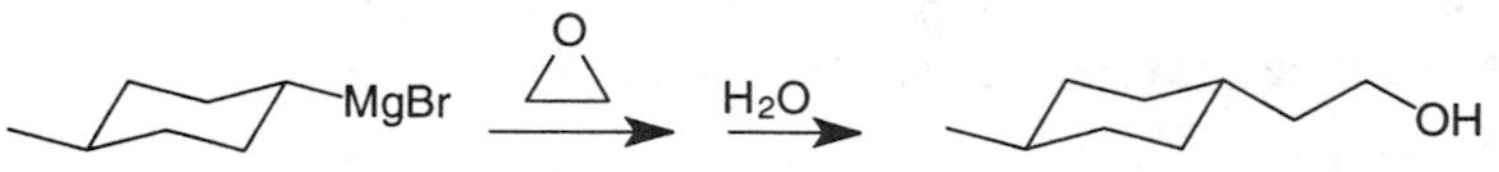

Pr 8.6   The fact that first-order kinetics are observed tells us that the rate-determining step requires only the substrate; that is, no external nucleophile assists in the cleavage of the C—Br bond. Because the solvolysis of a bromoalkane bearing the adjacent methoxy group is faster than that without the methoxy group, the methoxy group must participate in the rate-determining step in a way that makes cleavage of the C—Br bond faster. Here, the methoxy group is located at a site from which it assists the loss of the leaving group by intramolecular back-side displacement. The initial fragmentation of the C—Br bond to form a stable cation is assisted by the methoxy group. (This internal assistance of the loss of a leaving group is called *anchimeric assistance*.)

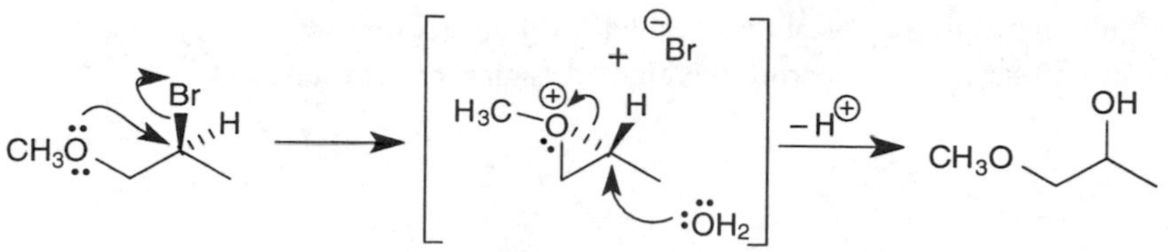

(a) This sequence includes internal back-side displacement of bromide ion as in an S_N2 reaction, but the rate-determining step involves only the substrate and is therefore unimolecular. The resulting three-membered-ring intermediate is opened by a second back-side attack, this time with water as the nucleophile, giving rise to the substitution product observed. This sequence therefore accomplishes two inversions of configuration (through two back-side attacks) and gives a product with the same absolute configuration as in the reactant (retention of configuration).

(b) These are typical S_N2 reaction conditions. With a more reactive nucleophile (such as cyanide ion) and a solvent less able to solvate bromide ion as it is lost, the rate of the simple S_N2 reaction can be higher than the intramolecular loss of bromide ion with anchimeric assistance. In this case, direct displacement through a single back-side attack gives product without an intervening intermediate, producing *(2R)*-2-cyano-1-methoxypropane.

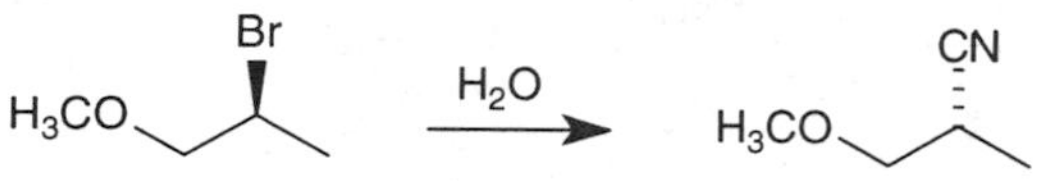

Pr 8.7 In virtually any synthetic plan, more than one correct answer will exist. The one correct answer given here uses the reactions we have just covered in this Chapter, but others are certainly possible.

(a) The Williamson ether synthesis is an excellent means for making ethers. Because this reaction takes place through an S_N2 pathway, treatment of the dianion of hydroquinone with methyl bromide will give the desired product; treatment of 1,4-dibromobenzene with sodium methoxide will not.

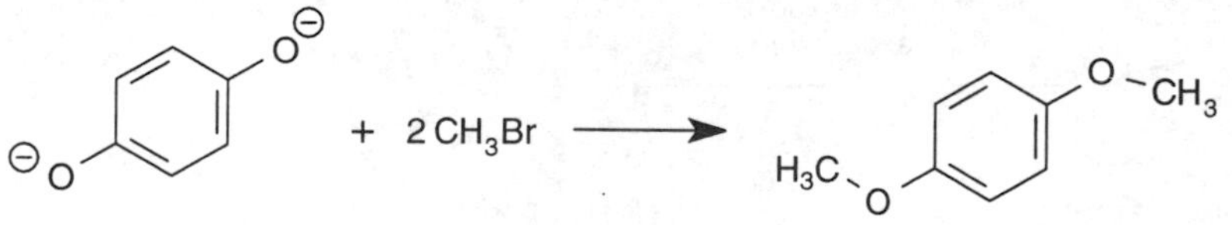

(b) Preparation of a cyclic ether by a Williamson ether synthesis requires an alkoxide and a leaving group on the same carbon skeleton. Treatment of 4-bromo-1-butanol with potassium *t*-butoxide produces the anion, which ring-closes by an S_N2 pathway. Because intramolecular S_N2 reactions that form five- and six-membered rings are faster than comparable bimolecular S_N2 reactions, a simple base such as NaOH could also be used.

(c) The synthesis of thioethers takes place in parallel with the Williamson ether synthesis. Here the required reagents are the anion of thiophenol and *n*-propyl bromide.

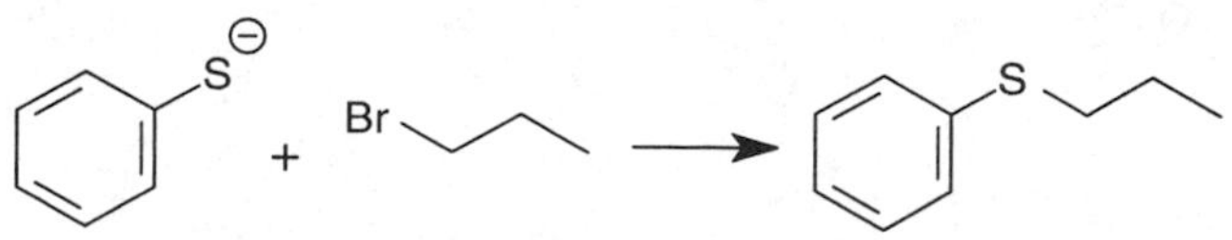

(d)   As in part *b*, an intramolecular cyclization is called for, but this time with a five-carbon skeleton; that is, by preparing the alkoxide anion of 5-bromo-1-pentanol.

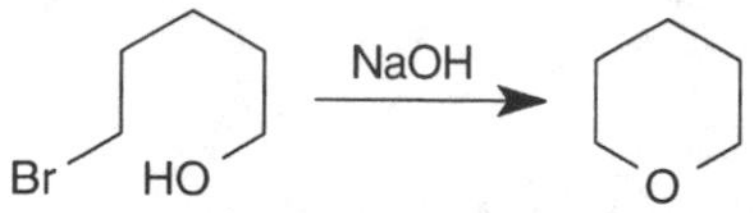

(e)   Organonitriles can be prepared by the S_N2 displacement of halides by potassium cyanide. Here, treatment of ethyl bromide with potassium cyanide will produce the desired nitrile.

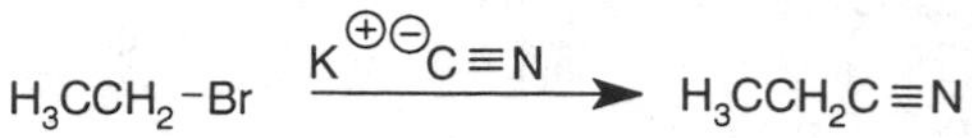

(f)   Benzylamine can be prepared by treating benzyl bromide either with excess ammonia or with phthalimide anion followed by hydrazine.

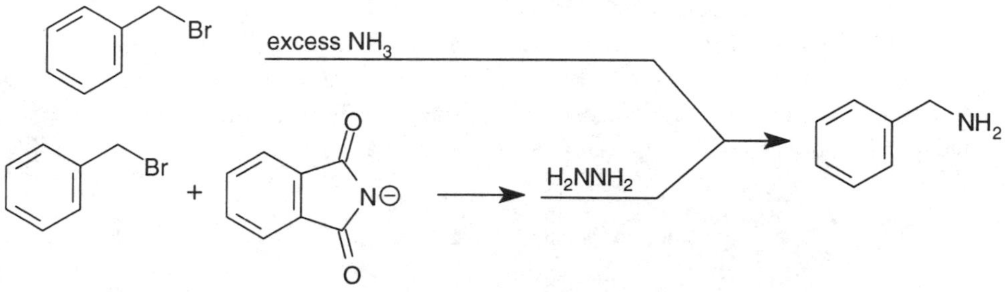

(g)   In parallel with part *b*, treatment of 4-bromo-1-butanethiol with potassium hydroxide produces the thiolate anion, which ring-closes by an S_N2 pathway.

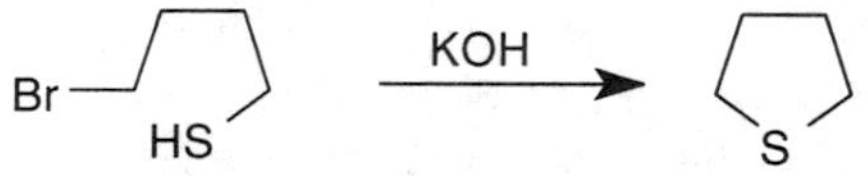

(h)   Ethylamine can be prepared by treating ethyl bromide either with excess ammonia or with phthalimide anion followed by hydrazine.

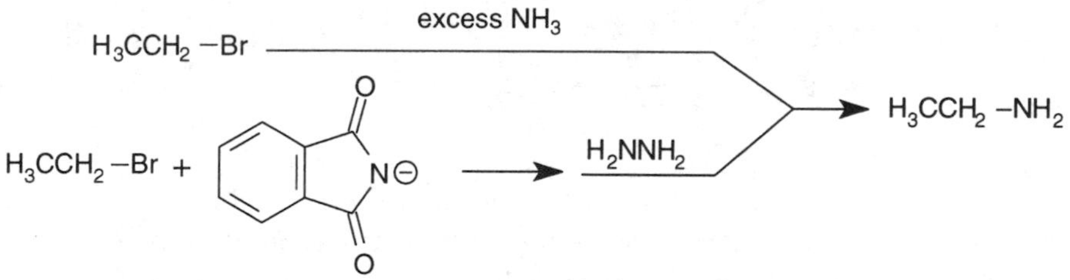

Pr 8.8   (a)   Treatment of 1,4-dibromobutane with NaOH would produce 4-bromo-1-butanol which would then undergo intramolecular cyclization under the reaction conditions.

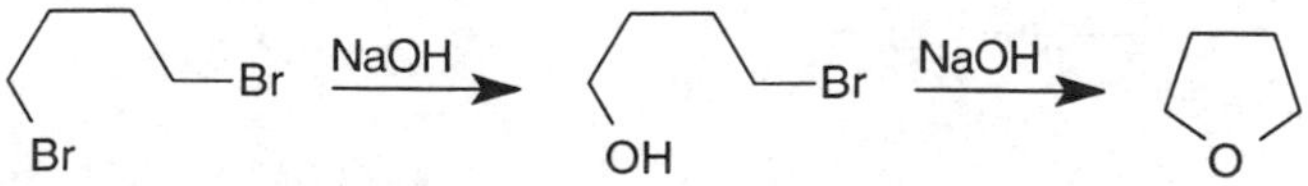

(b)

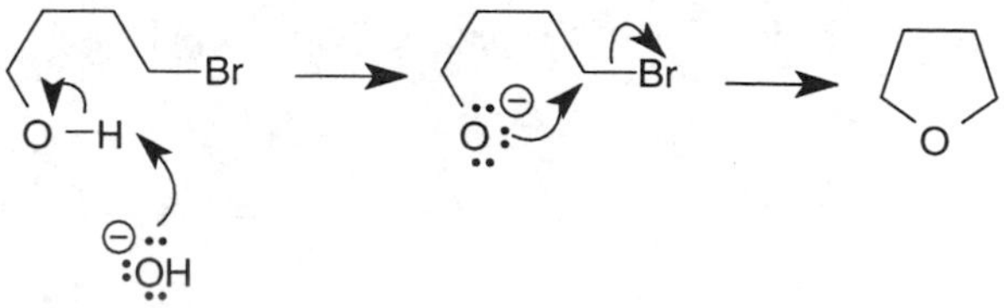

Pr 8.9 Preparation of an alcohol by treatment of a Grignard reagent with ethylene oxide extends the carbon chain by two carbons. The bromide precursor to the Grignard reagent is thus obtained by mentally deleting this two-carbon extension.

(a) *n*-Ethyl bromide + Mg gives ethylmagnesium bromide, which reacts with ethylene oxide to give 1-butanol.

(b) Cyclohexylmethyl bromide + Mg gives cyclohexylmethylmagnesium bromide, which reacts with ethylene oxide to give 3-cyclohexylpropanol.

(c) Bromobenzene + Mg gives phenylmagnesium bromide, which reacts with ethylene oxide to give 2-phenyl-1-ethanol.

(d) *t*-Butyl bromide + Mg gives *t*-butylmagnesium bromide, which reacts with ethylene oxide to give 3,3-dimethyl-1-butanol.

Pr 8.10 Each treatment adds two carbons to the skeleton. The desired product has four carbons more than the allowed starting material. Therefore, the chain extension with ethylene oxide will be used twice. The Grignard reagent obtained by metallation of bromocyclohexane produces 2-cyclohexylethanol upon treatment with ethylene oxide. This alcohol in turn can be converted to a Grignard reagent by treating first with PBr$_3$ and then with magnesium metal in ether. Treatment of this Grignard reagent with ethylene oxide provides the desired alcohol after neutralization.

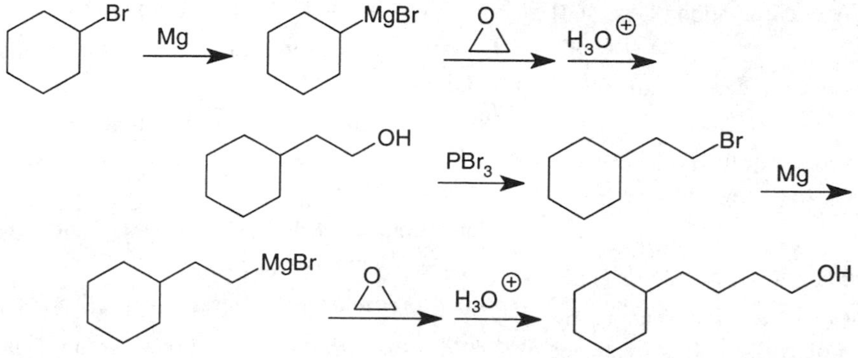

Pr 8.11 In a Gabriel synthesis, the anion of phthalimide is treated with the desired alkyl bromide, and the resulting primary amine is freed by treatment with hydrazine. In the problem, the following are the required alkyl bromides.

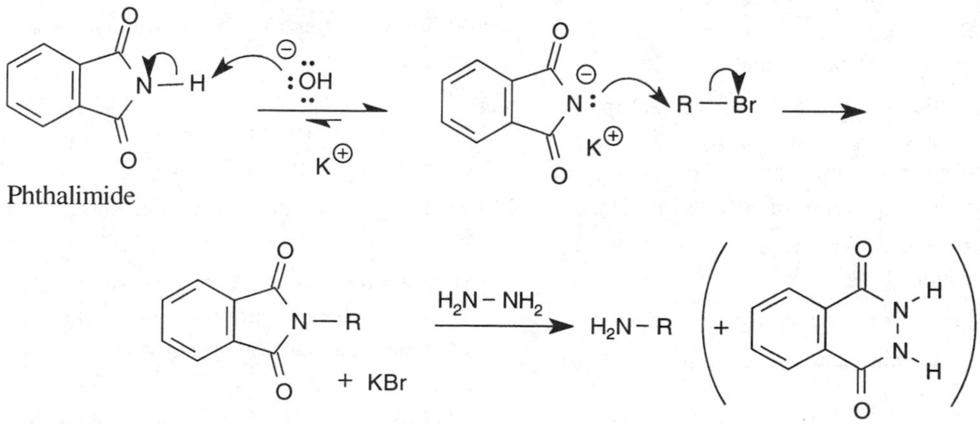

Phthalimide

(a) *n*-Butyl bromide

(b) Cyclohexyl bromide

(c) 2-Phenylethyl bromide

Pr 8.12   The Gabriel synthesis is a synthetic route to primary amines. It also requires an S_N2 alkylation of the phthalimide anion, which is possible only for methyl, primary, and secondary halides or tosylates.

(a)   Because this is a tertiary amine, it cannot be prepared directly by a Gabriel synthesis.

(b)   Although aniline is a primary amine, the carbon–nitrogen bond is at an aromatic carbon, which cannot be produced in the S_N2 first step of a Gabriel synthesis.

(c)   This primary amine can be prepared by a Gabriel synthesis.

(d)   This primary amine can be prepared by a Gabriel synthesis. The halide ion can be displaced from a primary carbon by nitrogen in the first step.

(e)   This primary amine can be prepared by a Gabriel synthesis. The halide ion can be displaced from the secondary carbon by nitrogen in the first step.

(f)   This compound has an amino group attached to a heteroaromatic ring. Displacement from this aryl halide precursor is impossible by an S_N2 process.

Important New Terms

Acetylide anion: *see* **Alkynide anion** (8.4)

Alkoxide: an anion obtained by deprotonation of the –OH group of an alcohol (8.3)

Alkynide anion: an anion formed by deprotonation of a terminal alkyne; $RC{\equiv}C^-$ (8.4)

Chlorosulfite ester: an intermediate in the conversion of an alcohol to an alkyl halide with thionyl chloride (8.3)

Cyanide ion: $^-C{\equiv}N$ (8.4)

Ethylene oxide: (C_2H_4O); the simplest epoxide (8.4)

Gabriel synthesis: the synthesis of a primary amine by alkylation of phthalimide anion, followed by treatment of the resulting *N*-alkylphthalimide with hydrazine (8.3)

Grignard reagent: a reagent in which carbon is directly bound to magnesium (8.4)

Insertion reaction: a reaction in which an atom, often a metal, becomes bound to two atoms that were themselves originally covalently bound (8.4)

Isotopic labeling: the replacement of an isotope of highest natural abundance with another isotope at a specific position in a molecule; for example, replacement of ^1H by ^2H (D), or of ^{12}C by ^{13}C (8.4)

Lithium dialkylcuprate: $R_2Cu^- Li^+$; an alkylating agent for alkyl halides (8.4)

Mesylate: $ROSO_2CH_3$; a methanesulfonate ester (8.3)

Organocuprate: a reagent in which carbon is directly bound to copper (8.4)

Organolithium: a reagent in which carbon is directly bound to lithium (8.4)

Organomagnesium compound: *see* **Grignard reagent** (8.4)

Organometallic compound: a reagent in which carbon is directly bound to a metal atom (8.4)

Phosphine: a functional group containing trivalent phosphorus (PR_3) (8.3)

Phosphonium salt: a tetravalent phosphorus cation ($^+PR_4$); obtained by protonation or alkylation of a phosphine (8.3)

Phosphonium ylide: $R_3P^+{-}(CR_2)^-$; an α-deprotonated phosphonium salt (8.3)

Phosphorane: *see* **Phosphonium ylide** (8.3)

Protecting group: a functional group that masks the characteristic reactivity of another group to which it can later be converted (8.3)

Selectivity: the formation of one product in preference to other possible products (8.2)

Sulfonate ester: an alkylated derivative of a sulfonic acid: $RO{-}SO_2R'$ (8.3)

Thionyl chloride: $SOCl_2$; an effective reagent for the conversion of alcohols to alkyl chlorides or of carboxylic acids to acid chlorides (8.3)

Tosylate: a *p*-toluenesulfonate ester: $ROSO_2\text{-}p\text{-}C_6H_4(CH_3)$ (8.3)

Transmetallation: an exchange of metals between an organometallic compound and either a metal or a different organometallic compound (8.4)

Williamson ether synthesis: the reaction of an alkoxide ion with an alkyl halide or tosylate to produce an ether (8.3)

Ylide: a zwitterion bearing opposite charges on adjacent atoms (8.3)

Elimination Reactions

9

Key Concepts

Mechanisms of E1, E2, and E1CB reactions

Stereoelectronic requirement for an *anti*-periplanar transition state in E2 eliminations: consequences in the stereoselective formation of geometric isomers

Thermodynamic control in Zaitsev eliminations

Kinetic control in Hofmann eliminations

Synthesis of alkenes by dehydrohalogenation of alkyl halides: stereochemical and regiochemical control

Synthesis of alkynes by dehydrohalogenation of vinyl halides

Formation of benzyne by dehydrohalogenation of an aryl halide

Synthesis of aniline and phenol by addition of ammonia or water to benzyne

Synthesis of alkenes by dehydration of alcohols

Acid-catalysis in the formation of oxonium ions by protonation of alcohols as a key step in alcohol dehydration

Cationic rearrangements in E1 reactions

Synthesis of alkynes by dehalogenation of vicinal dihalides

Synthesis of aldehydes and carboxylic acids by oxidation of primary alcohols

Chromate oxidation as a chemical (color) test for oxidizable groups

Synthesis of ketones by oxidation of secondary alcohols

Indirect routes for oxidation of alkanes in the laboratory

Answers to Exercises

Ex 9.1 (a) The E1 elimination proceeds through a carbocation intermediate. (Recall from Chapter 7 that this reaction is acid-catalyzed: acid is needed to convert the hydroxyl group to a better leaving group.) This exercise therefore asks: which reactant leads to the more stable carbocation? The reactant at the left is a secondary alcohol and forms a secondary cation as water is lost; the reactant at the right is a tertiary alcohol that forms a tertiary cation. Thus, dehydration is easier with the substrate at the right.

(b) This reaction also proceeds through an E1 mechanism with the formation of a carbocation intermediate in the rate-limiting step. Upon loss of bromide ion, the structure at the left forms a secondary cation. The structure at the right also forms a secondary cation, but inductive electron withdrawal by the carbonyl group on the adjacent carbon intensifies the positive charge and destabilizes the cation. Elimination is therefore faster with cyclohexyl bromide.

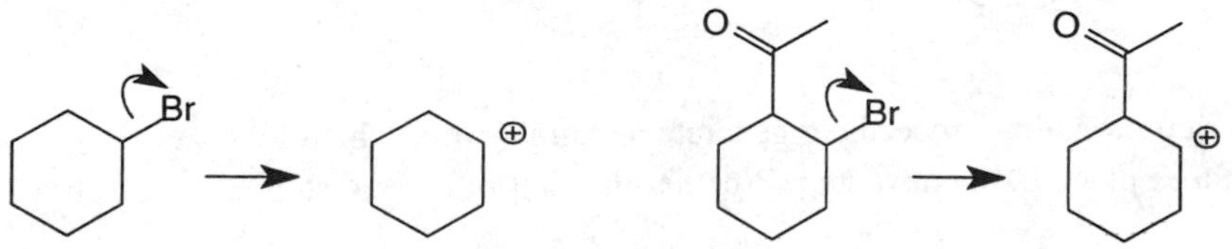

153

(c) Deprotonation with base of the structure at the right in the Exercise (shown here in the lower reaction) produces an enolate anion, whereas deprotonation with strong base of cyclohexyl bromide (at the left in the Exercise) would produce a simple alkyl carbanion (upper reaction). These anions are the conjugate bases of the starting materials, and the enolate anion is present at much higher concentration at equilibrium than is the simple carbanion because of the greater acidity of a C—H bond α to a carbonyl group. Because these anions participate in the rate-determining step in an E1cB pathway, the rate of reaction is faster for the lower reaction. In practice, simple alkyl halides, as cyclohexyl bromide in the upper reaction, undergo elimination instead by E2 (and E1) mechanisms because the anion is sufficiently unstable that loss of bromide ion occurs at the same time that the proton is removed by base.

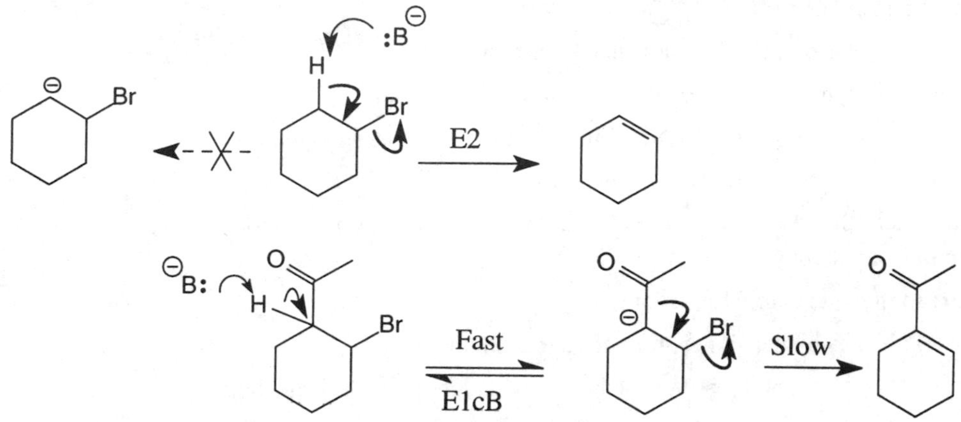

(d) In the structure at the left in the Exercise, four protons are available for concerted (E2) elimination, as H^+ and Br^- are simultaneously lost. In the structure at the right, elimination is blocked because of the absence of a proton on the adjacent carbon that is *trans* to bromine (in an *anti*-periplanar relation) while maintaining a chair conformation for the six-membered ring. E2 elimination is therefore faster with cyclohexyl bromide.

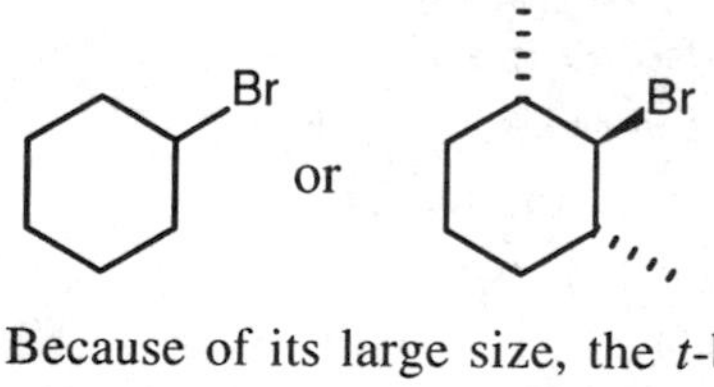

(e) Because of its large size, the *t*-butyl group acts as a conformational anchor, with the dominant conformer being that in which this group is equatorial. Therefore, the structure at the left in the Exercise has chlorine in an axial position, in a conformation from which the loss of HCl through a concerted E2 transition state is easy. The structure at the right in the Exercise has chlorine in an equatorial position, which is not *anti*-periplanar to a hydrogen on either adjacent carbon. As a result, no concerted elimination is possible from this conformation.

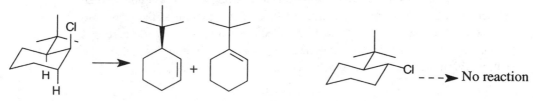

Ex 9.2 In the E1cB reaction, deprotonation proceeds to generate an anion, from which loss of the leaving group takes place in the next step. Whether the first or second step is

rate-determining depends on the relative heights of the corresponding activation energy barriers. In curve *a*, the first step is rate-determining; in curve *b*, the second step is. The Hammond postulate specifies that the transition state will closely resemble the chemical species lying closest to it in energy. The anion is a higher-energy species than either the reactant or the product, and is hence closer to both of the relevant transition states. Any factor that will stabilize the intermediate anion will therefore accelerate the reaction, irrespective of whether the first or second step is rate-determining.

Curve *a*:

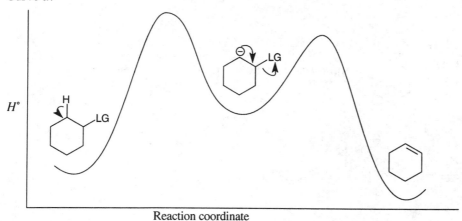

Reaction coordinate

Curve *b*:

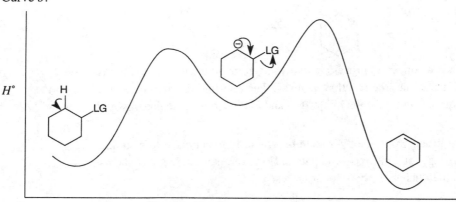

Reaction coordinate

Ex 9.3   This elimination takes place by deprotonation at the α-carbon, producing an enolate anion. Loss of hydroxide ion then takes place in an exothermic second step, producing the observed enone.

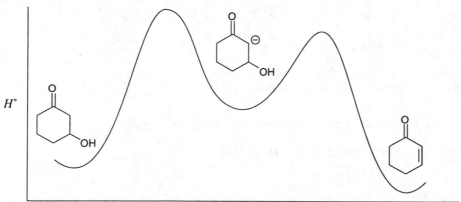

Reaction coordinate

Ex 9.4   The E1 elimination begins with the loss of the leaving group, producing a carbocation in the rate-determining step. Under neutral conditions the leaving group is hydroxide, a very poor leaving group. Thus, the E1 elimination is very slow under neutral conditions. However, the OH group can be readily converted to a much better leaving group by protonation by acid. Thus, loss of water from the protonated alcohol, present in an equilibrium with its neutral conjugate base when acid is present in catalytic amounts, produces the same carbocation in a much more rapid process in which water is the leaving group. Carbocation formation is followed by rapid deprotonation to form the alkene.

Acid-catalyzed:

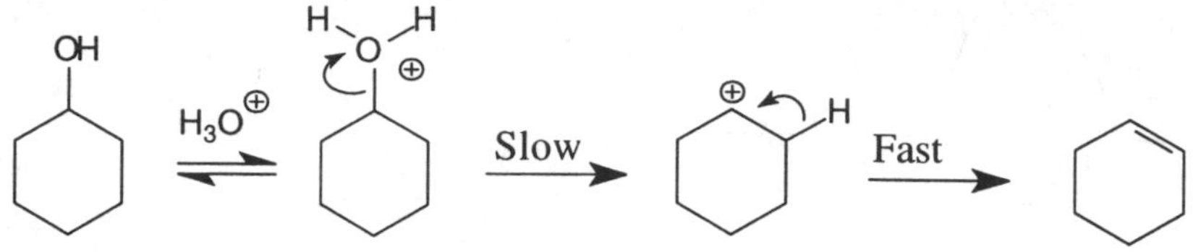

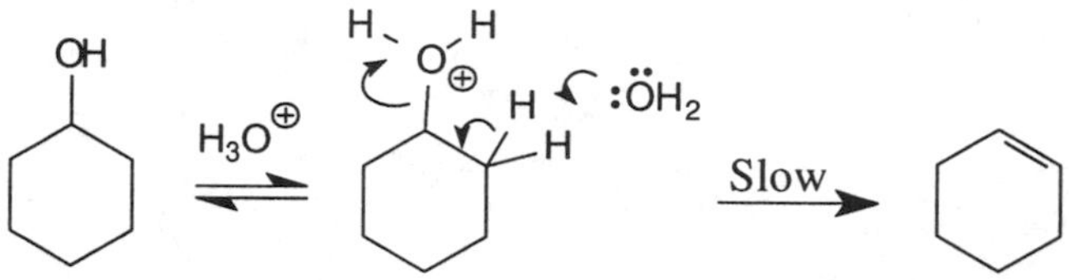

Base-catalyzed:

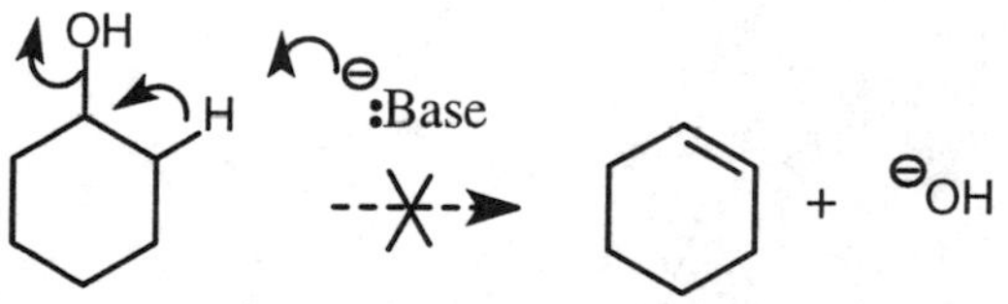

The E2 elimination is a concerted process in which there are no intermediates. It has a single transition state along the reaction pathway. The rate-limiting step of an E2 reaction in acid requires simultaneous loss of H^+ and H_2O. No intermediates would be formed.

In base, the leaving group would be hydroxide ion. In an E2 reaction of this simple system, a strong base is required for deprotonation. However, hydroxide ion is such a poor leaving group that this pathway is not observed.

Ex 9.5   In these examples of E2 elimination reactions, a *trans anti*-periplanar relation between bromine and a hydrogen on an adjacent atom is required as the reactants approach the transition states.

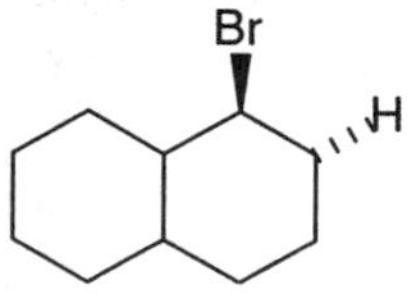

All hydrogens shown can be lost in an E2 elimination from an accessible conformation.

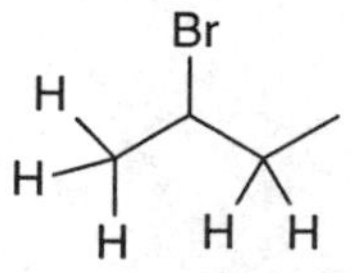

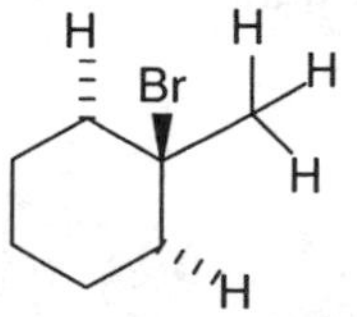

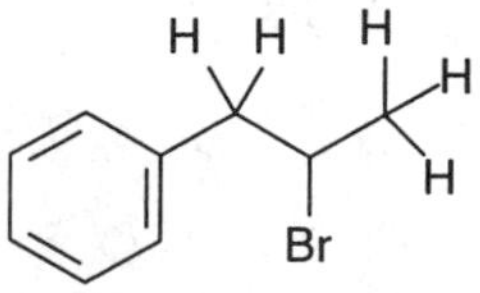

Ex 9.6   Treatment of ethanol with sodium metal produces sodium ethoxide and hyrogen gas.

$$CH_3CH_2OH \ + \ Na \ \longrightarrow \ CH_3CH_2O^{\ominus} \ ^{\oplus}Na \ + \ \tfrac{1}{2}H_2$$

Ex 9.7   Often, basicity, a measure of the affinity for a proton, parallels nucleophilicity, the affinity for a partially positively charged carbon atom. Both properties increase with increasing electron density. Thus, an anionic reagent is more basic and more nucleophilic that its neutral conjugate acid. Charge-intensive reagents that have negative charge localized on a single atom generally have a higher basicity-to-nucleophilicity ratio than do charge-delocalized anions. Charge-intensive reagents are sometimes called hard; charge-delocalized reagents are sometimes called soft.

(a) Charge is more concentrated in the anionic amide ion ($^-NH_2$), which is a much stronger base than a neutral ammonia molecule.

(b) Charge is more concentrated in the anionic hydroxide ion (^-OH), which is a much stronger base than a neutral water molecule.

(c) Charge is more dispersed on the larger sulfur atom of the thiolate anion than on the smaller oxygen atom of the hydroxide ion. Thus, the latter is the stronger base.

(d) The negative charges of both phenoxide and the enolate anion are delocalized, making both relatively poor bases but good nucleophiles. The negative charge in phenoxide ion is more delocalized than that in and enolate anion and, therefore, the latter is more basic.

(e) Because charge is delocalized over two oxygen atoms in a carboxylate anion, it is a poorer base than is the alkoxide anion, in which charge is localized on a single oxygen atom.

Ex 9.8   The Hofmann elimination proceeds through an E2 mechanism with the loss of the more sterically accessible proton, often giving the less stable product. The Zaitsev elimination can proceed through either an E1 or E2 mechanism but gives the more stable alkene.

(a) Zaitsev elimination. The product is a trisubstituted alkene. A Hofmann elimination would produce a less stable monosubstituted alkene.

(b) Hofmann elimination. The bulkier base attacks at the less sterically hindered proton at C-1, producing a less stable alkene product.

(c) Zaitsev elimination. This hydrolytic elimination proceeds through a cation, deprotonation of which gives the observed disubstituted conjugated alkene.

(d) Hofmann elimination. Attack of hydroxide at the unsubstituted position takes place in this E2 elimination because of the requirement for an *anti*-periplanar transition state.

(e) Zaitsev elimination. Under these weakly basic conditions, the cleavage of the carbon–iodine bond to generate a carbocation is rate-determining. An E1 elimination produces the more stable alkene—here, trisubstituted.

Ex 9.9   The regiochemistry in an E2 elimination is controlled by the requirement for an *anti*-periplanar transition state. When both regioisomers are accessible through such a transition state, the one formed from the transition state with fewer destabilizing interactions is the dominant product.

(a) Because the methyl group at C-2 is *trans* to the bromine at C-1, elimination through an E2 pathway is blocked toward that side. However, a proton at C-5 can be lost with bromine in an *anti*-periplanar arrangement, producing 3-methylcyclopentene.

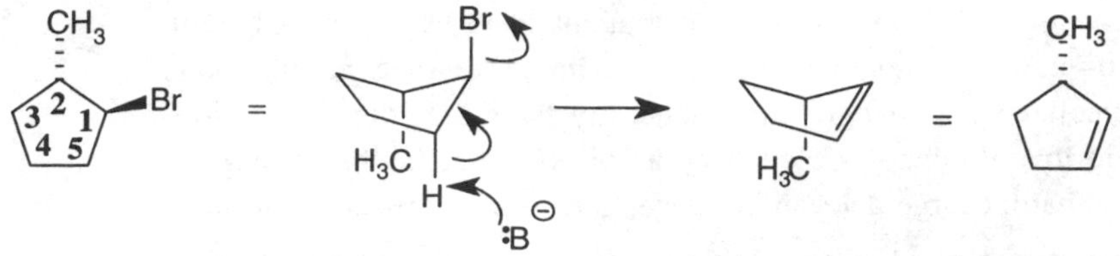

(b) Here, a hydrogen at C-2 occupies the position *trans* to bromine. Because elimination toward that site leads to the more stable product (1-methyl-cyclohexene), it is favored, if a small base is used, over the alternative dehydro-bromination in which a hydrogen at C-6 is removed.

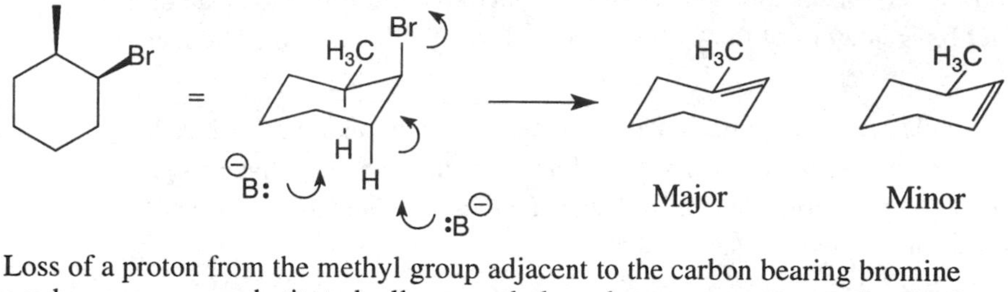

(c) Loss of a proton from the methyl group adjacent to the carbon bearing bromine produces a monosubstituted alkene and thus there are no stereochemical consequences for this elimination pathway.

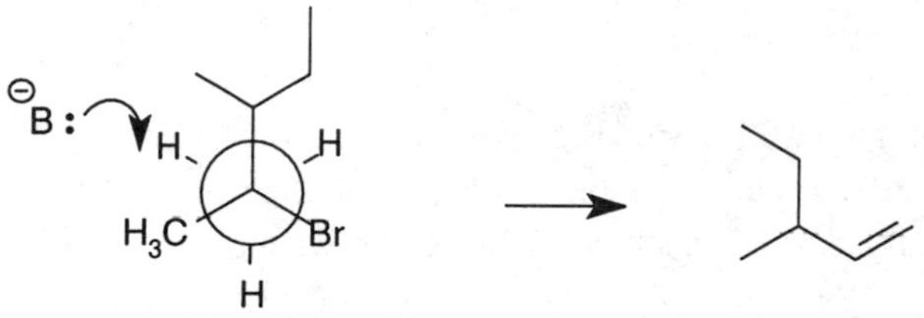

Loss of a proton from C-3 can occur via an *anti*-periplanar transition state through only one conformation, shown below, producing the *E* isomer of 3-methyl-2-pentene.

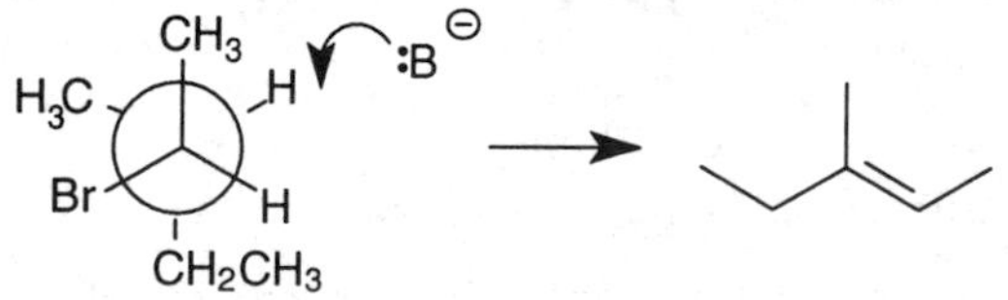

Ex 9.10 (a) This tertiary bromide is likely to react through an E1 elimination pathway. Deprotonation of the carbocation produced in the rate-determining step will take place so as to produce the more highly substituted alkene, 2-methyl-2-butene.

(b) As in part *a*, loss of bromide ion will produce an intermediate carbocation, from which deprotonation takes place to give the more stable isomer, here a tetrasubstituted alkene.

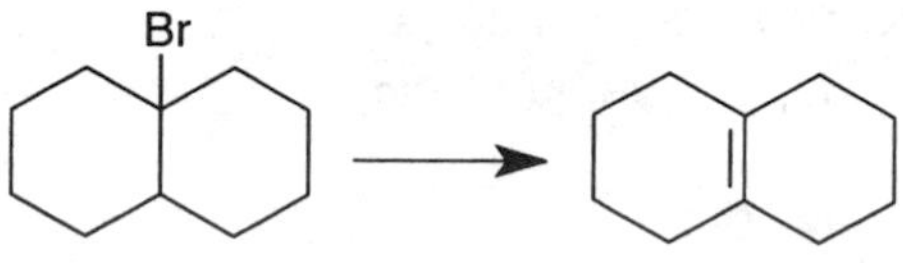

(c) In this secondary alkyl chloride, either an E1 or E2 elimination is possible. Under neutral conditions where no base assistance is possible, an E1 elimination is likely to occur. Here the originally produced carbocation can rearrange to a more stable tertiary carbocation upon shift of the tertiary hydrogen from C-3 to C-2. One of the C-2 protons is then lost to produce the most stable alkene, 2-methyl-2-butene. However, rearrangement of the cation does not affect the outcome of the reaction, as either cation will form the same, trisubstituted alkene.

(d) As in part *c*, this reactant is a secondary alkyl chloride from which E1 or E2 elimination can occur. Under neutral conditions where an E1 elimination proceeds through a carbocation intermediate, the more stable product, 1-methylcyclohexene, is favored over the less stable one.

Ex 9.11 This reaction begins by protonation of the hydroxyl group with acid. Loss of water takes place at the same time that a methyl group migrates from C-2 to C-1, leading to a stable tertiary carbocation. Deprotonation then leads to the observed product.

Ex 9.12 The hydroxide ion is a much poorer leaving group than is chloride ion, in part because chlorine is more electronegative than oxygen and can better accommodate the negative charge of the anion. As a result, hydroxide ion is a much stronger base than is chloride ion. In addition, chlorine is a third-row element and more polarizable than oxygen, reenforcing this trend. In addition, C—X bonds are weaker than C—O bonds (C—Cl 81, C—Br 68, C—I 51 versus C—O 86).

Ex 9.13 Because *cis*-stilbene is less stable than the *trans* isomer, its formation must be the result of a kinetically controlled reaction rather than a thermodynamic one. The preparation of *cis*-stilbene from the indicated optically active stereoisomer (i.e., it is not a *meso* compound) requires that the loss of the two bromines take place through an *anti*-periplanar transition state. A likely mechanism would involve insertion of Zn to one of the C—Br bonds, followed by concerted or bromide ion-induced cleavage through an *anti*-periplanar transition state.

Ex 9.14 A proton at C-1 can be oriented *anti*-periplanar to the bromine at C-2 so that elimination produces an allene, 1,2-butadiene. The protons on C-1 of 1,2-butadiene are vinylic and can be removed by treatment with a strong base such as sodium amide. The hybridized orbital bearing the lone pair in the resulting anion is coplanar with the *p* orbital of C-2 (that forming the C-2—C-3 π bond). Thus, the anion is delocalized over C-1 and C-3. Protonation on C-3 produces a terminal alkyne whose C-1 proton is more acidic than that of the allene.

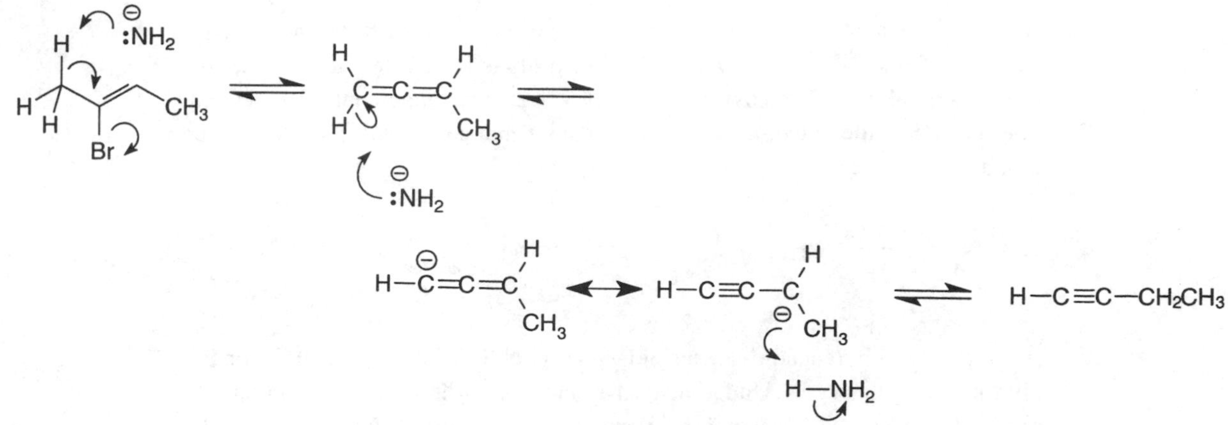

Concerted elimination of a C-3 hydrogen with bromine produces 2-butyne. Equilibration of this alkyne with 1,2-butadiene takes place by deprotonation at C-1 to form a delocalized anion, followed by reprotonation at C-3. Thus, all three species, 2-butyne, 1,2-butadiene, and 1-butyne are in equilibrium, but because the terminal alkyne is the most acidic, its anion dominates. 1-Butyne is therefore the major product obtained after neutralization.

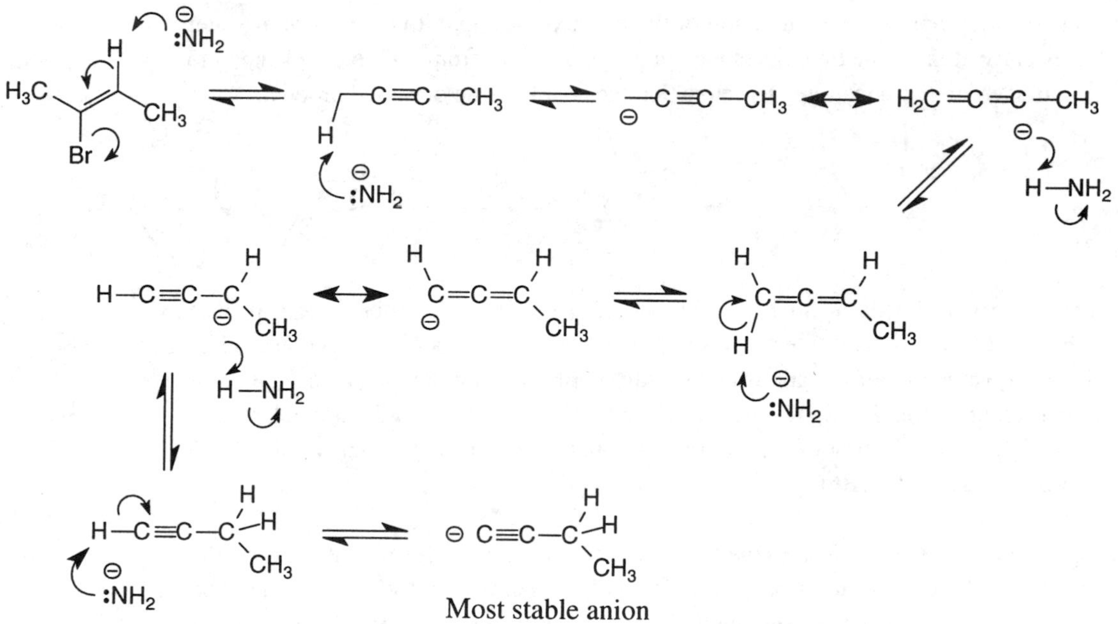

Most stable anion

Ex 9.15   Deprotonation at the *ortho* position of chlorobenzene and loss of chloride ion to form benzyne take place simultaneously in a concerted reaction. Because this alkyne is bent far from the ideal bond angle of 180° for *sp*-hybridized carbon atoms, it is highly reactive and is attacked by nucleophiles much more easily than is a normal alkyne. The phenyl anion produced by this attack by hydroxide ion is a highly unstable reactive intermediate, and protonation to produce phenol takes place almost as fast as it can collide with a water molecule.

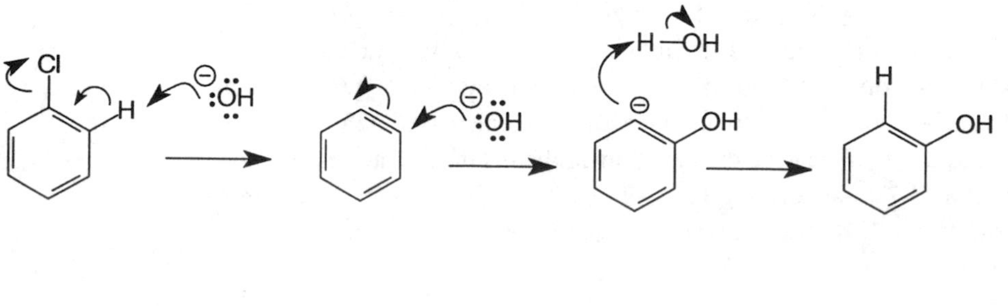

Alternatively, the formation of the C—OH and C—H bonds may take place simultaneously, as shown in the lower pathway. Because protonation of the anion would be so fast, there is little experimental data to permit a distinction between the stepwise and concerted pathways.

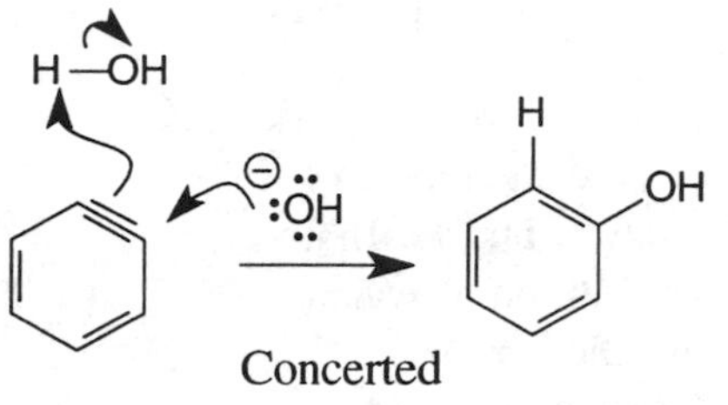

Concerted

Ex 9.16 Chromate oxidations in the presence of water convert primary alcohols to carboxylic acids and secondary alcohols to ketones. Tertiary alcohols and ketones resist oxidation, although under forcing conditions (high concentrations of oxidant at high temperatures), carbon–carbon bond fragmentation reactions begin to intervene. Aldehydes also are converted to carboxylic acids by aqueous chromate, but in the absence of water (for example, with CrO_3 and pyridine in CH_2Cl_2), the oxidation of an aldehyde is relatively slow.

(a) Cyclohexanone
(b) Butanoic acid
(c) No reaction
(d) 2-Butanone
(e) No reaction
(f) No reaction

Ex 9.17 The mechanism of the oxidation of the hydrate (geminal diol) of an aldehyde (lower reaction) is essentially identical with that for the oxidation of the alcohol to the aldehyde (upper reaction). Both oxidation steps include the formation of a chromate ester, followed by loss of a proton and a Cr(IV) species. [Remember that Cr(IV) is not a stable oxidation state of chromium: two Cr(IV) disproportionate to form Cr(III) and Cr(V).]

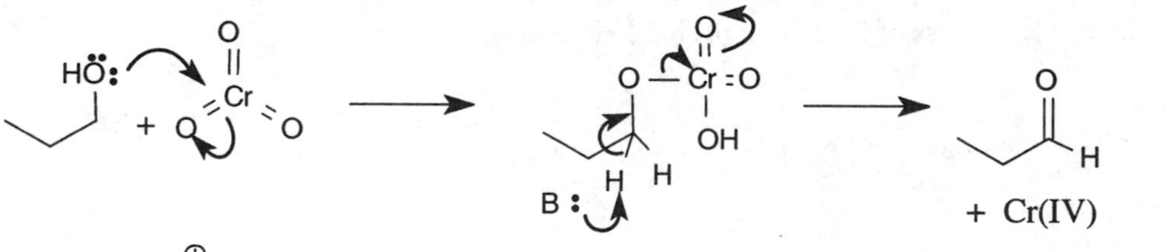

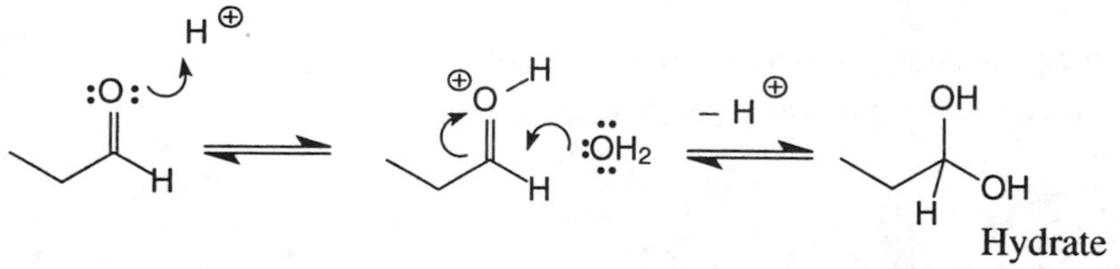

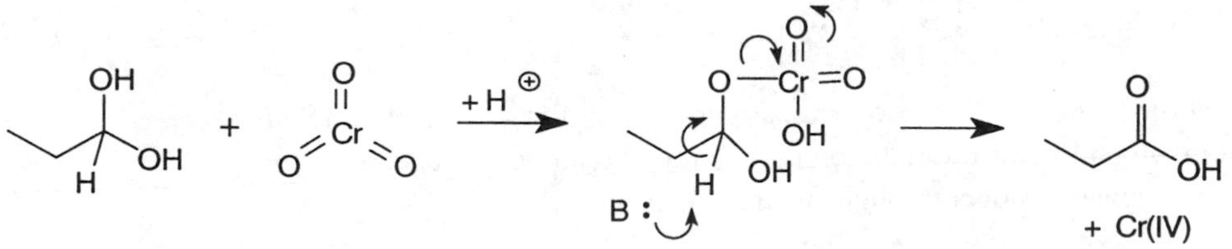

The mechanism of hydrate formation (center reaction) is treated in detail in Chapter 12.

Ex 9.18 Many routes are possible, but only one is shown here for each transformation.

(a) Free radical halogenation is selective and converts methylcyclohexane to 1-methylcyclohexyl bromide. Treatment with strong base effects dehydrohalogenation to form the elimination product. Only a small amount of the less stable exocyclic isomer is formed in this elimination.

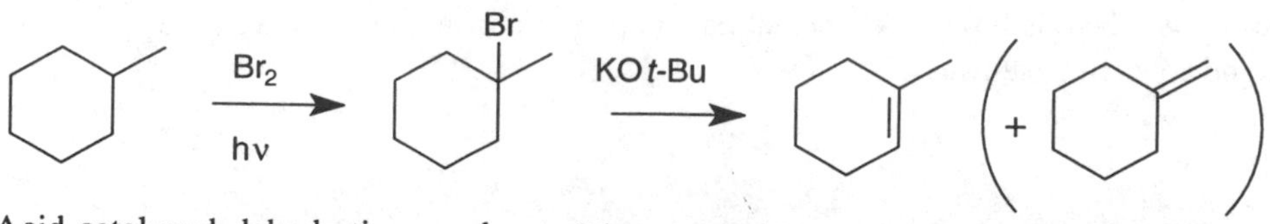

(b)   Acid-catalyzed dehydration produces 1-butene, acid-catalyzed hydration of which produces 2-butanol. Both steps require an acid catalyst, but the first effects the loss of water, whereas the second results in the addition of water. The difference is the result of concentration and conditions: the loss of water requires low initial concentrations of water and preferably some means, such as a Dean–Stark trap, for removal of water as it is formed. The addition of water is best effected with water as solvent, containing, if needed, a cosolvent such as dioxane or THF to help solubilize the starting alkene.

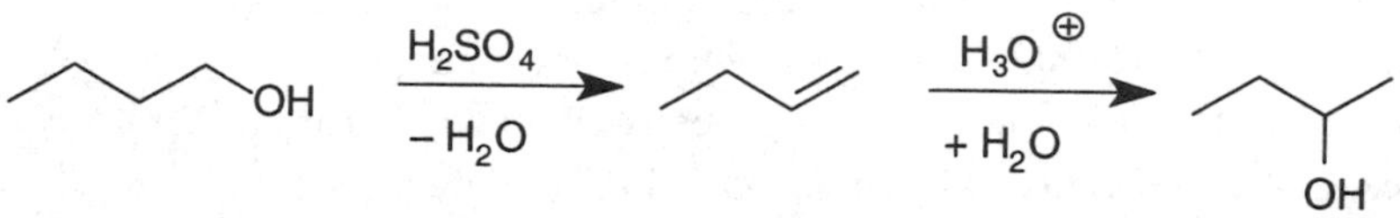

(c)   Free radical halogenation introduces a halide (for example, an alkyl bromide) that can be hydrolyzed to the corresponding alcohol under S_N1 conditions. Chromate oxidation of this secondary alcohol produces the ketone.

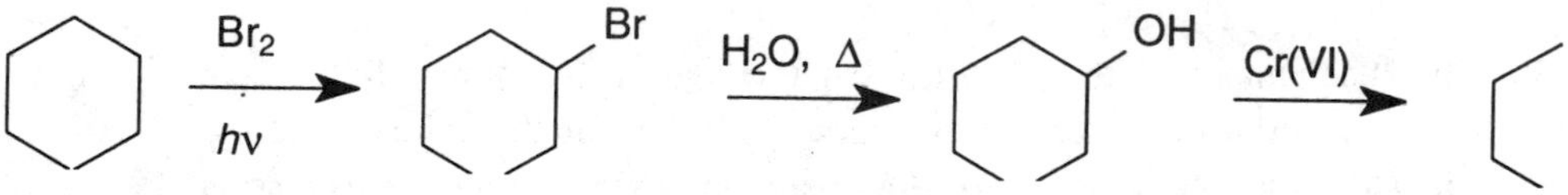

Answers to Review Problems

Pr 9.1   (a)   The reactant possesses neither a highly acidic hydrogen nor a structure leading to a particularly stable cation upon loss of bromide ion. Thus, elimination proceeds by concerted loss of HBr through an E2 mechanism.

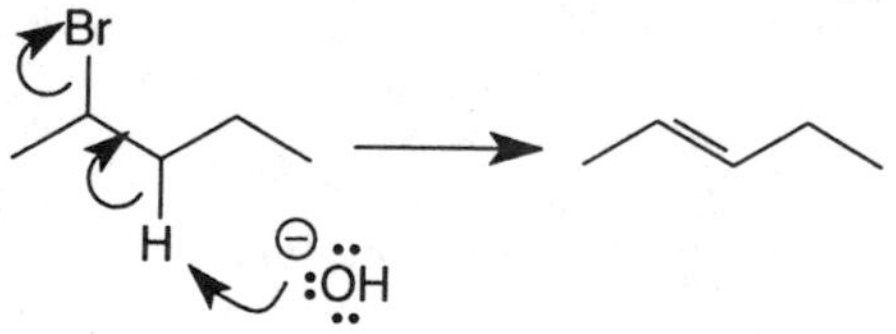

(b)   The loss of bromide ion produces a very stable tertiary benzylic cation. This stability influences the energy of the transition state, making cation formation feasible in this E1 reaction.

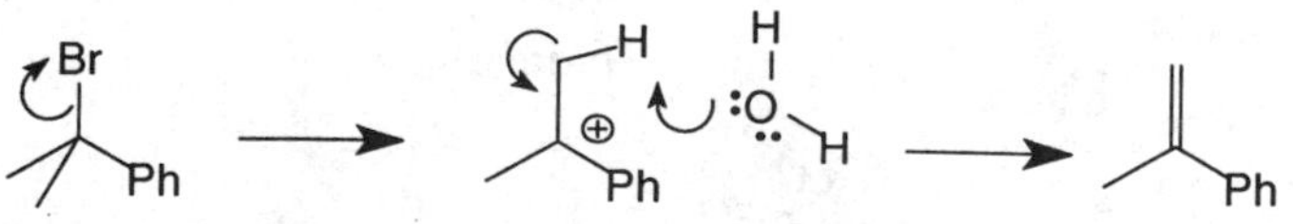

(c)   Deprotonation at the position α to the carbonyl group forms a stable enolate anion, from which loss of bromide ion in the rate-determining step leads to the observed conjugated product through an E1cB mechanism.

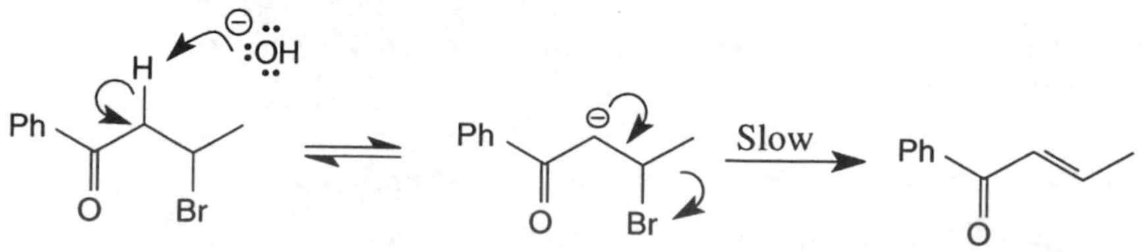

Pr 9.2   (a)   2-Bromo-3-methylbutane. E2 elimination from 1-bromo-3-methylbutane (left) can give only a Hofmann elimination product. There are two modes of elimination from 2-bromo-3-methylbutane (right): to form either 3-methyl-1-butene (Hofmann product) or 2-methyl-2-butene (Zaitsev product).

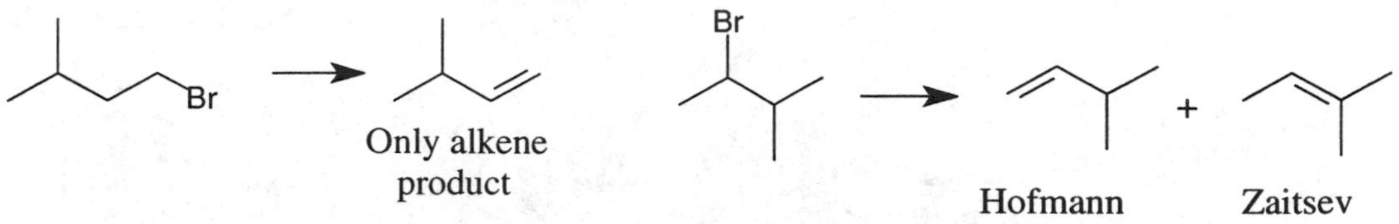

Only alkene
product

Hofmann   Zaitsev

(b)   2-Methyl-2-butanol. Alcohol dehydration and nucleophilic substitution take place by an E1/S_N1 mechanism under the conditions specified. Formation of the intermediate carbocation is therefore rate-determining. A tertiary cation is formed by protonation of the alcohol and loss of water from 2-methyl-2-butanol (left) and a secondary cation is produced in the same sequence with 3-methyl-2-butanol (right). Even though the cation formed from the latter alcohol can rearrange to a tertiary cation, the initial dehydration is slower.

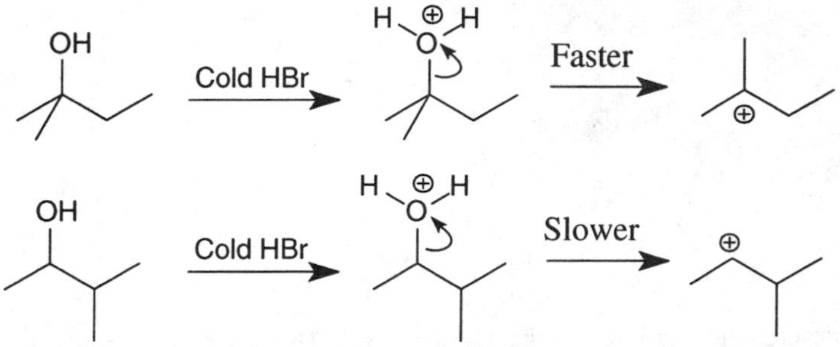

(c)   2-Methyl-1-phenyl-2-propanol. Dehydration from 2-methyl-1-phenyl-2-propanol (left) proceeds through an E1 reaction pathway in which a stable tertiary carbocation is formed. The formation of a trisubstituted conjugated double bond is favored because of the ease of deprotonation to form the conjugated product. The dehydration of 4-phenyl-1-butanol (right) requires concerted loss of H^+ and H_2O from the protonated alcohol to avoid the formation of an unstable primary carbocation, with bromide ion acting as a weak base (path a) to initiate deprotonation. Because bromide ion is a better nucleophile than a base, direct S_N2 displacement by bromide ion (path b) is more likely.

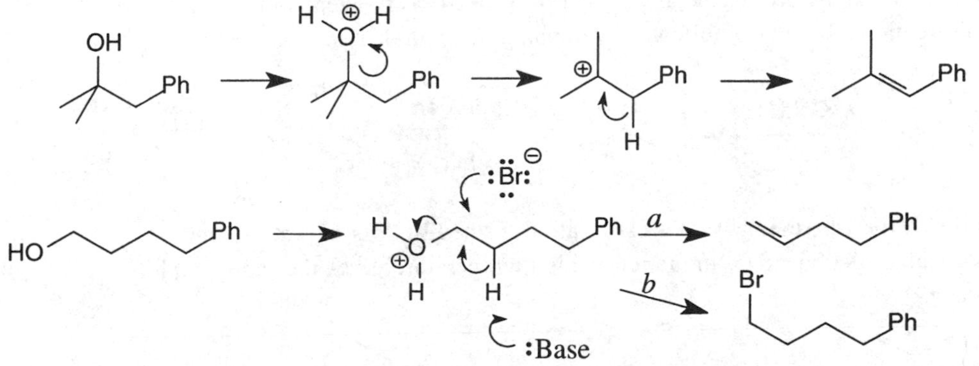

(d)   2,3-Dimethylcyclohexanol. The cation formed by dehydration of 2,3-dimethyl-cyclohexanol (left) has two different adjacent hydrogens that can be lost in producing two isomeric alkenes. The cation formed from 1,4-dimethylcyclo-hexanol (right) is symmetrical, and the adjacent hydrogens are equivalent. Hence, one alkene is produced.

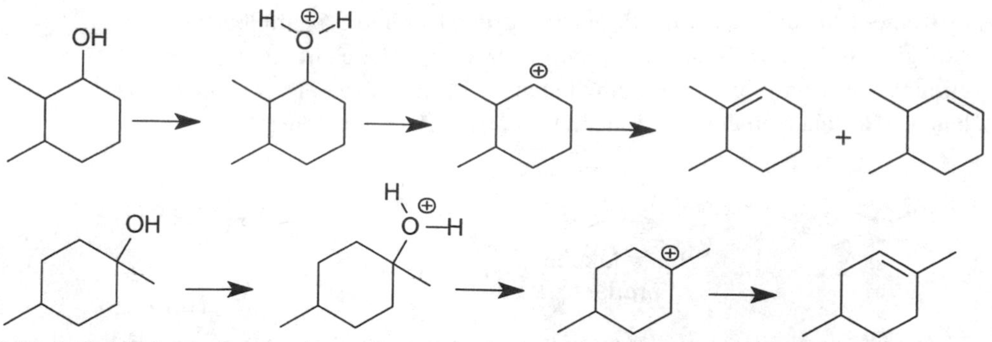

(e) 4-Methyl-2-chloropentane. A concerted elimination from 4-methyl-2-chloro-pentane can give either Hofmann (4-methyl-1-pentene) or Zaitsev (*E*- or *Z*-4-methyl-2-pentene) product, but a concerted elimination from 2,4-dimethyl-1-chloropentane can give only a single product (2,4-dimethyl-1-pentene).

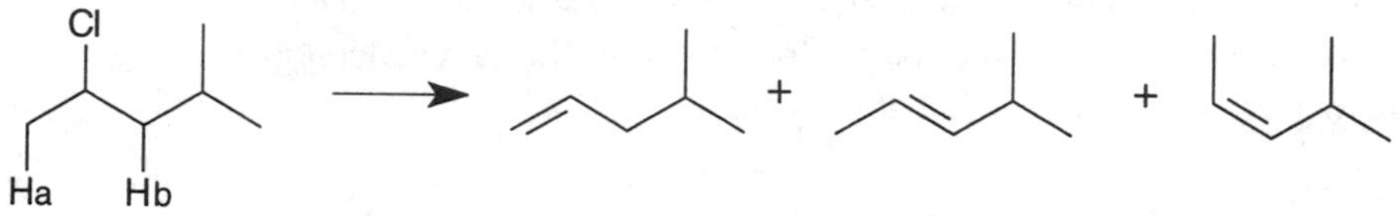

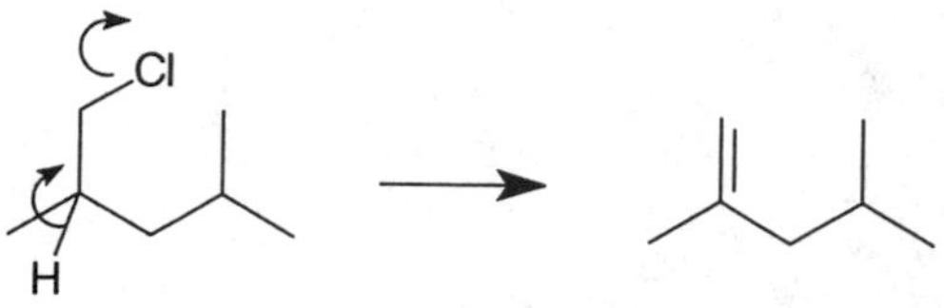

Pr 9.3 (a) These are conditions for heterolytic cleavage of the C—Br bond. The resulting cation can be either deprotonated or captured by water. At high temperature, elimination is favored, producing the more stable alkene.

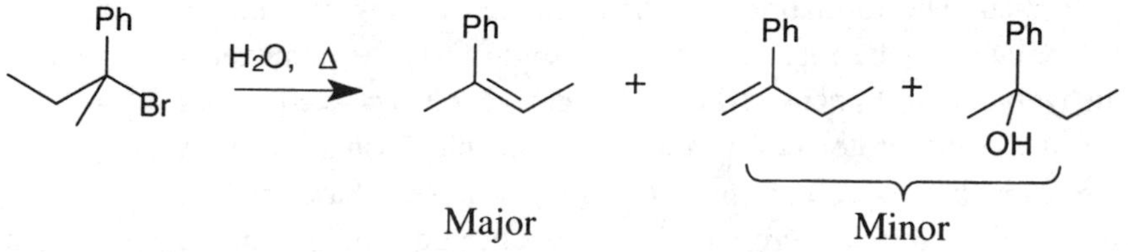

(b) At low temperature, the rate of deprotonation of the cation formed by the loss of water from the protonated alcohol is low. Instead, rearrangement to the more stable cation takes place, followed by trapping by bromide ion.

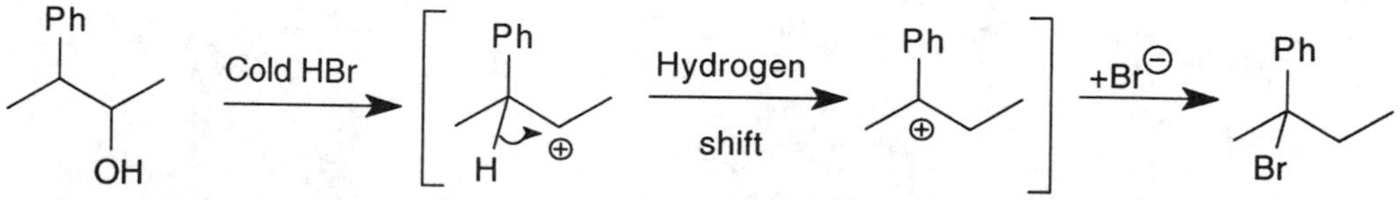

(c) Through the intermediacy of a carbocation formed by loss of water from the protonated alcohol, E1 elimination at high temperature gives the more stable alkene.

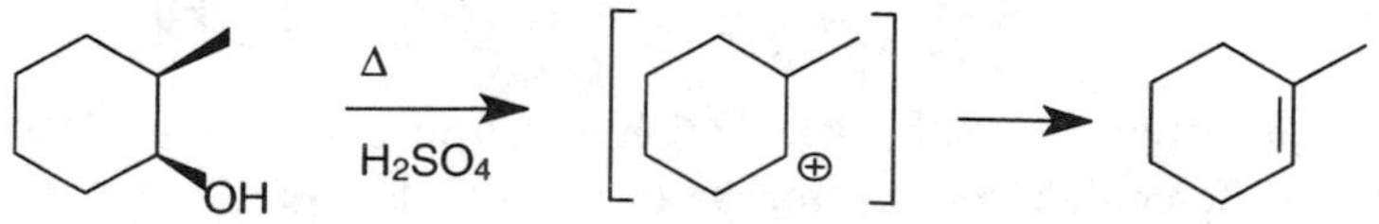

(d) These are conditions that favor both E2 elimination and S_N2 substitution. The proton at the unsubstituted position, rather than the proton on the carbon bearing the methyl group, must be removed by base to satisfy the requirement for an *anti*-periplanar transition state in the elimination. The formation of the ether through a Williamson ether synthesis also takes place, although the presence of a methyl group on the adjacent carbon makes the transition state somewhat crowded.

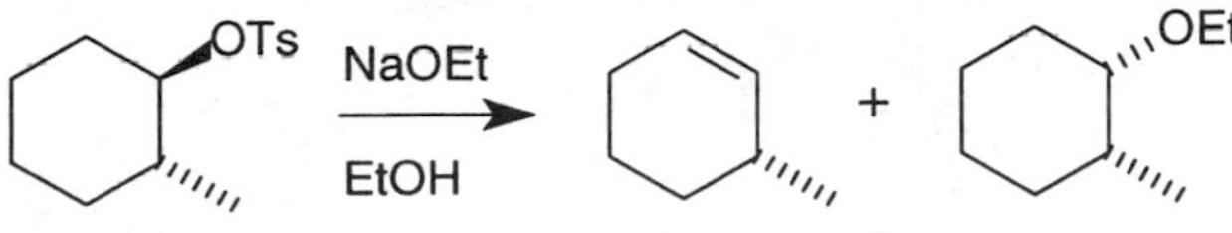

(e) Sodium hydroxide is a better base than a nucleophile, and its use favors elimination over substitution. Because the base is small, it can attack at C-1 or C-3 without difficulty. Because attack at C-3 produces the more stable alkene, the *trans*-2-alkene is the favored product.

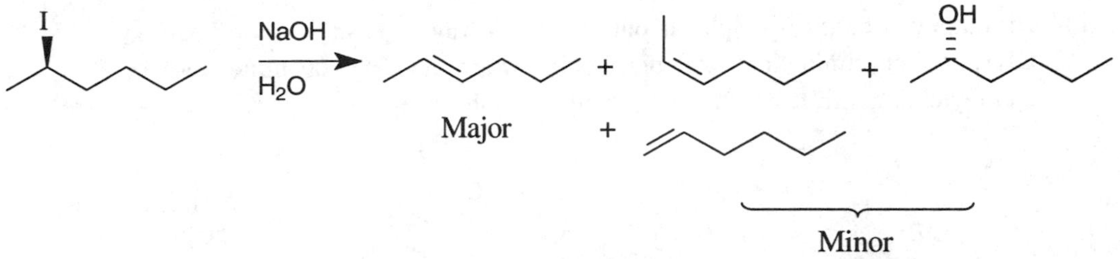

(f) The proton α to the carbonyl group is clearly the most acidic and is attacked preferentially by base, either by pre-equilibration to an enolate anion in an E1cB mechanism or directly to alkene product through an E2 route.

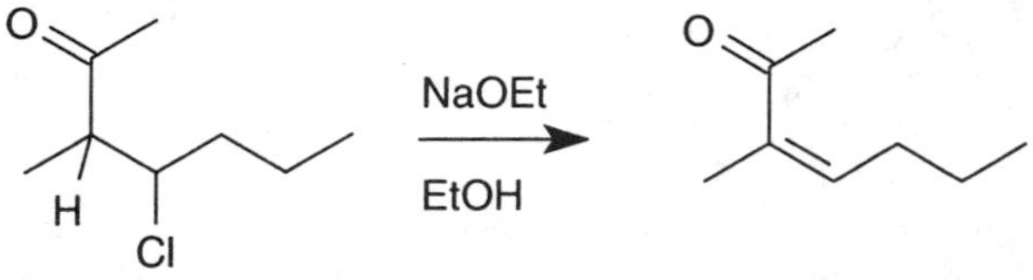

Pr 9.4   For each of these reactions, a choice must be made between a Hofmann elimination to form a 1-alkene or a Zaitsev elimination to form a 2-alkene.

(a) With a poorer leaving group, the C—X bond is less stretched in the transition state of the reaction of butyl chloride, and the acidity of the adjacent C—H bond is more important. This leads to more of the Hofmann elimination product.

(b) With a better leaving group, the C—X bond is more stretched in the transition state, placing more partial positive charge at the 2-position. Because this reaction is then more like an E1 elimination, product stability is more important and more Zaitsev product is obtained.

(c) These are E1 conditions—namely, a poor nucleophile at high temperature with a reactant possessing a leaving group at a tertiary position. E1 conditions give more of the more stable Zaitsev product. than do E2 conditions

(d) With a bulkier base, deprotonation in the E2 elimination is favored at the less sterically hindered site, producing more of the Hofmann product.

(e) With a smaller base, deprotonation is less sensitive to steric hindrance and, hence, more sensitive to product stability. Under these conditions, more Zaitsev product is obtained.

Pr 9.5 (a) In the Newman projection obtained by viewing down the C-2—C-1 bond, iodine is *anti*-periplanar to methyl and concerted elimination is blocked. In the view down the C-6—C-1 bond, hydrogen is *anti*-periplanar to iodine so that elimination takes place smoothly, producing 3-methylcyclohexene.

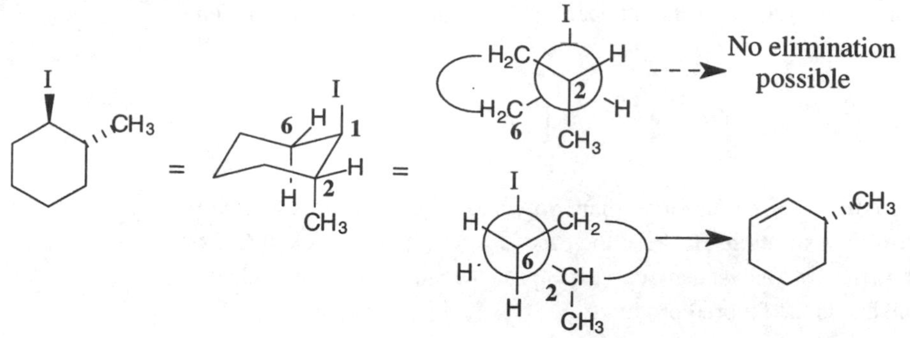

(b) Bromine can be *anti*-periplanar only to the hydrogen on the carbon bearing CH$_3$, so that elimination takes place only toward the CH$_3$-substituted carbon atom, yielding 3-trideuteromethyl-1-methylcyclohexene.

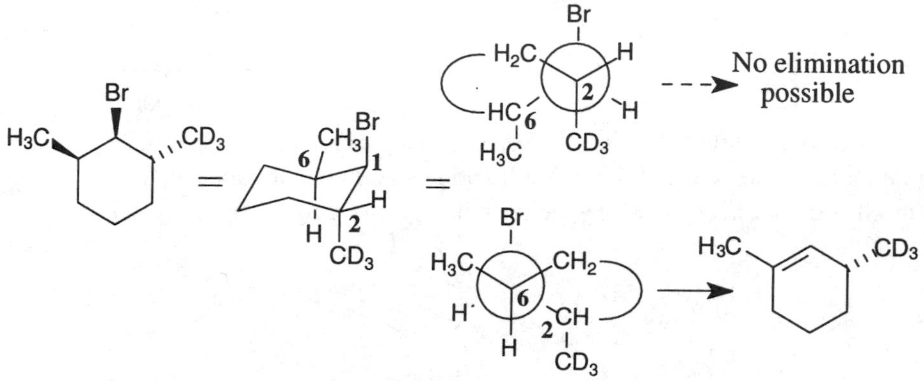

Pr 9.6 **(a)** This is an acid-catalyzed E1 elimination in which deprotonation of the tertiary carbocation gives a highly substituted alkene. All deuteriums present in the reactant are retained in the major product.

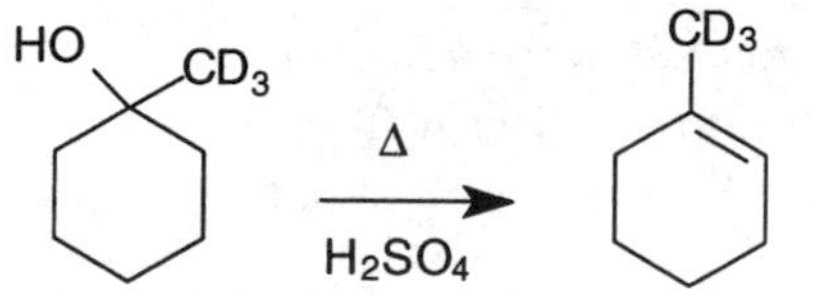

(b) This problem is identical to that in part *a*, except with a different deuterium labeling pattern. The same route described in part *a* leads to an alkene with one fewer deuterium than in the starting material.

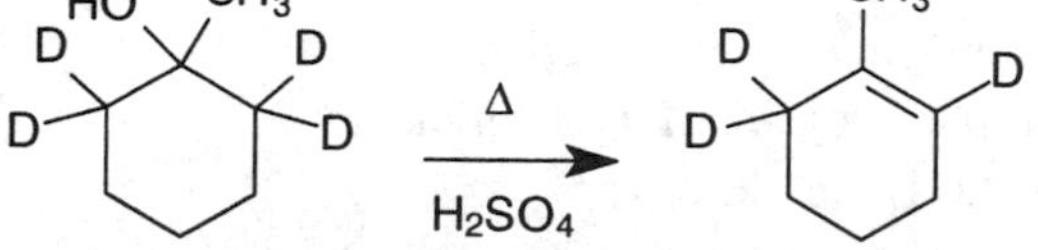

(c) In this E2 reaction, iodine can be oriented *anti*-periplanar to hydrogen, but not to deuterium. Thus, HI is formally lost instead of DI, giving a cyclohexene with two deuteriums. Substitution at this secondary site also produces some ether by a Williamson ether synthesis.

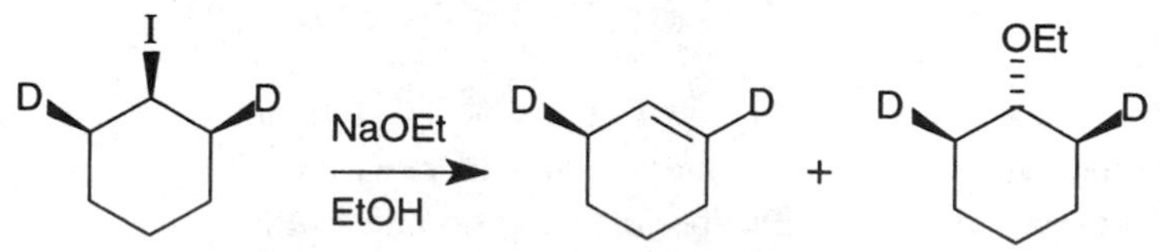

(d) The most acidic position is that α to the carbonyl group. Through an E1cB mechanism, the α-deuterium is removed to form an anion from which iodide is expelled to produce the conjugated enone product. (Williamson ether product is likely to be less important here than in part *c* because of steric blockage of the back face of the C—I bond by the adjacent acetyl group.)

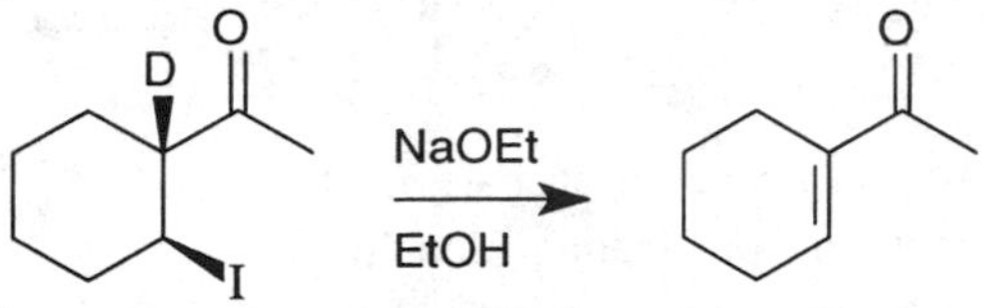

(e) Elimination of HI gives 3,5-dideuterobenzyne, which is trapped by ammonia to give the two regioisomers shown, with both deuteriums retained in the product.

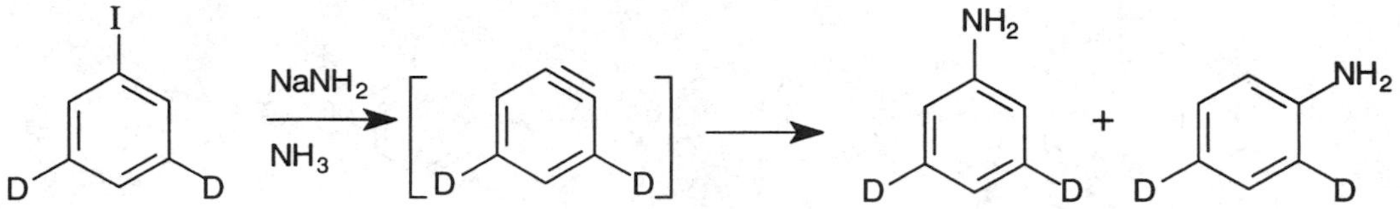

(f) Elimination of DI produces 3-deuterobenzyne, which is rapidly hydrated to 2-deuterophenol and 3-deuterophenol from the two possible orientations for water addition to the triple bond.

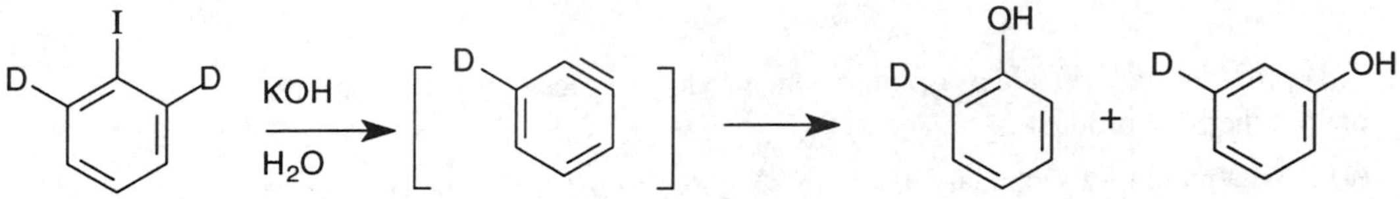

(g) Treatment with cold HBr constitutes reaction conditions in which S_N1 reactivity competes with elimination. The secondary cation initially formed by loss of water from the protonated alcohol rearranges by the shift of a deuterium to form a tertiary benzylic carbocation that is trapped by bromide ion. Although deprotonation of the rearranged cation to form elimination product also takes place, bromide ion is a good nucleophile and a relatively poor base, making the elimination product less important than if the reaction had been conducted at a higher temperature.

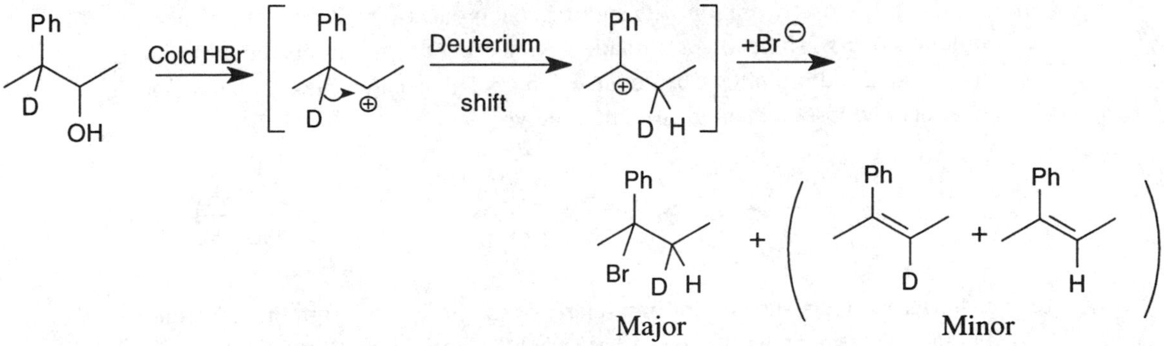

Pr 9.7   (a) This S_N1 substitution proceeds through a carbocation intermediate formed by protonation of the alcohol OH to an oxonium ion, from which water is lost. The resulting carbocation rearranges by a 1,2-hydrogen shift to form a much more stable tertiary benzylic cation, which is then trapped by bromide ion.

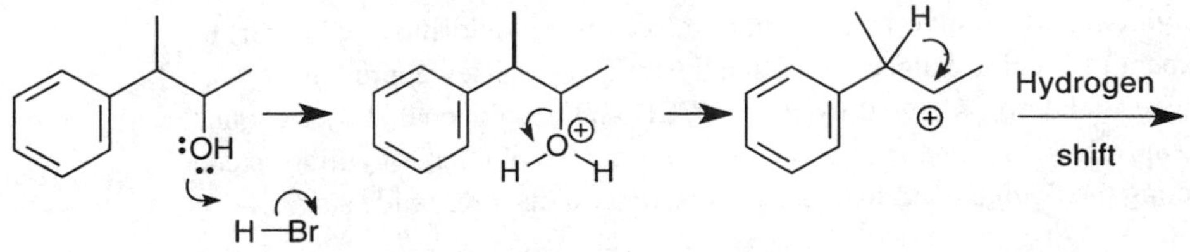

(b) This E1 reaction starts in the same way as the S_N1 reaction in part *a*. Thus, protonation of the alcohol again gives an oxonium ion, from which loss of water produces a secondary carbocation. This time, however, a methyl group migrates to produce the tertiary benzylic cation. Deprotonation under these elimination conditions provides the observed product, a conjugated alkene.

Pr 9.8 This problem requires a description of the product expected at each step in order to predict the final product.

(a) Treatment of a secondary alcohol with phosphorus tribromide converts it to the corresponding alkyl bromide (Chapter 8). Base-catalyzed elimination with a small base such as sodium ethoxide takes place with a predominantly Zaitsev orientation to form *trans*-2-butene as the major product.

(b) Acid-catalyzed dehydration with sulfuric acid produces the Zaitsev elimination product, *trans*-2-butene. Bromination with bromine in carbon tetrachloride forms the 2,3-dibromide. Then double dehydrohalogenation takes place upon treatment with sodium amide to form 2-butyne.

(c) Cold HBr effects substitution in preference to elimination, forming 2-bromobutane. Treatment with a large base such as potassium *t*-butoxide gives a Hofmann elimination product: namely, 1-butene.

(d) Thionyl chloride converts secondary alcohols to chlorides (Chapter 8). Treatment of 2-chlorobutane with a small base gives Zaitsev elimination. *trans*-2-Butene is then brominated by Br_2 in CCl_4 and dehalogenated by treatment with zinc. (Notice that the last two steps do not accomplish anything except protecting the double bond from further reactivity at the dibromide stage).

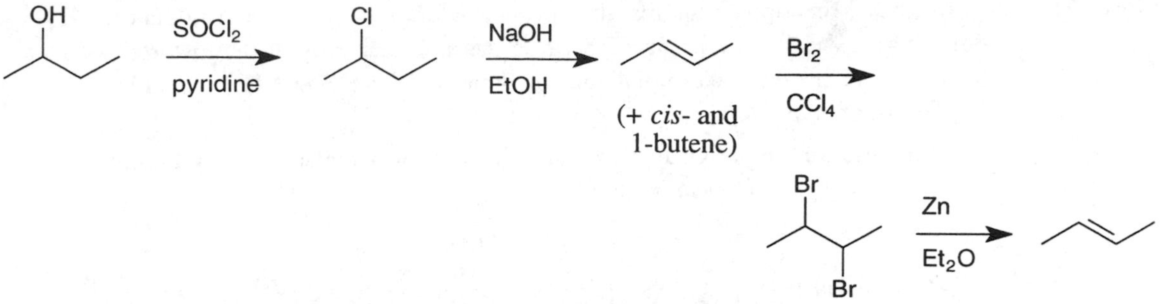

Pr 9.9   The color of a solution of chromic acid changes from bright red-orange to deep green as it is consumed in effecting oxidation reactions. Chromic acid oxidizes primary and secondary alcohols and aldehydes; it does not oxidize tertiary alcohols or ketones at room temperature.

The Lucas reagent (Chapter 3) is useful in classifying alcohols as primary, secondary, or tertiary. It employs an S_N1 substitution to produce the corresponding alkyl chloride, which is generally insoluble in the aqueous solvent in which the reagent is delivered. One therefore sees phase separation for a positive test, with two layers appearing immediately for a tertiary alcohol, within a minute or so for secondary alcohols, and much more slowly, if at all, for primary alcohols.

This question asks you to classify each substrate according to these tests.

(a)   This is a secondary alcohol, which causes a color change with chromic acid and a moderately fast phase separation with the Lucas reagent.

(b)   This reagent is also a secondary alcohol, and gives the same test results as in part *a*.

(c)   As in parts *a* and *b*, this cyclic alcohol is secondary and gives the same test results.

(d)   This tertiary alcohol gives no color change with chromic acid, but an immediate reaction with the Lucas reagent.

(e)   This is a primary alcohol, which causes a color change on treatment with chromic acid, but no reaction with the Lucas reagent.

(f)   As in part *d*, this tertiary alcohol gives the same test results.

(g)   Because ketones resist both oxidation and substitution, no positive test is indicated with either chromic acid or the Lucas reagent.

(h)   Aldehydes are oxidizable and therefore give a color change with aqueous chromic acid. They resist substitution, however, and so give negative results with the Lucas reagent.

Pr 9.10   There are many possible answers to this problem. The three listed here use reagents described in this chapter. All of the reactions shown here take place by dehydrobromination: 1) of a vicinal dibromide, 2) of a geminal dibromide, and 3) of a vinyl bromide.

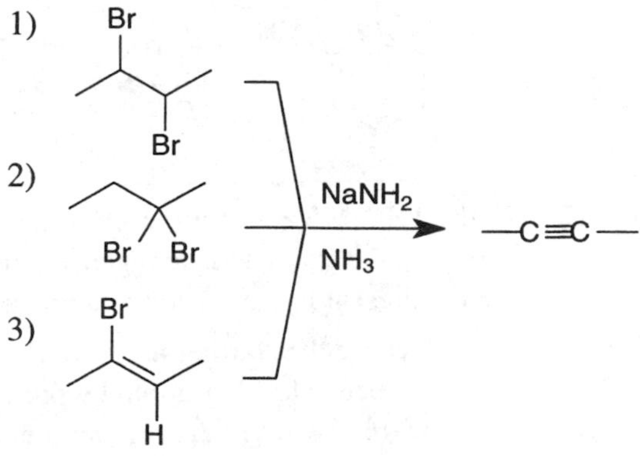

Pr 9.11 Each of these reactions requires the formation of a benzyne intermediate. Regioisomeric products are produced because the high reactivity of the distorted triple bond in a benzyne makes it possible for H and X (X = OH or NH$_2$) to add in either possible orientation.

(a) In this reaction, the original *para* iodine substituent is replaced by an amino group at the *para* and *meta* positions.

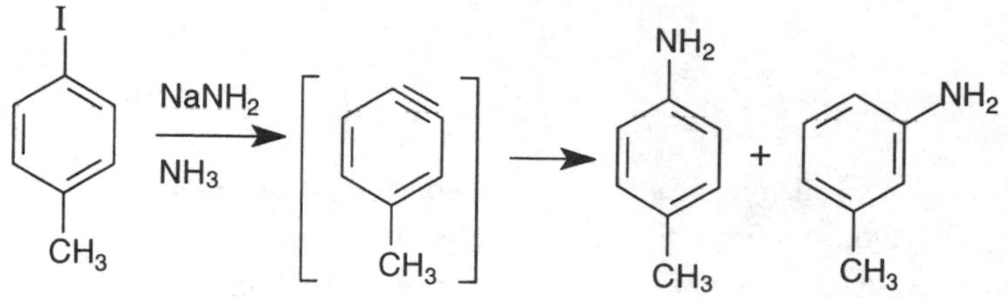

(b) Except for a small isotope effect, it is equally probable to lose HI or DI to produce two isomeric benzynes. Four possible phenols are produced by hydration of these two intermediates.

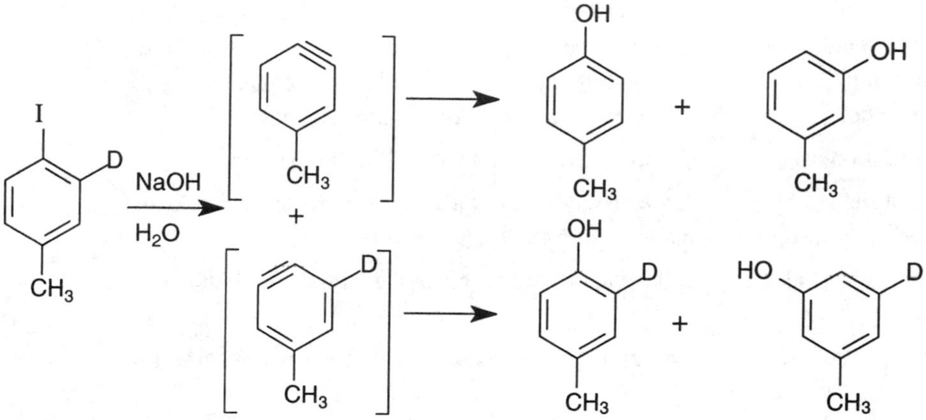

(c) Regardless of whether the 2- or 6-proton is lost, the same benzyne is formed. It can be captured by ND$_3$ in either orientation.

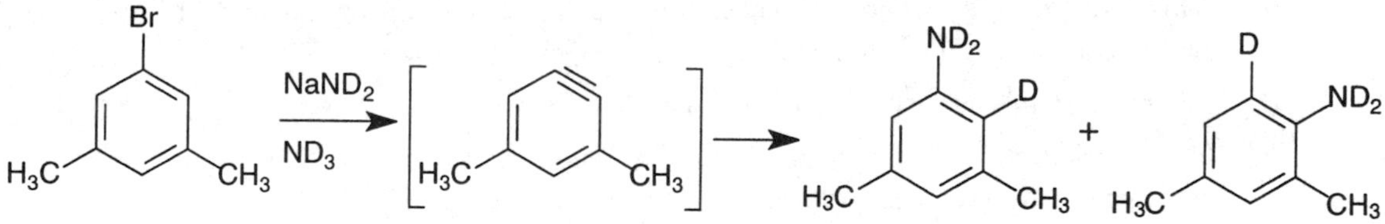

(d) Because one *ortho* position is blocked by a methyl group, elimination can take place in only one direction, producing one benzyne that is hydrated to either of the observed phenols.

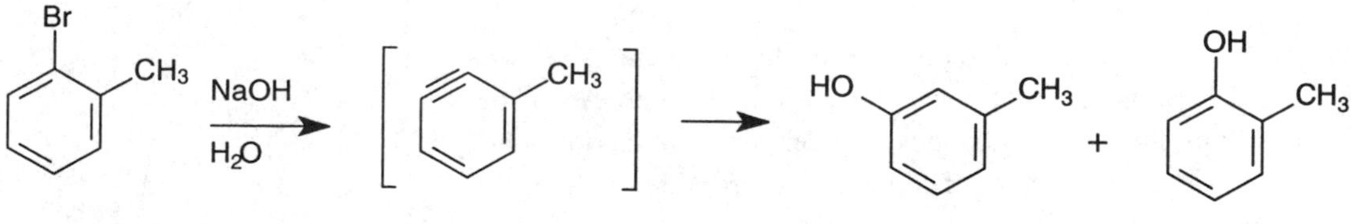

Pr 9.12 In the ^{1}H NMR spectrum each of the possible elimination products, vinyl protons appear in the characteristic region downfield from the aliphatic signals. Such protons are not present in the substitution product.

(a) Because the starting material is an alcohol, begin by converting the OH group to a better leaving group by protonation. Water is then lost from the oxonium ion in the rate-determining step. The resulting cation is either trapped by

bromide ion (path *a*) to give the observed substitution product, deprotonated at
C-1 (path *b*) to produce 1-alkene, or at C-3 (path *c*) to produce 2-alkene.

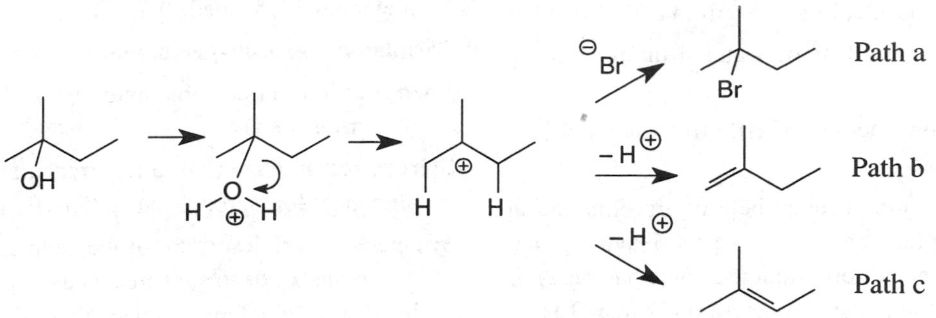

Two elimination products are therefore possible: 2-methyl-1-butene, the
Hofmann elimination product, and 2-methyl-2-butene, the Zaitsev elimination
product. In the first, there are two vinyl protons, and in the second, only one.
Therefore, the integration of the vinylic signal in the ^{1}H NMR spectra of the
two isomeric products is different.

(b) With a potential leaving group at a secondary position, either an S_N1 or an S_N2
(or E1 or E2) mechanism is possible. (This is as complicated as it gets!)
Because there is relatively little steric hindrance at C-2, the concerted
displacement (path *a*) is likely. In competition with this pathway, an E2
elimination can take place by the attack of methoxide as a base on one of the
hydrogens at C-3 to yield *cis-* and *trans*-2-butene, and at C-1 (path *b*) to yield
1-alkene. The choice between the geometric isomers of 2-butene depends on
which of the prochiral protons is removed, as shown in the Newman projections
shown directly below each possible regioisomer.

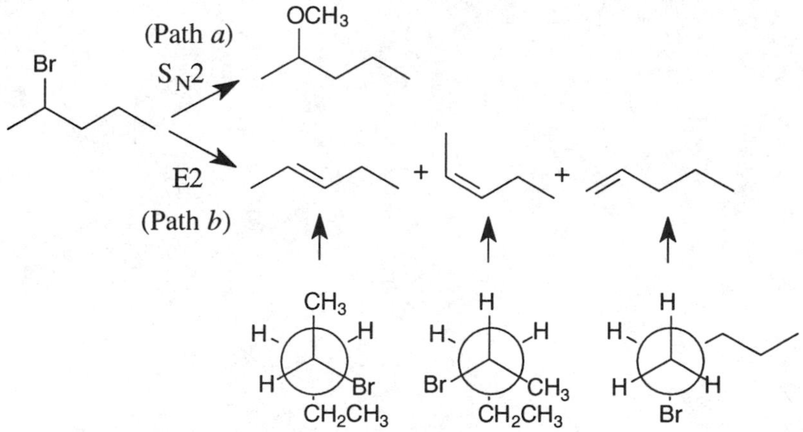

There are thus three possible olefinic products, *cis-* and *trans*-2-pentene, and 1-
pentene 1-Pentene can be differentiated from the other two by integrating the
vinylic region (three protons vs. two). *cis*-2-Pentene has a much smaller vinylic
coupling constant (4–6 Hz) than does the *trans* isomer (12–14 Hz).

Important New Terms

*Anti***-periplanar**: descriptor of the geometric relationship in
which the bonds to substituents on adjacent atoms of a σ
bond are coplanar, with a dihedral angle of 180°; the
preferred geometry for an E2 elimination (9.2)

Aryl halide: a functional group in which a halogen is
attached to an arene ring (9.8)

Benzyne: C_6H_4; an unstable species with a triple bond in a
benzene ring; a highly reactive ring compound related to
benzene in having two hydrogen atoms removed from
adjacent ring positions (9.8)

Chromate oxidation: the oxidation with Cr^{6+}, often of
alcohols to aldehydes, ketones, or carboxylic acids, that

is accompanied by a color change of the inorganic reagent from red-orange to green (Cr^{3+}) (9.9)

Dehalogenation: the formal loss of X_2 from a dihalide (9.3)

Dehydration: the formal loss of water, usually from an alcohol (9.2)

Dehydrobromination: the loss of HBr from an alkyl bromide (9.3)

E1cB reaction: a unimolecular, heterolytic elimination reaction taking place by the loss of the leaving group from the deprotonated form (anionic conjugate base) of the neutral substrate in the rate-determining step (9.1)

E1 reaction: a unimolecular, heterolytic elimination reaction taking place by breaking of the carbon-leaving-group σ bond, with the formation of a carbocation, in the rate-determining step (9.1, 9.4)

E2 reaction: a bimolecular, concerted elimination reaction in which bonds to both the proton and the leaving group are broken in the rate-determining step (9.1)

Elimination: a chemical reaction in which two groups on adjacent atoms are lost as a double bond is formed (9.1)

Hofmann elimination: a kinetically controlled elimination reaction in which the less substituted alkene is preferentially formed (9.3)

Periplanar: *see Anti*-periplanar and *Syn*-periplanar (9.3)

Phenyl anion: an unstable anion formed by deprotonation of benzene (9.8)

Stereoelectronic control: a requirement for precise orbital alignment for a proposed reaction (9.3)

***Syn*-periplanar**: descriptor of the geometric relationship in which the bonds to substituents on adjacent atoms of a σ bond are coplanar, with a dihedral angle of 180°; a possible geometry for an E2 elimination, although less preferred than the *anti*-periplanar alignment (9.3)

Vinyl halide: a functional group in which a halogen is attached to an alkenyl carbon (9.7)

Zaitsev's rule: an empirical prediction of preferential formation of the thermodynamically more stable, more highly substituted regioisomer in an elimination reaction (9.3)

Addition to Carbon–Carbon Multiple Bonds

10

Key Concepts

Mechanism of stepwise electrophilic addition to alkenes

Synthesis of alkyl halides by hydrohalogenation of alkenes

Synthesis of alcohols by alkene hydration

Ion pairing between carbocations and counterions

Factors influencing regiochemical control in alkene hydration: Markovnikov's rule

Skeletal rearrangements in hydrohalogenations and hydrations of alkenes

Regiochemical control in hydrohalogenation of alkynes

1,2 and 1,4 (conjugate) addition

Stereochemical control in hydration reactions: random in electrophilic additions; *syn* addition in hydroboration–oxidations

Methods for reversing Markovnikov regiochemistry: radical hydrohalogenation and hydroboration–oxidation

Blocking cationic rearrangements through oxymercuration–demercuration

Synthesis of vicinal dihalides by halogenation of alkenes

Stereochemical control (*trans* addition) in halogenation through intermediate cyclic halonium ions

Halogenation–dehalogenation as an alkene protecting group

Bromination as a chemical (color) test for multiple bonds

Synthesis of cyclopropanes by carbene addition in the Simmons–Smith reaction

Stereospecificity in singlet carbene additions

Synthesis of epoxides by peracid oxidation of alkenes

Oxidative degradation as a tool for structural simplification in natural product chemistry

Ozonolysis as a route for fragmentation of alkenes to the related carbonyl compounds

Permanganate oxidation as a chemical (color) test for alkenes and oxidizable groups

Cationic polymerization: using carbocations as electrophiles

Radical addition to alkenes: cyclization and polymerization

Answers to Exercises

Ex 10.1 This exercise requires you to identify the bonds of the starting materials that are broken and the bonds of the products that are made in the reaction. The difference in the total bond strengths of the bonds broken and the bonds made = $\Delta H°$.

(a)

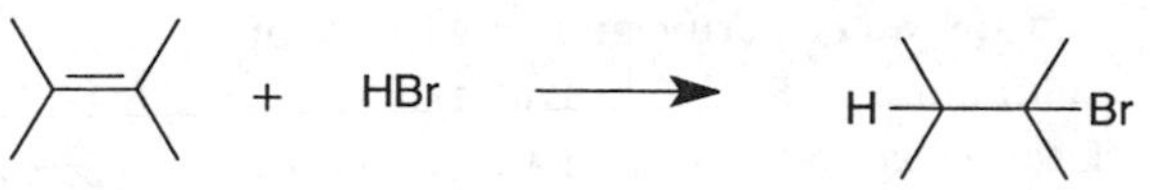

Broken			Made		
C—C π	63		C—H	99	
H—Br	87		C—Br	68	
	150			167	

$$\Delta H° = 150 - 167 = -17 \text{ kcal/mole}$$

173

(b)

Broken
C—C π   63
O—H   111
            174

Made
C—H   99
C—O   86
            185

$$\Delta H° = 174 - 185 = -11 \text{ kcal/mole}$$

(c)

Broken
C—C π   63
Br—Br   46
            109

Made
C—Br   68
C—Br   68
            136

$$\Delta H° = 109 - 136 = -27 \text{ kcal/mole}$$

Ex 10.2 In each of these reagents, identify the most polar bond and determine which part is (or becomes) partially positively charged (the electrophile) and which is (or becomes) partially negatively charged (the nucleophile).

(a) There is only one bond in HBr. It is highly polar, with electron density shifted toward bromine. Thus, H^+ is the electrophilic part.

(b) In benzenesulfonic acid, the most polar bond is the O—H bond of the sulfonic acid. Like a mineral acid, such as HBr considered in part *a*, this compound is a strong acid, and the O—H bond is polarized so that hydrogen bears partial positive charge and oxygen bears partial negative charge. Therefore, benzensulfonic acid also delivers H^+ as an electrophile.

(c) The direction of polarity in acetic acid (CH_3CO_2H) is the same as that in asulfonic acid, with the O—H bond polarized so as to produce H^+ and ^-OCOR. However, acetic acid has a higher pK_a than a typical sulfonic acid. Therefore, it is a weaker acid and a poorer source of protons.

(d) Molecular chlorine is itself nonpolar, but the electrons in the σ bond connecting the chlorine atoms are highly polarizable. Interaction of molecular chlorine with an electron-rich, nucleophilic species (like the π cloud of an alkene) causes a shift of electron density so that the chlorine atom closer to the nucleophile becomes electrophilic as the remote chlorine departs as chloride, with the electrons originally in the σ bond.

(This reaction occurs when Cl_2 is added to an aqueous NaOH solution, producing NaOCl, sodium hypochlorite, found in common bleach as well as in swimming pools that use chlorine for water purification. Reaction of NaOCl with amines produces chloramines, R_2N–Cl, and it is these compounds that cause burning of the eyes and that give the water its "chlorine" smell.)

(e) Because of the larger size of bromine, the σ bond in molecular bromine is even more polarizable than that connecting chlorine atoms in Cl_2 (discussed in part *d*). As with molecular chlorine, when molecular bromine approaches a species with surplus electron density, the σ bond electrons shift toward the remote bromine atom as the nearer bromine becomes electrophilic. Because the bromine–bromine bond is more polarizable and weaker than the chlorine–chlorine bond, Br_2 is a more reactive electrophile than is Cl_2.

(f) Both bonds in H—O—Cl are polar, and this molecule is a weak acid (H^+ and ^-OCl). However, the bond between hydrogen and oxygen is much stronger than that between oxygen and chlorine (111 kcal/mole versus 59 kcal/mole). Therefore, the O—Cl bond is broken when HOCl reacts with an alkene. Which atom, oxygen or chlorine, is the electrophilic end of the O—Cl bond is determined by the higher electronegativity of oxygen. Thus, as a nucleophile approaches HOCl, the σ electrons of the O—Cl bond shift more toward the more electronegative oxygen atom, making chlorine electrophilic. Notice that this reactivity is contrary to that expected from the stability of Cl^- and OH^-, as judged by the pK_as of their conjugate acids HCl and H_2O. However, because reactions of HOCl as an electrophile are generally quite exothermic, use of the Hammond postulate implies a transition state that more closely resembles the starting material than the products.

(g) A carboxylic peracid is unlike a simple carboxylic acid in that the anion obtained by removal of a proton from oxygen is not resonance-stabilized. Therefore, peracids are not particularly acidic. On the other hand, the oxygen–oxygen bond of a peracid is very weak (~36 kcal/mole). Heterolytic cleavage of this bond produces HO^+ and a resonance-stabilized carboxylate anion. Thus, the hydroxyl oxygen of a carboxylic peracid acts as an electrophile.

(h) The O—H bonds in water are polar, with hydrogen bearing partial positive charge and oxygen partial negative charge. Water is less acidic than are mineral acids and, thus, is less active as an electrophile.

Ex 10.3 Hydrochlorination takes place by protonation to form a carbocation in the rate-determining step. Because this step is endothermic, we deduce (from the Hammond postulate) that the transition state resembles the cation and is significantly influenced by the stability of the intermediate cation. Therefore, the compound that produces the more stable cation is more rapidly hydrochlorinated.

(a) *cis*-2-Butene. Protonation of butene produces a secondary carbocation more easily than protonation of ethene can produce a primary cation.

(b) Stilbene (1,2-diphenylethene). Protonation forms a secondary benzylic cation that is more stable than the secondary cation formed by protonation of *cis*-2-butene.

(c) 1,3-Butadiene. Protonation produces a secondary allylic cation, which is stabilized by resonance delocalization of positive charge.

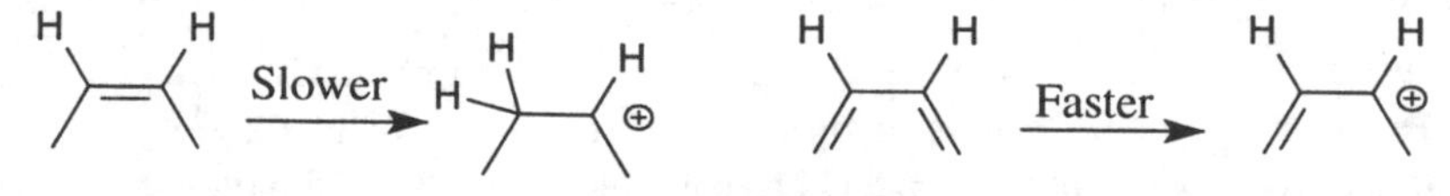

(d) 2,3-Dimethyl-2-butene. Upon protonation, this alkene produces a tertiary carbocation that is more stable than the secondary carbocation formed by protonation of *cis*-2-butene.

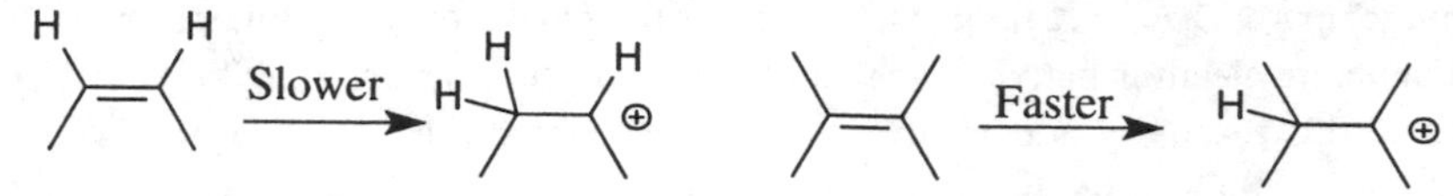

(e) *cis*-3,4-Dimethyl-3-hexene produces a tertiary carbocation upon protonation. This species is much more stable than the primary carbocation that would be formed by protonation of ethene.

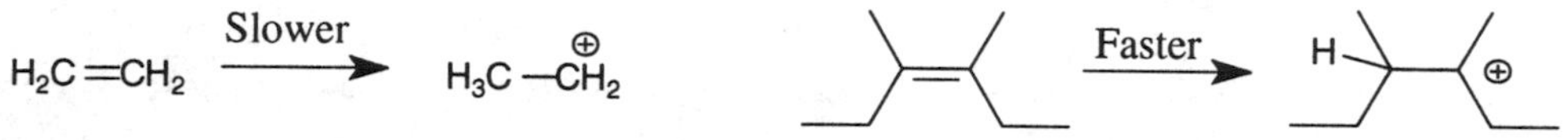

Ex 10.4 In each of these compounds, the two possible sites for protonation are not equivalent, and a different cation is produced by forming a C—H bond at each sp^2-hybridized carbon of the starting alkene. The site that is protonated in the product is the one that results in the more stable carbocation.

(a) Protonation of styrene is at the less substituted position, so as to produce the more stable benzylic cation.

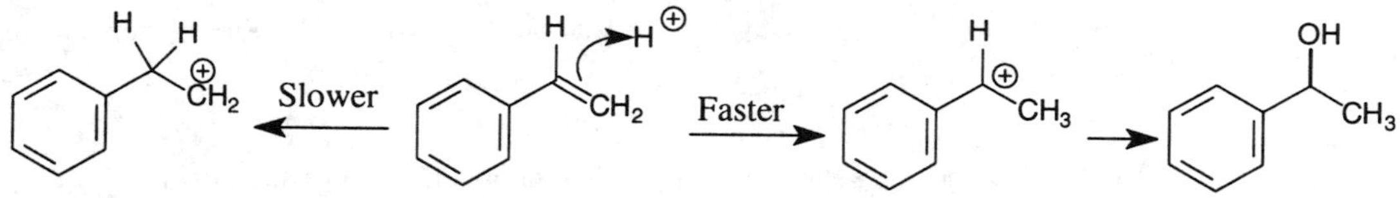

(b) Protonation at the less substituted site forms the more stable tertiary carbocation.

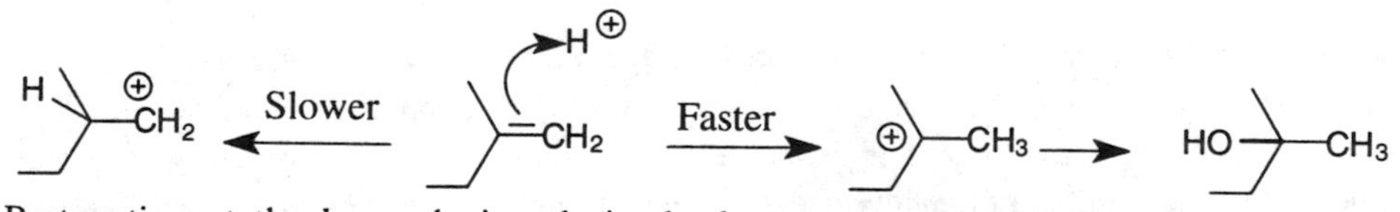

(c) Protonation at the less substituted site leads to a more stable tertiary carbocation.

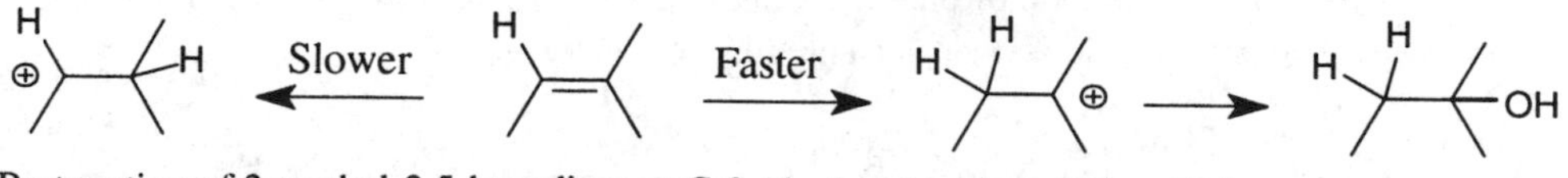

(d) Protonation of 2-methyl-2,5-hexadiene at C-3 (the less highly substituted site of the more highly substituted double bond) produces the most stable (tertiary) cation from this nonconjugated diene.

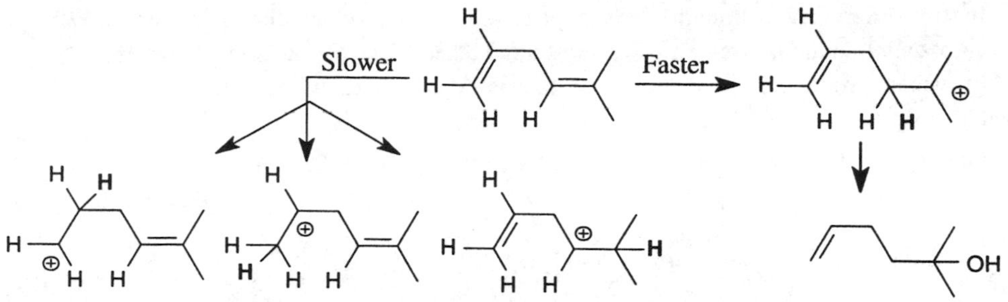

10.5 The addition of HBr to an alkene begins with initial protonation to form the most-stable cation.

(a) Protonation of 1,3-pentadiene at C-1 gives a secondary allylic cation (see Exercise 10.3, part c) that is captured by bromide ion at C-2 or C-4 to give the 1,2- and 1,4-adducts, respectively.

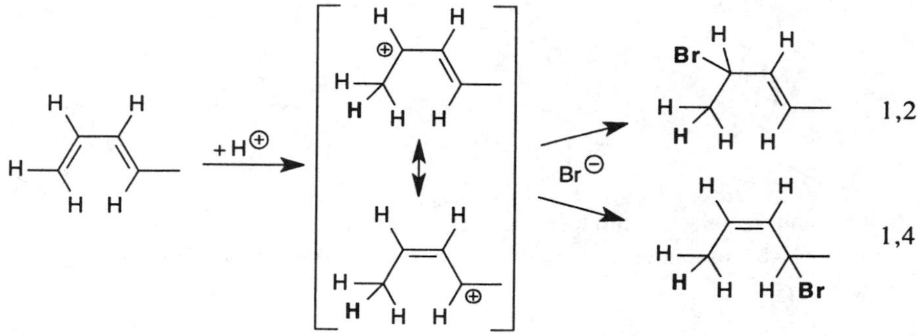

(b) Protonation of 4-methyl-1,3-pentadiene at C-1 leads to a tertiary allylic carbocation, which is captured by bromide at C-2 or C-4 to produce the indicated allylic bromides.

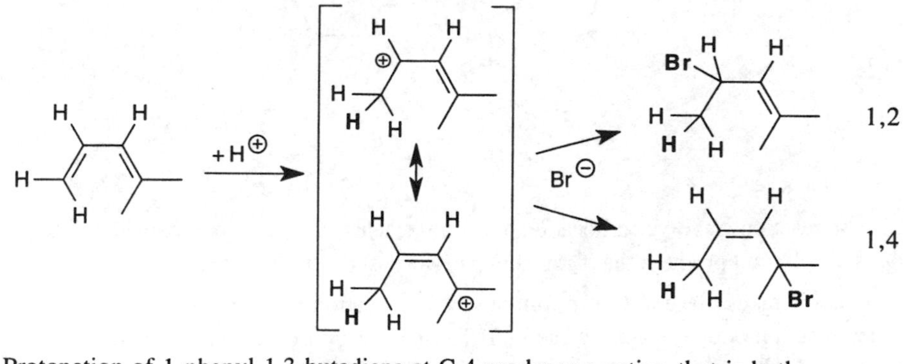

(c) Protonation of 1-phenyl-1,3-butadiene at C-4 produces a cation that is both benzylic and allylic. Trapping at the two possible sites that bear formal positive charge produces the two isomers shown.

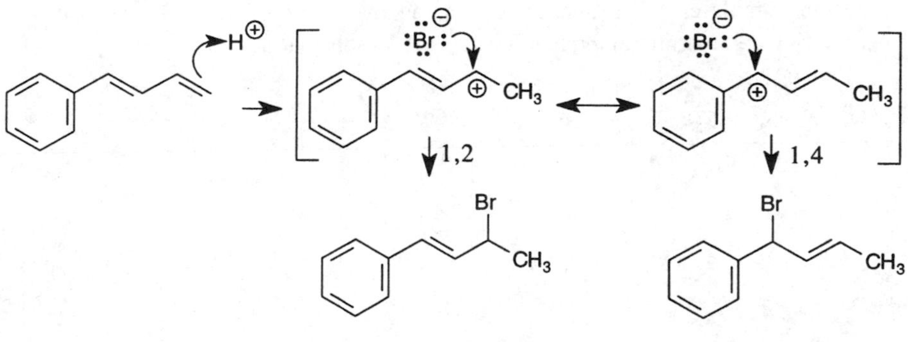

Ex 10.6 In this Exercise, protonation takes place so as to produce the more stable carbocation in an unsymmetrically substituted alkene. The cation is then captured by water, producing an oxonium ion that is deprotonated to yield the product alcohol.

(a) Protonation takes place so as to produce a teriary carbocation.

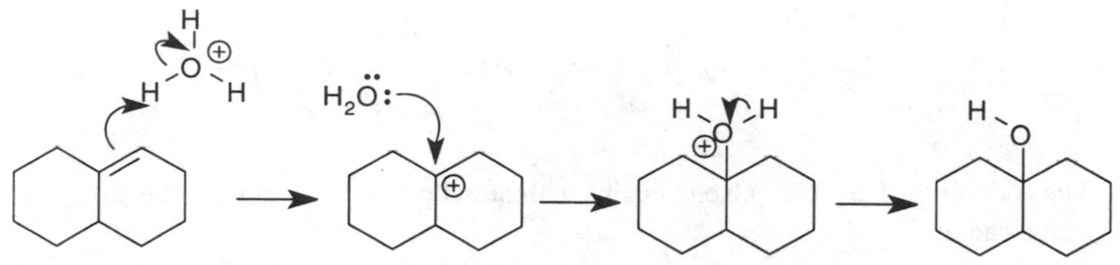

(b) Protonation at either sp^2-hybridized carbon in this symmetrically substituted alkene leads to the same intermediate tertiary carbocation.

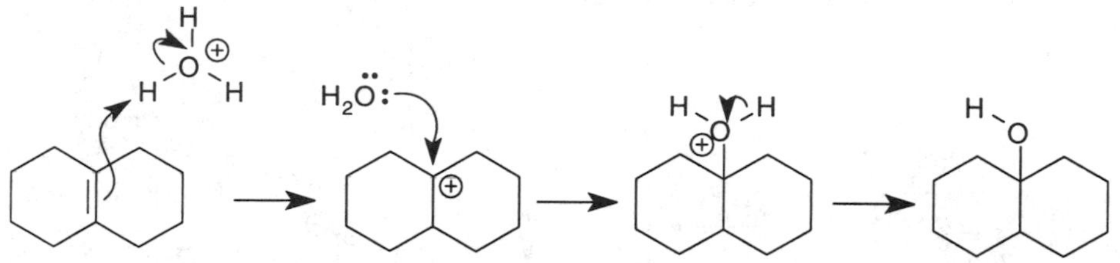

(c) Protonation takes place to produce a tertiary carbocation, but attack by water can take place from either the top or bottom face, leading to two isomeric alcohols.

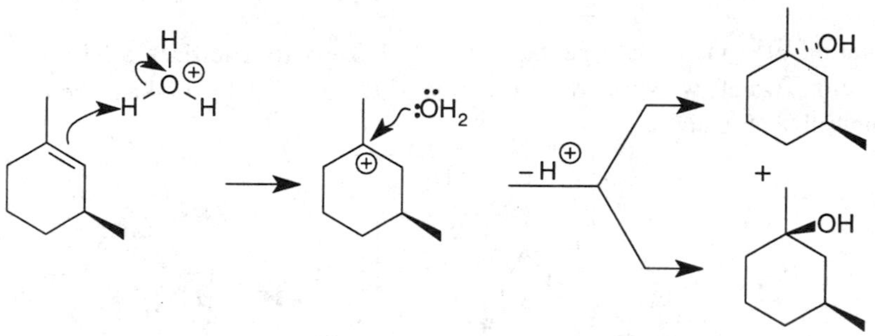

Ex 10.7 Both of these reactions involve carbocationic rearrangements. The product obtained in part *a* is an addition product; that formed in part *b* is a substitution product.

(a) Protonation takes place at C-1 to form a secondary cation rather than at C-2, which would produce a primary cation. Because a 1,2 shift of a methyl group from C-3 yields a more stable tertiary cation, this shift takes place rapidly. The resulting rearranged cation is captured by water, forming an oxonium ion. Transfer of a proton from the positively charged oxygen to a water molecule completes the sequence, producing the observed alcohol and regenerating the hydronium ion that is a catalyst for this hydration.

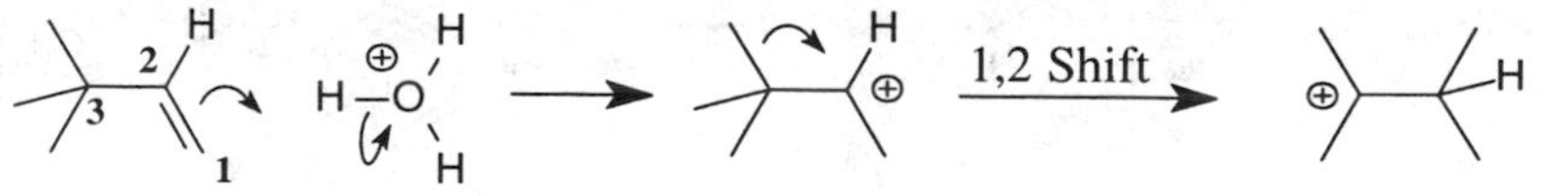

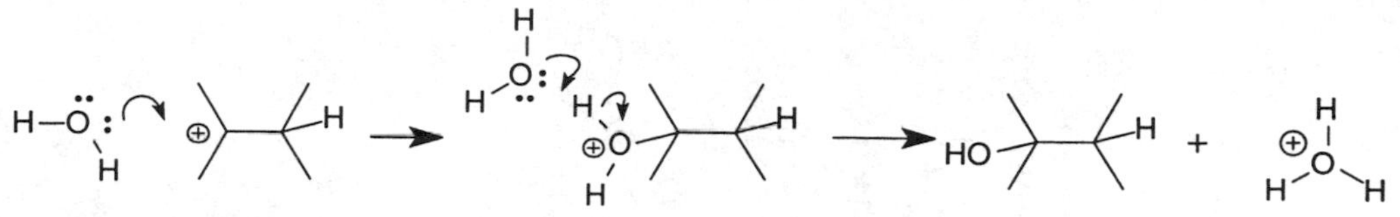

(b) Protonation of one of the lone pairs of electrons of the oxygen of the alcohol produces an oxonium ion. Direct loss of water from this ion would produce a primary carbocation. However, with simultaneous migration of a CH_2 group of the ring toward the electron-deficient center, a secondary cation is formed. This migration results in the incorporation of the CH_2 group to the larger ring. Thus, the five-membered ring of the starting alcohol is transformed to a cyclohexyl cation that is trapped by bromide to form the observed product.

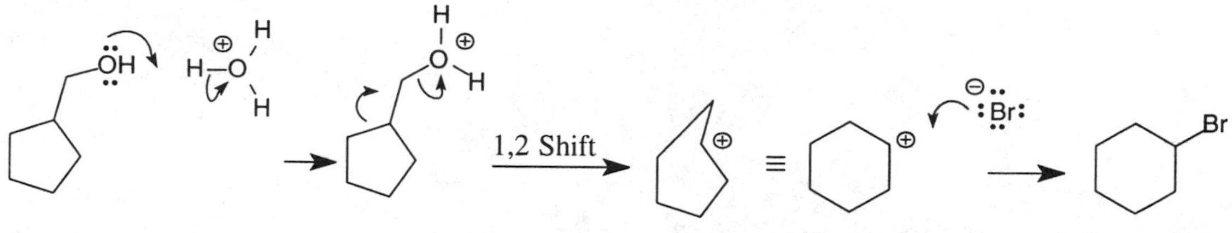

10.8 (a) Deuteration takes place with equal facility at either carbon and on either face of the double bond of cyclohexene, generating in each case the same carbocation. This intermediate can be captured by chloride on either face of the planar sp^2-hybridized carbocation with equal ease. Although the product has two chiral centers, the product mixture is optically inactive, because of the presence of equal amounts of the enantiomers of the two diastereomers in which D is *cis* and *trans* to Cl.

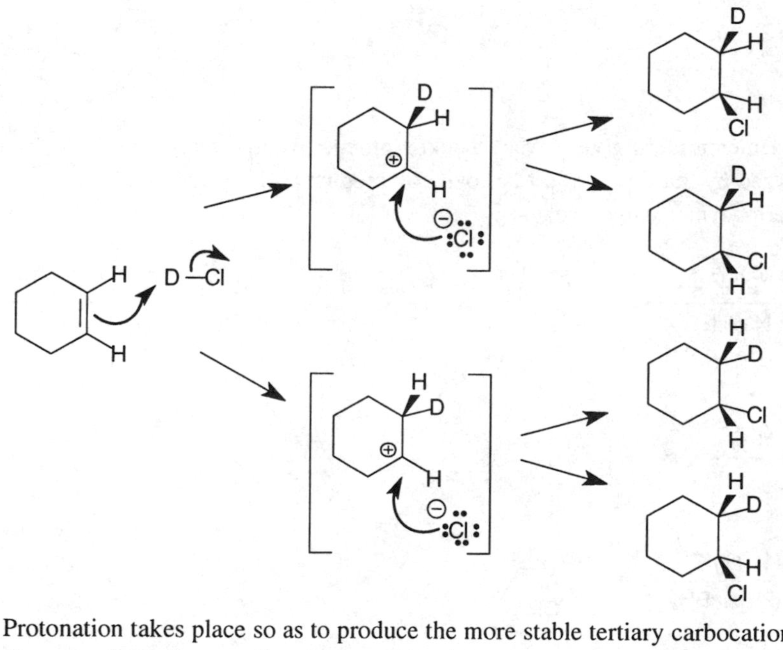

(b) Protonation takes place so as to produce the more stable tertiary carbocation. Capture of this intermediate by bromide yields a product that lacks a chiral center and is therefore optically inactive.

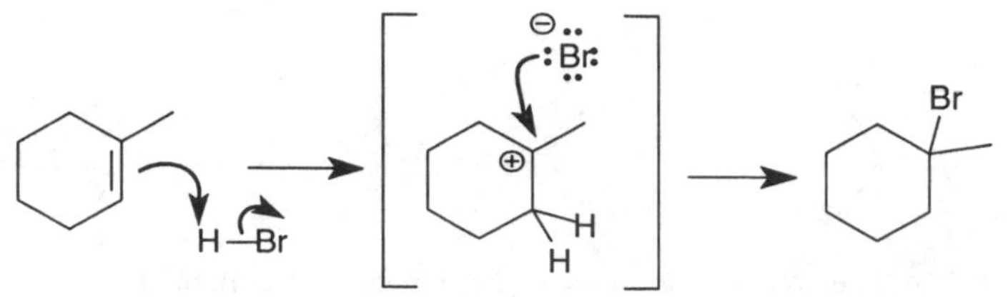

(c) Deuteration takes place with equal ease at C-2 on either face of the double bond, as in part *a*. Trapping by water, followed by deprotonation of the oxonium ion, gives a racemic mixture of *cis*- and *trans*-2-deutero-1-methylcyclohexanols. This problem is completely parallel to part *a*.

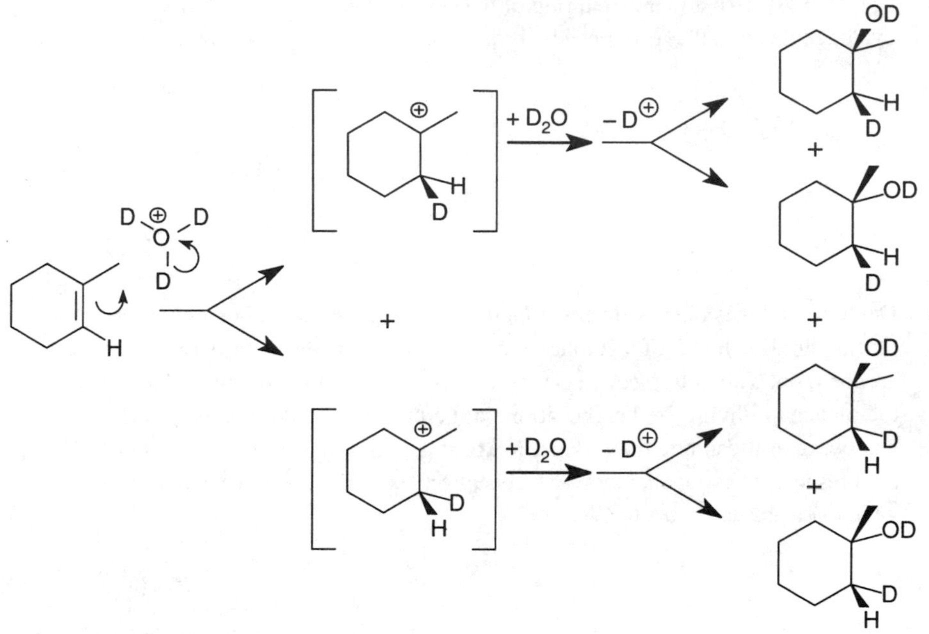

Ex 10.9 Oxymercuration-demercuration gives a Markovnikov product without rearrangement. Acid-catalyzed hydration gives a Markovnikov product with a rearranged skeleton, when thermodynamically favorable.

(a)

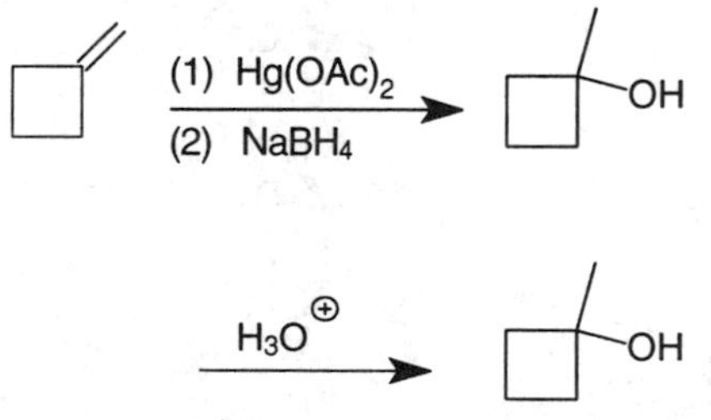

(b)

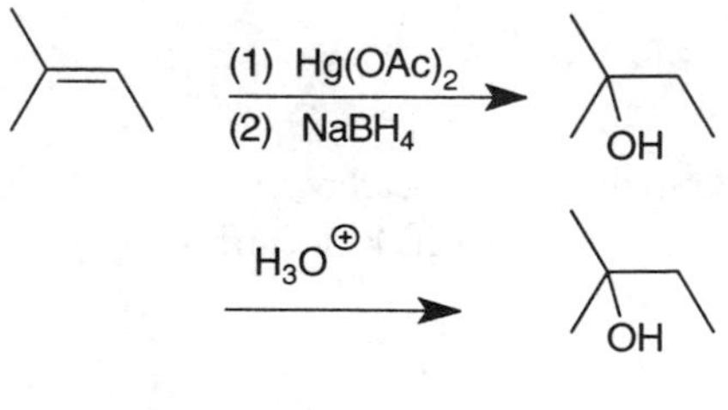

(c)

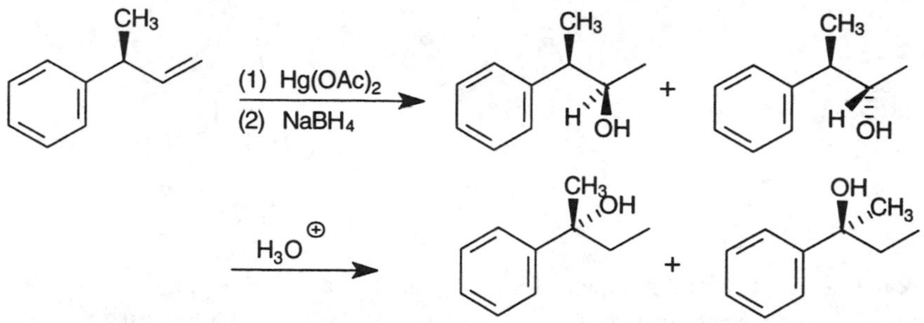

(d)

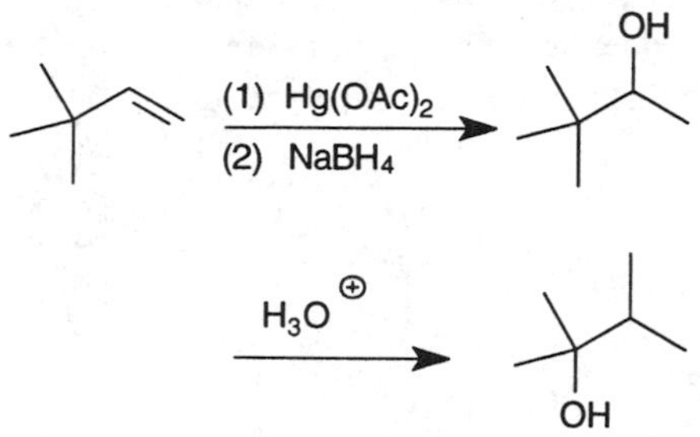

Ex 10.10 The hydration of an alkyne produces an enol with a regiochemistry dictated by Markovnikov's rule.

(a) Here protonation at the terminal carbon leads to an enol with the OH group at C-2. Tautomerization produces a methyl ketone.

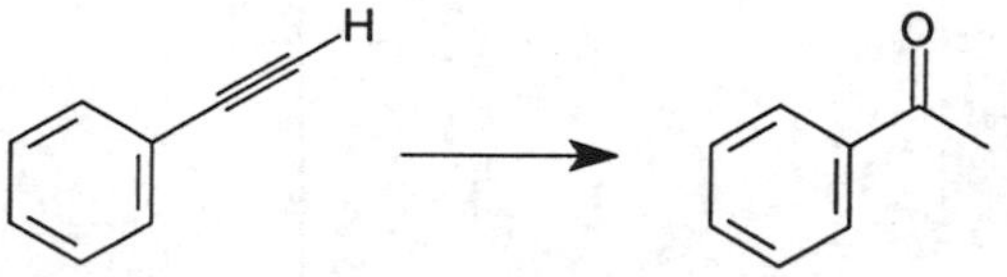

(b) In this symmetrical alkyne, there is no preferred site for protonation, but the same product (3-hexanone) is formed after tautomerization of the enol in any case.

(c) Both C-2 and C-3 are suitable sites for the intial protonation, leading to a product mixture of 2-hexanone and 3-hexanone, respectively, after tautomerization.

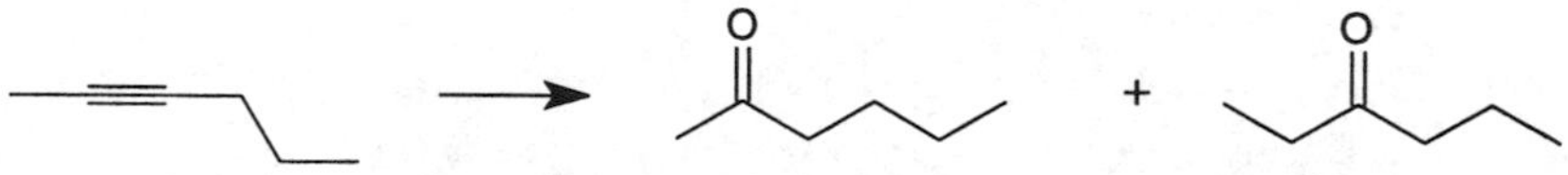

Ex 10.11 Protonation takes place at C-1 in accord with Markovnikov's rule, producing a secondary vinyl cation that is captured by water in a nucleophilic attack. Acid-catalyzed tautomerization, after deprotonation of the resulting oxonium ion, leads to the observed ketone.

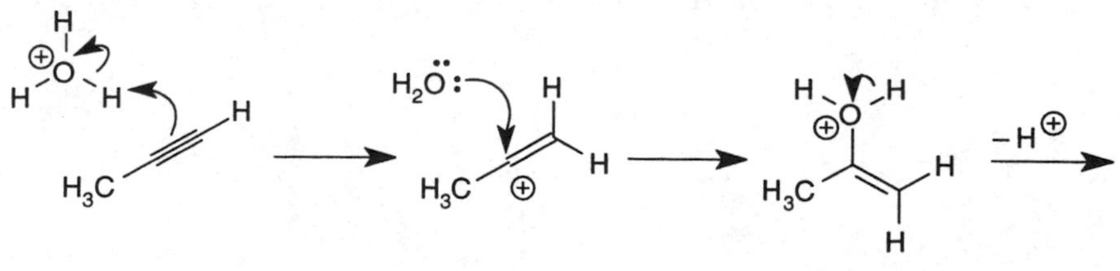

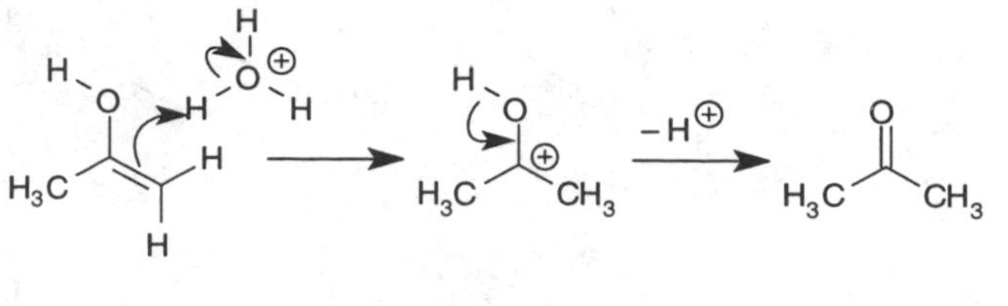

Ex 10.12   Shown here are the cyclic bromonium ions formed by attack on the top and bottom faces of *trans*-2-butene. Each of these bromonium ions can be opened with equal ease by attack at the upper or lower carbon atom, as drawn. Although the product is drawn to show the four possible stereochemical modes, all four structures are the same, a *meso* compound. (*Anti* addition to a *trans*-alkene produces *meso*-adducts; *anti* addition to a *cis*-alkene produces a *d,l*-pair.)

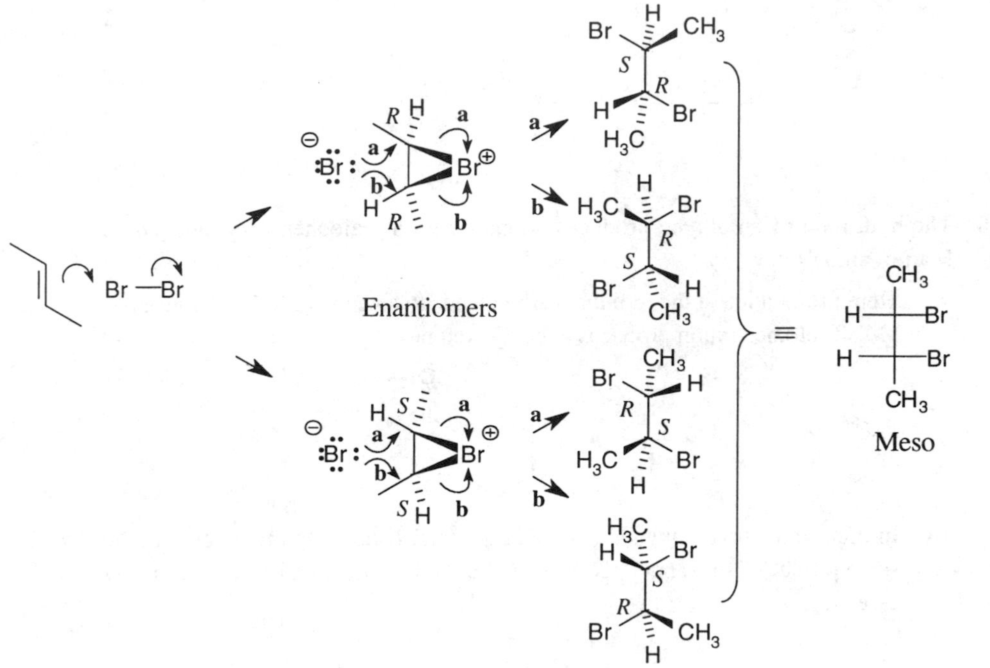

Ex 10.13   The difference between the bonds broken and those made in a reaction represents $\Delta H°$. (Recall that the value for a π bond is the difference between the bond strength for the double bond and the single bond.)

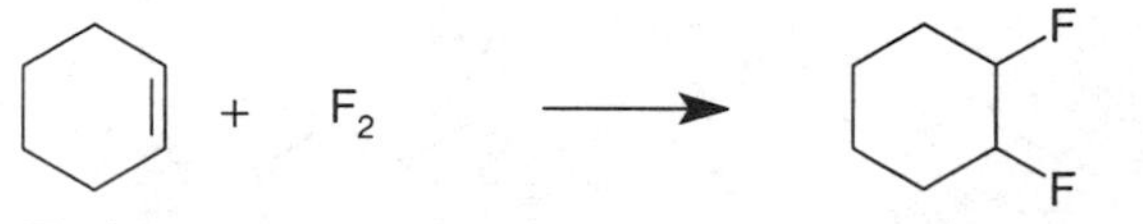

Broken

C—C π	63
F—F	38
	101

Made

C—F	108
C—F	108
	216

$$\Delta H° = 101 - 216 = -115 \text{ kcal/mole}$$

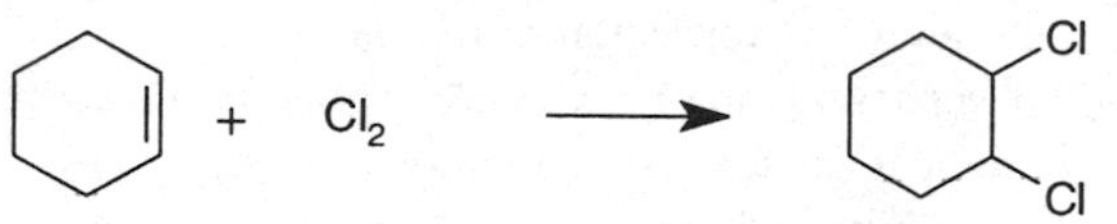

Broken		Made	
C—C π	63	C—Cl	81
Cl—Cl	58	C—Cl	81
	121		162

$$\Delta H^\circ = 121 - 162 = -41 \text{ kcal/mole}$$

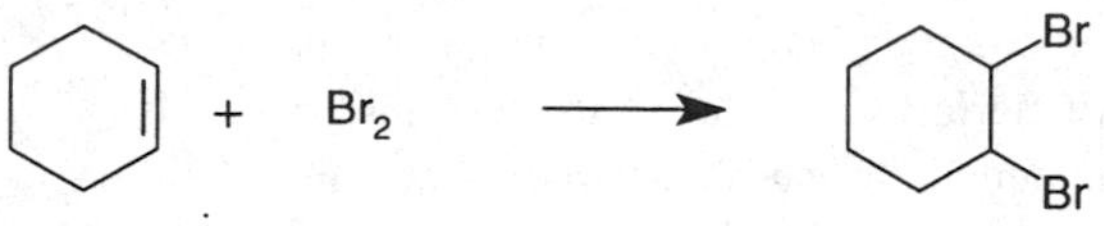

Broken		Made	
C—C π	63	C—Br	68
Br—Br	46	C—Br	68
	109		136

$$\Delta H^\circ = 109 - 136 = -27 \text{ kcal/mole}$$

Ex 10.14 In each example, a cyclic onium ion is formed because the electrophilic atom bears a lone pair of electrons that interacts with the developing carbocationic center. In these unsymmetrical reagents, the less electronegative atom acts as the electrophile.

(a)

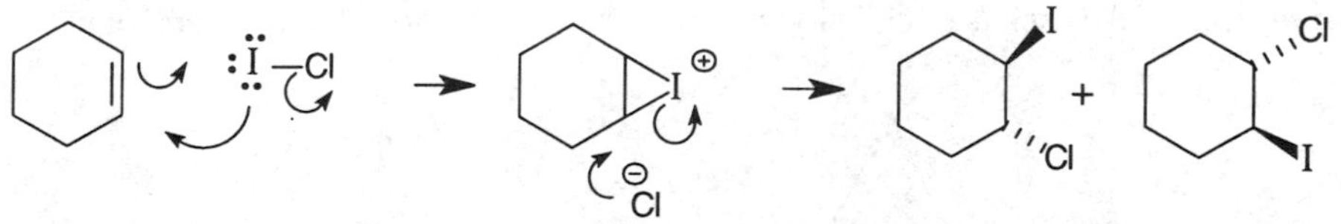

(b)

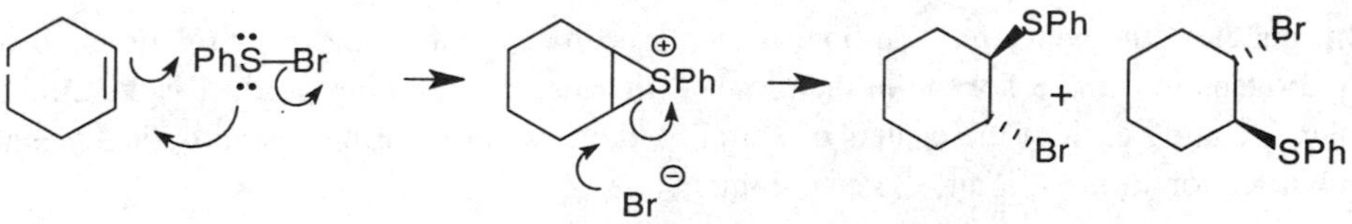

(c)

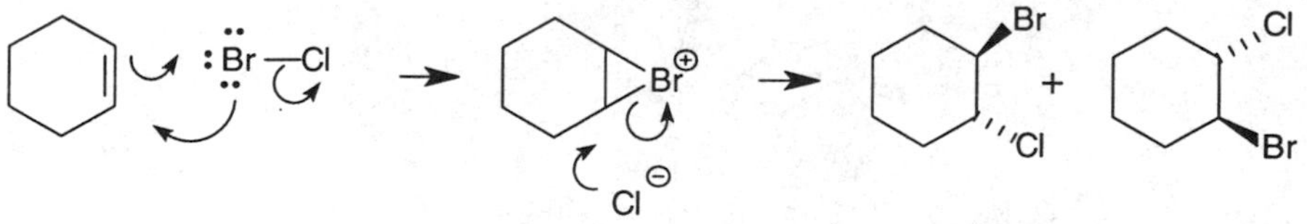

(d)

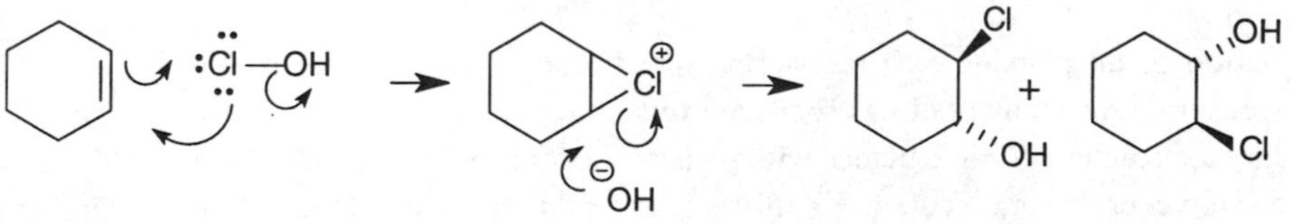

Ex 10.15 Markovnikov products are formed in polar solvents; anti-Markovnikov products in ether. This question thus asks whether the intended product results from Markovnikov or anti-Markovnikov addition.

(a) Markovnikov addition (through a tertiary carbocation); polar solvent.

(b) Markovnikov addition (through a tertiary carbocation); polar solvent.

(c) Anti-Markovnikov addition (through a brominated tertiary radical); ether.

(d) Anti-Markovnikov addition (through a brominated benzyl radical); ether.

Ex 10.16 The regularity of this polymer follows from the preferred formation of the more stable benzyl radical in the propagating polymer chain. Homolytic attack of the radical initiator on styrene produces a benzyl radical (rather than the alternative primary radical), thus ensuring the placment of the phenyl group at alternating positions along the alkyl chain.

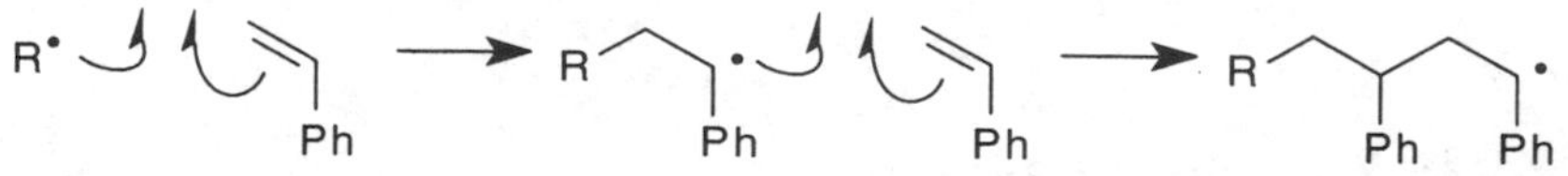

Ex 10.17 The addition of a singlet carbene (such as dichlorocarbene) to a double bond is stereospecifically *syn*. When the starting material is symmetric about a mirror plane through the double bond, a *meso* compound is formed; otherwise, a *d,l*-pair is produced.

(a) Concerted addition of dichlorocarbene retains the *cis* relationship of the alkenyl hydrogens in cyclohexene in the bicyclic product. Because the resulting product has a mirror symmetry plane relating each of the centers of chirality, it is an optically inactive *meso* compound.

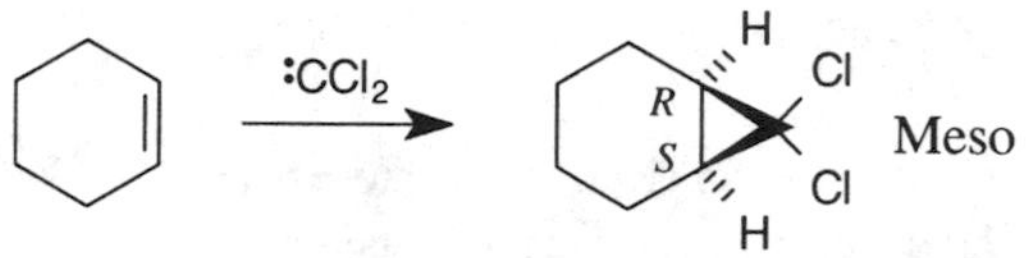

(b) As in part *a*, the concerted addition of dichlorocarbene retains the *cis* relationship of the alkenyl hydrogens in 2-butene in the product. Because the resulting product has a mirror symmetry plane relating each of the centers of chirality, it is also an optically inactive *meso* compound.

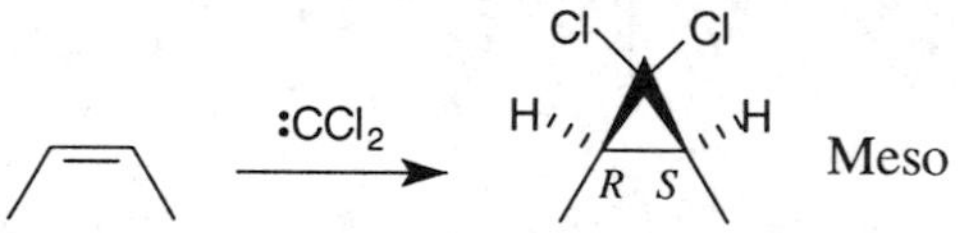

(c) In this case, the concerted addition of dichlorocarbene retains the *trans* relationship of the alkenyl hydrogens in *trans*-3-hexene in the product. Because the resulting product lacks a mirror symmetry plane relating each of the centers of chirality, the product is formed as an optically inactive racemic modification of the *R,R* and *S,S* enantiomers.

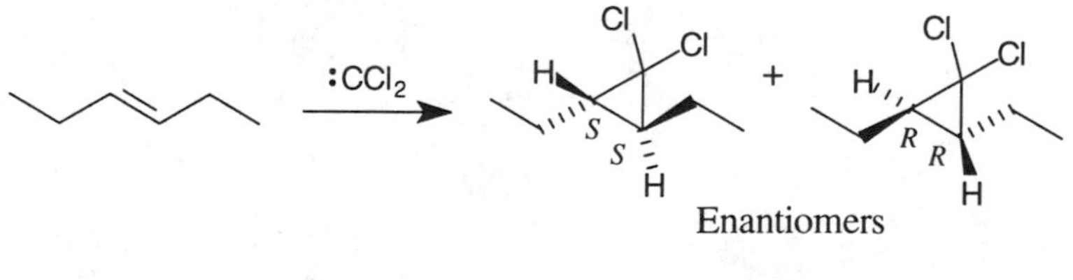

Enantiomers

Ex 10.18 The formation of an epoxide by the reaction of an alkene with a peracid is a concerted reaction, as shown below. There are two chiral centers in the epoxide produced by the reaction of *cis*-2-butene with peracetic acid: one center is *R* and the other is *S*. However, the molecule has a mirror plane of symmetry relating these two centers and, thus, the molecule is a *meso* isomer. It is therefore not optically active.

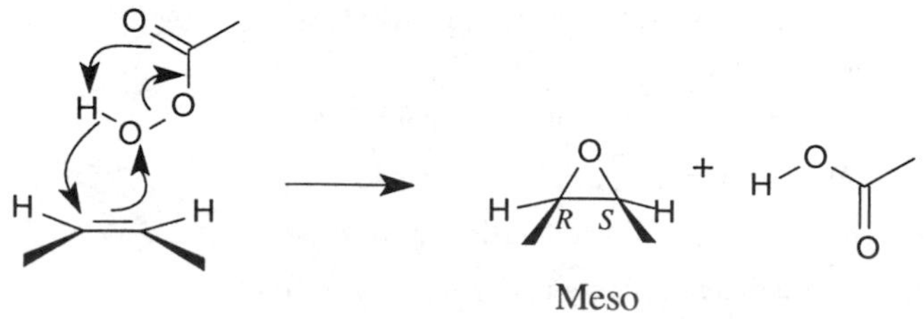

Meso

Ex 10.19 In each case, the starting material is converted to a Grignard reagent from which a two-carbon extension is obtained upon treatment with ethylene oxide.

(a)

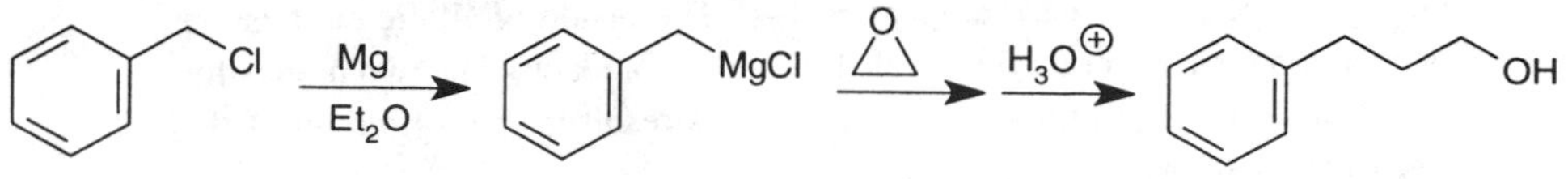

(b)

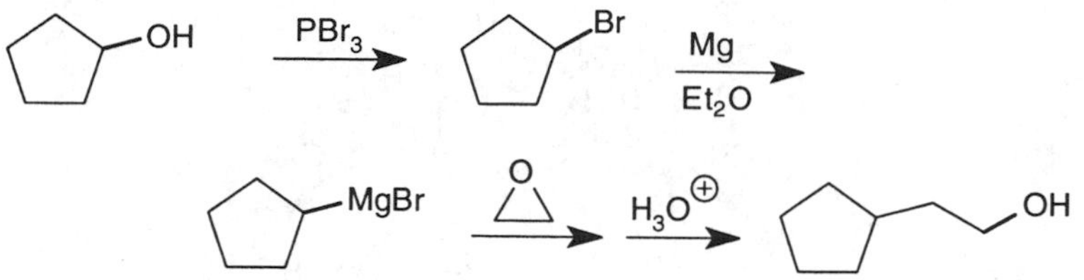

(c)

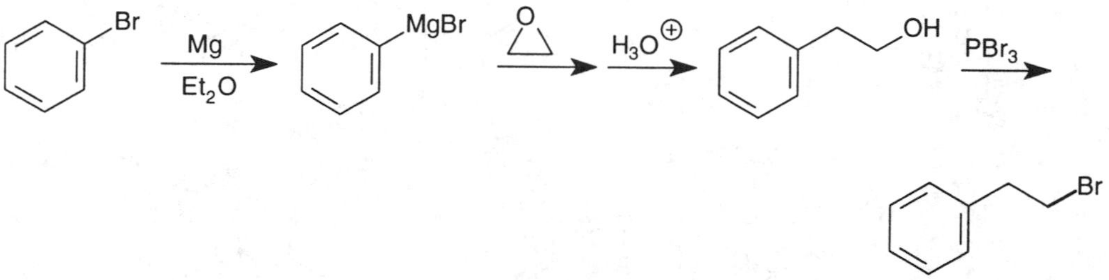

Ex 10.20 Grignard reagents are prepared by insertion of magnesium to a C—Br bond. The reaction of a nucleophilic species such as a Grignard reagent with ethylene oxide proceeds by S_N2-like cleavage of one of the carbon–oxygen bonds of the epoxide. The resulting alkoxide is converted to the product alcohol upon neutralization with acid.

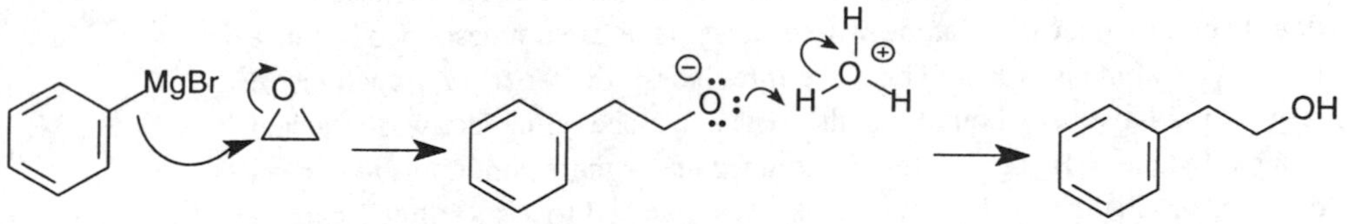

Ex 10.21 Acid-catalyzed hydration gives a Markovnikov hydration product; hydroboration-oxidation gives an anti-Markovnikov hydrate. This Exercise thus asks whether the observed product is a Markovnikov or anti-Markovnikov adduct, and, if the latter, whether the product has a *cis* relation between the hydroxyl group and added proton, as is obtained from hydroboration-oxidation.

(a) Markovnikov; acid-catalyzed hydration.

(b) Not at all. This is an anti-Markovnikov adduct, but from an *anti* addition, which can be achieved with neither acid-catalysis (wrong regioisomer) nor hydroboration-oxidation (wrong stereoisomer).

(c) Anti-Markovnikov, *syn* addition; hydroboration-oxidation.

(d) Anti-Markovnikov, no stereochemical marker to indicate *syn* or *anti*; hydroboration-oxidation.

(e) Not as a single stereoisomer. This is a *syn* Markovnikov adduct, resulting from the addition of both the proton and the hydroxyl group exclusively to one face of each of the sp^2-hybridized carbon atoms of the original alkene. Addition of water in the presence of acid to *cis*-1-phenyl-2-deuteropropene produces four stereoisomers (two enantiomers each of two diastereomers). (For practice, draw these four isomers.)

(f)   Not as a single product. This is one of several possible Markovnikov addition products. The presence of deuterium has little effect; protonation takes place with essentially equal ease at C-1 and C-4. The resulting allylic cations are then captured by water to yield both 1,2- and 1,4-adducts. The two modes for the initial protonation under acidic conditions result in the formation of the four isomers shown.

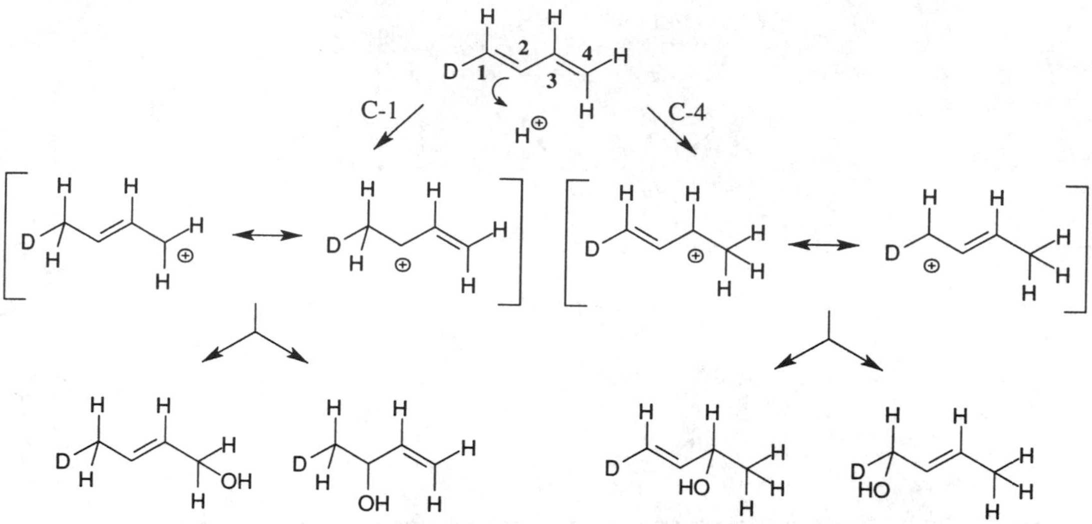

Ex 10.22   Potassium permanganate effects cleavage of an alkenyl double bond, producing the corresponding carbonyl compound. When an aldehyde is formed, further oxidation takes place, genereating a carboxylic acid.

(a)   The production of a ketone indicates that the carbonyl carbon was originally doubly bound to a carbon atom, whose substituents are not described because the other fragment is lost. There are three possibilities: (i) two cyclohexyl rings could have been bound together; (ii) a cyclohexyl moiety was attached to a methylene (CH_2) group that is oxidized under these conditions to carbon dioxide; or (iii) the cyclohexyl moiety was attached to a substituted carbon (CR_2) group that is converted under these conditions to an undefined carbonyl group.

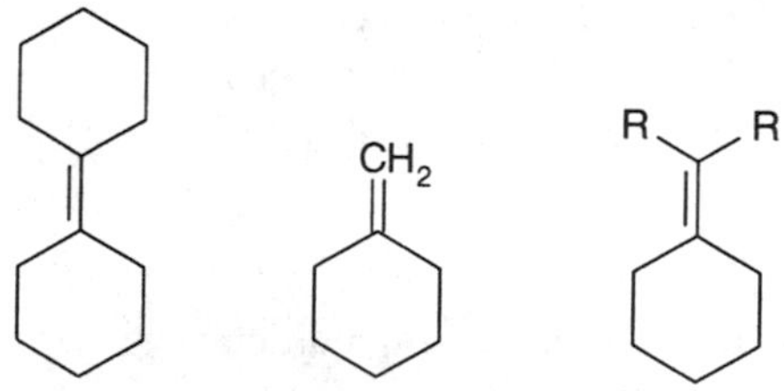

(b)   Either the two carbonyl carbons were attached to each other in a cyclic alkene or each of the two carbonyl carbons was attached to some other carbon moiety that, as in part *a*, was lost by further oxidative degradation under the reaction conditions.

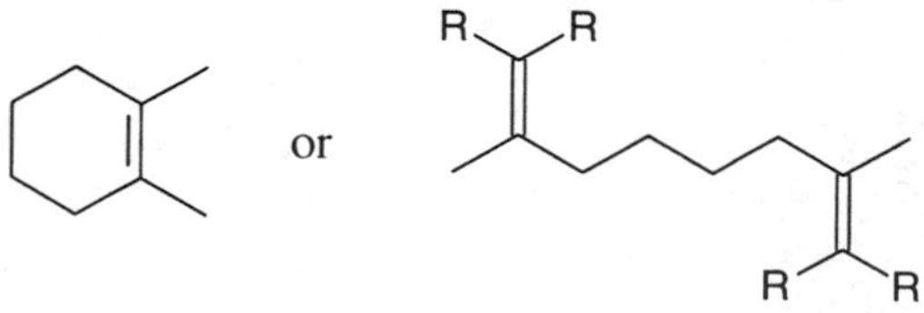

(c)   Oxidation of either *cis-* or *trans-*2-butene with $KMnO_4$ will produce acetic acid.

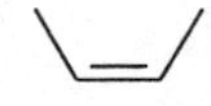

(d)   As in part *b* of this Exercise, the two carbonyl groups can be generated by oxidation of a single alkene.

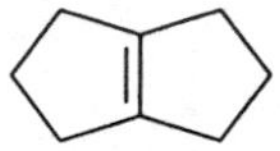

Ex 10.23   Ozonolysis converts an alkene to a pair of carbonyl groups.

(a)

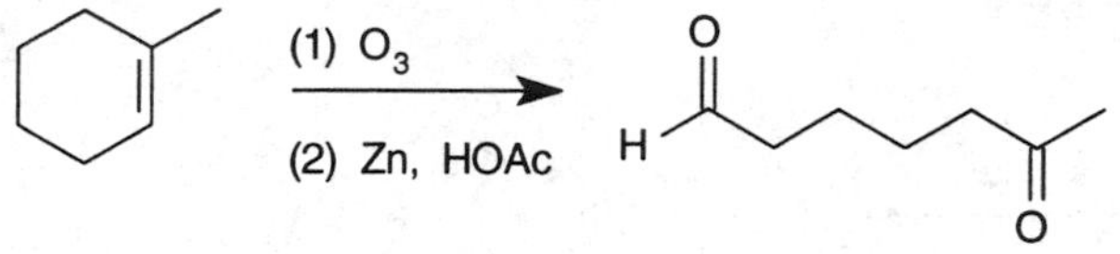

(b)

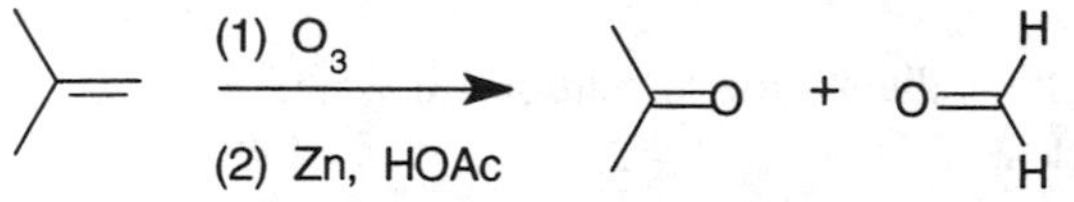

(c)

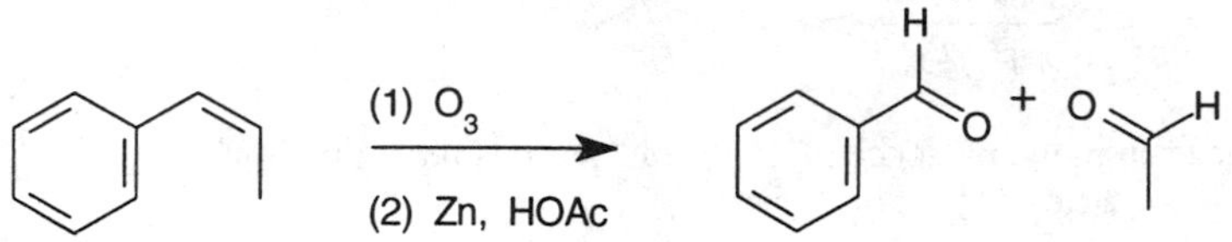

(d)

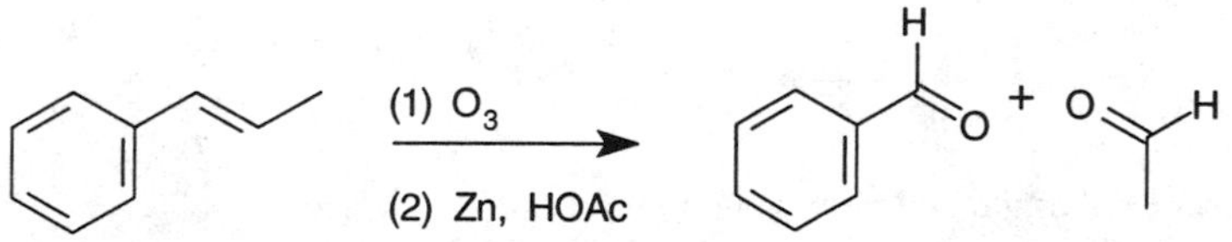

(e)

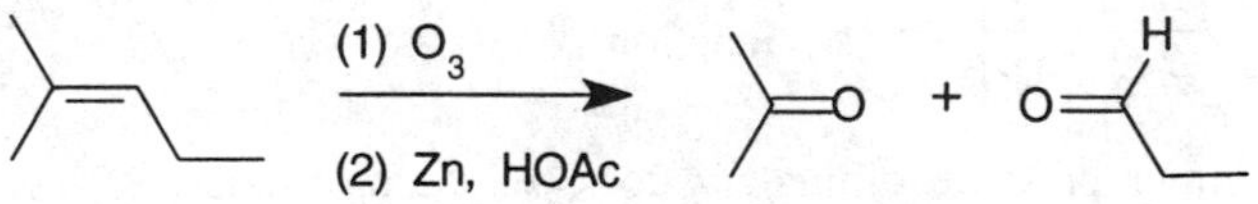

Ex 10.24   This problem asks us mentally to deconstruct the six-membered ring produced in a Diels-Alder reacion to the diene and dienophile from which it was constructed.

(a)

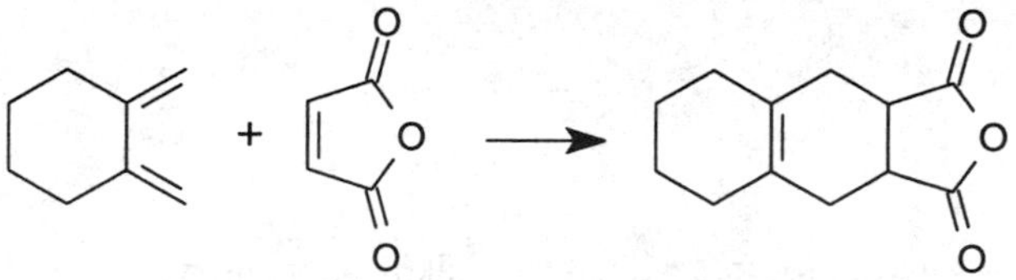

(b)

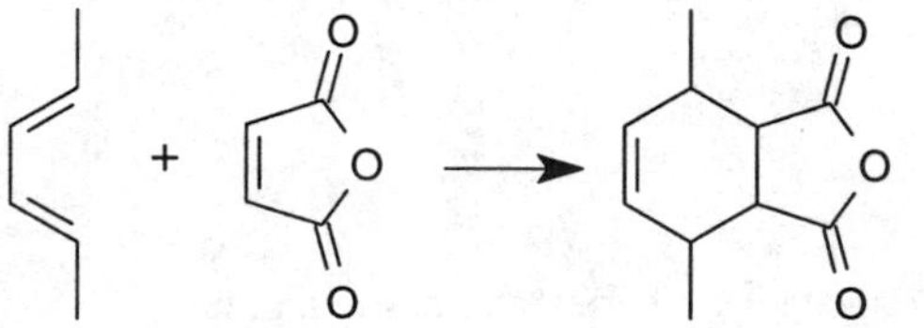

(c)

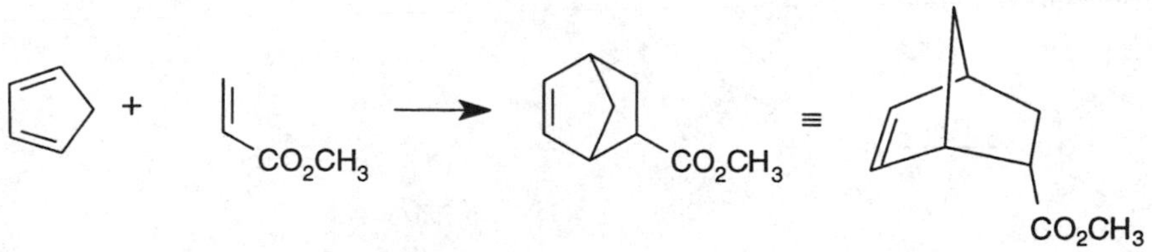

Ex 10.25 Identifying the preferred site for reduction of several possible functional groups by different reducing agents is the goal of this Exercise.

(a) Catalytic hydrogenation of the isolated alkene is preferred.

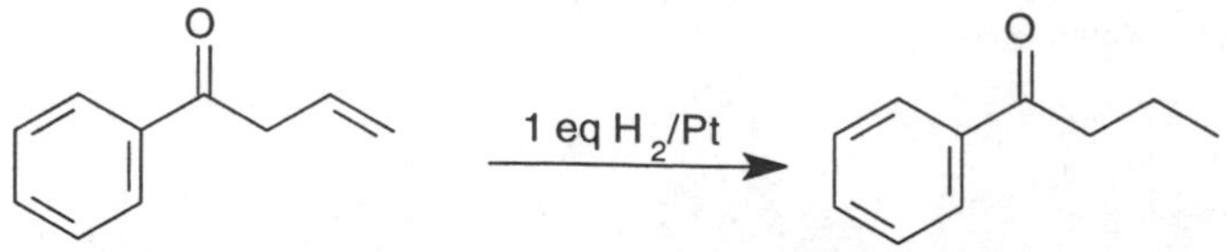

(b) The high electron affinity of a carbonyl group makes it the preferred site for a dissolving metal (sodium) reduction.

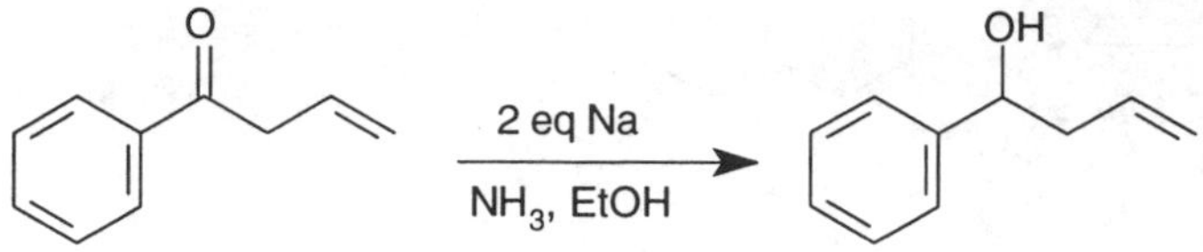

(c) An alkyne triple bond is reduced by sodium under conditions in which an alkene double bond remains unaffected.

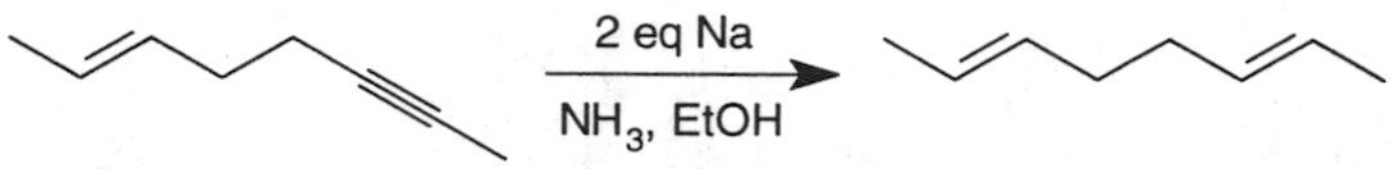

(d) Catalytic hydrogenation in the presence of a "poison" permits the reduction of an alkyne in the presence of an alkene.

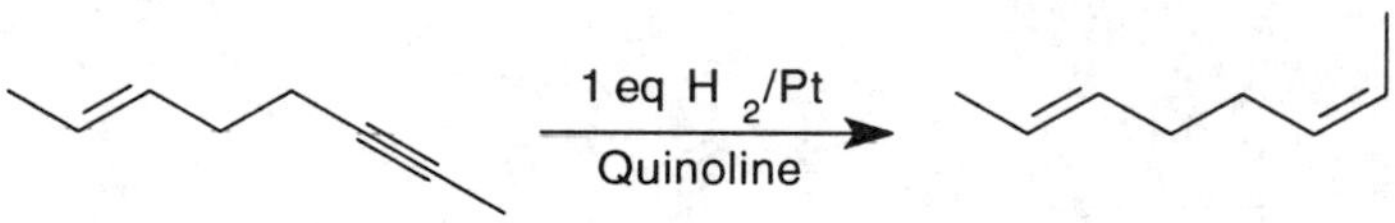

(e) Double bonds resist dissolving metal reduction: no reaction.

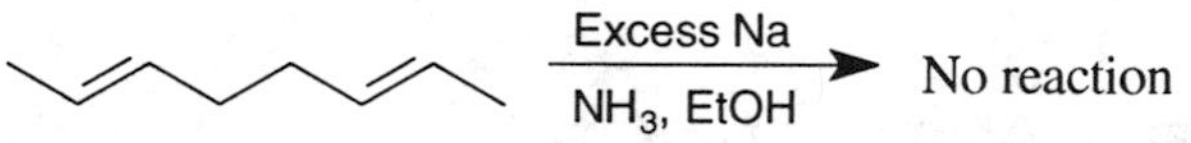

(f) Catalytic hydrogenation under pressure ultimately does reduce aromatic compounds. After aromaticty destroyed, complete reduction to the corresponding alkane skeleton takes place.

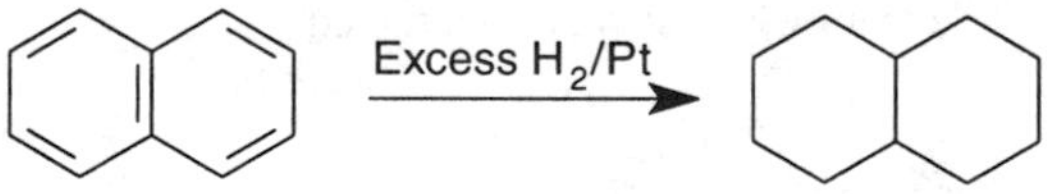

Answers to Review Problems

Pr 10.1 (a) Hydrohalogenation in a polar solvent proceeds with Markovnikov orientation.

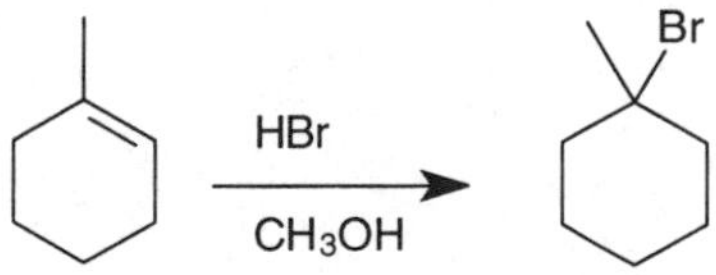

(b) Addition of HI proceeds with the same regiochemical orientation as in a polar HBr addition.

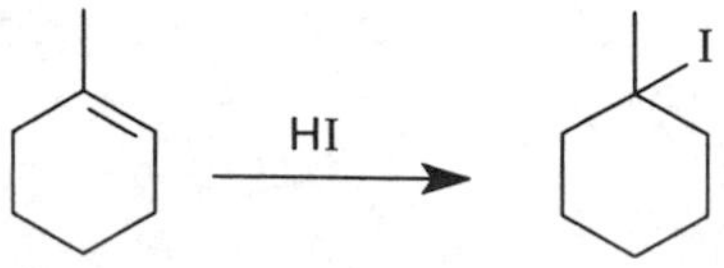

(c) In ether, traces of peroxides initiate radical addition, which yields anti-Markovnikov product. Both *cis* and *trans* products are formed upon trapping the intermediate alkyl radical with bromine, on the same and opposite faces, respectively, of the radical.

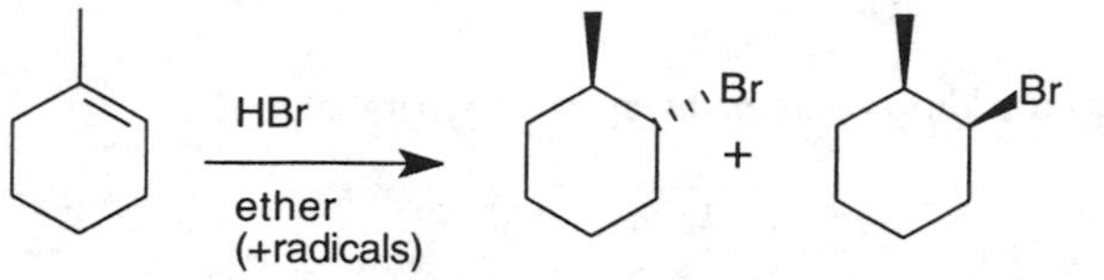

(d) Hydroboration–oxidation yields an anti-Markovnikov hydration product. The first step, hydroboration, is a *syn* addition and takes place from the less hindered face of a double bond. Because the replacement of the carbon–boron bond with a carbon–oxygen bond occurs with complete retention of stereochemistry, the OH group is introduced exclusively *cis* to the added hydrogen. Thus, in this example, the OH group appears in the product in a geometry *trans* to the methyl group.

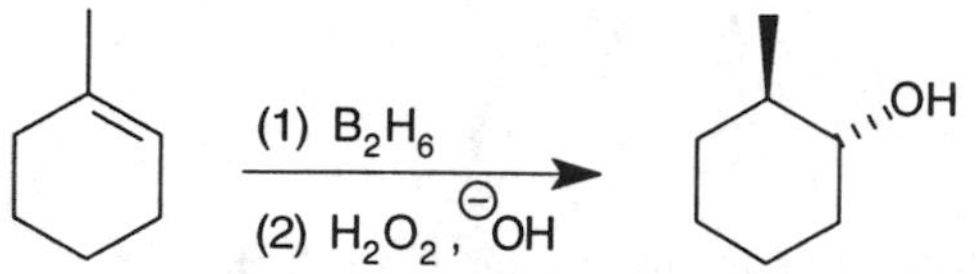

(e) Oxymercuration–demercuration effects a Markovnikov hydration, avoiding carbocation rearrangements that often accompany the more frequently used aqueous hydration. Here, no rearrangement would take place even with aqueous acid, so the same product is formed as in dilute acid.

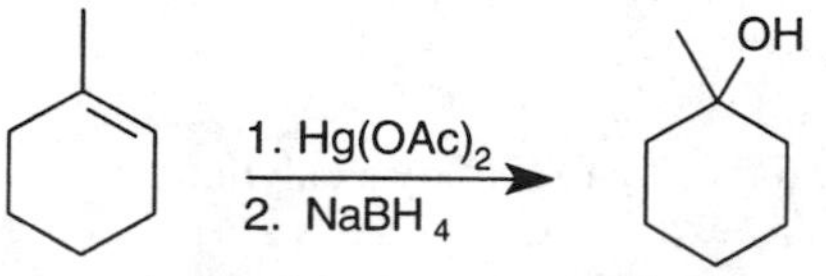

(f) Bromination takes place as an *anti* addition, producing a racemic (*d,l*) mixture.

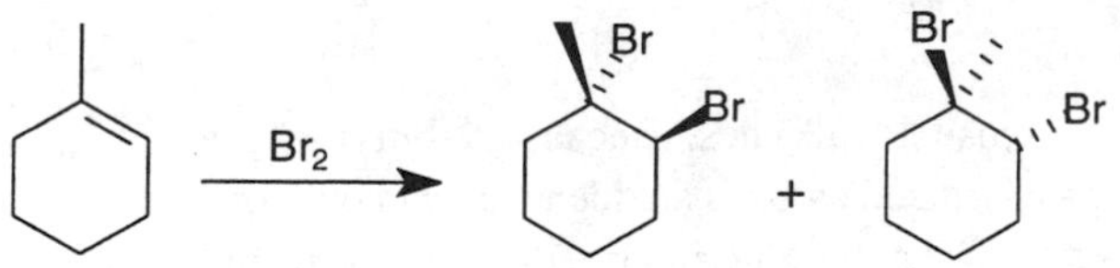

(g) Carbene additions are stereospecific in the singlet state.

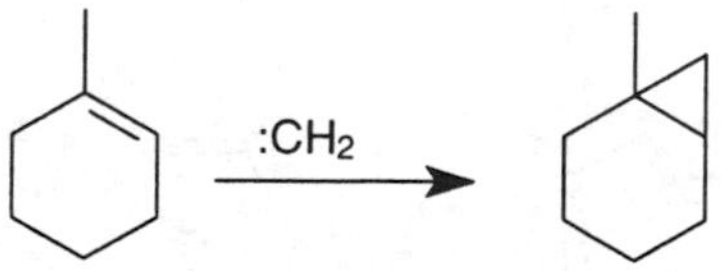

(h) Peroxidation occurs upon treatment with *m*-chloroperbenzoic acid.

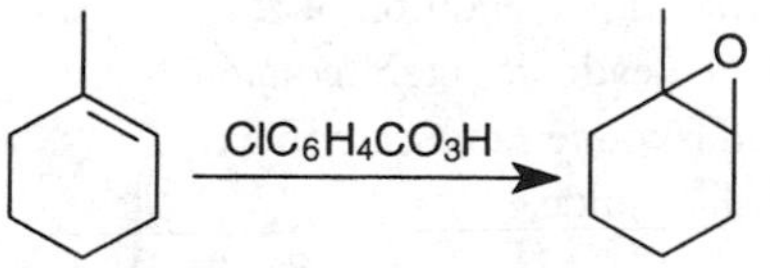

(i) Ozonation, followed by reduction with zinc in acetic acid, cleaves double bonds, converting the carbons at the ends of what was a double bond to carbonyl groups.

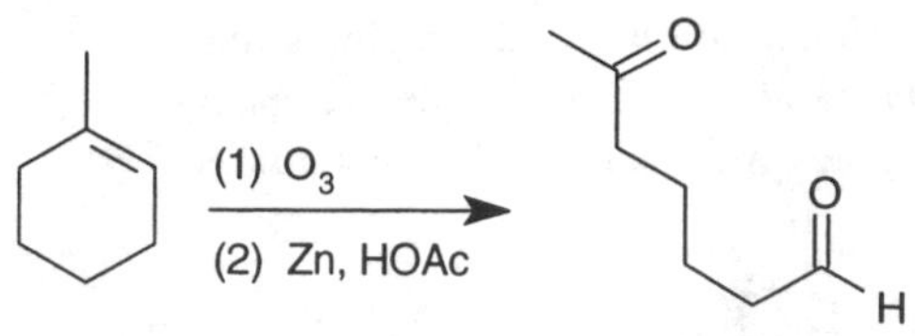

(j) Chlorination takes place by *anti*-addition, in a pathway exactly parallel to that in bromination in part *f*.

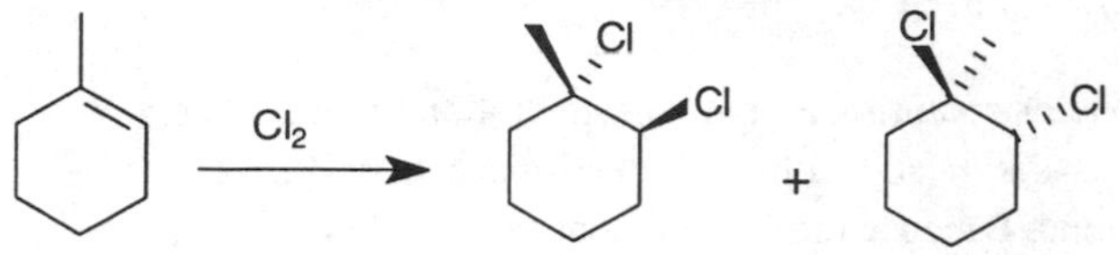

(k) Hot permanganate effects oxidative cleavage, with further oxidation of an initially formed aldehyde to a carboxylic acid.

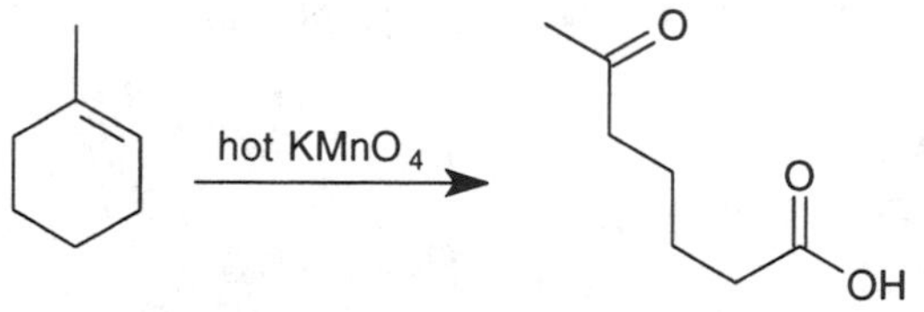

(l) In dilute sulfuric acid, the carbocation formed upon protonation of an alkene reacts with water to form the hydrate shown.

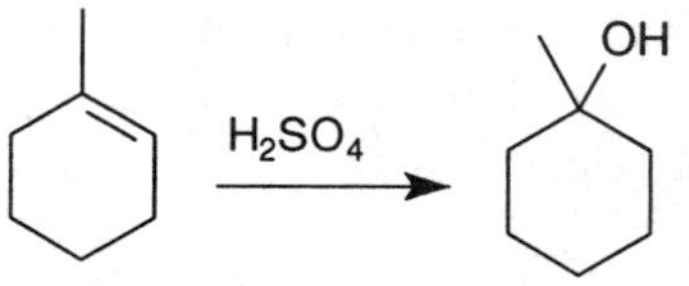

Ex 10.2 (a) Alkynes are hydrogenated first to *cis*-alkenes and then to alkanes by catalytic hydrogenation.

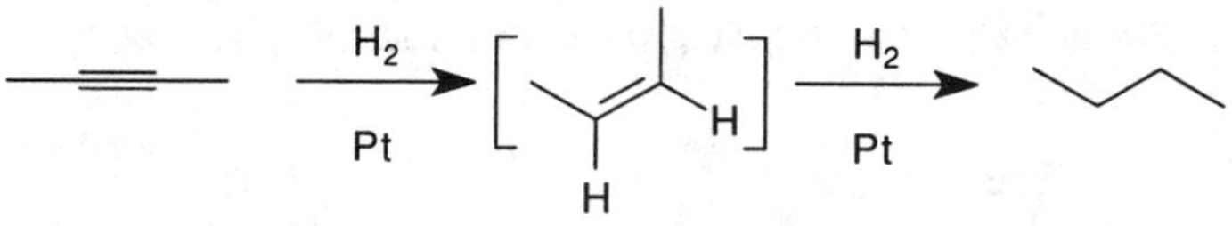

(b) HBr adds in a Markovnikov sense to alkynes. Because 2-butyne is a symmetrical alkyne, the two *sp*-hybridized carbons are identical, and only one vinyl bromide product is possible. The cationic intermediate is linear and both *cis* and *trans* products are formed.

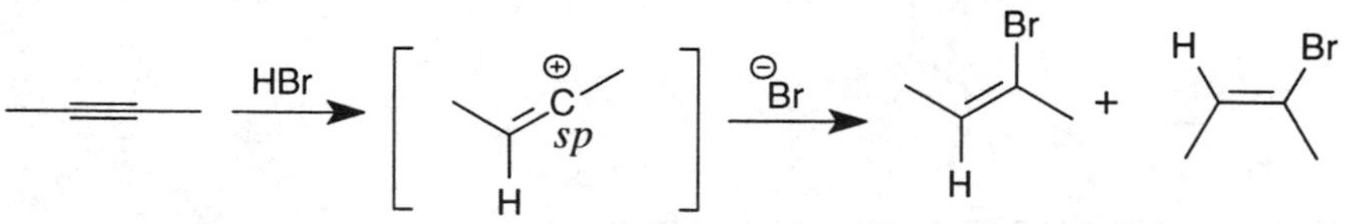

(c) The bromine atom present in the vinyl bromide formed upon addition of the first equivalent of HBr stabilizes cationic charge in the cationic intermediate for the second addition. Thus, the second bromine adds to the same carbon as the first, producing a geminal, rather than a vicinal, dibromide.

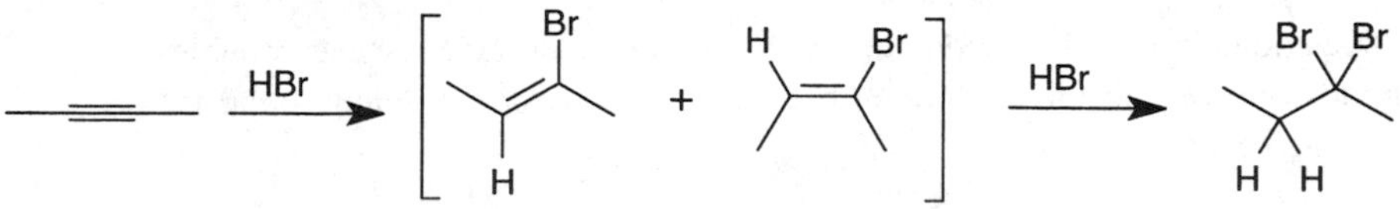

(d)   Molecular bromine adds to alkynes in a *trans* fashion.

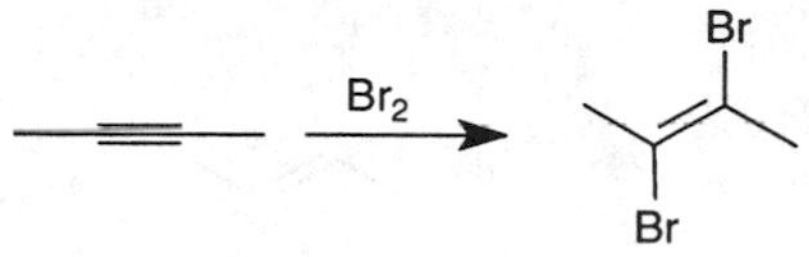

(e)   A tetrabromide is formed from alkynes by the reaction with two equivalents of molecular bromine.

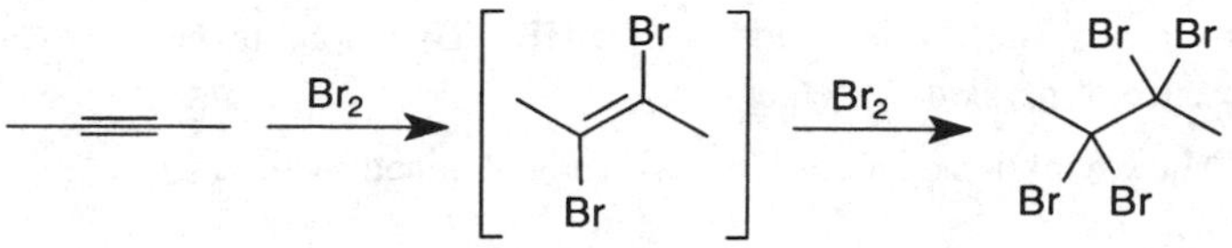

(f)   Aqueous sulfuric acid effects hydration of alkynes with a Markovnikov orientation. The enol formed tautomerizes to the corresponding carbonyl compound.

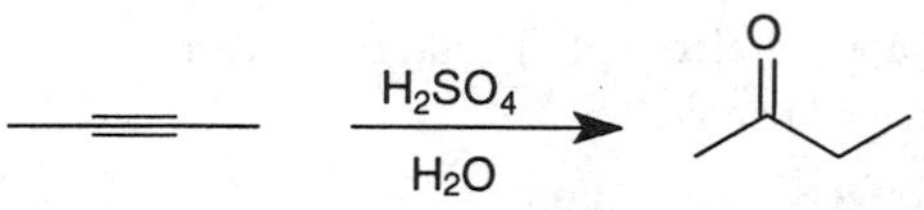

(g)   Ozonation, followed by reduction, cleaves triple bonds, converting each of the *sp*-hybridized carbons of the alkyne to the corresponding carboxylic acid.

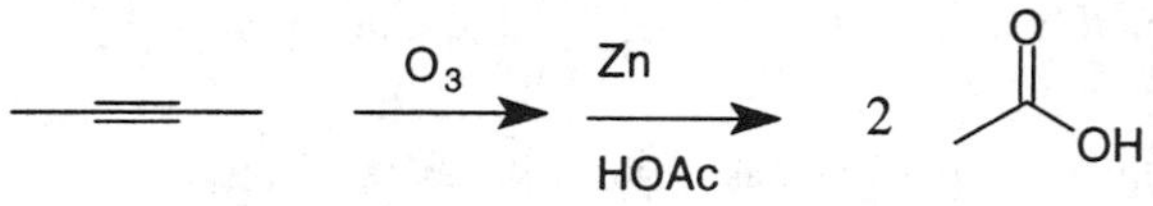

(h)   Hot permanganate gives the same products with alkynes as does ozonolysis.

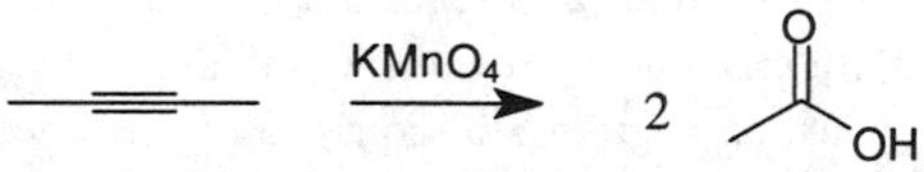

Pr 10.3   (a)   A diene is completely hydrogenated upon treatment with excess hydrogen in the presence of a noble metal catalyst.

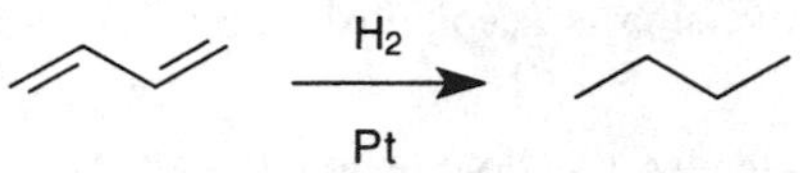

(b)   The cyclic bromonium ion formed by the electrophilic attack of bromine on this conjugated diene is opened by attack by bromide ion at C-2 or C-4. The former reaction leads to a 1,2-dibromide; the latter to a 1,4-dibromide.

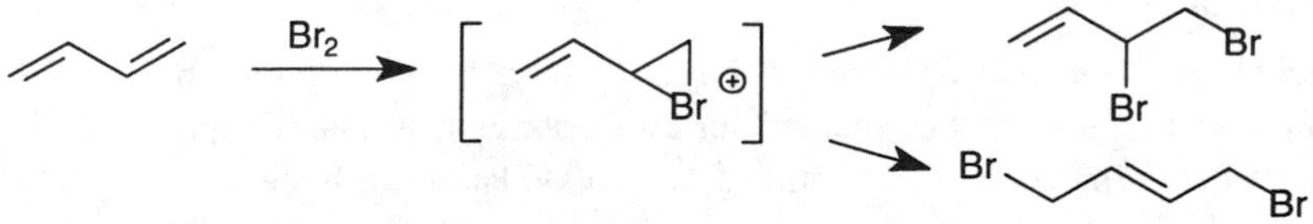

(c)   Dilute H_2SO_4 effects hydration of the diene. Because the intermediate cation is allylic, it is captured by water in both a 1,2 and a 1,4 fashion.

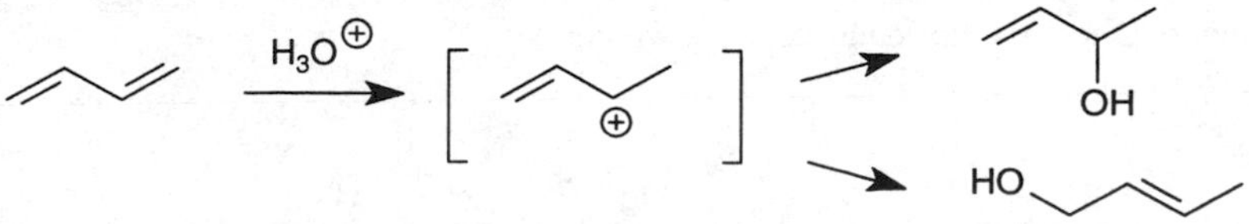

(d)   Electrophilic addition of HCl is governed by the same features as the electrophilic hydration reaction discussed in part *c*.

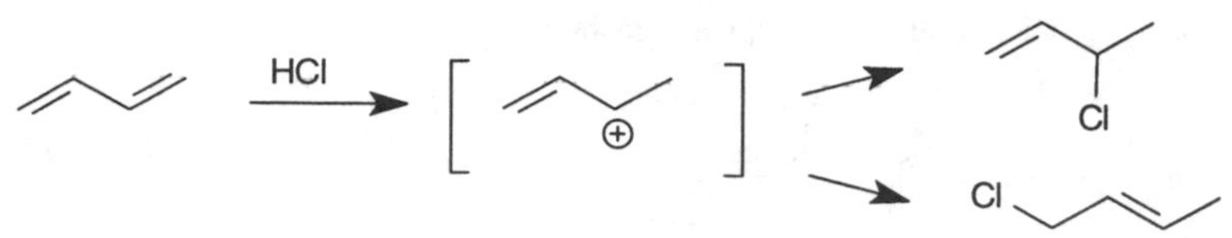

Pr 10.4 Several different routes for the indicated conversions are possible. One possibility that involves the reactions considered in this chapter is given here.

(a) This reaction calls for anti-Markovnikov addition of HBr. This is achieved with HBr in the presence of peroxides in ether.

(b) This product is the Markovnikov adduct of HBr, which is obtained by the use of HBr in a polar solvent.

(c) This anti-Markovnikov hydration product is formed by hydroboration–oxidation. The required reagents are B_2H_6, followed by basic aqueous hydrogen peroxide.

(d) This Markovnikov hydrate is formed by normal electrophilic addition, with either dilute aqueous acid or oxymercuration–demercuration.

(e) Catalytic hydrogenation (H_2 on Pt or Pd) effects this reaction.

(f) Double dehydrobromination of 1,2-dibromobutane yields the desired product. The dibromide is formed from 1-butene upon treatment with Br_2 in CCl_4. The elimination takes place upon treatment with $NaNH_2$ in liquid ammonia (Chapter 9).

(g) 2-Butyne could be formed by the same reactions as in part *f* if 1-butene were converted to 2-butene. This can be attained by the reaction in part *b*, followed by Zaitsev elimination from this product to yield 2-butene. Zaitsev elimination from 2-bromobutane is induced, for example, by treatment with a large base, such as potassium *t*-butoxide. Thus, an appropriate sequence of reagents would be: (1) HBr; (2) KO-*t*-Bu in *t*-BuOH; (3) Br_2 in CCl_4; (4) $NaNH_2$ in NH_3.

(h) *s*-Butyllithium is the metallation product of 2-bromobutane. When 2-bromobutane is prepared as in part *b*, the desired conversion is accomplished by treatment with lithium metal in ether.

(i) Reaction with hot potassium permanganate cleaves the double bond of alkenes, forming the corresponding carboxylic acid fragments. Oxidative degradation of 1-butene with $KMnO_4$ gives formic acid, which is further oxidized to carbon dioxide. Thus, the only (easily) isolated organic product is propanoic acid.

(j) 2-Bromobutane (from part *b*) is converted to the corresponding amine by treatment with a large excess of ammonia or by a Gabriel synthesis (Chapter 8). The latter reaction consists of treatment of an alkyl halide with the anion of phthalimide, followed by hydrolysis with aqueous acid or treatment with hydrazine.

(k) An additional three carbon atoms must be added in the conversion of 1-butene to 2-heptanone. One possible sequence is shown below.

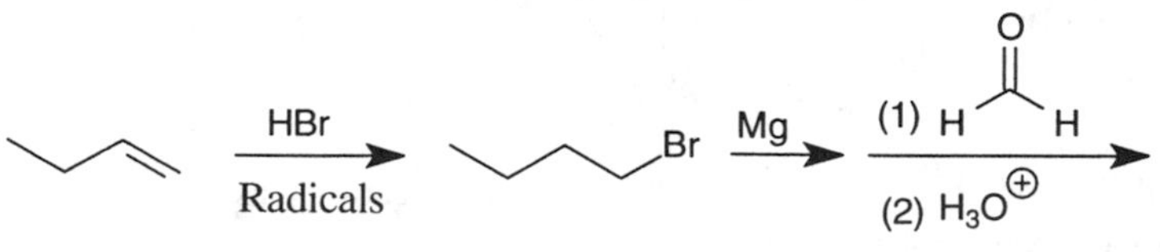

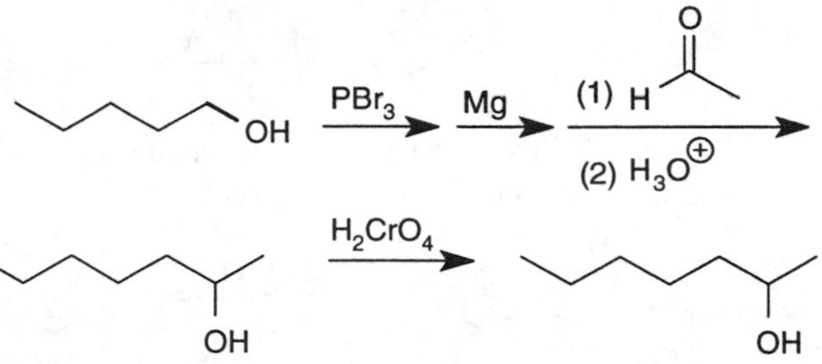

(l)   The sequence used to make 2-heptanone in part *k* can be modified to make hexanoic acid by substituting formaldehyde for acetaldehyde in the second Grignard reaction. Oxidation of 1-hexanol to hexanoic acid can be accomplished with H_2CrO_4.

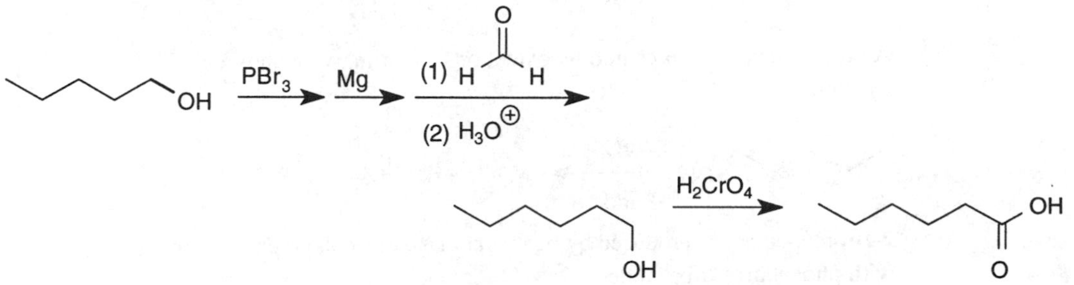

(m)   1-Hexanol can be prepared either by the route in part *l* or by treatment of a Grignard reagent with ethylene oxide. The necessary Grignard reagent is *n*-butylmagnesium bromide, which is prepared from 1-bromo-butane, synthesized as in part *a*.

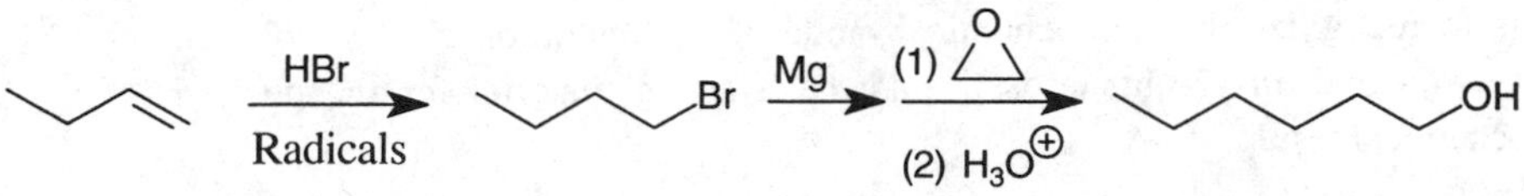

Pr 10.5   Again, there are several possible correct answers. One example is provided for each part of this exercise.

(a)   An alcohol can be converted directly to the corresponding halide upon treatment with thionyl chloride or phosphoryl chloride.

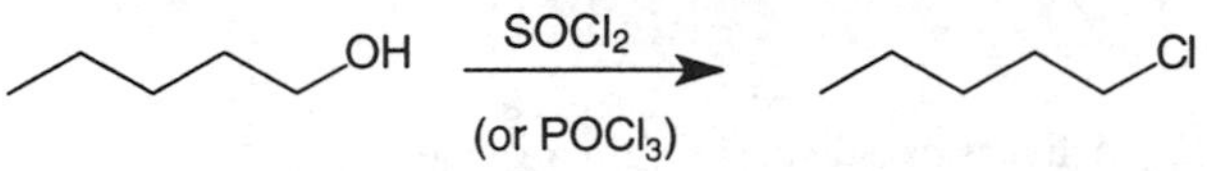

(b)   The dehydration of a primary alcohol is usually accomplished by heating with catalytic amounts of acid in an inert solvent. Benzene serves not only as a convenient solvent but also as a means for the azeotropic removal of water as it is formed in the reaction.

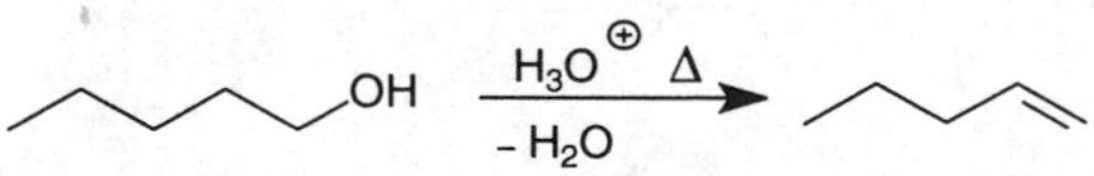

(c)   1-Pentanol can be isomerized to 2-pentanol by dehydration (as in part *b*), followed by Markovnikov hydration (treatment with dilute aqueous acid).

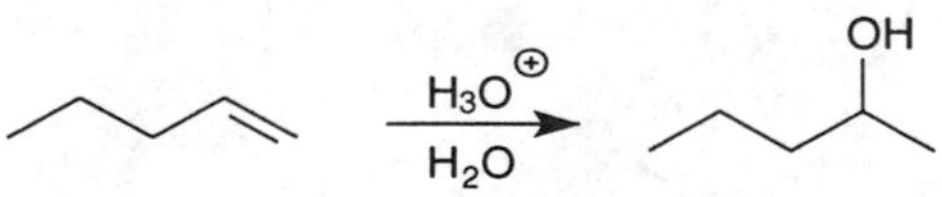

(d)   Dehydration of 2-pentanol (part *c*) under equilibrating conditions (heating in aqueous acid) produces the Zaitsev elimination product.

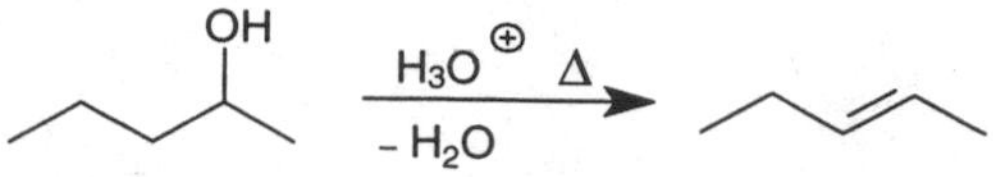

(e) 2-Pentanol (prepared as in part *c*) can be converted to 2-pentanone by oxidation with chromic acid.

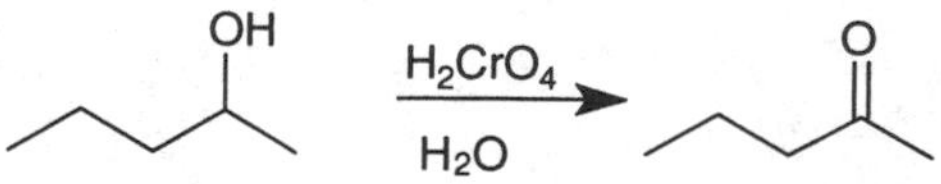

(f) 1-Pentanol is oxidized to the corresponding acid with chromic acid in an aqueous medium.

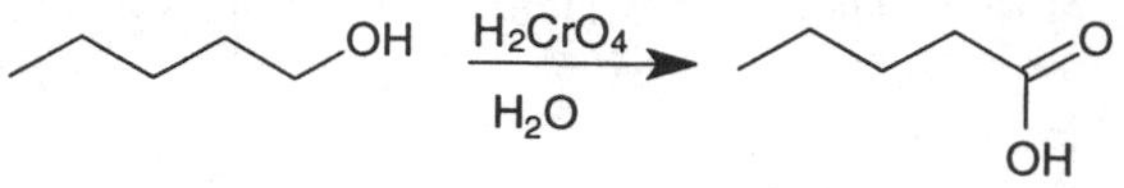

(g) An aldehyde can be prepared by oxidation of a primary alcohol with CrO_3 in pyridine.

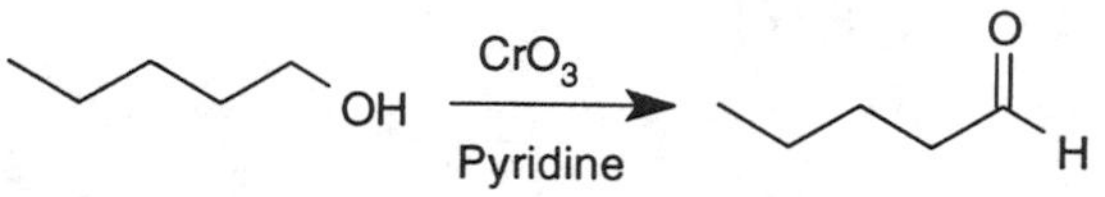

(h) 2-Bromopentane is produced by treatment of 2-pentanol, prepared as in part *c*, with phosphorus tribromide.

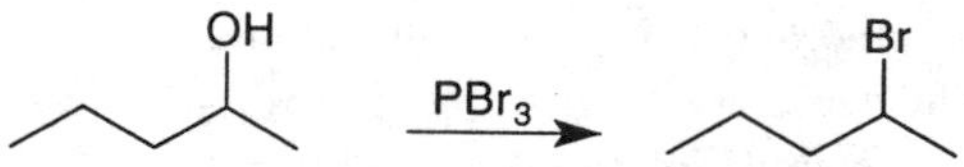

Pr 10.6   Each of the following examples consists of an addition reaction of 2-butene. Your task is to recognize the stereochemical mode of each addition so that you can choose *cis*- or *trans*-2-butene as the correct starting material for the desired stereochemical result.

(a) Because bromination is an *anti* addition, the *meso* compound is formed from *trans*-2-butene. (Electrophilic bromination of *cis*-2-butene gives rise to a racemic mixture of the *d,l* isomers.)

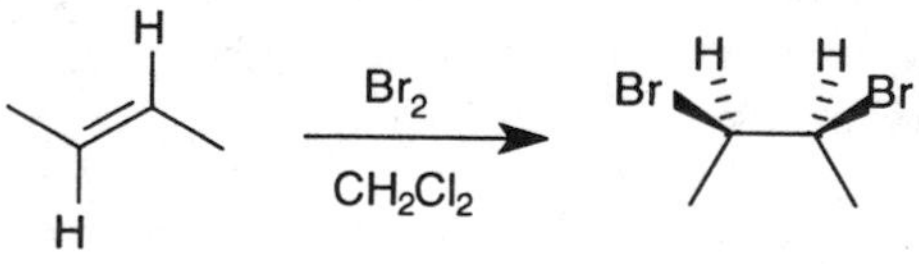

(b) Catalytic hydrogenation with D_2 delivers two deuteriums in a *syn* addition. In order to prepare the *d,l* mixture, D_2 is added to *trans*-2-butene. (Catalytic hydrogenation of *cis*-2-butene produces the *meso* compound.)

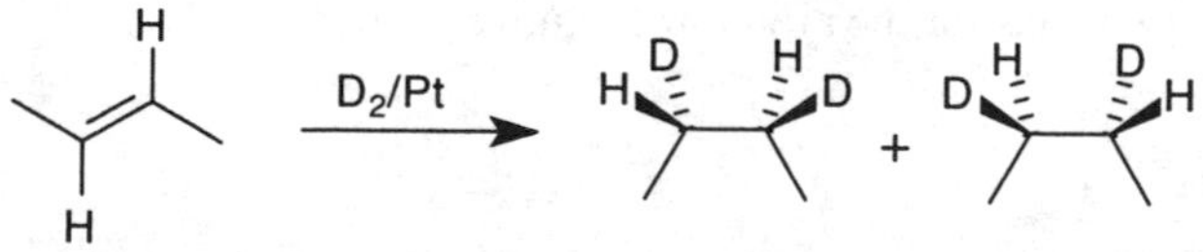

(c) Electrophilic addition of chlorine is through a chloronium ion, opening of which achieves an *anti* addition. A *d,l* racemic mixture is therefore obtained from *cis*-2-butene. (As in part *a*, chlorination of *trans*-2-butene produces the *meso* compound).

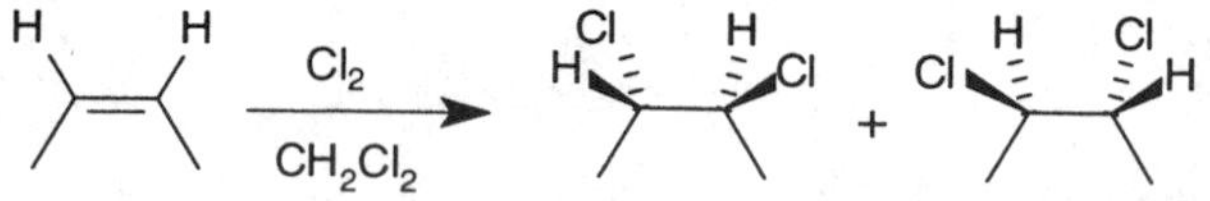

(d)   The addition of HBr in a polar solvent takes place through a carbocationic intermediate. A racemic mixture of 2-bromobutane is obtained from the 2-butyl cation. This intermediate is formed by protonation of either 1- or 2-butene, because the regiochemistry of the protonation is determined by cation stability. Therefore, this same product is formed from 1-butene, *cis*-2-butene, or *trans*-2-butene.

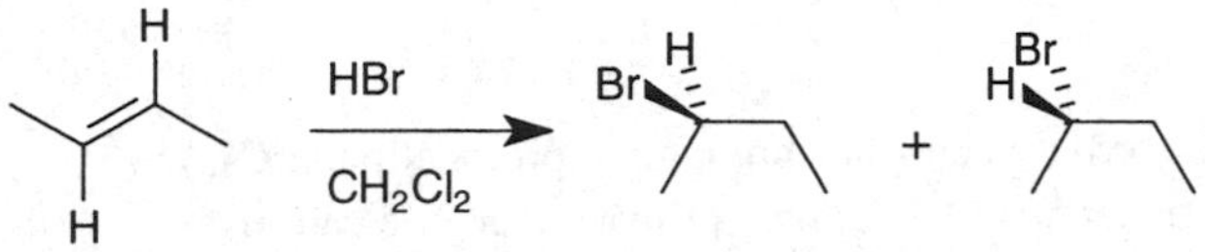

(e)   Carbene addition in the singlet state is a stereospecific *syn* addition. To obtain the *d,l* product, begin with *trans*-2-butene. (Carbene addition to *cis*-2-butene produces the *meso* compound).

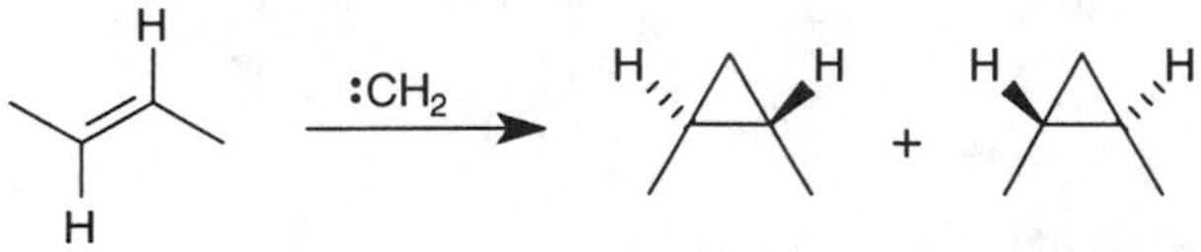

(f)   Like a carbene addition, epoxidation is stereospecifically *syn*. The *meso* epoxide is therefore obtained from *cis*-2-butene. (As with the carbene addition in part *e*, epoxidation produces the *d,l* epoxide of *trans*-2-butene.)

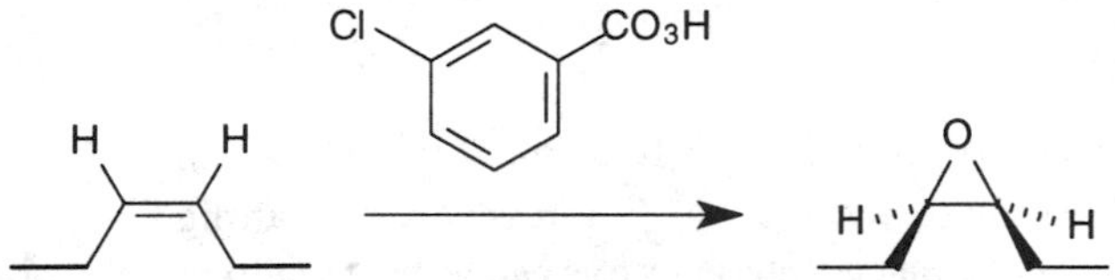

Pr 10.7   The rate of a chemical reaction is determined by the activation energy of the rate-determining step. In an acid-catalyzed hydration of an alkene, the rate-determining step is the formation of a carbocation intermediate. Therefore, two factors are important: the stability of the intermediate cation (the more stable the cation, the faster the reaction), and the stability of the starting alkene (the less stable the reactant, the faster the reaction). In general, the difference in stability between 3°, 2°, and 1° cations is greater than the difference in stability between pairs of various alkenes.

(a)   2-Methyl-1-pentene (forms a tertiary cation) > 1-hexene > 2-hexene. Both hexenes can form secondary cations. However, the activation energy to form the this cation is higher for 2-hexene because this alkene is more stable than is 1-hexene.

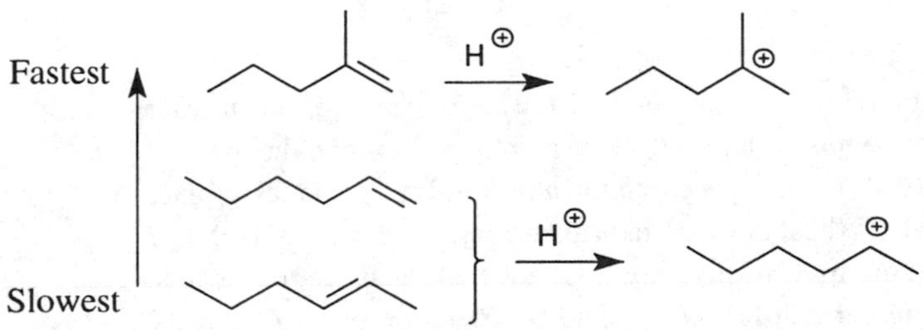

(b)   2-Methylpropene (forms a tertiary cation) > *cis*-2-butene > *trans*-2-butene. Both 2-butenes form the same secondary cation. However, the activation energy to form the this cation is higher for the *trans* isomer because this alkene is more stable than is the *cis* isomer. Thus, the activation energy for protonation of the former is higher.

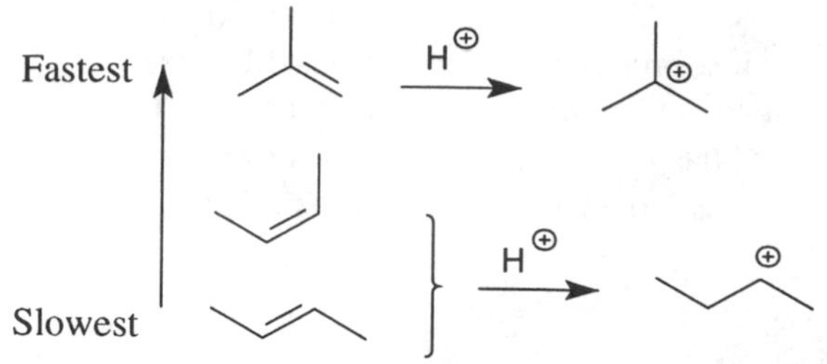

Fastest / Slowest diagram (c)

(c) 2-Phenyl-1-butene (forms a tertiary benzylic cation upon protonation at C-3) > 1-phenyl-1-butene (protonation at C-2 forms a primary benzylic cation) > 1-phenyl-2-butene (forms a mixture of secondary cations by protonation at either C-2 or C-3, although a hydrogen shift from C-1 gives a rearranged cation which is the same as that obtained by protonation of 2-phenyl-1-butene).

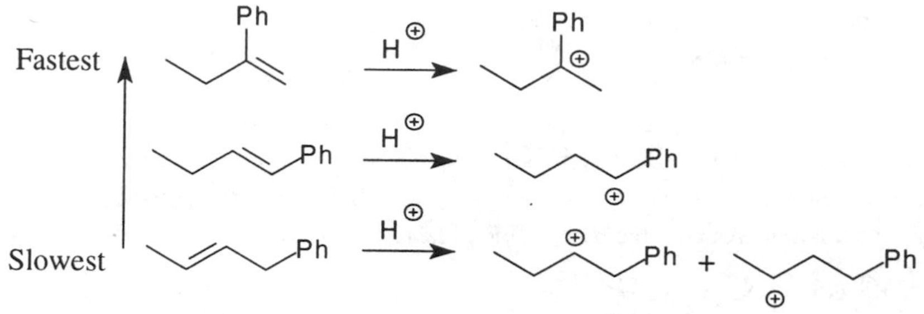

Pr 10.8 In aqueous acid, C-1 of 1-pentene is protonated to form the 2-pentyl cation. Deprotonation by an E1 mechanism (Chapter 9) at C-1 regenerates starting material, whereas that at C-3 produces 2-butene. Steric factors can be best analyzed in the Newman projections obtained by viewing down the C-2–C-3 bond of the cationic intermediate. The *trans* isomer is formed from the conformer represented at the left, in which the methyl group at C-2 is directed toward the region of space opposite to that of the ethyl group at C-3. The *cis* isomer is formed from the conformer represented at the right, in which the alkyl groups on C-2 and C-3 are directed toward the same region of space.

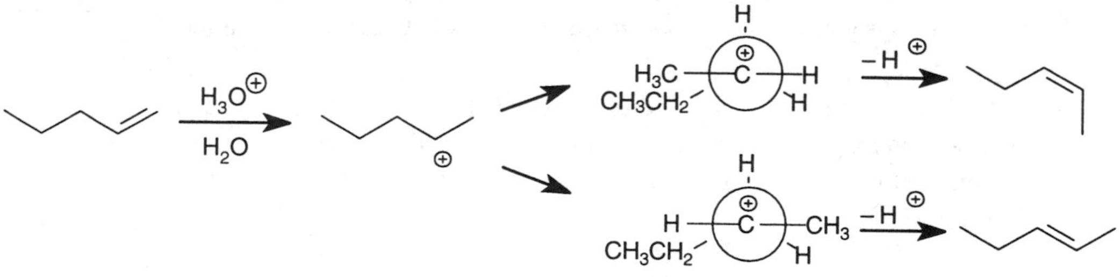

Pr 10.9 After initiation by the thermal decomposition of the peroxides present in trace amounts, radical addition begins with the attack of a bromine atom on the diene to form the more stable radical. As in an electrophilic attack, this addition takes place at C-1, producing an allylic radical in which radical character is present at both C-2 and C-4. Each of these atoms can therefore abstract hydrogen from a second molecule of HBr, forming either a 1,2- or 1,4-adduct. There are two 1,4-adducts, and the more stable *trans* isomer is formed in greater quantity than the *cis* isomer.

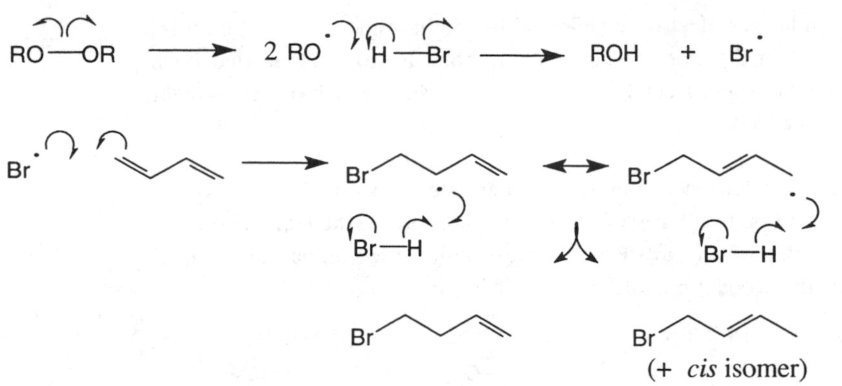

RO—OR ⟶ 2 RO·   H—Br ⟶ ROH + Br·

Br·

Br   Br   ↔   Br

Br—H   Br—H

Br   Br
(+ *cis* isomer)

Pr 10.10   Both electrophilic chlorination and chlorohydrin formation begin with an attack by Cl_2 as an electrophile to form a chloronium ion. Back-side attack on the chloronium ion by chloride gives the *meso* dichloride. The same *meso* compound is formed if the initial attack had taken place on the opposite face of the double bond or if chloride had opened the chloronium ion by attack at the other carbon. (Use molecular models to convince yourself that this is correct.)

When water acts as a nucleophile in the same way as chloride ion to open the chloronium ion, an oxonium ion is formed. Deprotonation of the resulting oxonium ion forms the chlorohydrin discussed in the Problem. The same stereochemistry [equal amounts of the (2-R,3-S) and the (2-S,3-R) enantiomers] is observed in the chlorohydrin as in the electrophilic chlorination. The only difference is that because the two attached groups in the chlorohydrin are not identical, the two centers of chirality do not reflect to each other through a mirror plane. Hence, the product is a racemic mixture, rather than a *meso* compound.

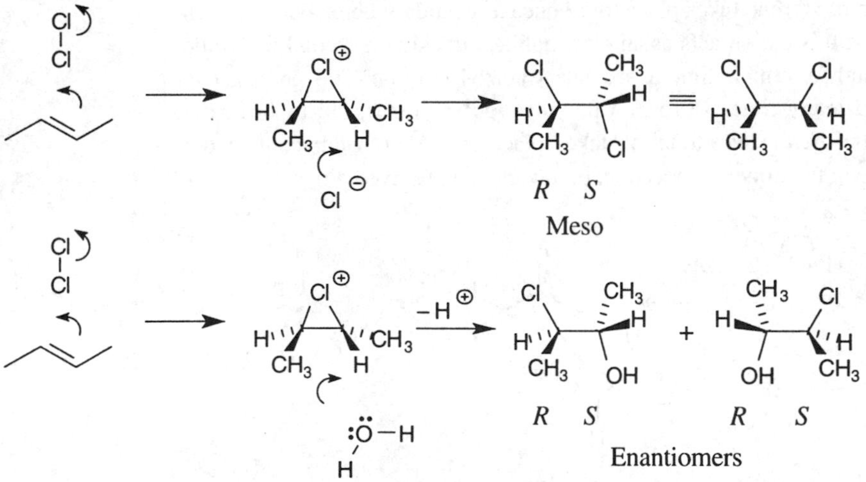

R   S
Meso

R   S   *R   S*
Enantiomers

Pr 10.11   Alkynes have a region of higher π electron density than do alkenes and, hence, are more easily approached by electrophiles. Furthermore, each π bond in an alkyne is individually less stable than that in a comparable alkene. However, the carbocation formed by electrophilic attack on an alkyne is vinylic and is less stable than the sp^2-hybridized cations formed in an electrophilic attack on a typical alkene. The Hammond postulate teaches us that in endothermic transformations, the transition-state energy is best approximated by the energy of the intermediate cation formed in the rate-determining step. The formation of an unstable vinyl cation is therefore slower than that of the cation obtained by protonation of an alkene, despite the higher electron density of the alkyne. In the product of addition of a single equivalent of bromine to an alkyne, a dibromoalkene, the π bond is quite electron-poor because of inductive electron withdrawal by the bromine atoms. Therefore, electrophilic attack on the electron-poor double bond is slower than that on a simple alkene, and the reaction can be stopped at the stage at which one equivalent of bromine has been added.

Pr 10.12 Both singlet carbene addition and epoxidation are stereospecific *syn* additions. Attack on the top and bottom faces of *trans*-2-butene leads to the (*R,R*) and (*S,S*) stereoisomers of both products, as shown. These enantiomeric products are formed in equal amounts, the product mixture is optically inactive.

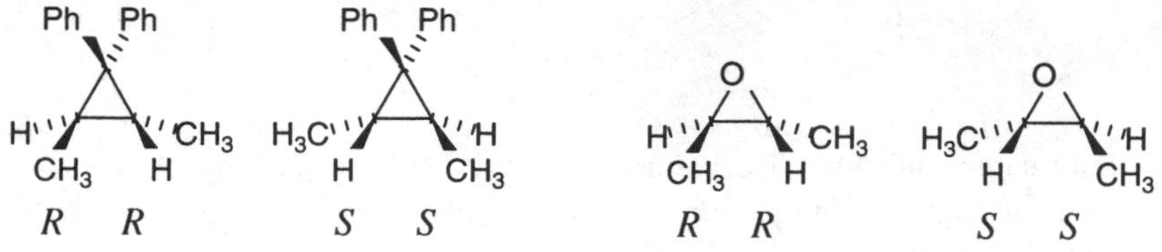

Pr 10.13 (a) Protonation of the alkene is at the 1-position to generate the more stable secondary carbocation. A rapid 1,2 shift of a methyl group from C-3 produces an even more stable carbocation that is tertiary and benzylic. This cation is then trapped by bromide ion to yield the observed product.

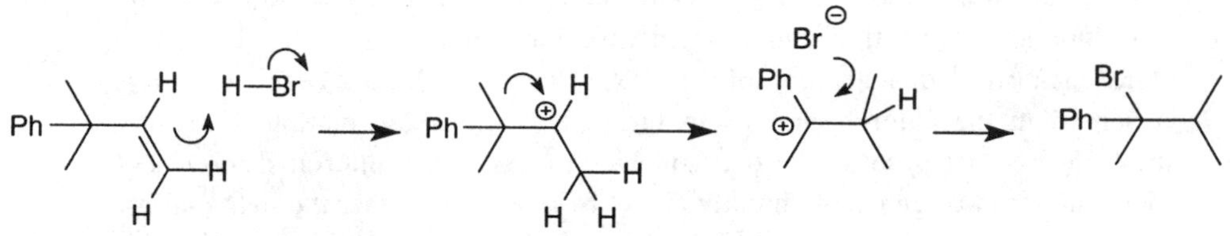

(b) Protonation of styrene takes place to produce a secondary benzylic cation. In neat styrene, this cation acts as an electrophile, attacking a second molecule of alkene and forming a dimeric secondary benzylic cation. This process can be repeated many times to form a polymer (polystyrene), but to form the product cited here, deprotonation takes place at the atom adjacent to the positive site in the dimer carbocation, producing the observed dimer.

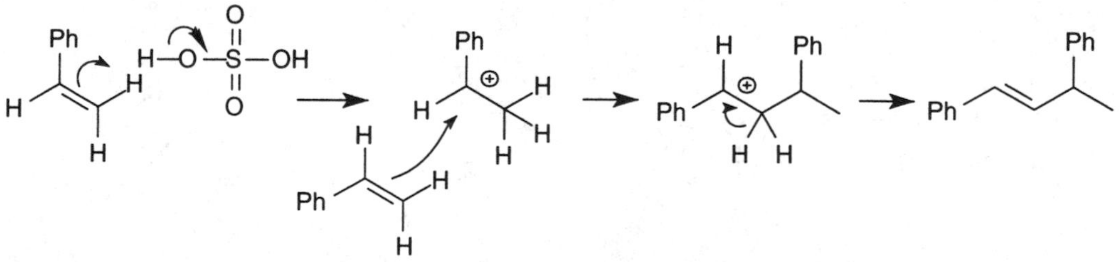

(c) Begin by drawing ozone with formal positive and negative charges (as in a Lewis dot structure) to see where electrophilic and nucleophilic attack might be likely. The observed molozonide product can be formed through a concerted six-electron transition state that shifts electron density from the alkenyl π bond onto the formally positively charged oxygen as the formally negatively charged oxygen traps the incipient carbocation.

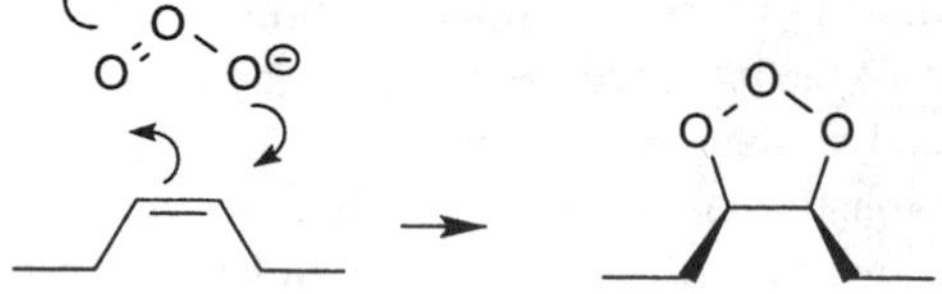

(d) Diborane is in equilibrium with BH$_3$. Monomeric trivalent boron compounds are highly electron deficient and actively seek centers of electron density such as that in a carbon–carbon double bond. The B—H bond in BH$_3$ therefore adds in concerted fashion across the C=C bond. The larger BH$_2$ group binds at the less sterically hindered site through a concerted transition state. This concerted reaction must be *syn*, but there is no stereochemical probe in the example here to test for that predicted stereocontrol.

(e) The addition of chlorine to an alkene proceeds through an intermediate chloronium ion.

(f) In the presence of peroxides, thermal or photochemical initiation produces trace amounts of alkoxy radicals. These species can abstract a hydrogen atom from H—Br, forming alcohol and a bromine atom, which initiates the addition. By attacking the less highly substituted end of the alkene, bromine forms a carbon–bromine bond and a tertiary radical. This radical then abstracts hydrogen from a second molecule of HBr, completing the addition and generating another bromine atom that can carry on the chain.

(g) The oxygen–oxygen σ bond in a peracid is both polar and polarizable. Formation of epoxides from the reaction of alkenes with peracids is believed to take place through a concerted cyclic transition state leading in one step to the products.

(h) Protonation of a terminal alkyne is at the less substituted carbon so that the more stable cation is formed. An enol is formed when the oxonium ion produced by trapping this cation with water is deprotonated. Acid-catalyzed tautomerization of the enol to a methyl ketone by protonation on carbon and deprotonation on oxygen completes the reaction. (Mercuric sulfate increases the rate of this reaction by providing Hg^{2+} ion as a Lewis acid, replacing protons in the first step. As discussed in the text, a mercuronium ion may also be an intermediate. For the purposes of a first course in organic chemistry, the simple hydration mechanism proposed here is adequate to rationalize the observed stereochemistry.)

Pr 10.14 The chemical tests with which you should now be familiar include: 1) the decolorization of the red solution of bromine in carbon tetrachloride upon addition to a multiple bond; 2) the color change of a red-orange chromic acid solution to deep green as it is reduced while an organic molecule is oxidized; 3) the fading of the intense purple color of potassium permanganate as it is reduced and an organic compound is oxidized; 4) the Lucas test, which distinguishes primary, secondary, and tertiary alcohols on the basis of their S_N1 reactivity and consequent solubility of the substituted alkyl chloride; and 5) the formation of a precipitate upon treatment of a terminal alkyne with basic aqueous silver ion. Many spectroscopic probes are possible to distinguish these pairs of compounds. One or two are given here.

(a) Cyclohexene decolorizes bromine; cyclohexane does not. Cyclohexene exhibits vinyl protons in its ^{1}H NMR spectrum; cyclohexane does not. Cyclohexene shows three signals in its ^{13}C NMR spectrum; cyclohexane shows one.

(b) Both hexene and hexyne decolorize bromine, but only 1-hexyne forms a precipitate with $Ag(NH_3)_2OH$. The unsymmetrical triple bond appears in the characteristic region of the infrared spectrum. The region from 1700 to 2900 cm^{-1} is blank in the IR spectrum of 1-hexene.

(c) Hexene decolorizes bromine; hexanol does not. Hexanol causes a color change of chromic acid; hexene does not. The broad IR band at around 3600 cm^{-1} characteristic of an OH stretch is present for hexanol and absent for hexene.

(d) Hexene decolorizes bromine; 2-bromohexane does not. Hexene exhibits vinyl protons in its ^{1}H NMR spectrum; bromohexane does not. The presence of bromine is clear from the pair of peaks reflecting the isotopic abundance of bromine and its +2 isotope in the mass spectrum.

(e) Both 2-hexene-1-ol and 1-hexanol cause a color change of a chromic acid solution, but only the alcohol containing an alkenyl group also decolorizes bromine. Both compounds show strong OH stretches in their IR spectra, but the unsaturation of 2-hexene-1-ol is evident from the signals assigned to the vinyl protons in its ^{1}H NMR spectrum, which are absent in that of 1-hexanol.

(f) A positive Lucas test is attained with 2-hexanol, but not with 2-bromohexane. The presence of the alcohol group is evident in the OH stretch region of the IR spectrum of hexanol, which is blank with 2-bromohexane. Bromine is quite evident from the pair of peaks in bromohexane's mass spectrum.

(g) When an aldehyde is oxidized by chromic acid, a color change is observed that is not seen with hexane. Hexanal shows a strong carbonyl stretch at about 1700 cm^{-1}, which is absent from the IR spectrum of hexane.

Pr 10.15 Because a *syn* diol is formed and because the hydrolysis takes place with retention of configuration, the C—O bonds in the osmate ester must be *cis* to each other. A *syn* addition usually implies that both bonds are formed simultaneously on one face of the double bond (as in hydroboration). This can be achieved in the six-electron transition state shown, in which the formal oxidation level of osmium changes, taking on electrons as the organic substrate is oxidized.

10.16 The formula of limonene tells us that the sum of the number of double bonds and rings in this compound is three. (A saturated acyclic C_{10}-hydrocarbon would have 22 hydrogens.) Because limonene takes up two equivalents of H_2, there must be one ring and two multiple bonds. The ozonolysis results tell us that these multiple bonds must be two double bonds, not a triple bond, because aldehyde and ketone products are formed rather than a carboxylic acid. The tricarbonyl compound has C=O groups present at sites that initially participated in double bonds. Because this ozonolysis product is acyclic, one of the original double bonds must be between two of the carbonyl groups present in the tricarbonyl compound. Formaldehyde is the other ozonolysis product, so the other acyclic double bond must contain a terminal CH_2 group.

We can therefore piece together the broken ring in one of three ways: by binding the aldehyde to the ketone at the left, or the aldehyde to the ketone at the right, or the two ketones together. The remaining carbonyl group would then be capped with the CH_2 group. This analysis yields the cyclic structures shown at the right.

There is no symmetry in any of the three possible structures, so all three would show ten ^{13}C NMR signals. They would have to be distinguished by a more detailed analysis of the chemical shifts and splitting patterns in their ^1H or ^{13}C NMR spectra or by analysis of their mass spectral fragmentation patterns.

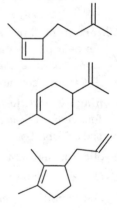

Pr10.17 Hydroboration-oxidation accomplishes a *syn* addition of H—OH with anti-Markovnikov regiochemistry. With an alkyne, an enol is produced. Tautomerization leads to a carbonyl compound with a regiochemistry opposite to that produced with acid-catalyzed hydration.

(a) Hexanal

(b) 3-Hexanone

(c) Phenylacetaldehyde ($PhCH_2CHO$)

Important New Terms

Acid-catalyzed: descriptor of a reaction that is accelerated in the presence of an acid but in which the acid is not consumed in forming the product (10.1)

1,2-Addition: a mode of addition in which two groups are bound to adjacent carbons in the product (10.1)

1,4-Addition: a mode of addition in which two groups are bound to the ends of a four-atom system in the product; *see also* **Conjugate addition** (10.1)

Alkylborane: a functional group in which carbon is attached to a trivalent boron atom (10.4)

***Anti* addition:** the formation of an addition product by delivery of an electrophile and a nucleophile to opposite faces of a double (or triple) bond (10.2)

Anti-Markovnikov regiochemistry: that taking place in the opposite sense from that predicted by Markovnikov's rule; an addition in which a proton is delivered to the more substituted carbon and the nucleophile to the less substituted carbon of an alkene (10.3)

Bromonium ion: a three-membered, cyclic, cationic intermediate in which bromine bears formal positive charge; formed by the addition of Br^+ (or a source of this species) to an alkene (10.2)

Chloronium ion: a three-membered, cyclic, cationic intermediate in which chlorine bears formal positive charge; formed by the addition of Cl^+ (or a source of this species) to an alkene (10.2)

Conjugate addition: the addition of a reagent across a four carbon conjugated π system, producing a 1,4 adduct and a double bond between C-2 and C-3 (10.2)

Cyclic bromonium ion: *see* **Bromonium ion** (10.2)

Cyclic chloronium ion: *see* **Chloronium ion** (10.2)

Cyclic halonium ion: *see* **Halonium ion** (10.2)

Dimer: a compound containing most or all of the atoms of two molecules of a starting material (10.2)

Electrophilic addition: a chemical reaction in which a $C=C$ π bond is replaced by two σ bonds upon reaction with an electrophilic reagent, often a proton, followed by addition of a nucleophile (10.1)

Epoxidation: the preparation of an epoxide from an alkene (10.4)

Epoxide: a three-membered ring functional group containing oxygen (10.4)

Halogenation: the formal addition of dihalogen to an alkene (10.2)

Halonium ion: a three-membered cyclic cationic intermediate in which a halogen bears formal positive charge; formed by the reaction of X^+ with an alkene; important for chlorine and bromine (10.2)

Hydration: the formal addition of water (10.1)

Hydroboration: the addition of a B—H bond to an alkene (10.4)

Hydroboration–oxidation: a reaction sequence used to achieve anti-Markovnikov hydration of an alkene; initiated by concerted *syn* addition of borane, followed by oxidation with basic hydroperoxide (10.4)

Hydrohalogenation: the formal addition of HX (X = halide) to a multiple bond (10.1)

Hydrogen peroxide: H_2O_2 (10.4)

Hydronium ion: H_3O^+ (10.1)

Markovnikov's rule: empirical generalization that the less highly substituted position in an unsymmetrically substituted alkene is attacked by the electrophile in an electrophilic addition (10.1)

Molozonide: a five-membered ring containing three oxygen atoms; produced by the direct addition of O_3 to an alkene (10.4)

Oxidative degradation: the cleavage of a carbon skeleton (often at a $C=C$ double bond) with the introduction of new carbon–oxygen bonds (10.4)

Oxirane: *see* **Epoxide** (10.4)

Oxymercuration–demercuration: a reaction sequence used to achieve Markovnikov hydration of an alkene without accompanying skeletal rearrangements; initiated by treatment with mercuric acetate in aqueous acid, followed by $NaBH_4$ (10.2)

Ozonation: the addition of O_3 to an alkene (10.4)

Ozone: an electrophilic allotrope of oxygen that exists in a zwitterionic form in which the central oxygen formally bears positive charge; O_3 (10.4)

Ozonide: a five-membered ring containing three oxygen atoms; produced by rearrangement of a molozonide in the addition of O_3 to an alkene (10.4)

Ozonolysis: a sequence in which a $C=C$ double bond is oxidatively converted to two carbonyl groups through sequential treatment with O_3, followed by Zn in acetic acid (10.4)

Peracid: RCO_3H; an oxygenated relative of a carboxylic acid (10.4)

Peroxide: a functional group containing an oxygen–oxygen σ bond; ROOR (10.4)

Prochiral: descriptor of an achiral center that can become a center of chirality either by replacement of one of two identical groups or by addition to a π system (10.1)

Protecting group: a functional group that can be formed reversibly and lacks the reactivity of another portion of the molecule (10.2)

Radical hydrobromination: an anti-Markovnikov hydrobromination of an alkene taking place through the radical addition of a bromine atom; initiated by peroxide decomposition (10.3)

Simmons–Smith reaction: formation of a cyclopropane through stereospecific carbene addition to an alkene through treatment of a vicinal dihalide with zinc–copper couple in the presence of an alkene (10.4)

Stereorandom: descriptor of a reaction without any stereochemical preference (10.1)

***Syn* addition**: the formation of product by delivery of an electrophile and nucleophile to the same face of a multiple bond (10.4)

Trimer: a compound containing most or all of the atoms of three molecules of a starting material (10.2)

Electrophilic Aromatic Substitution

11

Key Concepts

Mechanism of electrophilic aromatic substitution

Resonance stabilization of the cationic intermediate in aromatic substitution

Rationale for substitution (rather than addition) with aromatic π-systems

Methods for activating electrophiles for aromatic substitution

Synthesis of aryl halides by aromatic halogenation

Synthesis of nitrobenzenes by aromatic nitration

Synthesis of anilines by reduction of nitrobenzenes under acidic (Clemmensen reduction) or basic (Wolff-Kishner reduction) conditions

Synthesis of benzenesulfonic acids by aromatic sulfonation

Synthesis of alkyl aromatics by Friedel-Crafts alkylation or by reduction of Friedel-Crafts acylation products

Synthesis of aryl ketones by Friedel-Crafts acylation

Synthesis of benzoic acids by Friedel-Crafts alkylation or acylation, followed by permanganate oxidation

Synthesis of azo dyes with aryl diazonium salts

Ring substituents as electron donors or acceptors

Directive and kinetic effects of donor and acceptor substituents on further ring substitution

Rationale for the unusual directive effects of halogens

σ- and π-electronic effects

Incorporating directive effects to synthetic planning

Directive effects in electrophilic substitution in polycyclic aromatics

Answers to Exercises

Ex 11.1   Association of one bromine atom of Br_2 renders the other bromine atom electrophilic, and initiates electron flow from the π-system of the ring toward the electrophile, producing a C—Br σ bond. The resulting arenium ion (a cyclic pentadienyl cation) is rearomatized by loss of a proton.

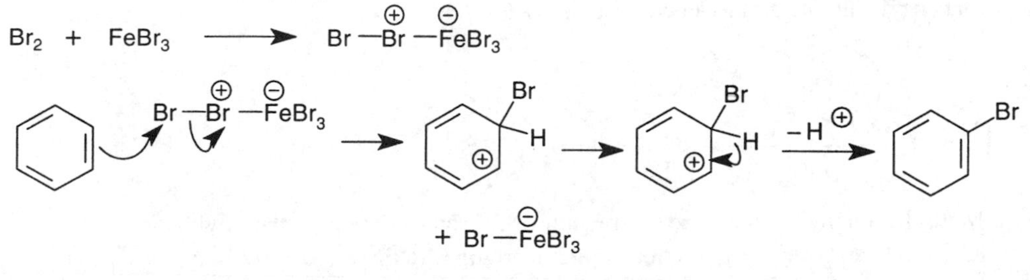

Ex 11.2   The NO_2^+ cation and CO_2 are isoelectronic; that is, they have the same bonding electronic structure, but contain different atoms (and hence, different charges). Both nitrogen and carbon can form four bonds. Nitrogen and carbon in NO_2^+ and CO_2 both have a σ bond and a π bond to each oxygen atom and, therefore, are *sp*-hybridized. Thus, both species are linear.

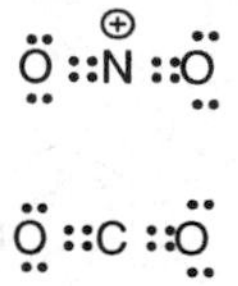

Ex 11.3   Upon complexation with a Lewis acid, 1-chloro-2,2-dimethylpropane undergoes a cationic rearrangement, with a methyl group shifting from C-2 to C-1 as the C—Cl bond is broken. The 2-methyl-2-butyl cation thus produced then attacks the ring, forming an arenium cation bearing a new C—C bond. Deprotonation restores aromaticity in the product.

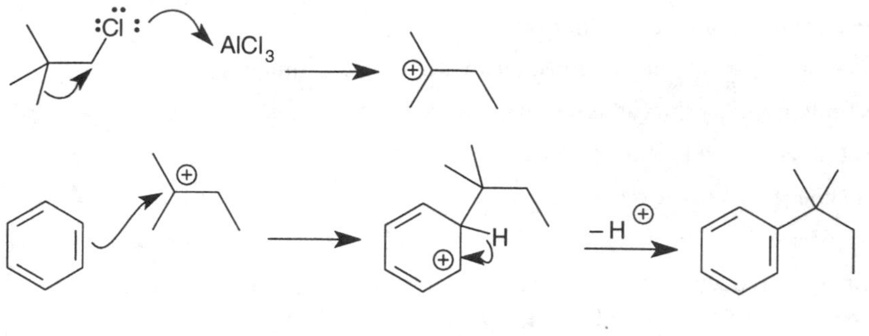

Ex 11.4   Friedel-Crafts alkylations proceed by the reaction of carbocationic electrophiles with aromatic systems. Because cations take part in these reactions, the product incorporates a rearranged skeleton whenever rearrangement to a more stable carbocation is possible.

(a)   *s*-Butylbenzene. AlCl$_3$ acts as a Lewis acid to break the C—Cl bond of *n*-butyl chloride to form a carbocation-like electrophile that rearranges by a 1,2 hydrogen shift to the secondary *s*-butyl cation. That cation then attacks the ring and effects an aromatic substitution.

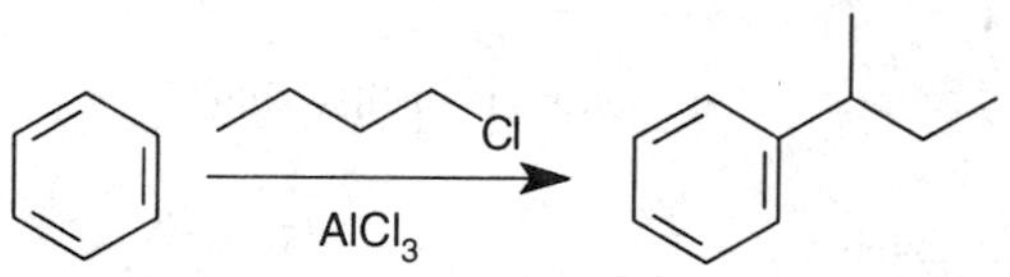

(b)   *s*-Pentylbenzene is also formed from this reactant, because the *s*-pentyl cation is formed directly upon interaction with the Lewis acid; no rearrangement is needed. The same intermediate then gives the same product by the route described in part *a*.

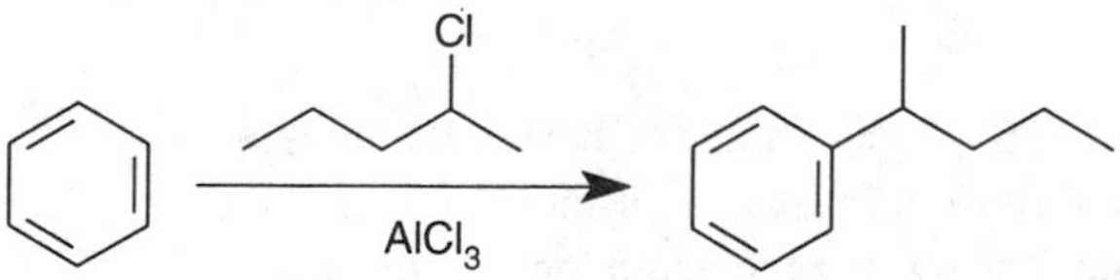

(c)   Diphenylmethane. The benzyl cation formed by interaction of the Lewis acid with benzyl chloride acts as an electrophile to attack benzene.

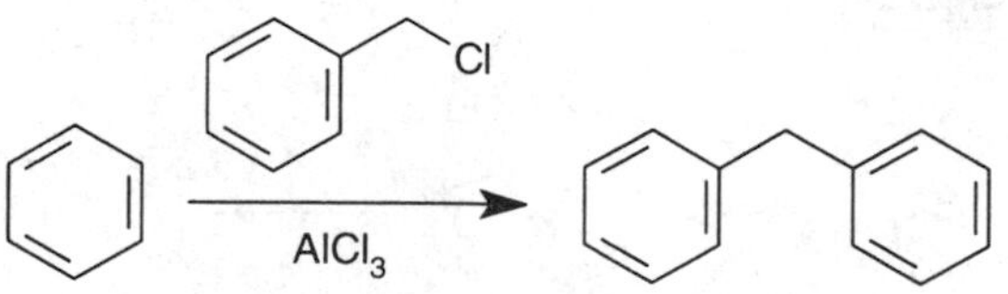

(d)   2-Methyl-2-phenylbutane. The interaction of 2-bromo-3-methylbutane with AlCl$_3$ produces a secondary cation, which rearranges to the 2-methyl-2-butyl cation that attacks the benzene ring to initiate electrophilic substitution.

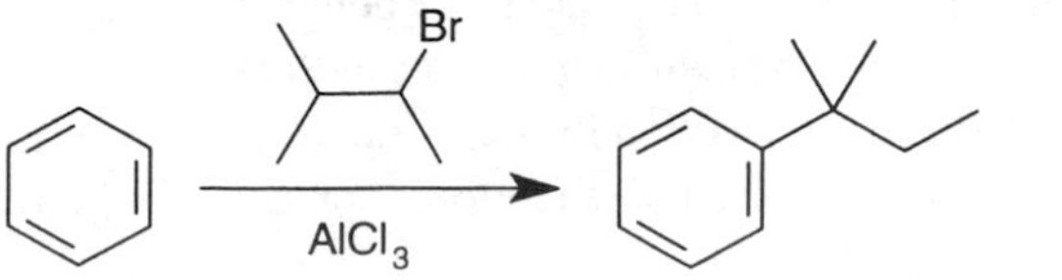

(e) Cyclohexylbenzene. Although these are not the standard conditions for a Friedel-Crafts alkylation, treatment of an alcohol with acid effects dehydration to a secondary cation, which participates in electrophilic substitution like those produced under standard Friedel-Crafts conditions.

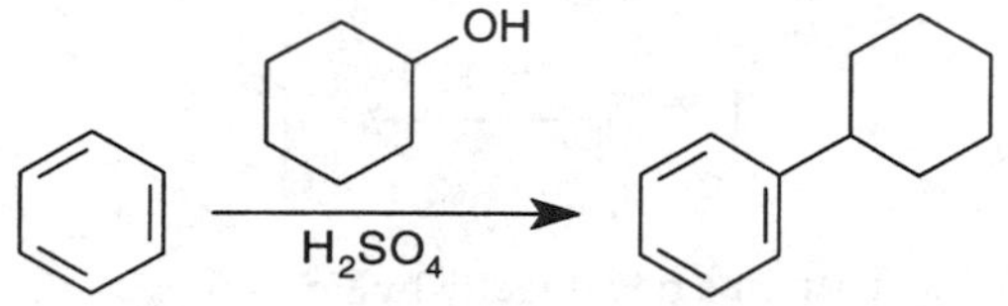

(f) 2-Methyl-2-phenylbutane. A hydrogen shifts from C-2 to C-1 in a cationic rearrangement when 1-bromo-2-methylbutane is treated with a Lewis acid (see Exercise 11.3).

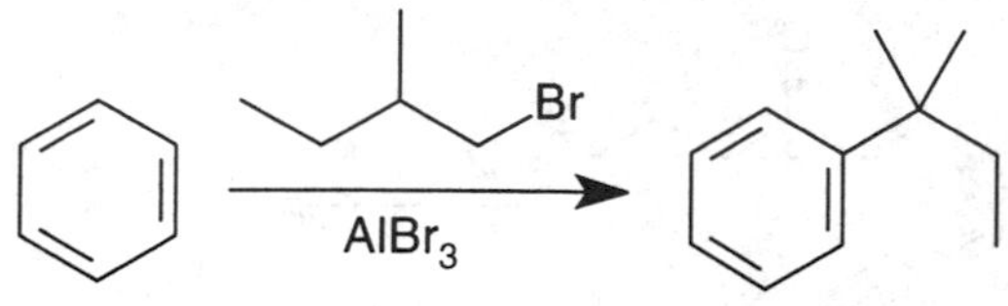

Ex 11.5 To convert benzene to a hydroxy-substituted azobenzene, we would hope to couple phenol with a benzene diazonium salt. To prepare the latter, we must first introduce nitrogen, most conveniently by nitration. Reduction, followed by treatment with nitrous acid, converts nitrobenzene produced in the first step first to aniline and then to a diazonium salt. This diazonium salt can serve as a precursor to the required phenol. By treatment with aqueous sulfuric acid, a portion of the benzene diazonium salt is converted to phenol.

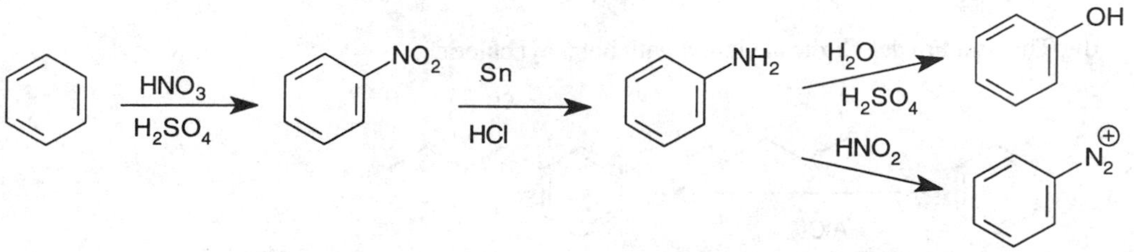

Because phenol is electron rich by virtue of the electron-releasing OH group, benzene diazonium salts are sufficiently electrophilic to initiate electrophilic attack, forming a new C—N bond connecting the two rings. Upon treating phenol with the remaining portion of the benzene diazonium salt, C—N bond formation takes place at the *p* position of phenol because the OH group is an *o,p*-director and because the attacking electrophile is large. Deprotonation of the resulting arenium cation restores aromaticity to the phenolic ring and completes the synthesis of the desired substituted azobenzene.

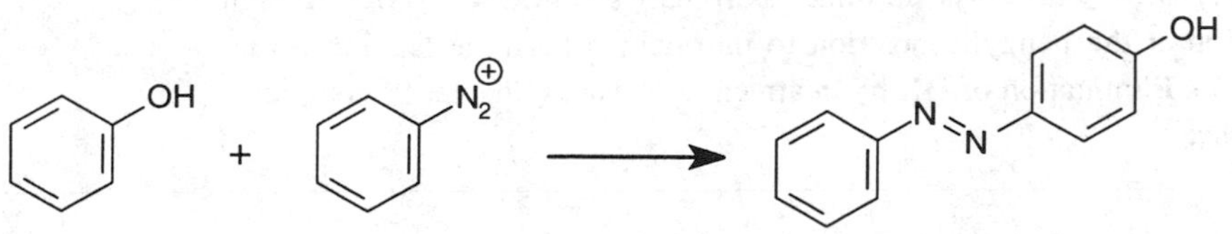

Ex 11.6 Several sequences are possible for each part of this Exercise. Only one sequence is given here for each part; it uses reactions presented in this and earlier chapters.

(a) Freidel-Crafts acylation is necessary to attach an alkyl chain in which the first carbon atom is a methylene group. Reduction of the resulting ketone, under either Clemmensen or Wolff-Kishner conditions, then produces the desired hydrocarbon.

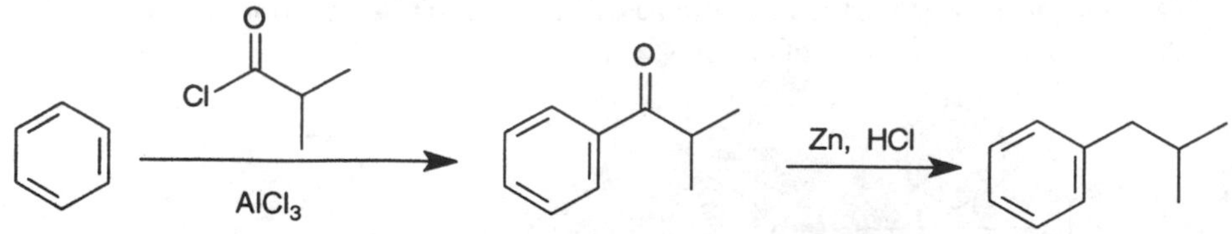

(b) A primary alkyl chain cannot be attached to a benzene ring directly by Friedel-Crafts alkylation because rearrangement of a straight-chain alkyl group would produce a secondary cation that leads to a branched alkyl substituent. It is therefore necessary to acylate benzene and then reduce the ketone formed.

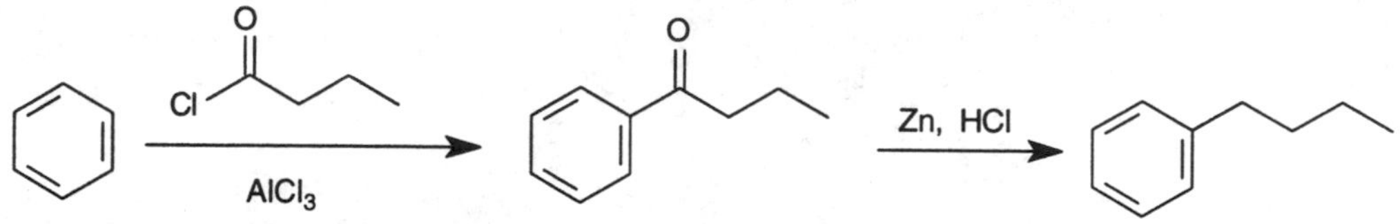

(c) This is the product expected from a Friedel-Crafts alkylation with a straight-chain pentyl group. Either a 1- or a 2-halopentane can be used as the alkylating agent. (But not 3-halopentane: there is no driving force for rearrangement to the 2-pentyl cation.)

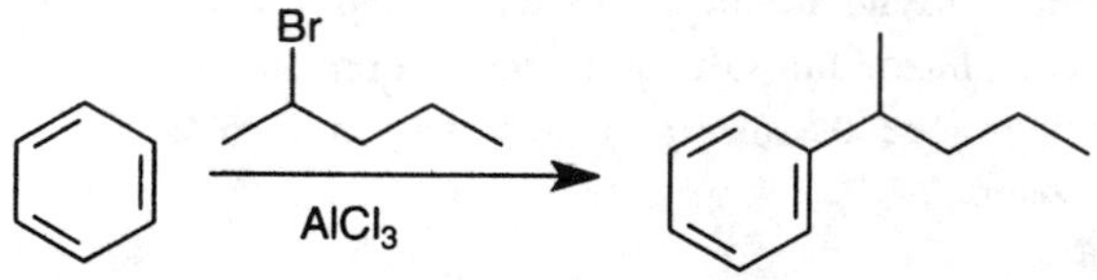

(d) This is a Friedel-Crafts acylation with butanoyl chloride.

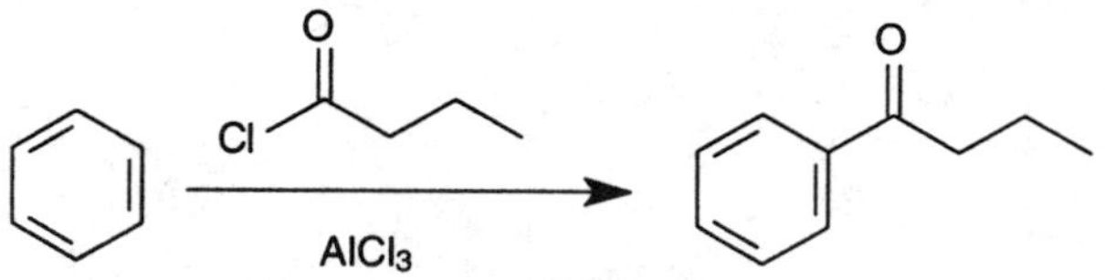

(e) This transformation can be effected by Friedel-Crafts alkylation with ethyl bromide.

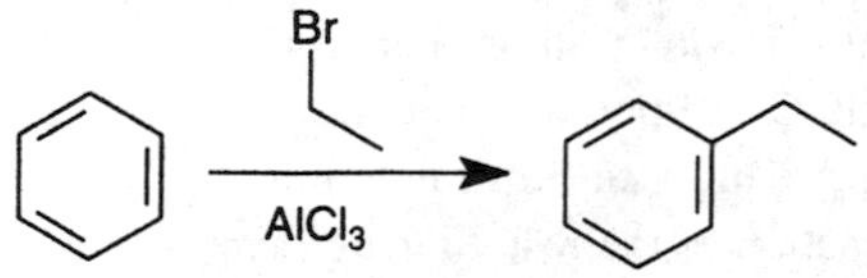

(f) This transformation requires attachment of a two-carbon chain to the aromatic ring, followed by side-chain functionalization. This can be achieved by Friedel-Crafts alkylation with ethyl bromide, as in part *e* followed by free radical substitution at the benzylic position to introduce a bromide leaving group (Chapter 7). Elimination of HX by treatment with base (Chapter 9) completes the sequence.

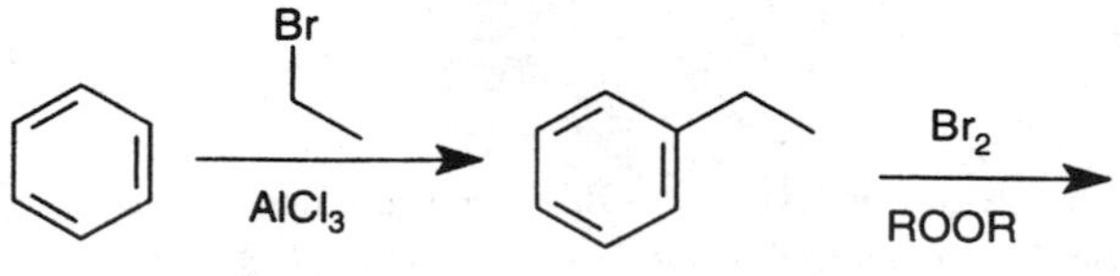

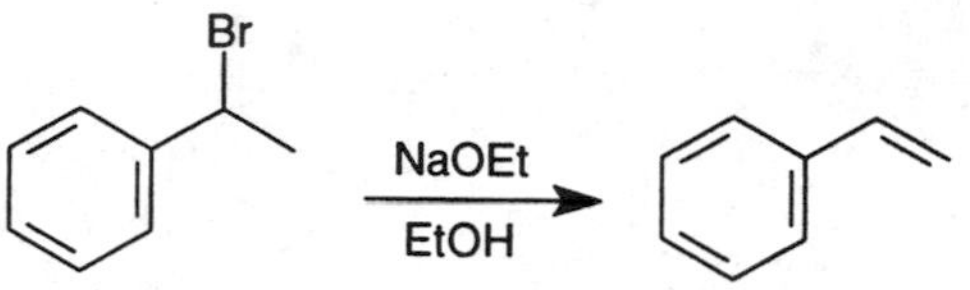

(g) This transformation is quite similar to that in part *h*, except that the functional group is at a nonbenzylic position of the alkyl side chain. This can be accomplished by an anti-Markovnikov hydration of the alkene formed as in part *f*. This is achieved with hydroboration (with diborane), followed by oxidation with hydrogen peroxide in aqueous base.

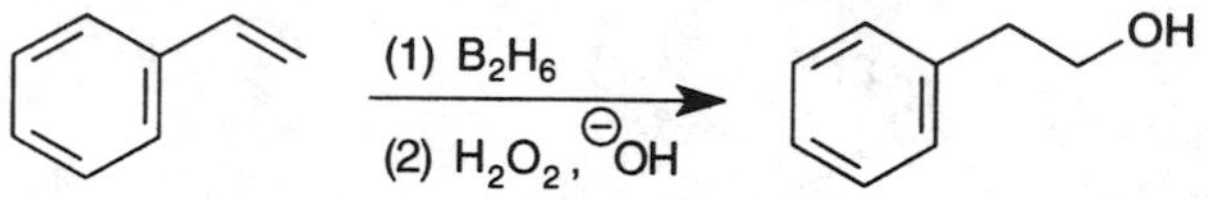

(h) This alcohol is isomeric to that in part *g* and can be obtained by hydrolysis of the benzylic bromide prepared in part *f*. (This substitution takes place by an $S_N 1$ mechanism, and the only function of the Na_2CO_3 is to prevent the reaction medium from becoming acidic.) Alternatively, acid-catalyzed hydration of the alkene produced in part *f* would give the desired alcohol, but because styrene is so easily polymerized, the yield of the desired alcohol may be lower by the latter route.

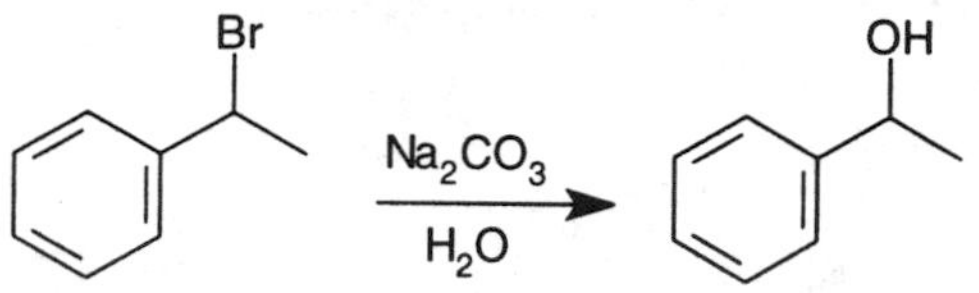

Ex 11.7 An alkyl group is a weakly activating substituent that directs electrophilic substitution to the *o,p*-positions. In *m*-xylene, each position that is *ortho* to one substituent is also either *ortho* or *para* to the other as well. Thus, each activated substituent is doubly activated. In contrast, in both *o*- and *p*-xylene, each position that is *ortho* to one substituent is *meta* to the other, and hence each is only singly activated. Accordingly, *m*-xylene is the most reactive substrate toward electrophilic chlorination. The *para* isomer undergoes substitution more slowly than the *meta* isomer because all four open sites are *ortho* to a methyl group. Hence, substitution is slowed because of steric interactions.

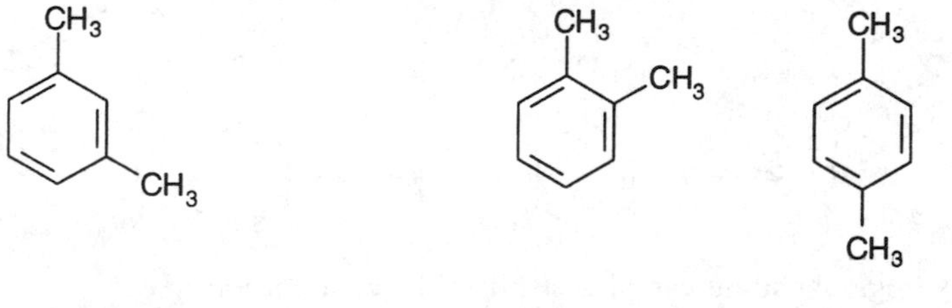

Ex 11.8 Like an alkyl group, a phenyl substituent is weakly activating and *o,p*-directing. This follows both from polarizability and from resonance stabilization of the arenium ion intermediate produced in the electrophilic halogenation. Thus, the phenyl ring will direct bromination to the positions *ortho* and *para* to the point of phenyl attachment. If *para* attack ensues, a particularly stable cationic intermediate is formed: not only are there the three resonance contributors available analogous to those shown in Figure 11.6, but as well there are three more structures in which the positive charge is delocalized to the attached phenyl ring.

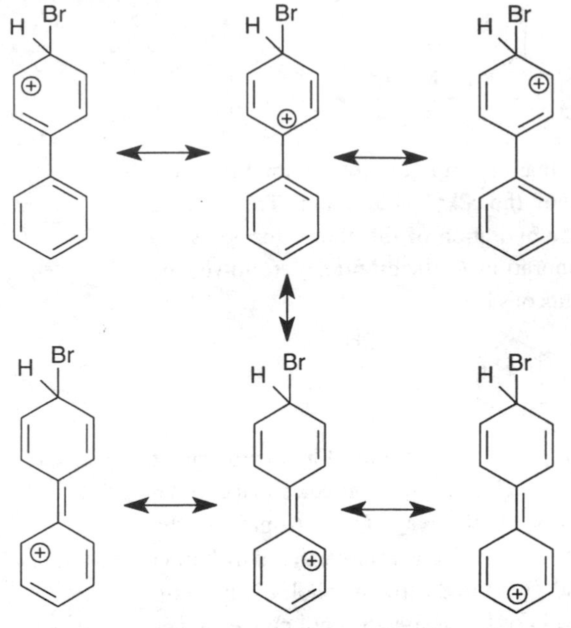

These same additional structures are also available to the arenium ion produced by *ortho* attack, but the two *ortho* positions are much more crowded than the sterically more accessible *para* position.

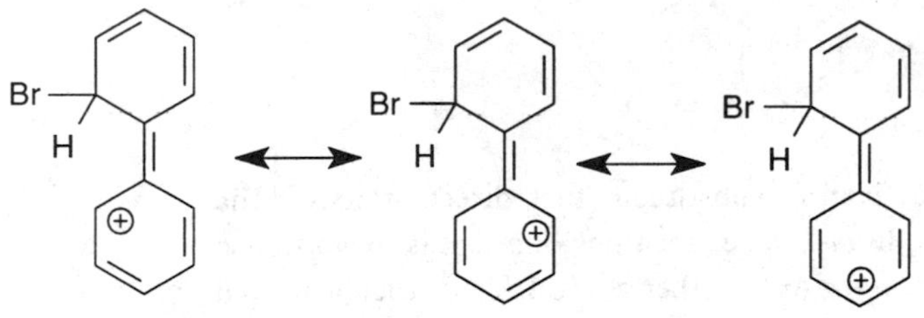

Ex 11.9 With phenol, *ortho* attack by an electrophilic chlorinating agent produces an arenium ion in which intramolecular hydrogen bonding between the phenolic OH and the σ-bound chlorine can take place. Thus, the resonance contributors analogous to those shown in Figure 11.7 are further stabilized by this noncovalent interaction.

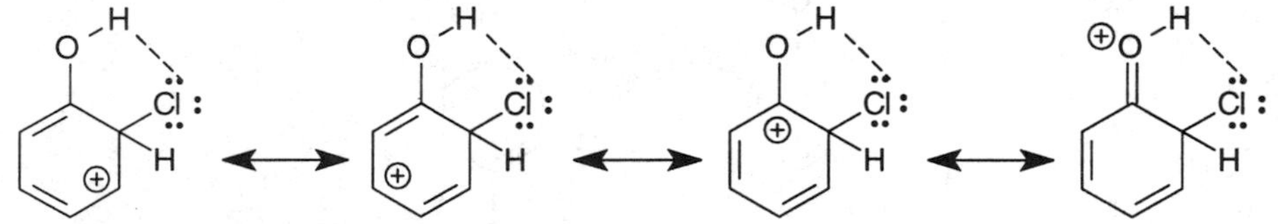

Analogous intramolecular hydrogen bonding can also stabilize the arenium ion formed upon electrophilic nitration at the *ortho* position.

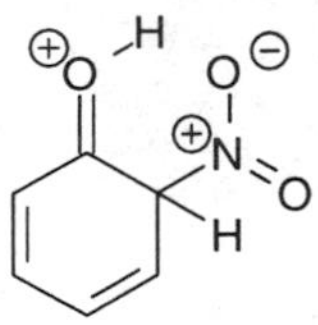

Ex 11.10 In addition to the resonance contributor shown at the left of each line (reproduced from Figure 11.10), two other contributors localize charge at the other *ortho* and *para* positions.

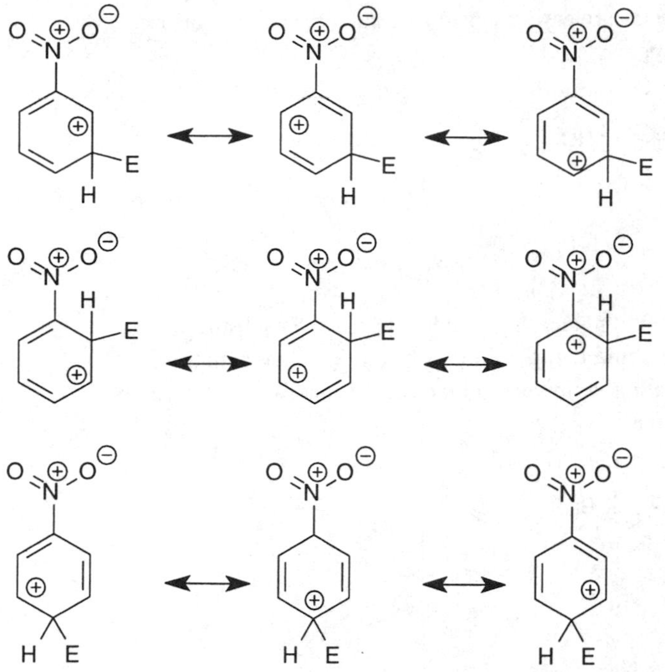

Ex 11.11 A fluorine substituent, like the other halogens, is *ortho, para* directing, but deactivating. Fluorine is electron-withdrawing regardless of its position relative to the positively charged carbons in the intermediate cation. However, lone-pair electron donation partially compensates when fluorine (or another halogen atom) is directly attached to a carbocation center, as shown in the resonance contributor in which fluorine bears formal positive charge. Analogous structures can be drawn for the cation resulting from attack of the electrophile at the *ortho* positions, but donation of electron density from fluorine is not possible in the cation resulting from attack at the *meta* position.

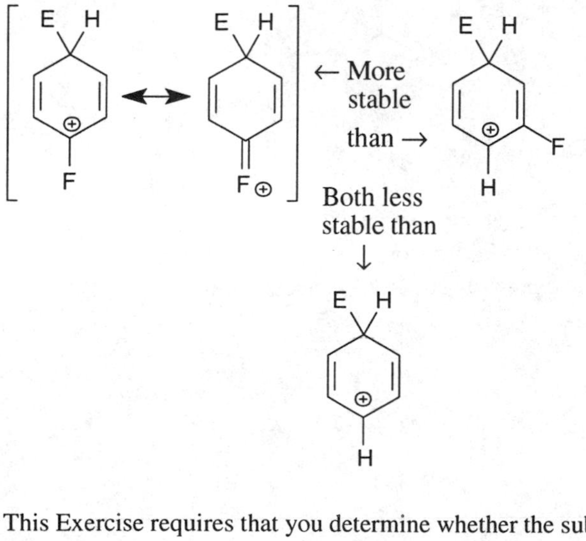

Ex 11.12 This Exercise requires that you determine whether the substituent originally present donates electron density to the intermediate cation, favoring *ortho, para* substitution, or withdraws electron density, favoring *meta* substitution.

(a) An alkyl group is a weak electron-releasing substituent and directs substitution to the *ortho* and *para* positions.

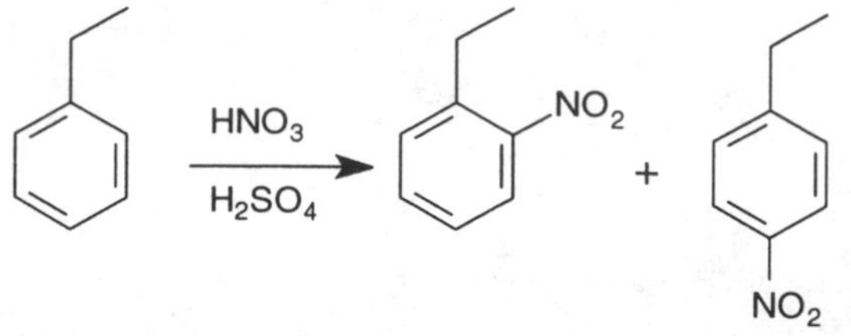

(b) An ether group is strongly electron releasing and directs aromatic substitution *ortho, para*. There may be some competition with multiple halogenation, although the second bromination is slower than the first because of inductive electron withdrawal by bromine.

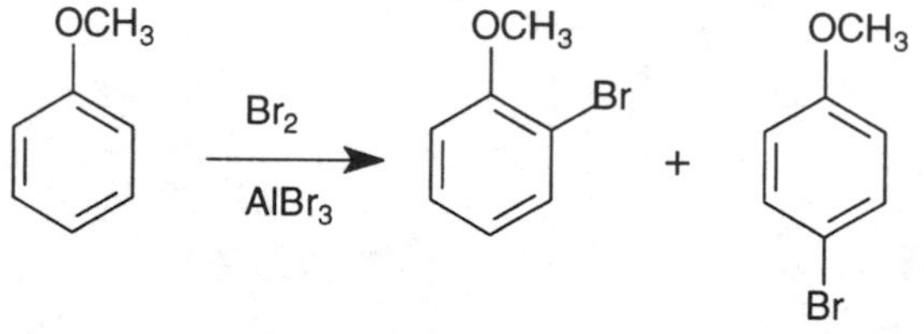

(c) A nitro group is strongly electron withdrawing; it directs substitution to the *meta* position.

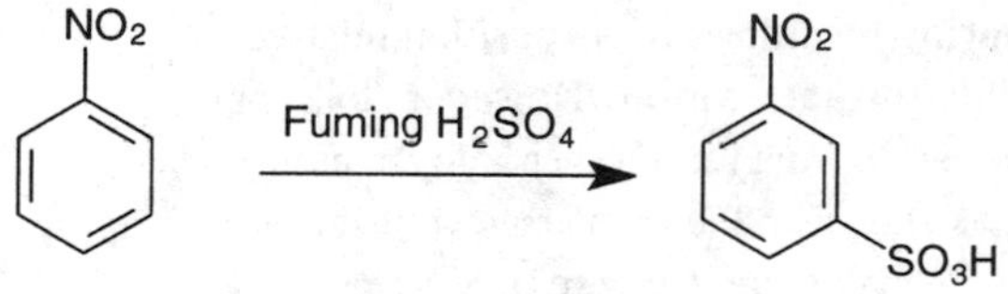

(d) The methyl group of toluene directs alkylation to the *ortho* and *para* positions. The carbon skeleton of the alkyl group is rearranged in this conversion.

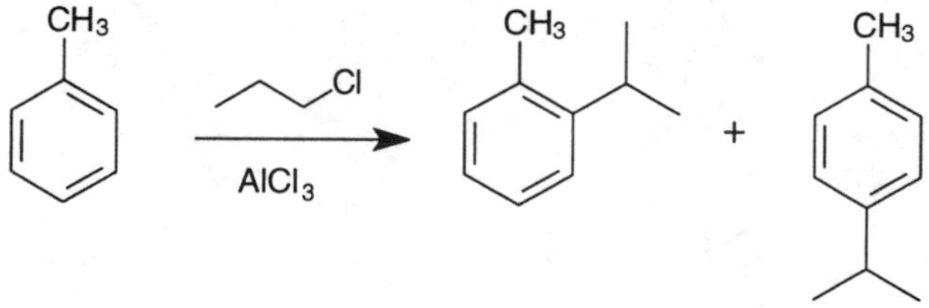

(e) The ester is bound to the aromatic ring at an oxygen atom with two nonbonded lone pairs of electrons. The ester is therefore electron releasing, although less so than an ether. *Ortho* and *para* substitution products are thus obtained.

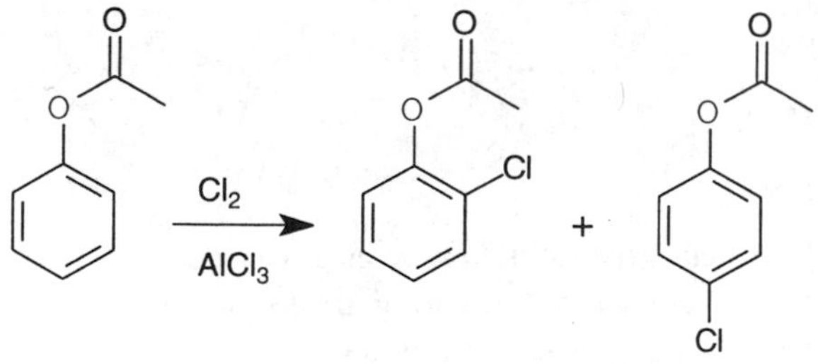

(f) The cyano group is electron withdrawing and directs to the *meta* position.

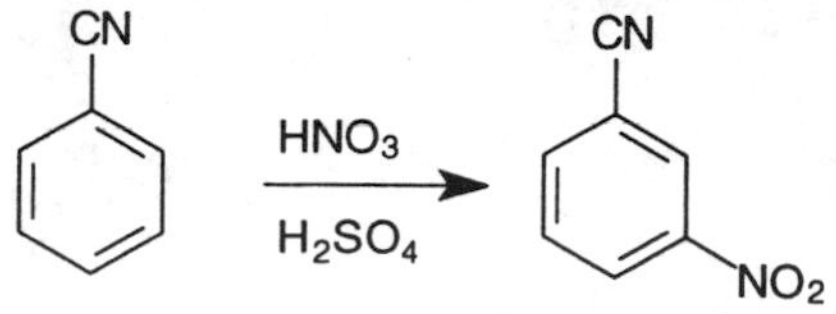

(g) Fluorine, like other halogens, directs to the *ortho* and *para* positions.

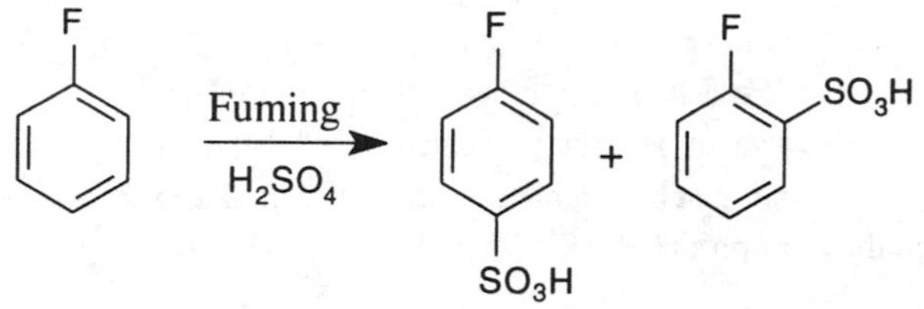

Ex 11.13 (a) A methoxy group is a strong electron-releasing group whose influence dominates over that of bromine. The sulfonic acid is therefore introduced at the position *para* to methoxy. Attack at the *ortho* position (in relation to methoxy) is favored electronically but disfavored sterically, because of the presence of two groups on adjacent ring positions in the reactant.

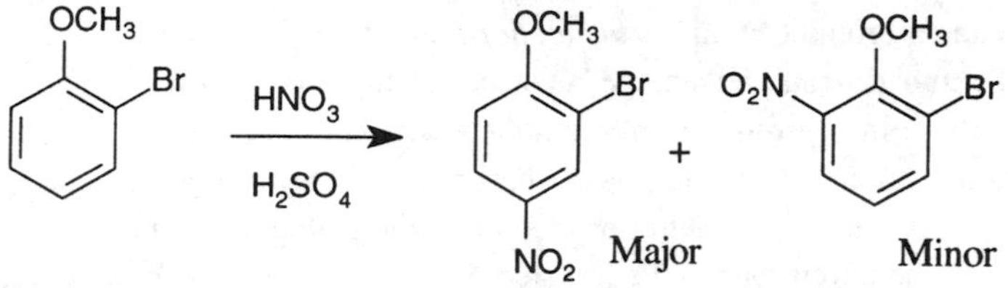

(b) The acetamide group is a moderate activator and directs *ortho, para*. Because its effect is stronger than that of a halogen which is deactivating, the product has an acetyl group in an *ortho* position to the acetamide group.

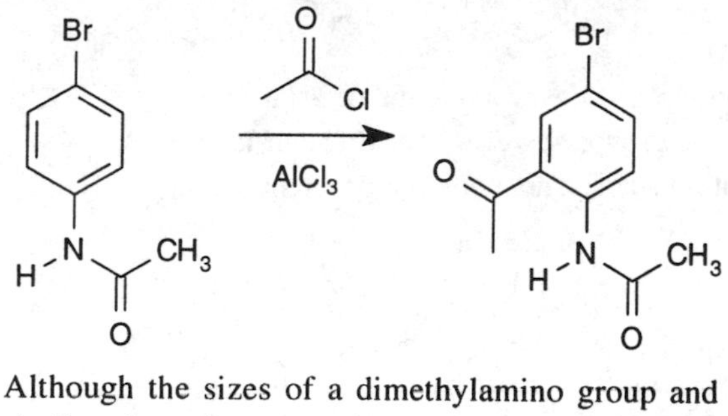

(c) Although the sizes of a dimethylamino group and an isopropyl group are similar, the amino group is a strong electron-releasing substituent that directs chlorination to the *ortho* position relative to this group. Multiple chlorination is likely in this highly activated ring. A Lewis acid is not required to activate the electrophile here. Sodium carbonate can be added to neutralize the HCl produced as the reaction proceeds.

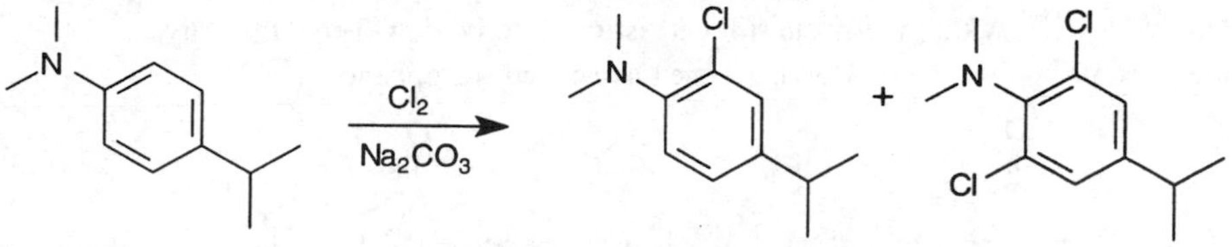

(d) An methoxy group is a strong electron-donating substituent, whereas a carboxymethyl group is an electron-withdrawing substituent. Bromination is therefore directed by the ether to the *ortho* position, the *para* position being blocked by the ester.

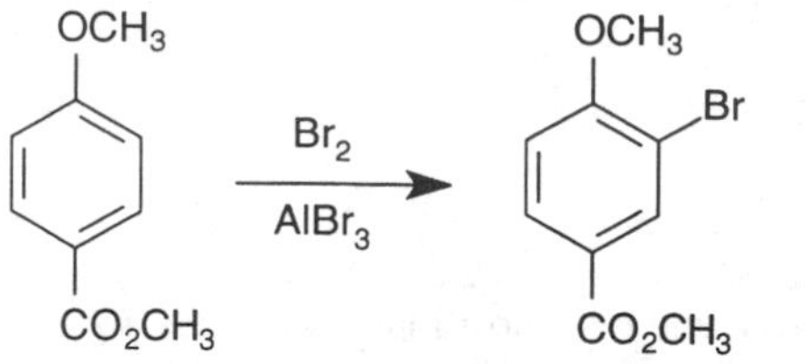

Ex 11.14 (a) Methyl is *ortho, para* directing, and nitro is *meta* directing. The methyl group must therefore be attached first. This is also necessary because Friedel-Crafts alkylation fails when strongly deactivating substituents are present. The *ortho* product must be separated from the *para* product by distillation.

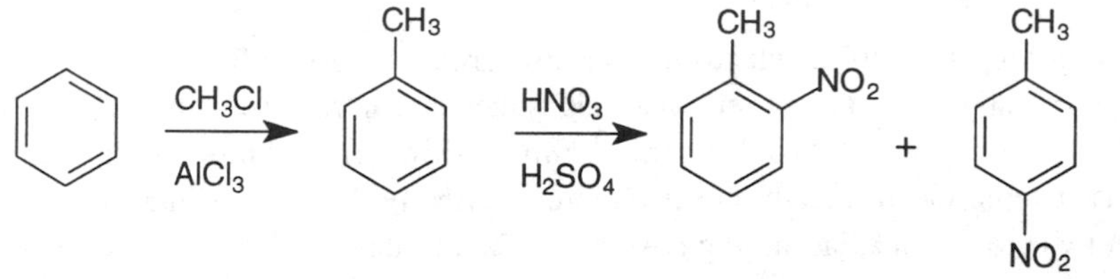

(b) Both an amino group and a bromine atom are *ortho, para* directing. To obtain a *meta* relation of the substituents, advantage must be taken of the *meta*-directing ability of the nitro group, a precursor to the amino group. Electrophilic bromination of nitrobenzene places the bromine and nitrogen substituents in the desired *meta* relation. After bromination, the aniline is then produced by reduction of the nitro group in the last step.

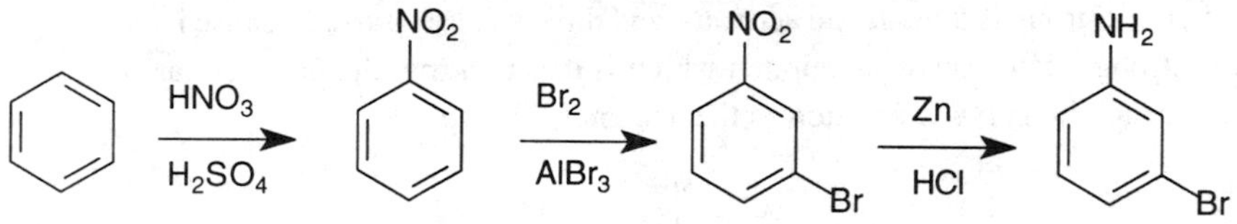

(c) As noted in part *b*, both the amino group and the bromine direct *ortho, para*. With the strong activation of the amino group, however, it is difficult to stop the bromination at the monosubstitution stage. This potential problem can be avoided by brominating first.

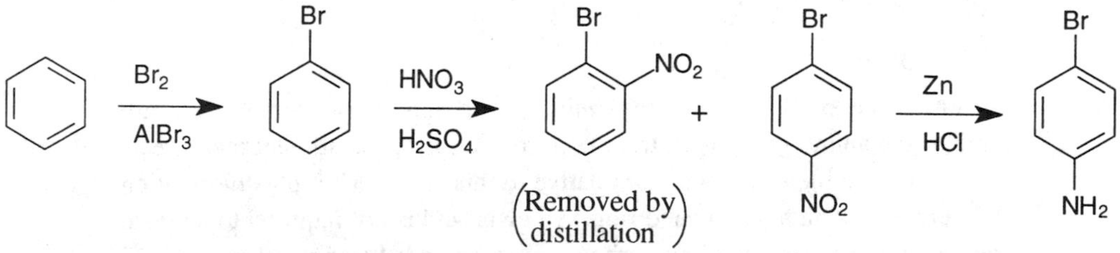

(d) Both the nitro and the acetyl groups are *meta* directing, but Friedel-Crafts acylation of nitrobenzene is not possible because the nitro group (the more strongly electron-withdrawing substituent) is too deactivating. Thus, the only route possible is nitration of acetylbenzene (also called acetophenone).

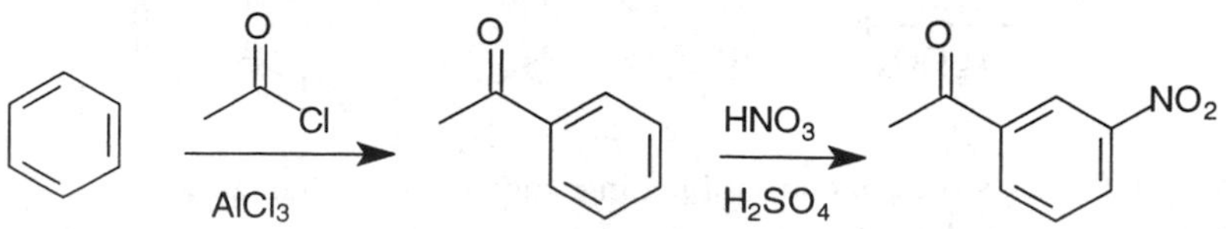

(e) Although the nitro group is a *meta* director, Friedel-Crafts alkylation fails with nitrobenzene. Instead, we must use reagents to selectively reduce *m*-nitroacetophenone, prepared as in part *d*, to the corresponding nitroethyl-benzene. This can be best accomplished selectively under Wolff-Kishner reduction conditions.

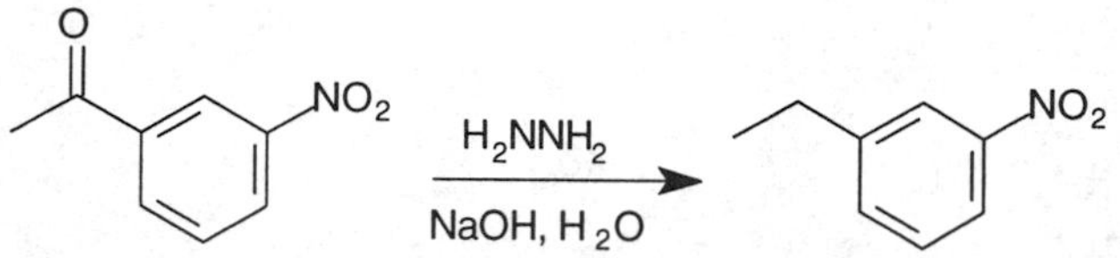

(f) Both the ethyl group and the amino group are *ortho, para* directing. We must therefore use the *meta*-directing effect of some precursor to one of these substituents. For example, alkyl groups can be formed by reduction of acyl groups (which direct *meta*), and amino groups can be formed by reduction of nitro groups (which also direct *meta*). The desired product can be obtained by reducing the nitro and ketone groups of the product obtained in part *d*.

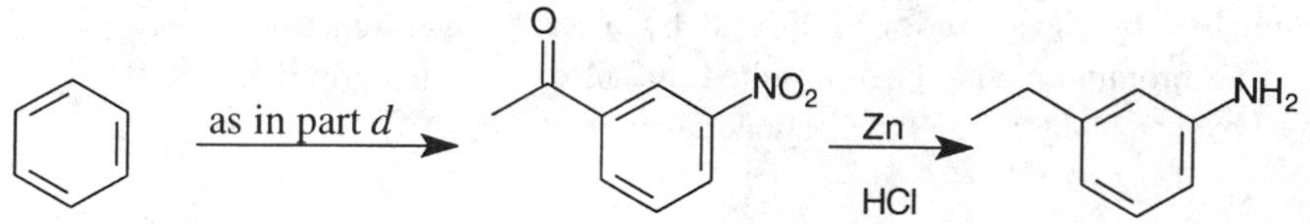

(g) Nitration of bromobenzene produces the desired *o*-bromonitrobenzene, along with substantial amounts of the *para* isomer.

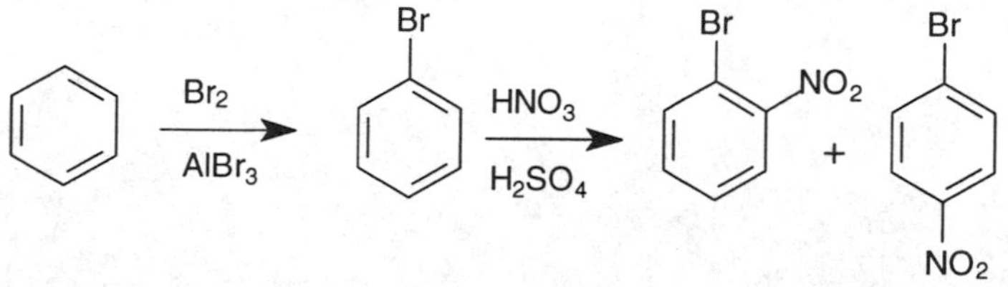

(h) The nitro group is *meta* directing and chlorine is *ortho, para* directing (despite its inductive withdrawal). Therefore, nitration must take place first.

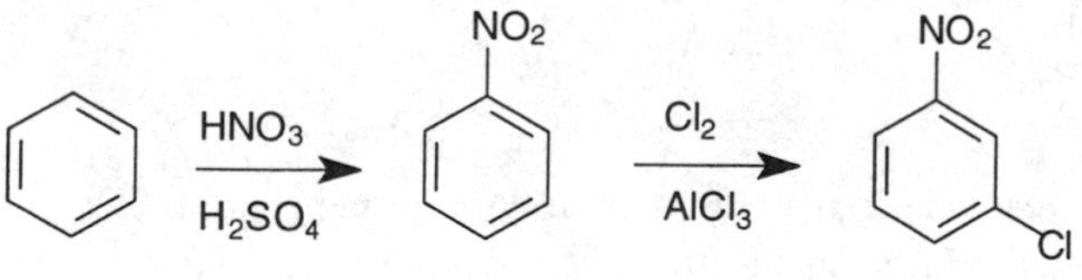

Ex 11.15 (a) From 1,3,5-tribromobenzene prepared as described in this exercise, electrophilic sulfonation gives the desired product.

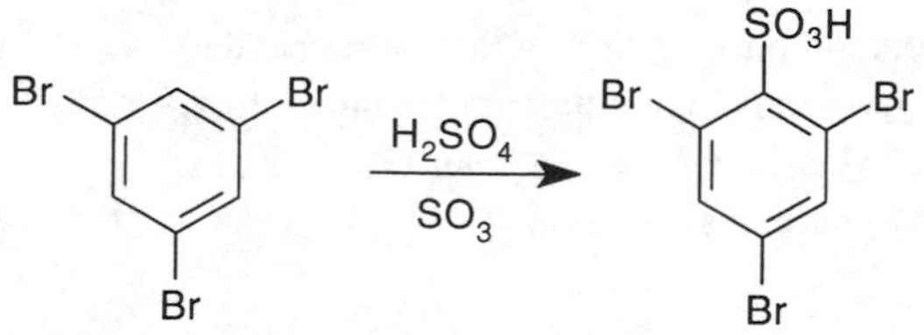

(b) Here incomplete bromination is required. This can be attained by blocking the *para* position with a group that can later be removed. For example, electrophilic bromination of *p*-nitroaniline takes place at the two positions *ortho* to the NH₂ group. Reduction of the nitro group, followed by diazotization of both amino groups, produces a 1,4-bis(diazonium) salt that gives the unsubstituted product upon treatment with H₃PO₂.

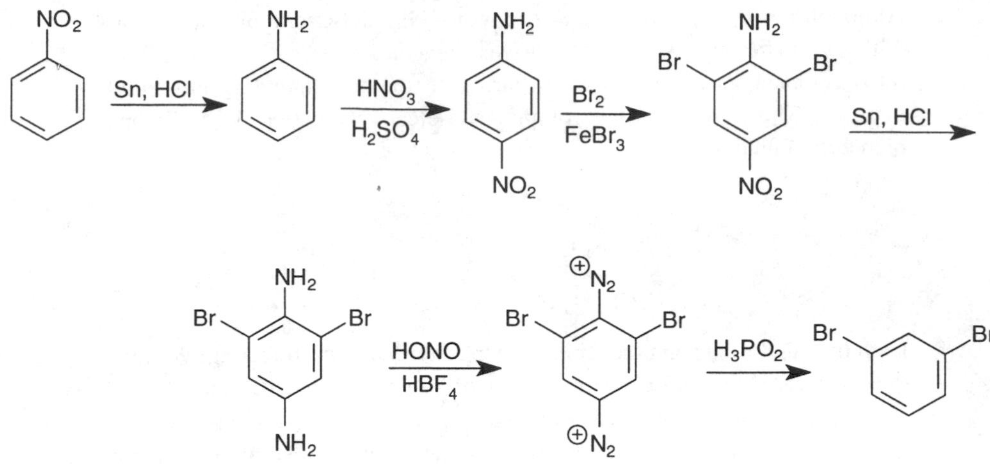

(c) This problem is similar to that posed in part *b*, except that a bromine atom must be installed after the *o,p* direction has taken place. This can be accomplished by diazotization, followed by a Sandmeyer reaction. The intervening bromine having been installed, the blocking nitro group can be removed by a pathway parallel to that undertaken in part *b*.

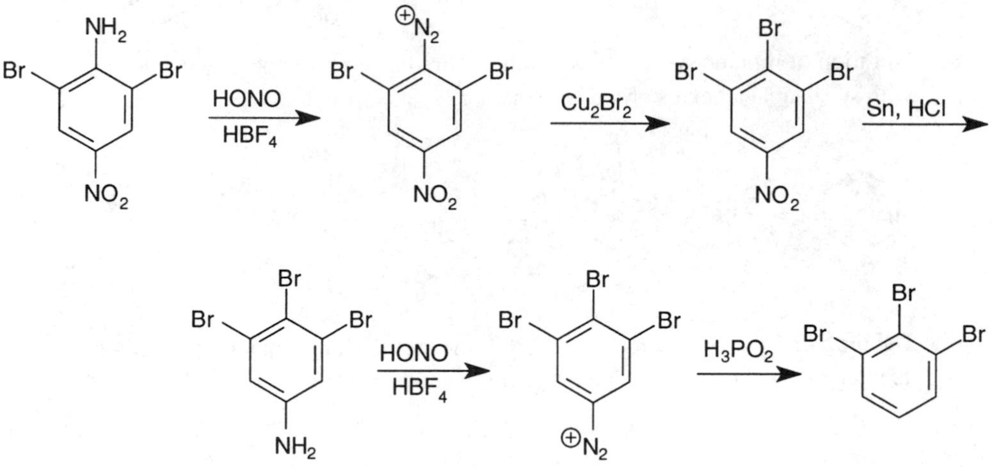

Ex 11.16 (a) Because there are more resonance structures maintaining benzenoid configurations in the cation produced upon attack at C-1, substitution is favored at that site. Also, the aromatic stabilization of two benzene rings is greater than that attained in one naphthalene ring.

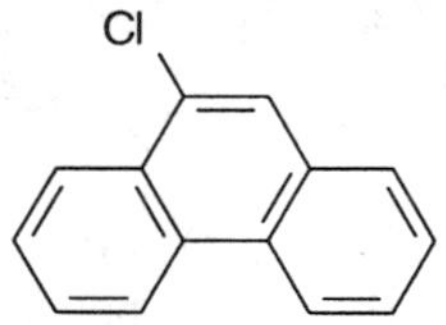

(b) Attack at the 9-position retains a benzenoid configuration in the two appended rings, whereas attack at C-1 or C-2 gives only naphthalenoid contributors. Because the resonance stabilization of two benzene rings is greater than that of a single naphthalene ring, the 9-substituted product is formed.

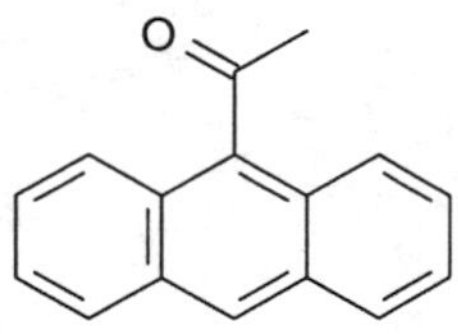

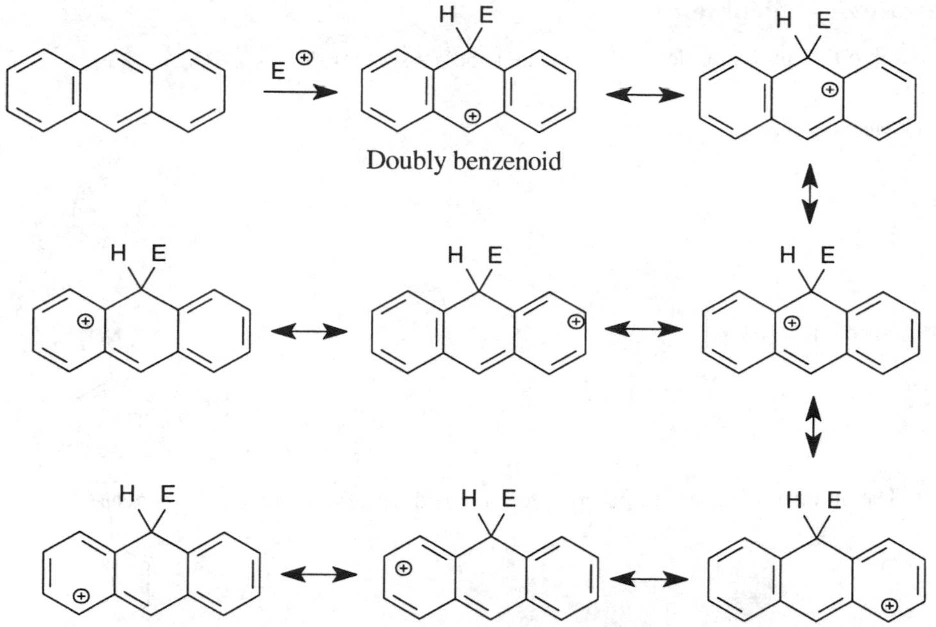

(c) The electronic effect of a nitro group is strongest at the positions indicated by arrows in the structure at the right. Because the effect of a nitro group is destabilizing, attack takes place at one of the other sites. The deactivation by the nitro group is greater on the ring to which it is attached, so attack takes place on the unsubstituted ring. Of the remaining sites, one is at an α and the other at a β position. Attack at the α site minimizes the destabilizing interaction with the nitro group and maximizes the benzenoid character of the intermediate cation, but at the expense of steric interaction with the $-NO_2$ group. These factors are off-setting, and both isomeric products are formed.

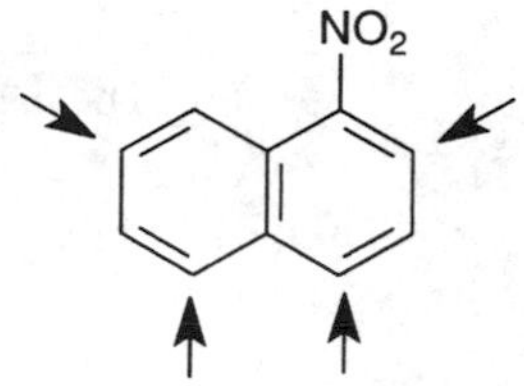

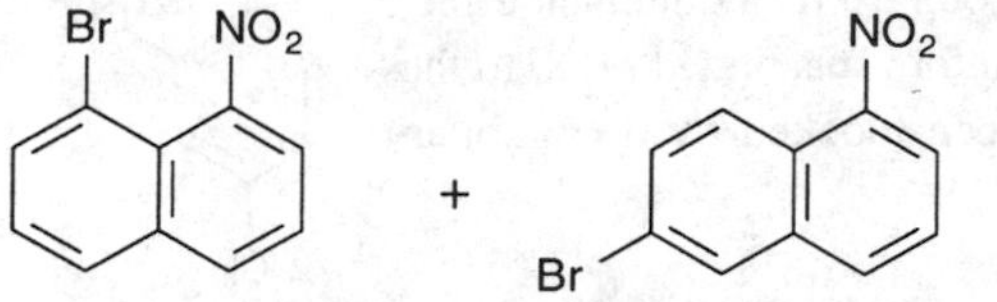

(d) The electronic effect of the methyl group is greatest at the positions indicated by arrows in the structure at the right. Because the methyl group is electron releasing, attack at these positions is favored. Of these positions, two are α positions, and one is on the ring bearing the methyl group. The maximum activating effect is felt at that site; that is, at an α position in the same ring as the substituent.

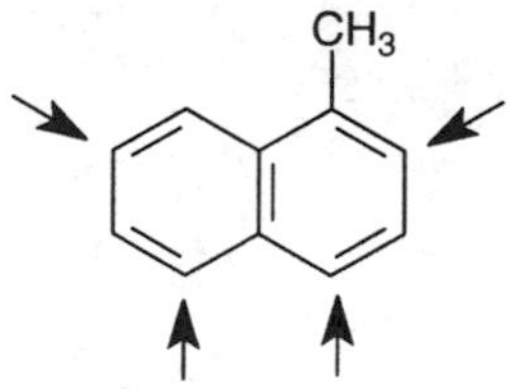

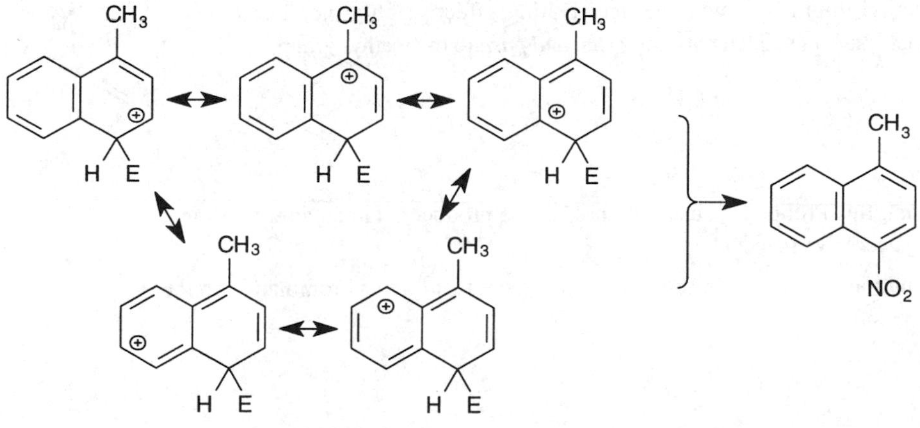

Answers to Review Problems

Pr 11.1   To solve this problem, determine whether the substituent directs *ortho, para* or *meta*.

(a) Methyl directs *ortho, para*.

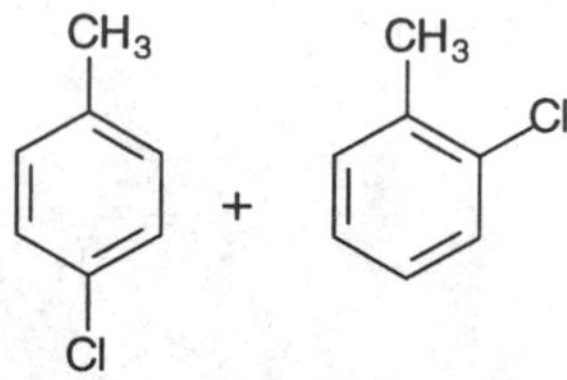

(b) Nitro directs *meta*.

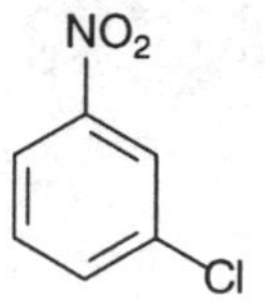

(c) The nitrogen of the amide group in acetanilide directs *ortho, para*.

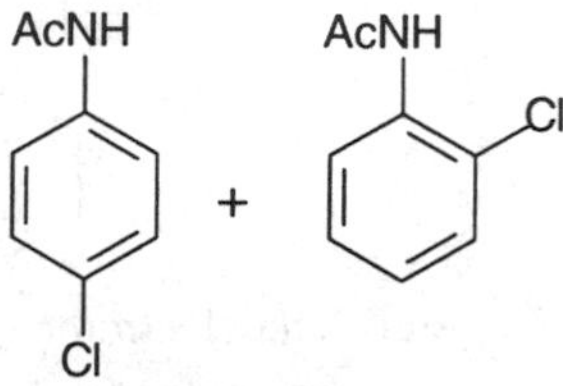

(d) The methoxy group in anisole directs *ortho, para*.

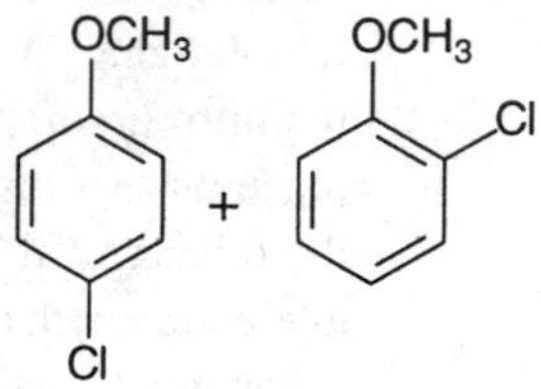

(e) Both the hydroxy group of phenol and chlorine direct *ortho, para*, but the greater electron release by an OH group exerts a dominant effect over that of chlorine. The product is therefore substituted *ortho* to the OH group, the *para* position having been blocked by the halogen substituent.

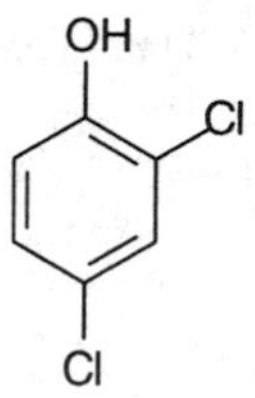

(f) The acetyl group directs *meta*.

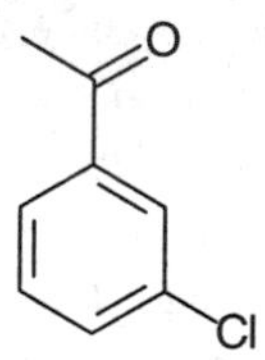

(g) Both methyl and chlorine direct *ortho, para*, but the activating effect of methyl dominates over the deactivating effect of chlorine. The products have the added chlorine *ortho* and *para* to the methyl group.

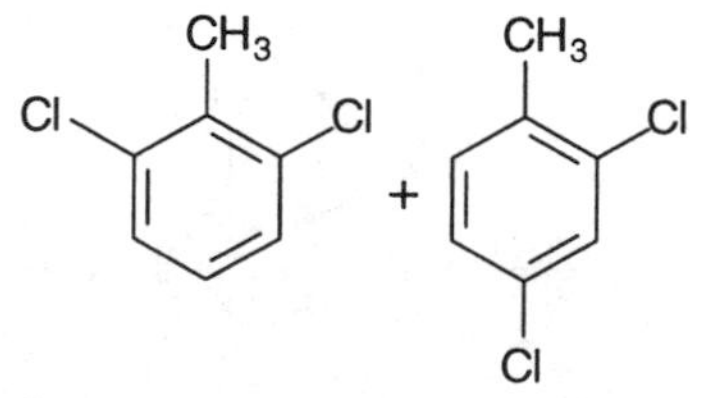

Pr 11.2   In answering this Problem, you should predict the product obtained after treatment with each successive reagent.

(a) Sulfonation of benzene introduces a substituent that directs bromination to the *meta* position.

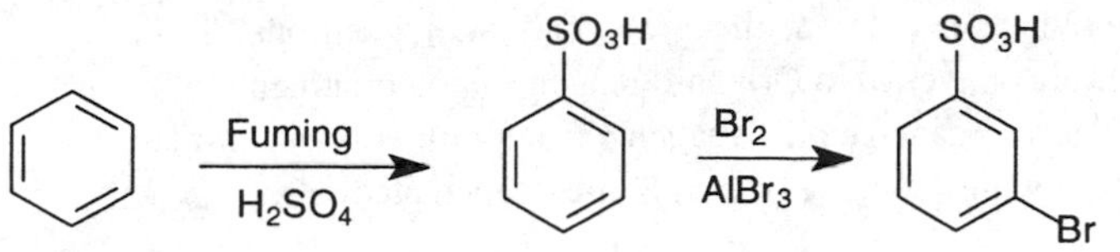

(b) Bromination of benzene introduces a substituent that deactivates, but directs *ortho, para*. The *ortho* and *para* sulfonic acids are therefore obtained by inverting the order of reagents from that used in part *a*.

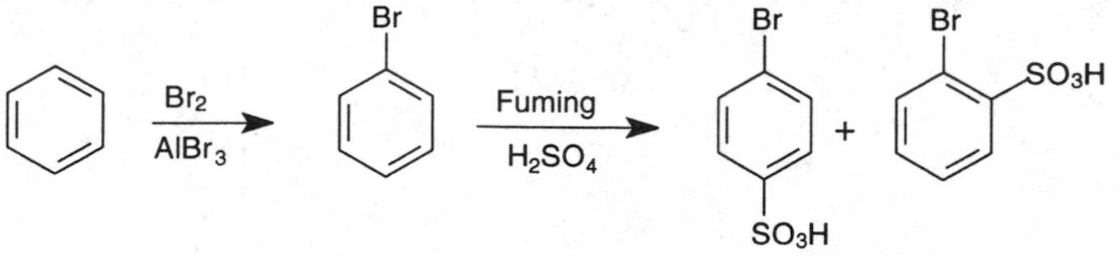

(c) Bromobenzene (prepared as in part *b* of this Problem) is acylated at the *ortho* and *para* positions. The resulting ketone group is reduced by tin in HCl to the corresponding alkyl group. Free-radical halogenation then permits selective bromination at the benzylic position.

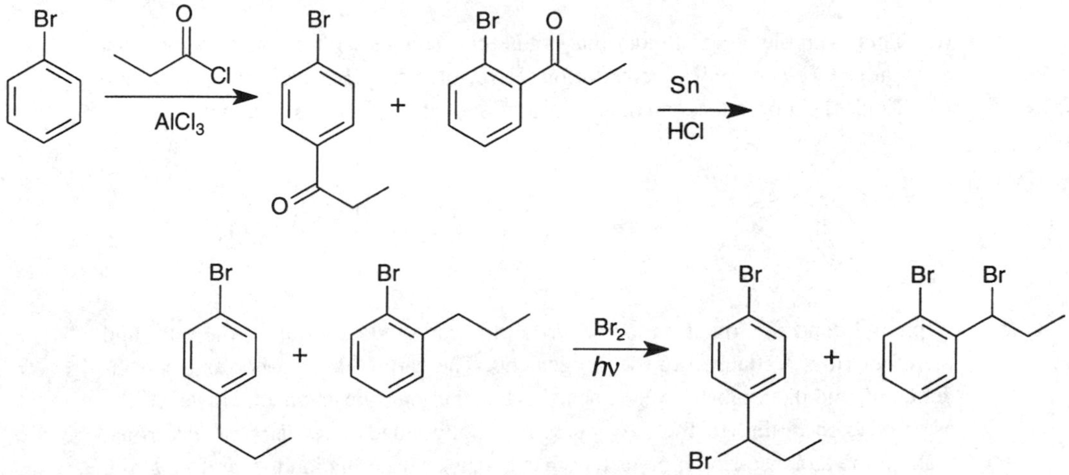

(d) Friedel-Crafts alkylation of phenol takes place at the *ortho* and *para* positions, and because a primary alkyl chloride is used, the alkyl chain in the product is rearranged. Base deprotonates the phenol, producing a highly nucleophilic phenoxide anion that reacts with 1-chloropropane in a Williamson ether synthesis to produce an arylalkyl ether.

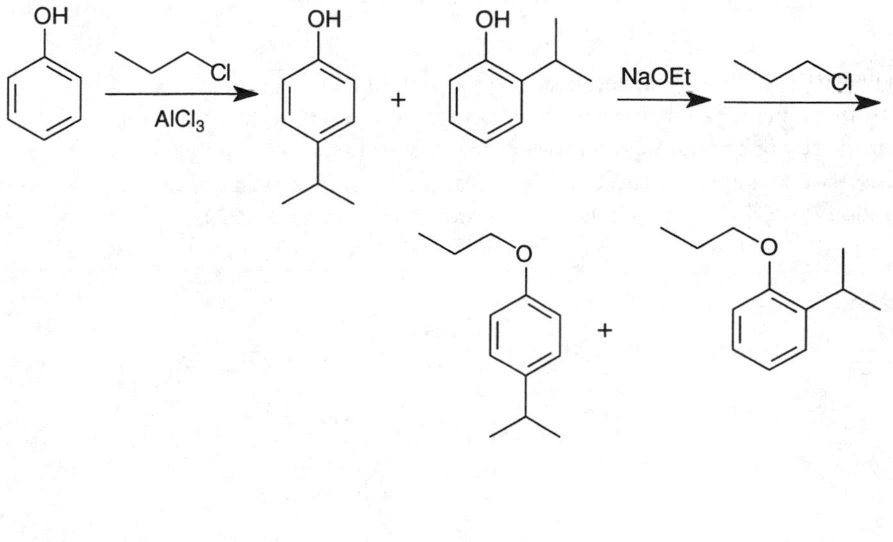

(e) Bromination of biphenyl takes place at the *ortho* and *para* positions. The resulting aryl bromides are converted to Grignard reagents upon treatment with magnesium in ether. These organometallic reagents react with ethylene oxide to form, after protonation with water, the two-carbon-extended alcohols.

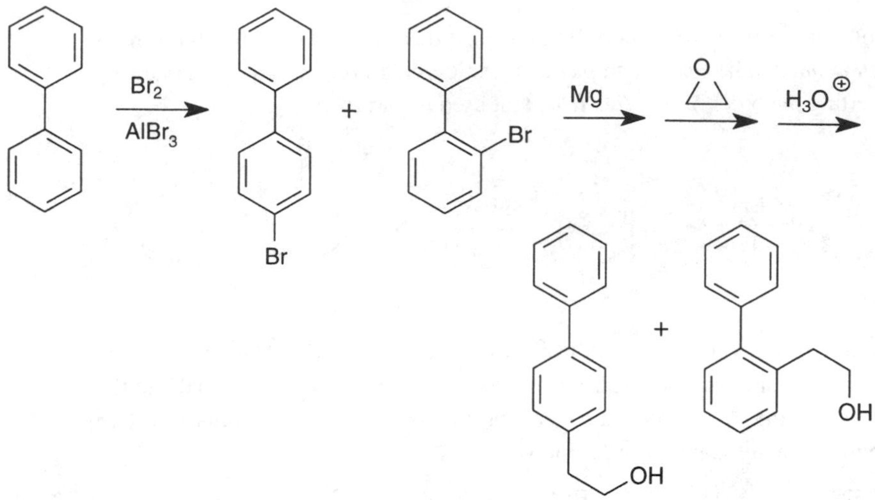

(f) This example is a reminder that not all compounds with aromatic rings must undergo electrophilic substitution. Here, electrophilic addition of bromine yields the dibromide, and then dehydrobromination produces an alkyne.

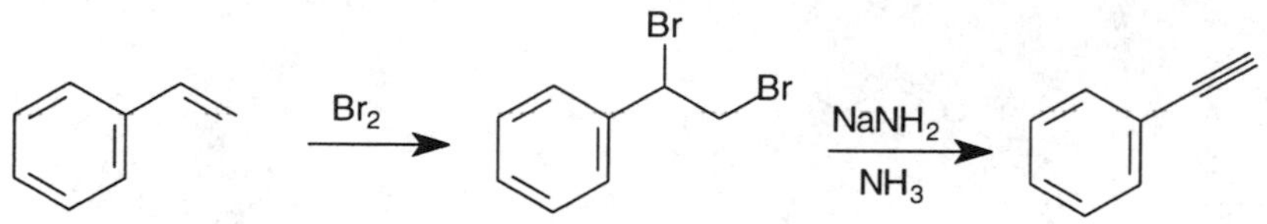

Pr 11.3 In phenyl benzoate, one ring is attached to the carboxylate group of the ester and the other ring is attached to the oxygen end. The carboxylate carbon is electron-deficient, and the aromatic ring attached to that carbon is deactivated. However, the ester oxygen bound to the ring bears two nonbonded lone pairs of electrons, making the attached ring more active than benzene. Bromination is therefore at the phenyl ring bound to oxygen of the ester group, at the *ortho* and *para* positions.

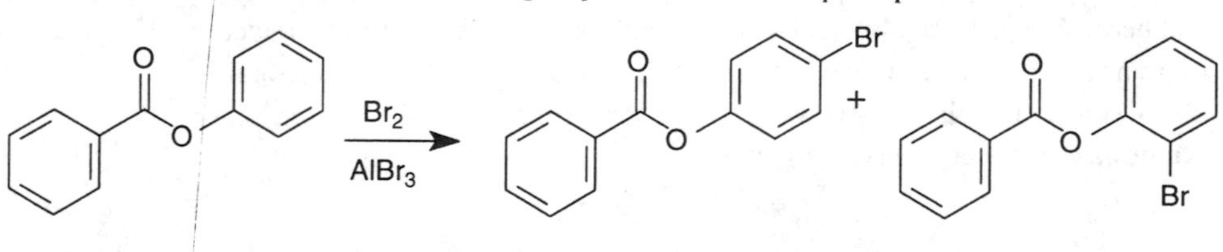

Pr 11.4 (a) The methyl group of toluene is an *ortho, para*-directing and activating substituent, but it can be oxidized by hot $KMnO_4$ to a carboxylic acid that is *meta*-directing and deactivating. Here, *para* substitution is required. It is necessary therefore to nitrate before the oxidation. Thus, the correct sequence is nitration ($HNO_3 + H_2SO_4$), followed by oxidization with hot $KMnO_4$.

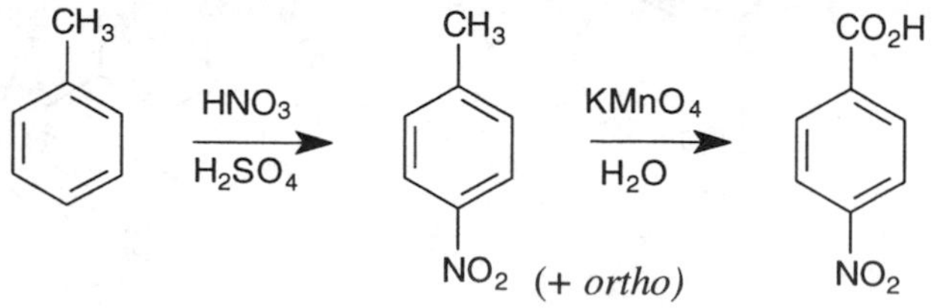

(b) To prepare the *meta* product, invert the order of the reagents used in part *a*.

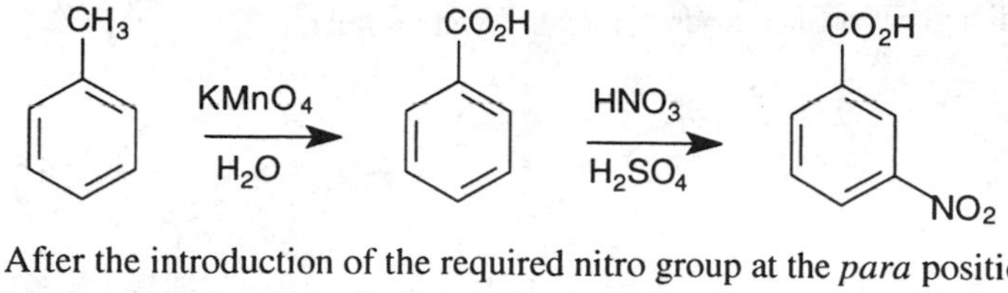

(c) After the introduction of the required nitro group at the *para* position (directed by the methyl group of toluene), the benzylic position can be functionalized by free-radical halogenation. The resulting benzyl bromide is then converted to the alcohol by hydrolysis through an S_N1 mechanism.

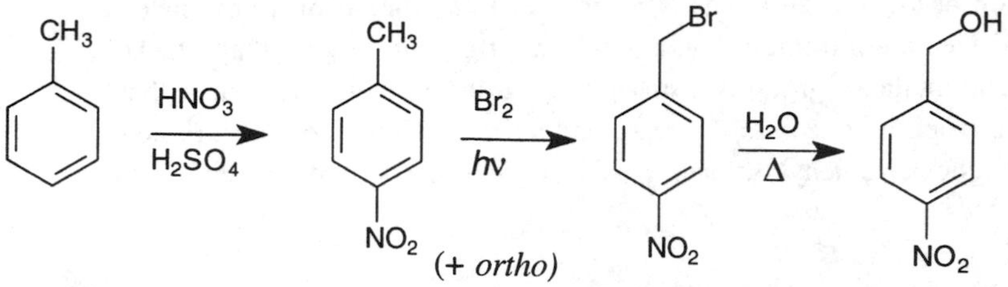

(d) *p*-Toluenesulfonic acid is prepared by electrophilic sulfonation of toluene.

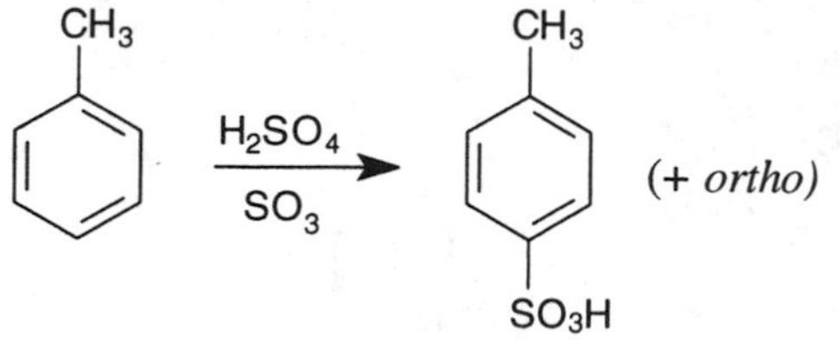

Pr 11.5   (a) This reaction begins with the formation of the active electrophile. $AlCl_3$ acts as a Lewis acid, accepting one of the lone pairs on the chlorine atom of the alkyl chloride. This complex undergoes rearrangement to a secondary cation, which attacks the benzene ring, forming a resonance-stabilized, cyclic, pentadienyl cation. Deprotonation of this cation by $AlCl_4^-$ completes formation of the indicated product.

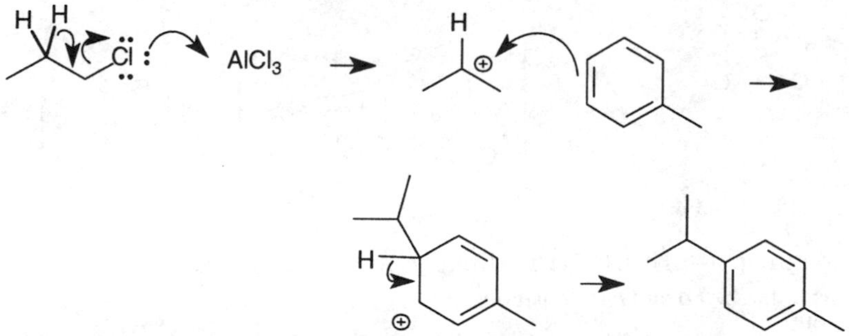

(b) An acylium ion is formed upon interaction of an acid chloride with $AlCl_3$. Reaction of the resonance-stabilized acylium ion with the aromatic ring produces a cyclic pentadienyl cation that is converted to product by a route parallel to that in part *a*.

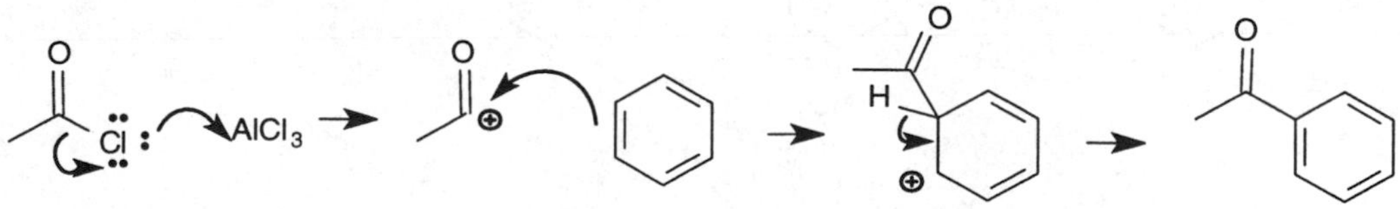

(c) Molecular bromine (Br_2) is activated as an electrophile by the interaction with a Lewis acid such as $FeCl_3$. Attack on the aromatic ring is directed to the *meta* position by the electron-withdrawing nitro group. Deprotonation of the resulting cyclic pentadienyl cation then leads to the observed product.

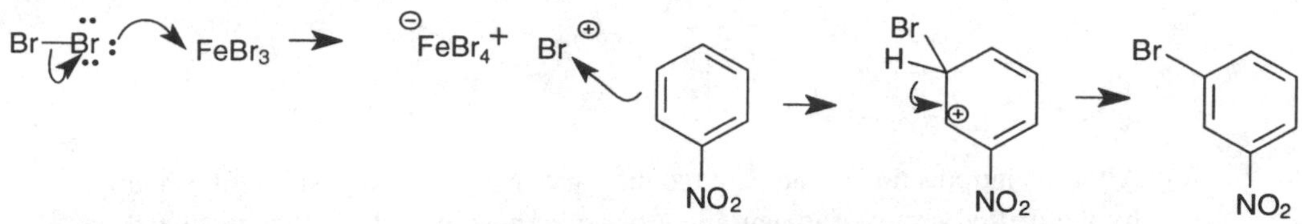

(d) Nitric acid can be protonated at any of the oxygen atoms, but only protonation as shown here (drawn as the lower left oxygen atom) is productive. Protonation at all oxygen atoms is reversible. Dehydration of a protonated nitric acid yields a nitronium ion, a highly electrophilic reagent that attacks anisole. The methoxy group is a strongly electron-releasing substituent that directs electrophilic attack to the *ortho* and *para* positions. Loss of a proton from the cyclic pentadienyl cation intermediate yields the desired product.

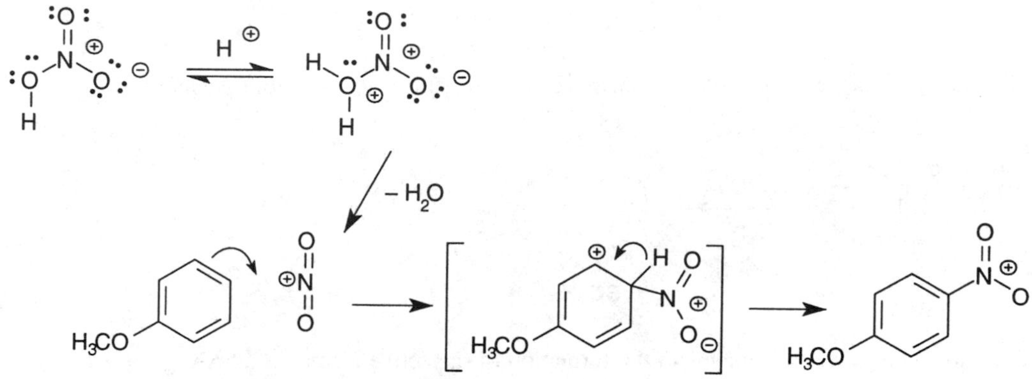

(e) Electrophilic aromatic sulfonation occurs with SO_3, the gas present in fuming sulfuric acid. The chlorine substituent in this Problem directs attack to the *para* position, forming the observed product after loss of a proton from the arenium cation intermediate.

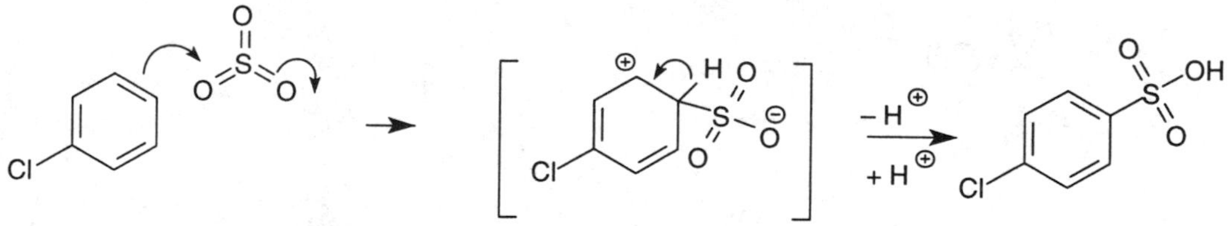

Pr 11.6 (a) The acetamido group ($-HNCOCH_3$) is strongly electron releasing and, therefore, directs to the *ortho* and *para* positions.

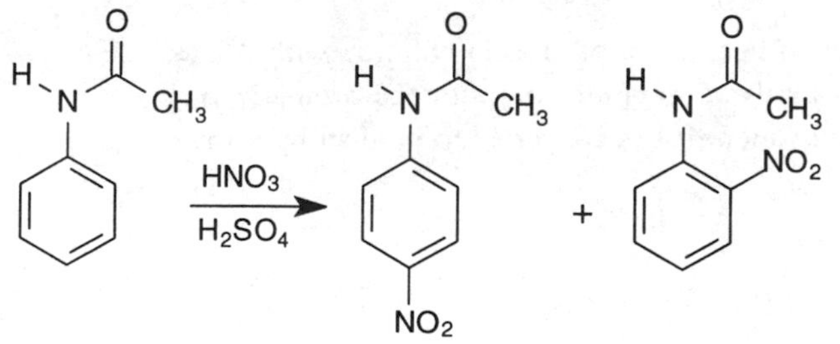

(b) The carboxylate group of methyl benzoate is electron withdrawing and, hence, *meta* directing.

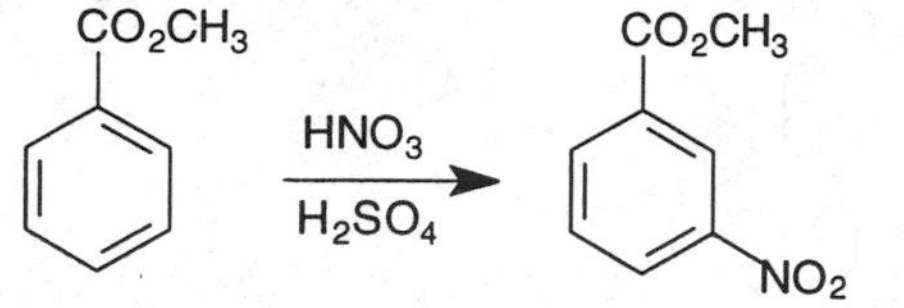

(c) Fluorobenzene, like the other halobenzenes, is deactivated but the halogen is *ortho, para* directing.

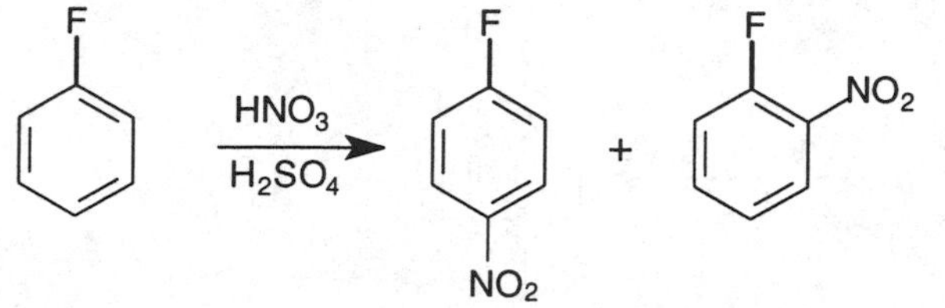

(d) Alkoxy groups are strongly electron-releasing and *ortho, para* directing.

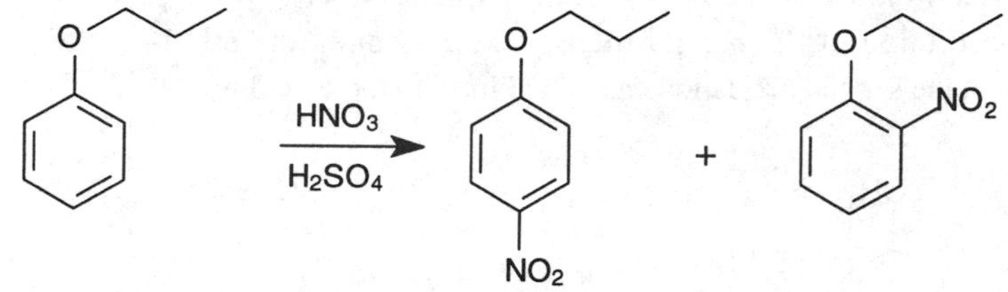

(e) 1-Methylnaphthalene undergoes electrophilic substitution at the *para* position in the same ring bearing the activating methyl group.

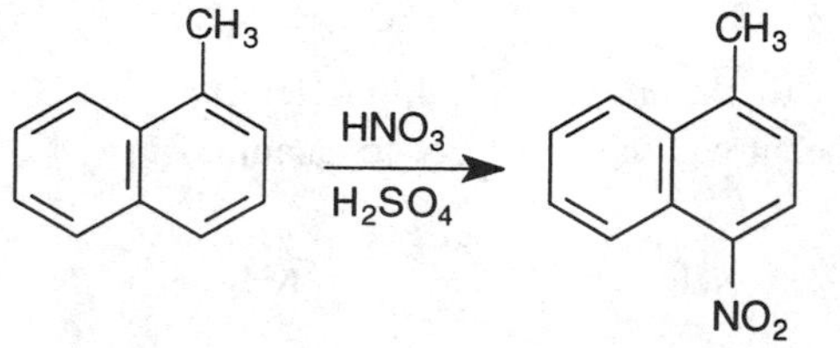

Pr 11.7 Several correct answers are possible for each part of this Problem. The one shown here employs the reactions presented in this chapter.

(a) To attach a straight alkyl chain, one should acylate and then reduce the resulting ketone.

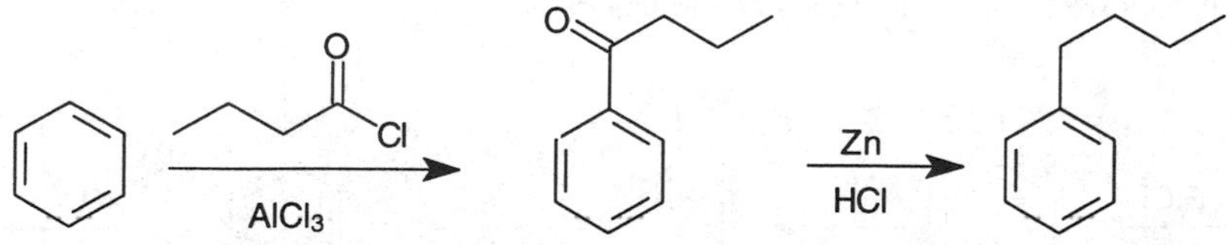

(b) A nitro group is a *meta* director, allowing chlorine to be introduced in a second step.

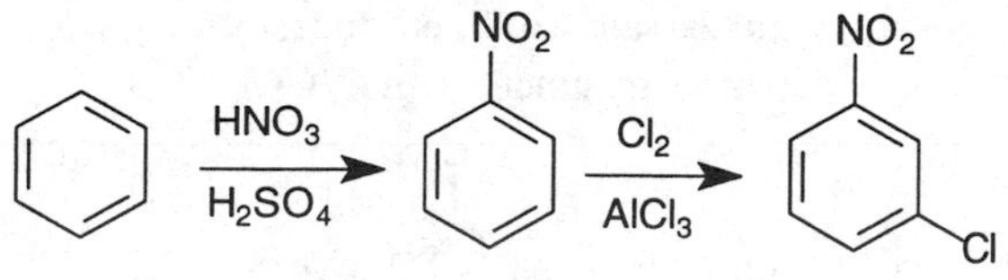

(c) Acylation produces an arene in which further substitution is directed to the *meta* position.

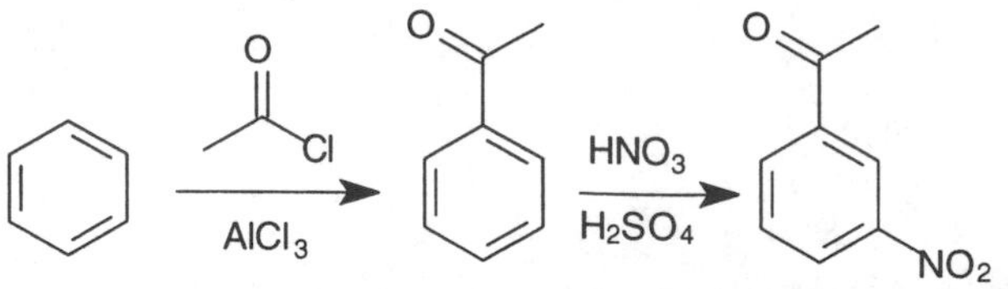

(d) Bromination introduces an *ortho, para* director, allowing regiospecific sulfonation at the desired *para* position.

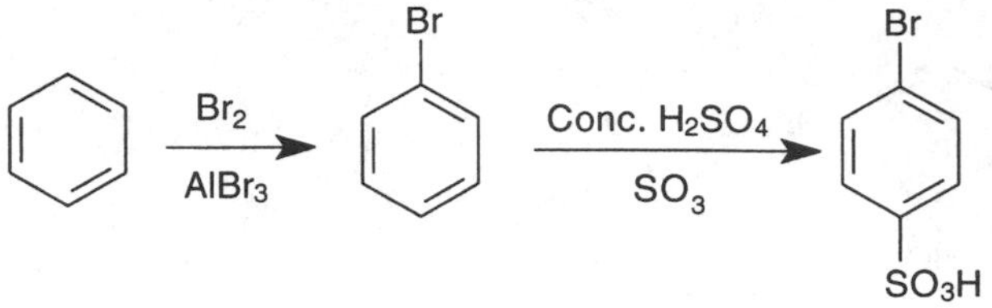

(e) Friedel-Crafts alkylation introduces a methyl group, which activates the ring toward further substitution at the *o,p* positions. The *p*-xylene (dimethyl-benzene) produced in the second alkylation can be oxidized to the diacid by hot aqueous KMnO$_4$.

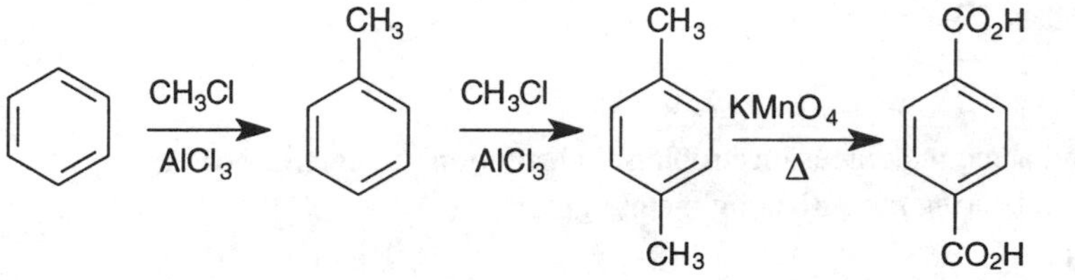

(f) Nitro groups direct further substitution to the *meta* position. After the introduction of the bromine substituent, the nitro group is reduced to an amino group.

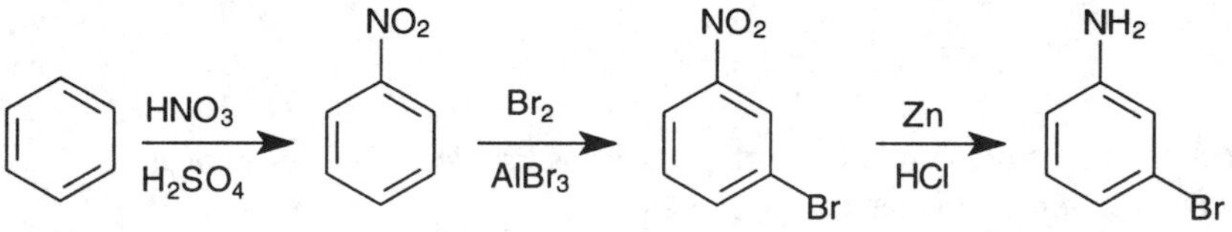

(g) Friedel-Crafts ethylation directs further functionalization to the *o,p* positions. After nitration, the benzylic position can be functionalized by free-radical bromination, followed by base-induced dehydrohalogenation.

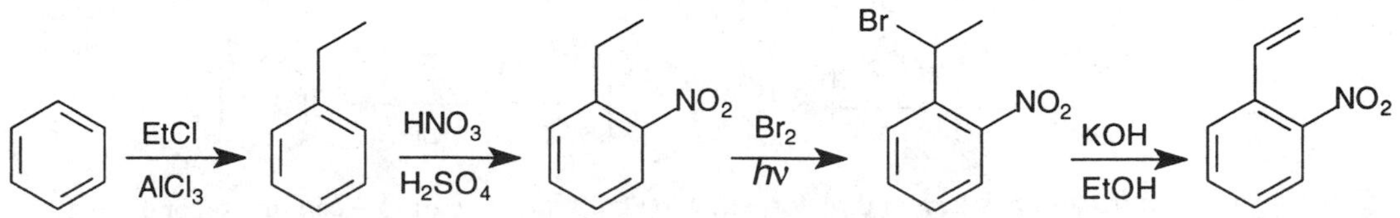

(h) Alkyation of benzene leads to toluene, which can be nitrated in the *para* position. Reduction of the nitro group with Zn and HCl produces *p*-methylaniline, which is methylated on nitrogen upon treatment with CH$_3$Cl.

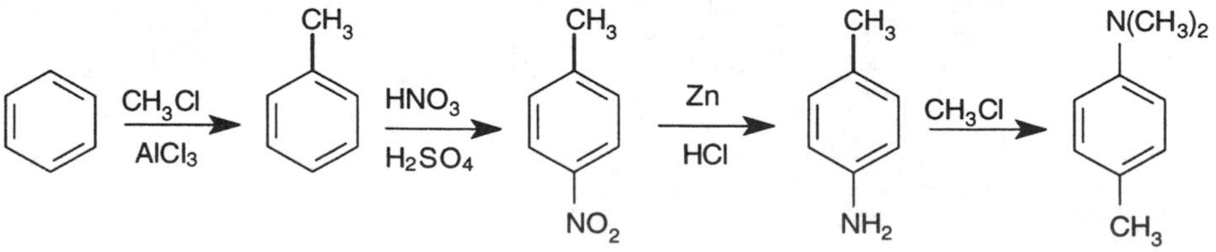

(i) Methylation of toluene can be followed by free-radical bromination to generate the benzylic bromide. (The two methyl groups are equivalent, so the bromination site is of no consequence.) Conversion of the bromide to the corresponding Grignard reagent, followed by treatment with ethylene oxide and hydrolysis, gives the desired alcohol.

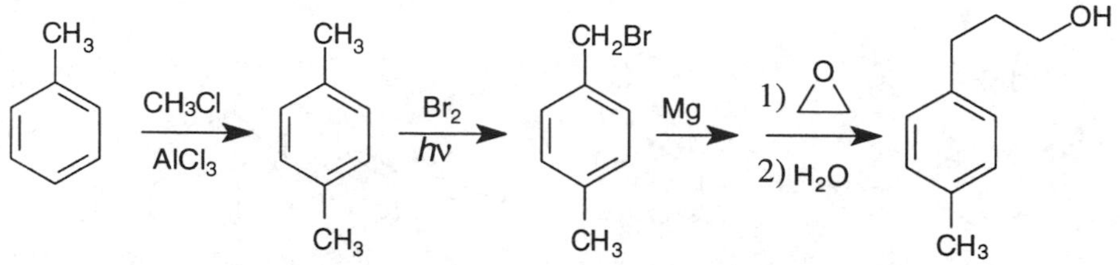

(j) Methylation of toluene allows for *o,p* functionalization by electrophilic halogenation. The benzylic position can be functionalized by a free-radical halogenation with chlorine.

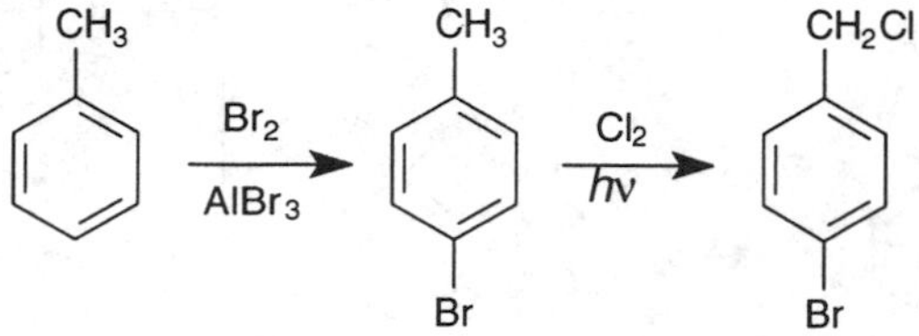

Pr 11.8   (a) The simple acetylation in this example tells us that an *o,p*-directing group was present, the dimethylamino group present in the product.

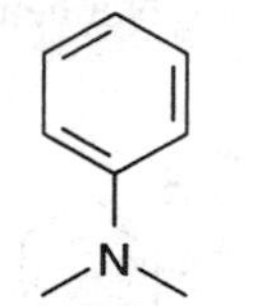

(b) The presence of the bromine in the *para* position after the bromination in the second step tells us that a *p*-directing substituent was attached at that point. The reduction in the first step tells us that an oxidized precursor to the *n*-alkyl group was present initially, logically a propionyl group, reduction of which gives the *n*-propyl group in the product.

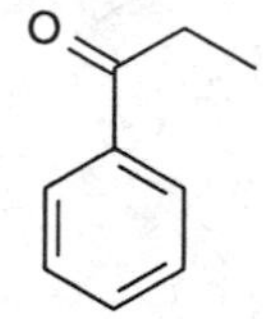

(c) The *meta* relation of the groups in the product requires that bromination in the first step was *meta*-directed. This requires the initial presence of an electron-withdrawing group. The reduction in the second step to form an *n*-propyl group implies that the original group was propionyl.

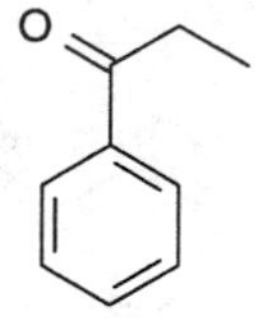

(d) The *ortho* relation in the product tells us that bromination in the first step was directed by an oxygen-containing *o,p* director. Because the methyl group was attached in the second and third steps by a Williamson ether synthesis, the original group must have been a phenolic OH group.

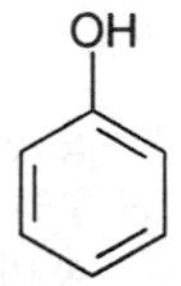

(e) There are two reduction steps here, as well as an electrophilic aromatic nitration. The last two steps effect the introduction of an amino group *meta* to the group obtained by reduction of the original substituent. An acyl substituent is such a group.

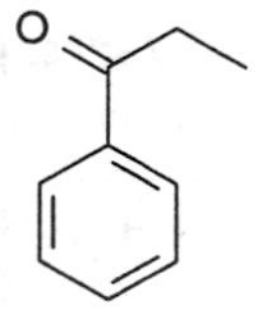

Pr 11.9   Friedel-Crafts acylation proceeds through an acylium ion, which attacks the ring as an electrophile at the site leading to the most stable carbocation. In naphthalene, the benzenoid character of the resonance-stabilized carbocations dictates the

preference for attack at C-1. Deprotonation of the indicated cations effects
rearomatization and produces the acylated substitution product.

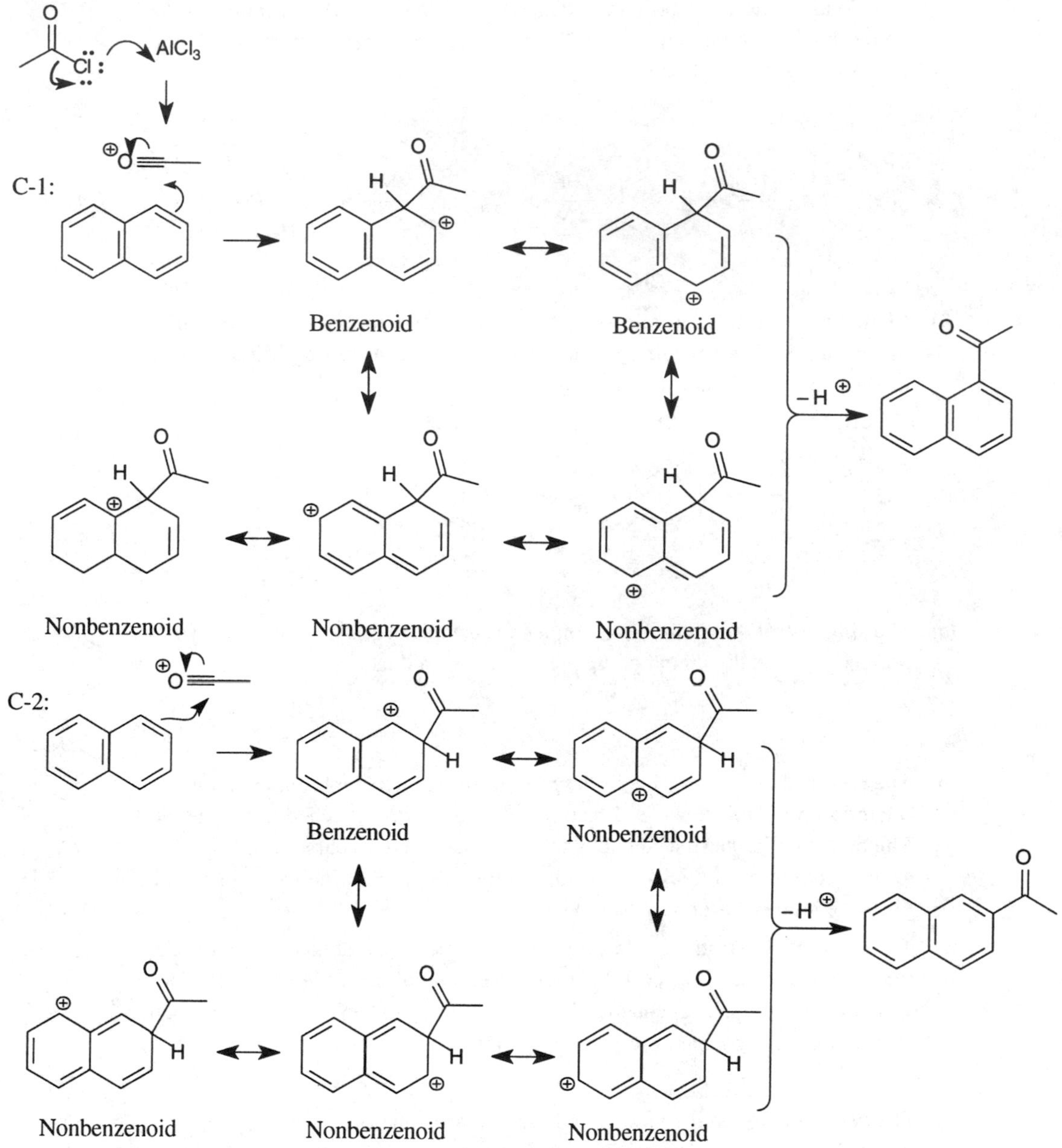

Pr 11.10   Under forcing conditions, phenol can act as an acid catalyst to form a *t*-butyl
cation. The strong electron-donation by the phenolic OH group directs electrophilic
attack by this cation to the *ortho* and *para* positions. Because the *para* position is
blocked in the starting material by a methyl group, the alkylation proceeds at both
ortho sites, producing BHT. Forcing conditions are required because of the steric
hindrance of two adjacent groups present after the first alkylation.

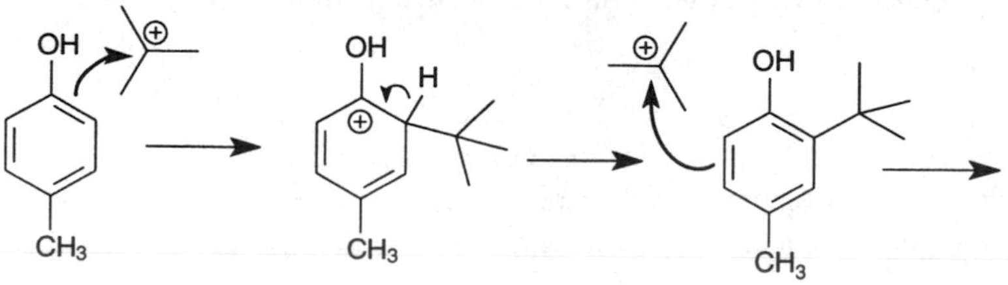

Important New Terms

Active electrophile: a more active than normal form of an electrophilic reagent often prepared by interaction of an electrophilic reagent with a Lewis or Brønsted acid (11.2)

Acylation: the replacement of H by an acyl group (11.2)

Acylium ion: $RC≡O^+$; a resonance-stabilized cation in which positive charge is distributed between carbon and oxygen (11.2)

Azo dyes: highly colored compounds containing the —N=N— linkage; among the first synthetic colorfast agents (11.3)

Benzenoid ring: a six-membered ring in a polycyclic aromatic compound that retains three formal double bonds (11.5)

Clemmensen reduction: the reduction of a ketone to a $-CH_2-$ group by treatment with zinc in HCl (11.2)

Diazo coupling: the connection of two aromatic rings through an azo linkage, usually by electrophilic attack on one ring by an aryl diazonium salt (11.3)

Diazonium salt: $[Ar–N≡N]^+ X^-$, prepared by treatment of a primary aniline with nitrous acid, HNO_2 (11.3)

Diazotization: the conversion of a primary amine to a diazonium salt (11.3)

Directive effect: a substituent effect that influences the regiochemistry of a reaction (11.4)

Electron acceptor: a group that withdraws electron density from an attached atom (11.4)

Electron donor: a group that releases electron density to an attached atom (11.4)

Electronic effect: the perturbation of molecular properties by shifts in electron density by a substituent (11.4)

Electrophilic aromatic substitution: *see* **Electrophilic substitution** (11.1)

Electrophilic substitution: the replacement of a substituent (usually hydrogen) on an aromatic ring upon interaction of an aromatic π system with an active electrophile (11.1)

Friedel–Crafts acylation: the reaction of a carboxylic acid chloride with an aromatic compound in the presence of a Lewis acid, resulting in the replacement of a hydrogen by an acyl substituent (11.2)

Friedel–Crafts alkylation: the reaction of an alkyl halide with an aromatic compound in the presence of a Lewis acid, resulting in the replacement of a hydrogen by an alkyl substituent (11.2)

meta **director**: a functional group that directs electrophilic aromatic substitution to the *meta* position (11.4)

Nitration: the replacement of H by an NO_2 group (11.2)

ortho,para **director**: a functional group that directs electrophilic aromatic substitution to the *ortho* and *para* positions (11.4)

Side chain oxidation: the conversion of an alkyl or acyl side chain on an aromatic ring to a $—CO_2H$ group upon treatment with hot aqueous $KMnO_4$ (11.3)

Substituent effect: the altered reactivity induced by the presence of a substituent group on a reactant, often affecting rates, stereochemistry, or regiochemistry (or all three) of a reaction (11.4)

Sulfonation: the replacement of H by an $—SO_3H$ group (11.2)

Wolff–Kishner reduction: the reduction of a ketone to a $-CH_2-$ group by treatment with basic hydrazine, NH_2NH_2 (11.2)

Nucleophilic Addition and Substitution at Carbonyl Groups

12

Key Concepts

Mechanism of nucleophilic addition to carbonyl groups

Mechanism of nucleophilic acyl substitution

Rationale for accelerating nucleophilic addition and substitution under acidic and basic conditions

Characteristics of the key tetrahedral intermediate in nucleophilic addition and substitution

Differentiating acid (or base) catalyzed reactions from acid (or base) induced reactions

Relative reactivity of various functional groups toward nucleophilic attack

Anions as activated nucleophiles

Structure and reactivity of complex metal hydrides

Comparison of selectivity of hydride reduction with those of catalytic hydrogenations, dissolving metal reductions, and biological cofactor reductions

Stereochemistry of catalytic hydrogenation and dissolving metal reduction of alkynes

Synthesis of alcohols by hydride reduction of aldehydes, ketones, and esters

Synthesis of amines by hydride reduction of amides

Formation of hydrates, hemiacetals, acetals, hemiketals, and ketals from carbonyl compounds

Acetals and ketals as protecting groups for aldehydes and ketones

Disproportionation of aldehydes to alcohols and acids in the Cannizzaro reaction

Formation of imines and Schiff bases with amino nucleophiles

Hydrazone formation as a chemical (color) test for aldehydes and ketones

Imine-enamine tautomerization

Synthesis of amines by reductive amination of aldehydes and ketones

Reagents and conditions for the interconversion of carboxylic acid derivatives: hydrolysis, esterification, amidation, dehydration to anhydrides, preparation of acid chlorides with thionyl chloride

Synthesis of carboxylic acids by nitrile hydrolysis

Comparison of esters and amides of sulfonic acids and phosphoric acids with those of carboxylic acids

Answers to Exercises

Ex 12.1 The addition of a halide ion to an aldehyde results in breaking of a $C=O$ π bond and the formation of a C—X bond. In addition, the relatively non basic halide ion is converted to an alkoxide ion, a process that is disfavored because of the conversion of a less to more basic species. In progressing from chloride to bromide to iodide ion, the effect of bond strength favors the aldehyde and halide ion over the addition product as the strength of the C—X bond progressively decreases. In addition, the halide becomes less basic (as can be determined from the increased acidity of the halogen acid), further favoring starting materials.

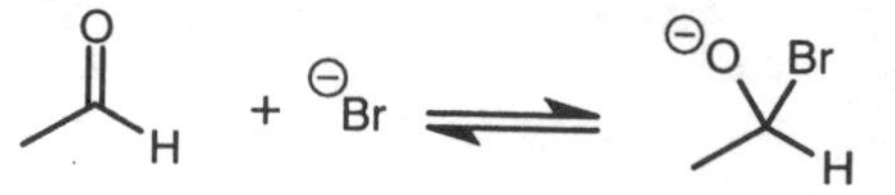

Bond energy $C{=}O\ \pi$ 91 $C{-}Br$ 68

pK_a $H{-}Br$ -9 $O{-}H$ 16

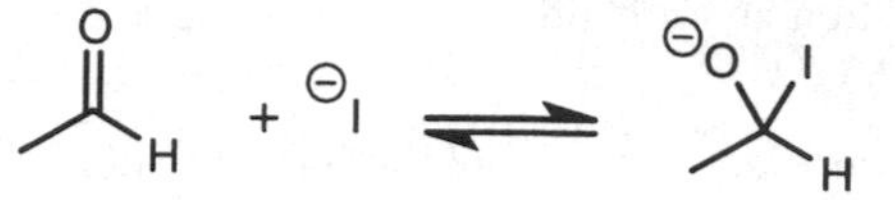

Bond energy $C{=}O\ \pi$ 91 $C{-}I$ 51

pK_a $H{-}I$ -10 $O{-}H$ 16

Ex 12.2   Hydroxide ion adds as a nucleophile to the acid chloride, forming a tetrahedral intermediate. Because chloride ion is a better leaving group than is hydroxide ion (compare the pK_as of HCl and HOH), Cl⁻ is lost as the $C{=}O$ bond is reformed. The reverse reaction (attack by chloride ion on the carbonyl group of a carboxylic acid) is so slow as to be insignificant, because the ultimate products, the carboxylate and chloride ions are much more stable than the starting acid chloride and hydroxide ions. This difference is due mainly to resonance stabilization in the carboxylate anion and the difference in basicity between hydroxide and chloride ions. The principle of microscopic reversibility requires that the same transition state be involved for both the forward and backward reactions. Thus, the activation energy for the reverse reaction is much larger (by ΔG) than that for the forward reaction.

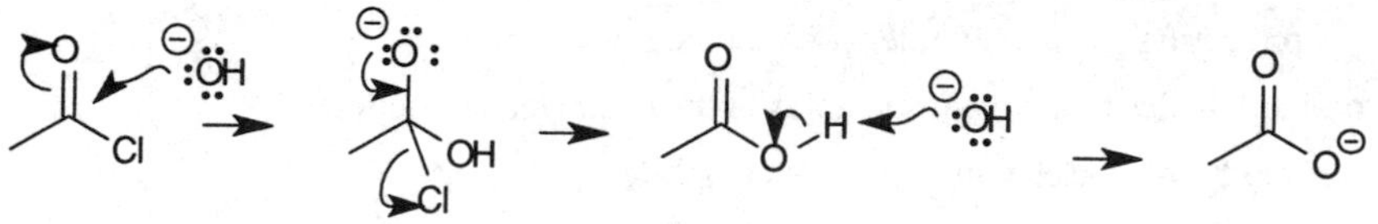

Ex 12.3   The ⁻AlH₄ ion delivers a hydride ion equivalent to the carbonyl carbon, forming a C—H bond, shifting an electron pair onto the carbonyl oxygen, and generating a trivalent aluminum compound. This process is greatly facilitated by the simultaneous association of the lithium cation with the carbonyl oxygen. Aluminum is in the same column of the periodic table as boron and is similarly electron deficient: trivalent aluminum compounds are highly electrophilic and actively seek centers of electron density as Lewis acids. The aluminum atom of AlH₃ associates with the negatively charged oxygen atom. The resulting anion now has Al—H bonds similar to those in the aluminum hydride and delivers a second hydride equivalent to another carbonyl group. This series of reactions is repeated until each of the four hydride equivalents of the original ⁻AlH₄ ion has been delivered. (Only the first stage is shown here.) Upon addition of water, the Al—O bonds in the resulting aluminate are replaced by H—O bonds. The final product mixture thus yields four equivalents of reduced alcohol for each mole of lithium aluminum hydride consumed, as well as inorganic salts and hydrates. Although a center of chirality is formed in the product, a racemic mixture results, because the rate of attack by the achiral ⁻AlH₄ ion is the same on each face of the planar carbonyl group.

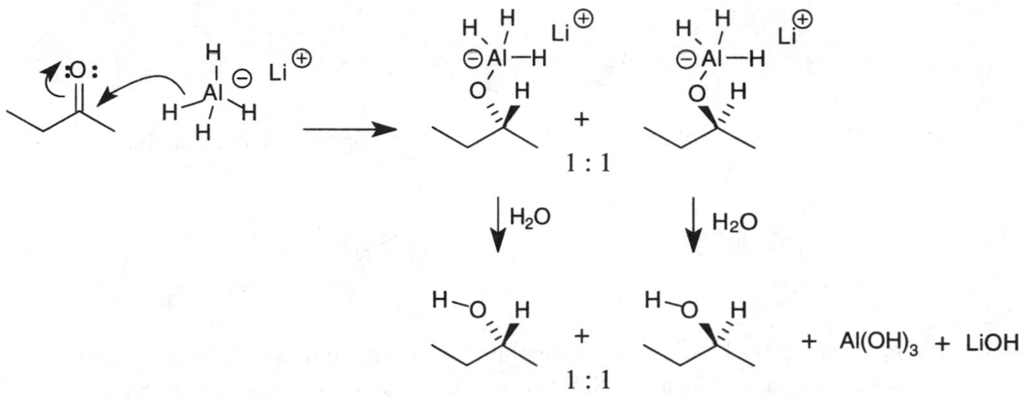

Ex 12.4 The reduction of acetamide (CH_3CONH_2) begins by delivery of a hydride ion equivalent to the amide carbonyl carbon. Cleavage of the C—O σ bond in the resulting tetrahedral intermediate is assisted by a lone pair of electrons from the adjacent amide nitrogen atom. Upon formation of a nitrogen–carbon bond and loss of a proton, an imine is produced. This functional group resembles the original carbonyl group electronically and is readily reduced by a second hydride ion equivalent, delivered from $^-AlH_4$ (or from other aluminum hydride species formed in the reduction) to the carbon end of the C=N π bond. The formation of this second C—H bond completes the conversion of what was originally a C=O double bond to a CH_2 group. Upon treatment with water, the aluminate is converted to the free amine product.

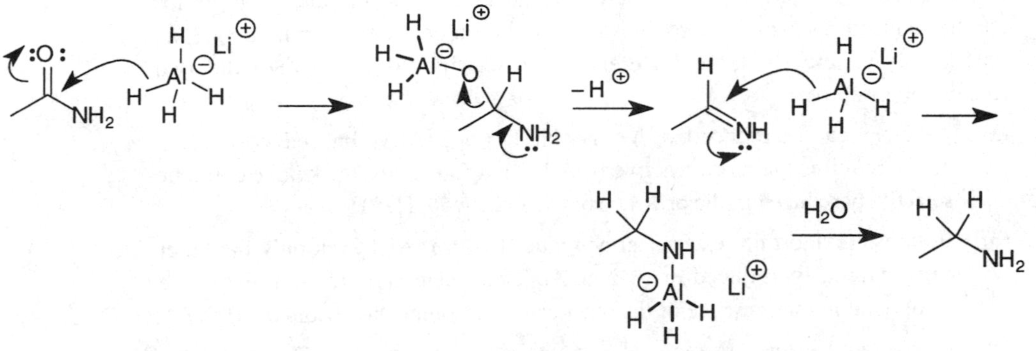

Ex 12.5 Lithium aluminum hydride reductions of amides preserve the carbon skeleton of the starting amide, simply effecting the net conversion of a C=O group to a CH_2 group.

(a) The symmetry of the amine permits only one amide starting material.

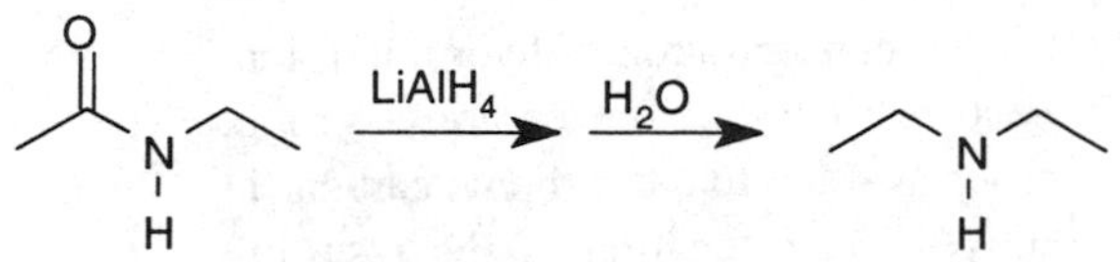

(b) As in part *a*, a highly symmetric amine can result from only one amide.

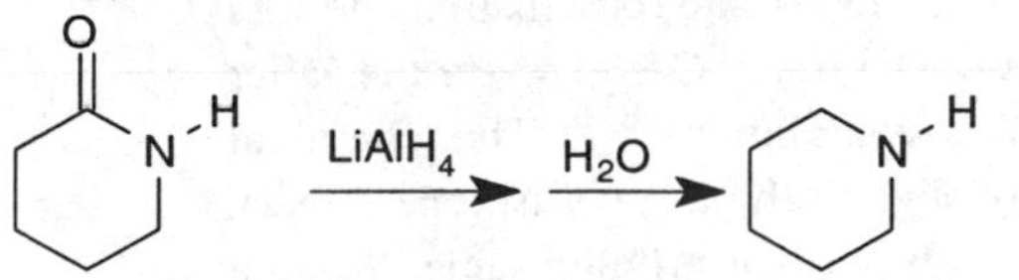

(c) Only one methylene group is adjacent to the amino group, permitting only one amide starting material.

(d)   Reduction of the tertiary amide show here affords *N*-ethyl-*N*-methyl aniline.

Ex 12.6   Lithium aluminum hydride reductions involve nucleophilic attack of a hydride equivalent to a carbonyl carbon. Substantial electron deficiency is needed at that site for complex metal hydride reduction. Because a carboxylic acid bears an acidic proton, it is rapidly deprotonated by a highly basic complex metal hydride. A carboxylic acid is thus converted to a carboxylate anion by treatment with $LiAlH_4$. The electron density in the carboxylate anion is much higher than in the neutral carboxylic acid, and, as a result, addition of a nucleophile is substantially more difficult.

Esters lack an acidic OH group and do not undergo a comparable acid-base reaction. The carbonyl group of an ester is more electrophilic than a carboxylate anion, so, esters are more easily reduced by complex metal hydrides such as $LiAlH_4$ than are carboxylic acids.

Ex 12.7   In this question, we are asked to determine whether each of the functional groups can be reduced by sodium borohydride or lithium aluminum hydride and, if so, whether a sufficiently large difference in reactivity exists between the two functional group types.

(a)   A ketone can be reduced with either $NaBH_4$ or $LiAlH_4$, but reduction of an amide requires the more reactive ($LiAlH_4$) reagent. Thus, the ketone could be selectively reduced in the presence of the amide with $NaBH_4$.

(b)   Both esters and thiol esters can be reduced with $LiAlH_4$, but only the latter can be readily reduced with $NaBH_4$. Thus, selective reduction of one compound in the presence of the other can be accomplished with $NaBH_4$.

(c)   Amides and esters are reduced with the same reagent ($LiAlH_4$). Selective reduction is therefore not possible.

(d)   An aldehyde is much more readily reduced than an ester. As in part *a*, $NaBH_4$ will reduce the aldehyde in the presence of the ester. $LiAlH_4$ would reduce both functional groups.

Ex 12.8   Even in neutral water there is a small (10^{-7} M) concentration of hydronium ion. Protonation of acetone enhances the electrophilicity of the carbonyl group, resulting in rapid addition of nucleophiles such as water to the carbonyl carbon (in this case, isotopically labeled water, $^{18}OH_2$). Deprotonation of the resulting tetrahedral species leads to a hydrate in which one of the oxygen atoms is ^{16}O and the other is ^{18}O. Reversal of the steps leading to the hydrate (protonation and loss of water) then proceeds with loss of either $^{18}OH_2$ or $^{16}OH_2$, where the latter process results in the formation of acetone containing ^{18}O. The chemical reactivities of $^{16}OH_2$ and $^{18}OH_2$ differ only slightly (by a small isotope effect), and the rate of loss of normal and ^{18}O-labeled water is nearly the same. When a large excess of heavy water is used (for example, as solvent), acetone with a high loading of ^{18}O is formed.

Ex 12.9 The conversion of a carbonyl compound to a hydrate is rapid, catalyzed by either acid or base, and involves formation of a C—O σ bond at the expense of a C=O π bond. The position of the equilibrium depends mainly upon the strength of the C=O π bond.

(a) Because aldehydes are somewhat less stable than ketones (having only one substituent group that participates in hyperconjugation), the equilibrium position for propanal lies further toward the hydrate than does that for acetone, but the carbonyl compound nonetheless is favored.

(b) Because of the three electron-withdrawing fluorine substituents, the carbonyl group of trifluoroacetaldehyde is even less stable than that of acetaldehyde, and, indeed, the equilibrium favors the hydrate.

(c) For carboxylic acid amides, there is a large degree of stabilization of the carbonyl system by delocalization of the nitrogen lone pair of electrons that must be disrupted in order to form the hydrate. Here, the amount of hydrate in equilibrium is so small that it has not been detected.

Ex 12.10 The Cannizzaro reaction is a disproportionation that takes place with aldehydes that have no enolizable α-hydrogen atoms.

(a) Lacking any α-hydrogen atoms, this aldehyde does participate in the Cannizzaro reaction.

(b) Because this compound has hydrogen atoms α to the carbonyl group and can therefore enolize, it is an inappropriate candidate for a Cannizzaro reaction. Instead is deprotonated to an enolate anion.

(c) The Cannizzaro reaction does not take place with ketones: the aldehydic hydrogen must be present in order to be transferred in the key reduction step.

(d) As in part *a*, this aldehyde lacks α-hydrogen atoms and can participate in a Cannizzaro reaction.

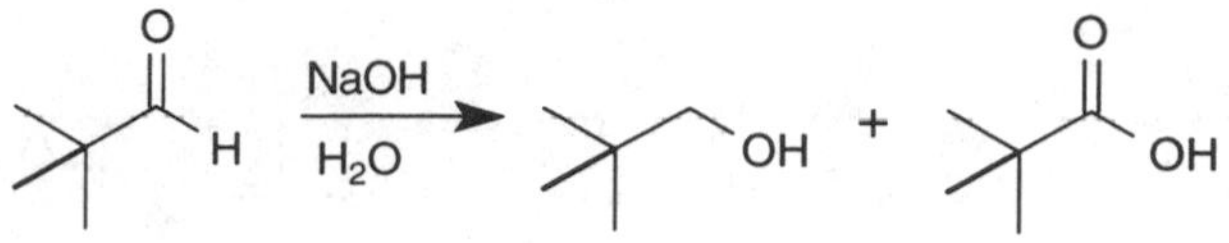

Ex 12.11 The formation of a ketal from a ketone and a 1,2-diol (or two equivalents of a simple alcohol) begins by protonation of the carbonyl oxygen. Nucleophilic attack by one oxygen atom of the diol, followed by deprotonation, results in a neutral hemiketal. Protonation of the hemiketal converts the hydroxyl group to a good leaving group, loss of which forms a stabilized cation to which the other oxygen of the original glycol adds as a nucleophile. Loss of a proton results in the product ketal. The steps for hydrolysis of the ketal are *exactly* the reverse of those shown below: that is, each reaction step arrow points in the other direction and each curved arrow showing the flow of electrons is reversed.

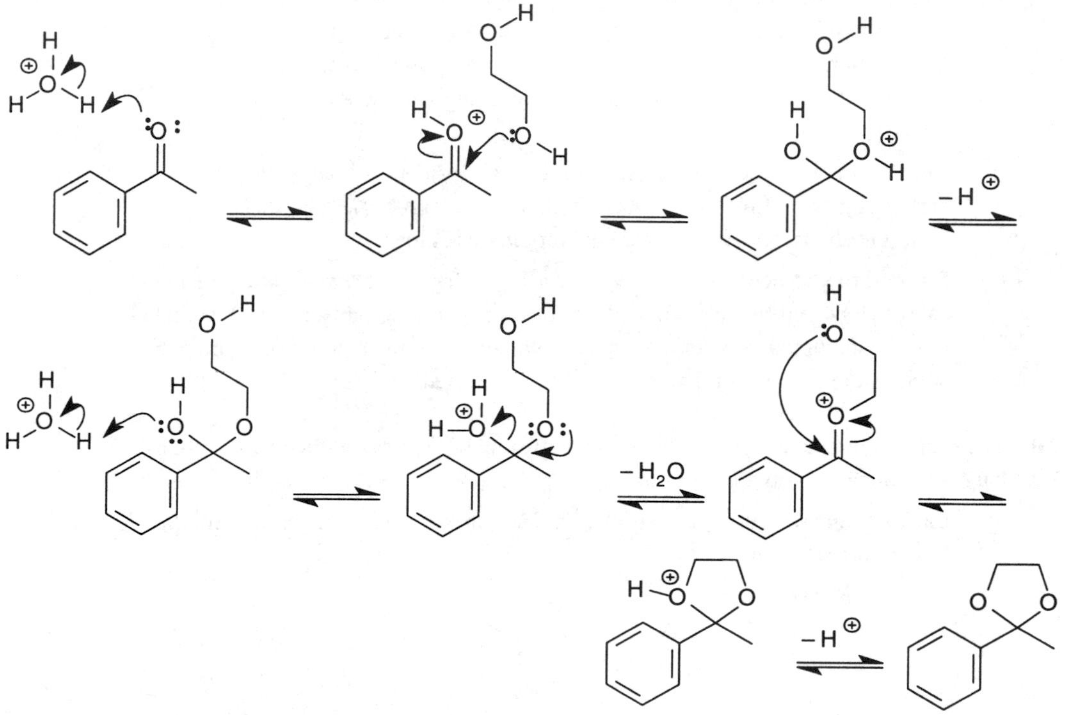

Ex 12.12 The formation of an imine from a ketone and an amine begins in the same way as ketal formation: by protonation of the carbonyl oxygen. Nucleophilic addition of an amine nitrogen followed by loss of a proton results in an intermediate (called an aminal) that is analogous to a hemiketal. Protonation of the hydroxyl group of the aminal sets the stage for loss of water, forming a protonated imine. Deprotonation of this species results in the product imine.

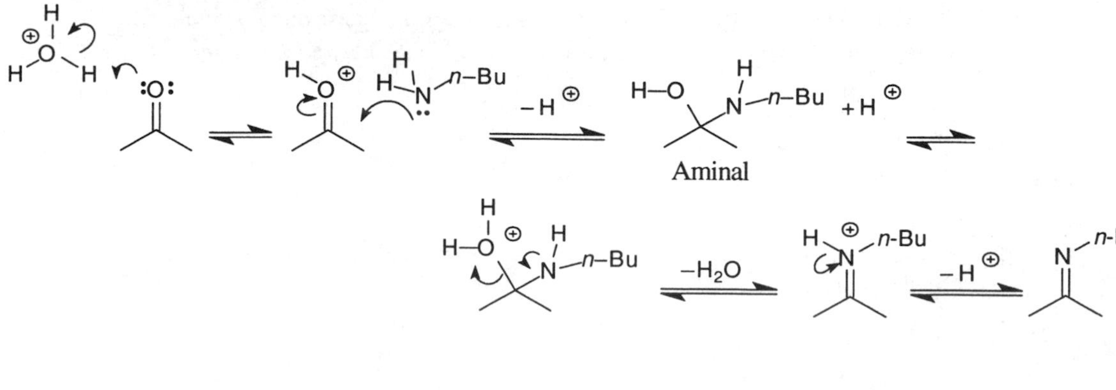

Aminal

Ex 12.13 The imine of a ketone is basic, and in moderately acidic solution (pH 5) it exists predominatnt in the protonated form as an iminium ion. Thus, a hydride ion equivalent is added from sodium cyanoborohydride to the imine carbon as the π electron density is shifted onto nitrogen, resulting in the formation of the product amine.

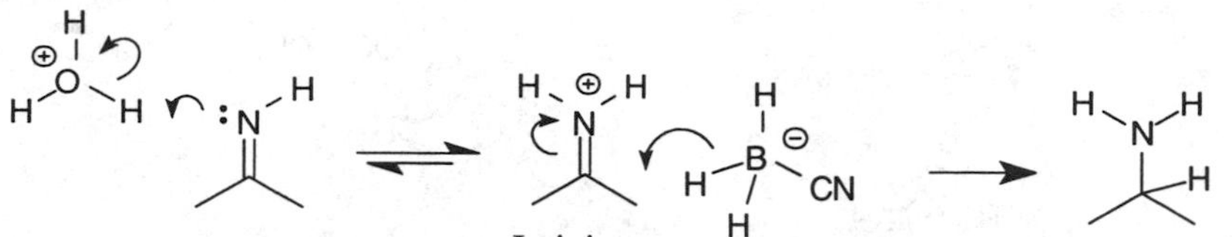

Ex 12.14 The mechanism for the formation of a hydrazone from hydrazine and a ketone is essentially identical to that in Exercise 12.12 for the corresponding reaction producing an imine from a ketone and an amine.

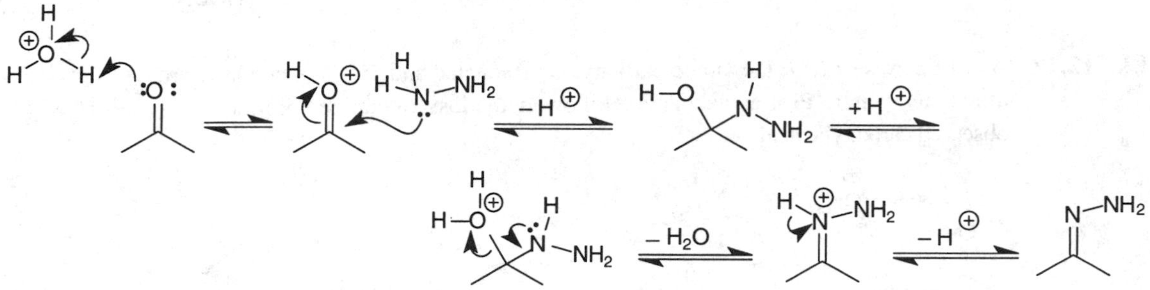

Ex 12.15 In each resonance contributor, a lone pair from the heteroatom attached to the carbonyl group is donated toward the carbonyl carbon as the pair from the C=O π bond shifts onto oxygen.

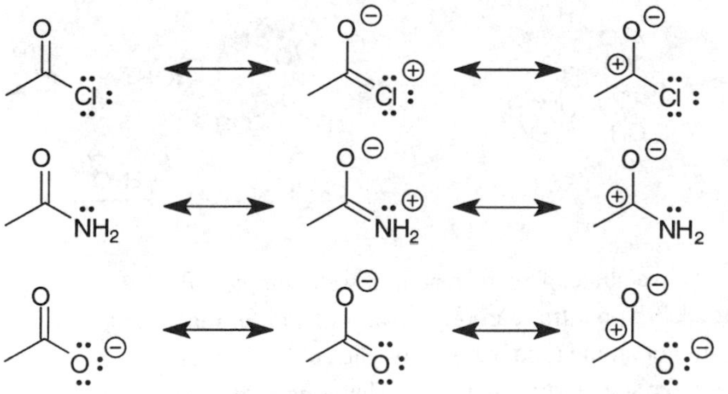

Ex 12.16 The reaction of a carboxylic acid and an alcohol in the presence of acid produces an ester and water in a reversible reaction. Removal of the water as it is formed with a Dean-Stark trap shifts the equilibrium position toward the side of the ester (Le Chatelier's principle). In essence, the reverse reaction becomes impossible when water is no longer present in the reaction flask.

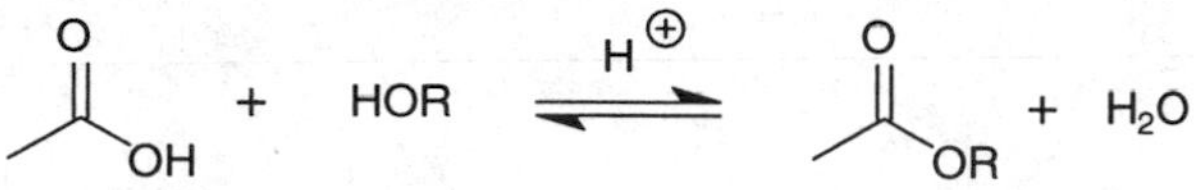

Ex 12.17 The reaction begins with protonation of the carbonyl oxygen, increasing its reactivity toward nucleophiles. Methanol then attacks, forming a tetrahedral intermediate. Upon proton exchange, the OEt group is converted to a leaving

group, and upon reformation of the carbonyl group, ethanol is lost, effecting the observed transesterification.

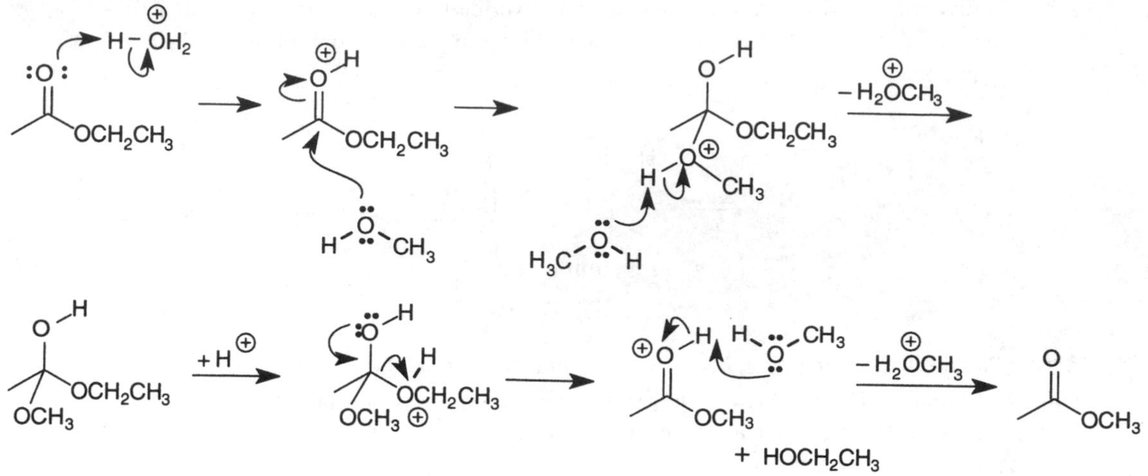

Ex 12.18 As in Exercise 12.17, the amide carbonyl is protonated and thus activated toward attack by water. Proton exchange, followed by loss of amine, produces the observed carboxylic acid.

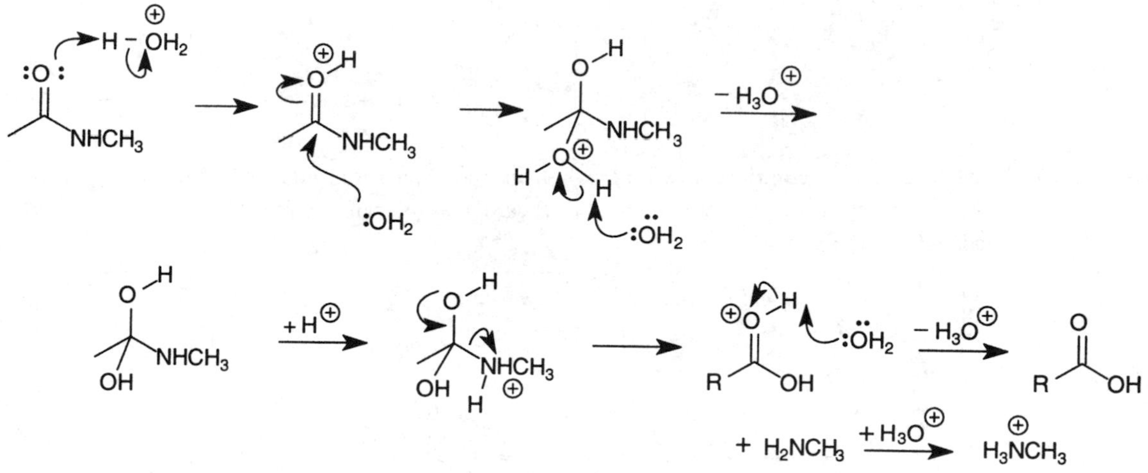

Ex 12.19 In base, the hydroxide ion (a more reactive nucleophile than neutral water employed in the previous two Exercises) attacks the neutral carbonyl group of the amide. From this tetrahedral intermediate, loss of amine produces carboxylic acid, whereas loss of hydroxide ion regenerates the amide starting material. The course of this reaction is thus controlled by the relative lability of these reagents to act as leaving groups.

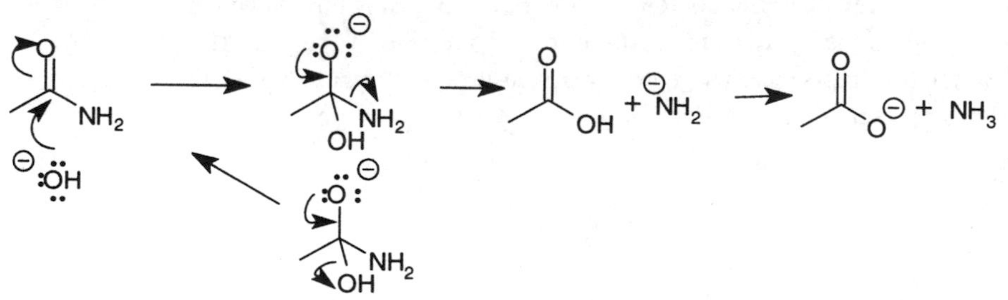

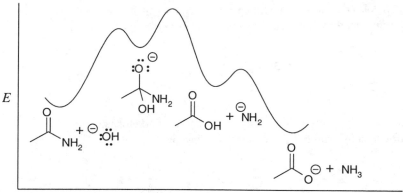

Reaction coordinate

Ex 12.20 Because formation of the tetrahedral intermediate in nucleophilic acyl substitution is normally rate limiting, factors that increase the stability of the starting carbonyl group increase the activation energy and decrease the reaction rate.

(a) Poor overlap between chlorine and carbon orbitals provides only limited resonance stabilization of carboxylic acid chlorides as compared with esters. Acid chlorides are therefore more reactive than carboxylic acid esters.

(b) The electron-withdrawing chlorine substituent of methyl α-chloroacetate increases the electrophilicity of the carbonyl group, increasing the rate of reaction of this ester compared with methyl acetate.

(c) As with chlorine (see part *a*), sulfur's orbitals are substantially larger than carbon orbitals. Thus, there is little resonance stabilization of thiol esters as compared to their oxygen analogs, and therefore the former are more reactive.

(d) The difference in resonance stabilization between primary and tertiary amides is small. On the other hand, steric hindrance in the transition state for addition of a nucleophile to tertiary amides is substantially greater, so reaction of the more substituted amide is slower.

Ex 12.21 The preparation of a carboxylic acid by hydrolysis of a nitrile formed by displacement of an alkyl halide requires a viable S_N2 reaction—that is, the alkyl halide must be primary or secondary. Conversely, formation of a carboxylic acid by reaction of a Grignard reagent with CO_2 can be carried out only when there are no functional groups present that react readily with a Grignard reagent.

(a) Because second-order substitution reactions are not possible on aryl halides, benzoic acid can be synthesized only from bromobenzene via a Grignard reagent.

(b) The presence of the hydroxyl group precludes the use of the Grignard route to form 4-hydroxycyclohexanecarboxylic acid.

(c)   A tertiary alkyl halide is not a substrate for an S_N2 reaction. Thus, a Grignard route is used for the synthesis of 1-methylcyclohexanecarboxylic acid.

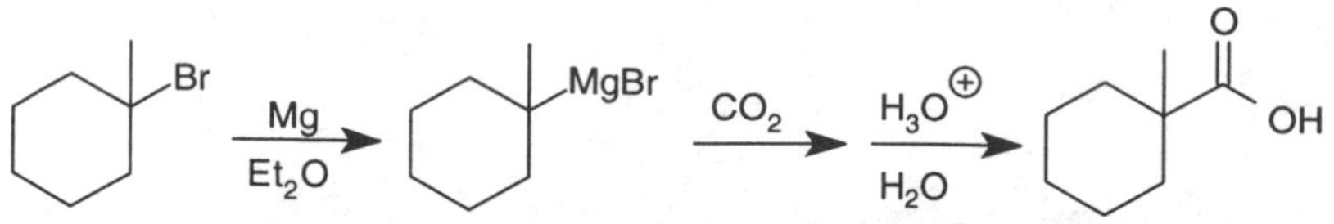

(d)   Because ketones react with Grignard reagents, the synthesis of 5-ketohexanoic acid is best carried out by synthesis of the corresponding nitrile.

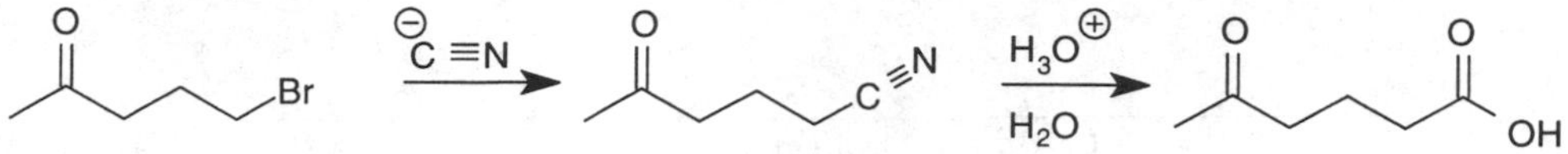

Ex 12.22 Protonation at nitrogen increases the electrophilic reactivity of a nitrile. A water molecule adds to the carbon atom of the protonated nitrile as the electron density of the π system is shifted to the positively charged nitrogen atom. Deprotonation of the oxonium ion, followed by protonation on nitrogen, leads to an amide. Activation of the amide by protonation on oxygen sets the stage for nucleophilic attack by a second molecule of water. The tetrahedral intermediate formed is deprotonated at oxygen and reprotonated on nitrogen, forming a tetrahedral species from which ammonia is readily lost. The protonated carboxylic acid thus formed is deprotonated, completing the reaction.

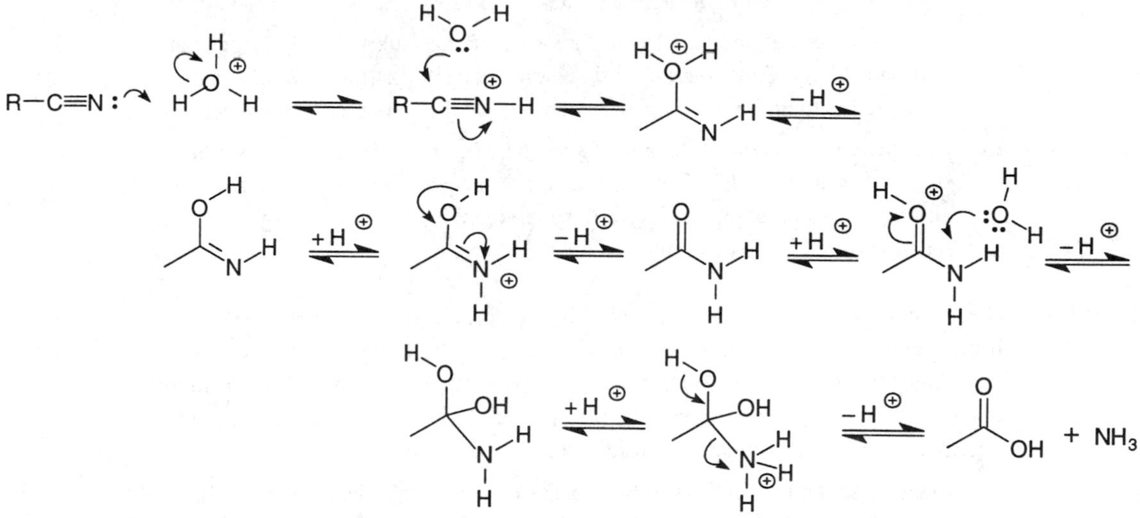

Ex 12.23 This Exercise reviews some of the reactions of sulfonic acids and sulfonyl chlorides from earlier chapters and places them in context with the new reactions considered here.

(a)   Conversion of toluene to *p*-toluenesulfonyl chloride is accomplished by electrophilic sulfonation (fuming H_2SO_4), followed by conversion to the sulfonyl chloride by treatment with thionyl chloride, a reaction parallel to that involved in the preparation of a carboxylic acid chloride from a carboxylic acid.

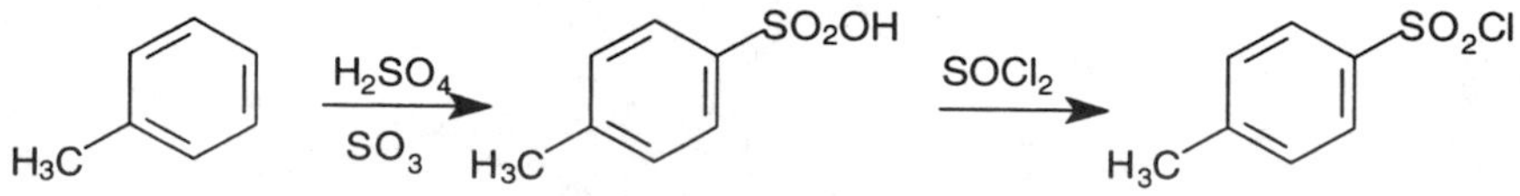

(b) The transformation of an alcohol to an iodide with inversion of configuration is an S_N2 reaction that requires conversion of the alcohol to a tosylate (a better leaving group) by reaction with *p*-toluenesulfonyl chloride, followed by an S_N2 displacement by iodide ion in a polar, nonaqueous solvent.

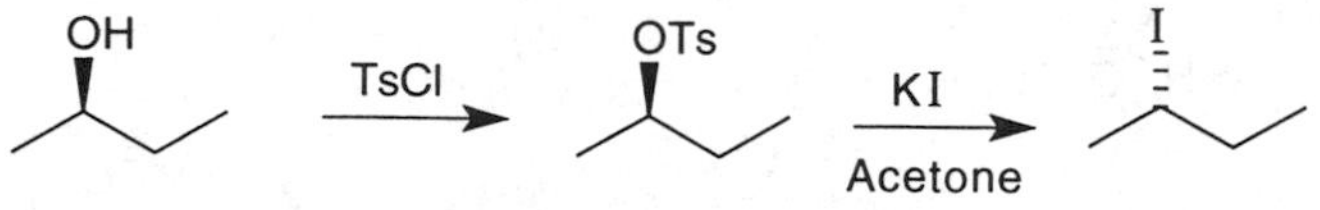

(c) The product is a tosylate ester which requires the alcohol and tosyl chloride. The alcohol is prepared from the ketone by reduction of cyclohexanone, using a complex metal hydride.

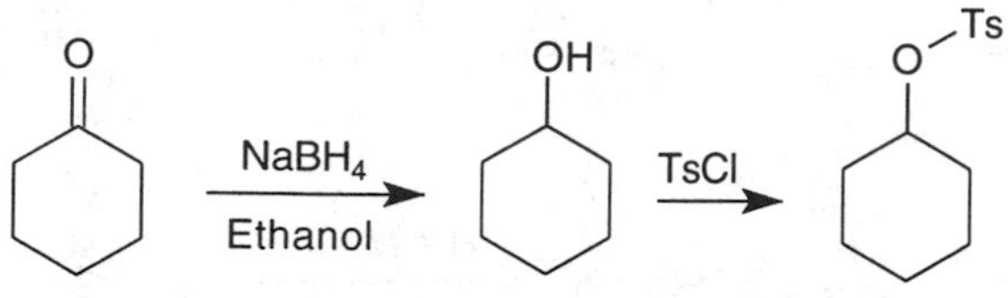

(d) The electron-withdrawing ester group directs electrophilic substitution to the *meta* position on the aromatic ring of a benzoate ester. Sulfonation provides the sulfonic acid, which is activated toward nucleophilic substitution by conversion first to the sulfonyl chloride and then to the sulfonamide by reactions parallel to those of carboxylic acid derivatives discussed in this chapter.

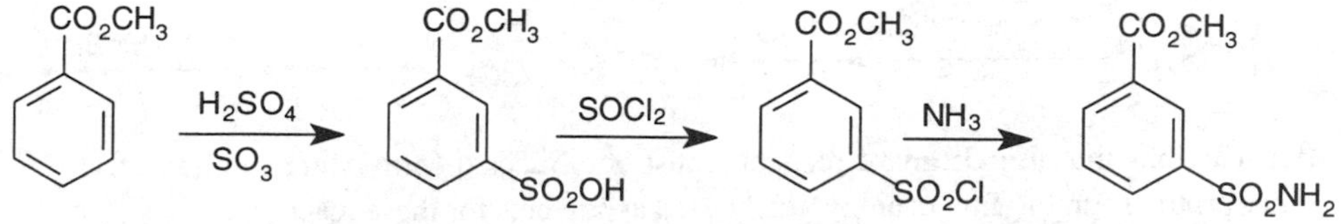

Ex 12.24 The S=O bond in thionyl chloride is polarized in the same way as the carbonyl C=O double bond. It is especially reactive because of the presence of two adjacent highly electronegative chlorine atoms. Nucleophilic attack by alcohol forms a new O—S bond as π electron density shifts onto oxygen. Upon deprotonation to relieve the positive charge of the oxonium ion and reformation of the S=O double bond as chloride is expelled, a thionyl ester is formed. The C—O bond is particularly polarized so that S_N2 attack by chloride occurs, releasing a stable molecule (SO_2) and a free chloride ion as the alkyl chloride is formed.

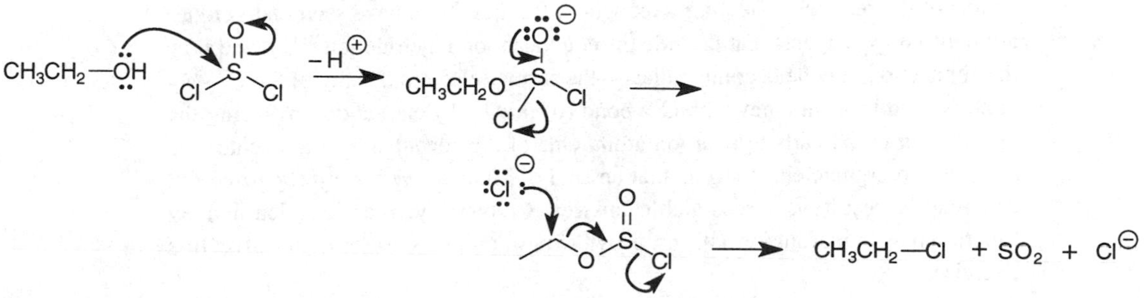

Ex 12.25 The first step in the Strecker synthesis is similar to reductive amination: an intermediate imine is formed by reaction of ammonia with the aldehyde, as in Exercise 12.12. Addition of cyanide ion to the protonated imine forms the α-aminonitrile. Hydrolysis of the nitrile follows the same pathway as detailed in the answer to Exercise 12.22.

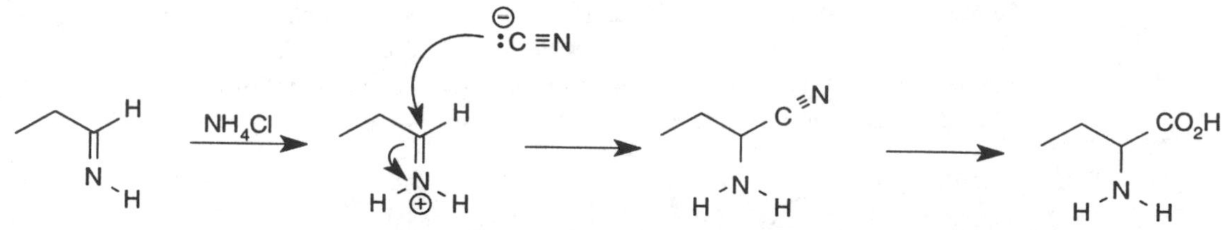

Ex 12.26 We know that at least one of the C—C bonds to the carbinol carbon of a product alcohol is formed during a Grignard synthesis.

(a)

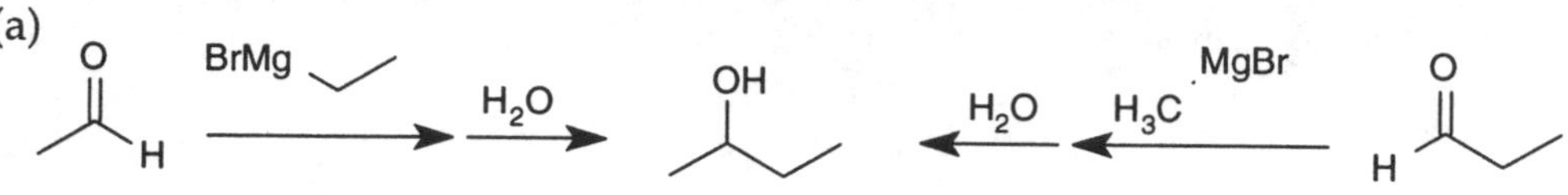

(b)

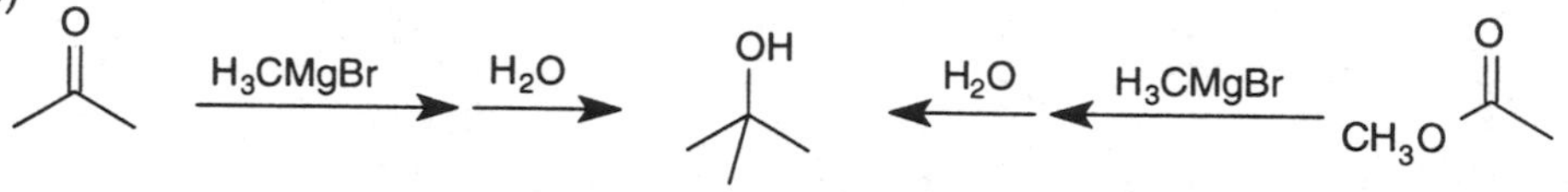

(c)

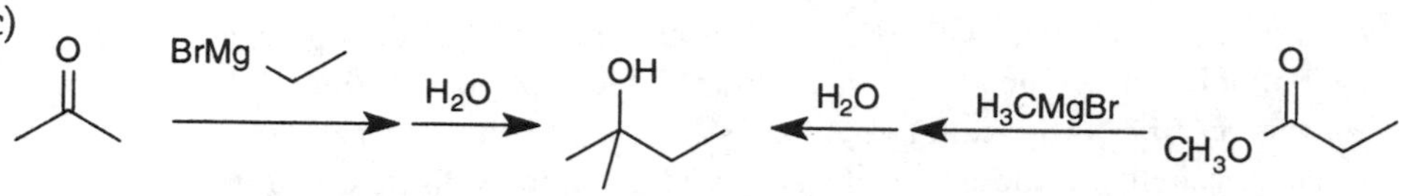

(d)

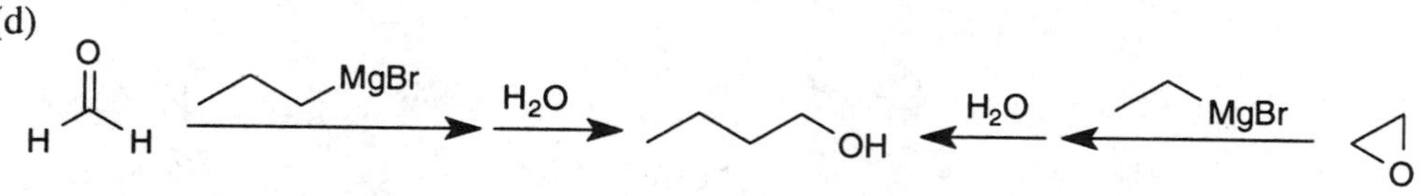

All reactions that use Grignard reagents must be protected form water (and any other proton source). Ethers are typically used as solvents for these reactions.

Ex 12.27 This Exercise is an excellent example of one that appears difficult if not impossible upon "quick" inspection (that is, have I seen the answer somewhere before?), but becomes relatively easy if we stop to analyze what happens, before we pick up the electron-pushing blue pencil. Because the carbon atoms of the Grignard reagent emerge in the products as H_2C=CH_2, we can presume that one hydrogen atom has been lost, while, conversely, the product alcohol has gained two hydrogen atoms relative to the starting ketone. "Elementary, Dear Watson;" a hydrogen atom has been transferred from the Grignard reagent to the ketone in the course of the reaction, effecting a reduction. But how? We have seen that ketones are reduced by reagents that provide the equivalent of a hydride ion (H^-), and now the connection may be becoming clear—the surplus electron density of the C—Mg bond is used to form a new C=C π bond (forming ethylene) and transferring the hydride ion to the carbonyl carbon atom. One of the advantages of a mechanistic approach to organic chemistry is that an analysis such as we have just carried out can readily result in a reasonable answer. Conversely, reactions learned by functional group transformations provide new clues as to how to solve this Exercise.

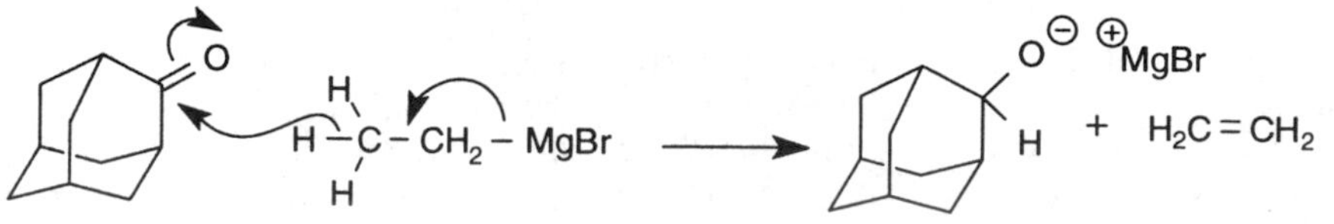

Ex 12.28 There is a simple, graphical way to arrive at the starting materials necessary for the synthesis of an alkene by a Wittig reaction: erase the middle of the double bond, and add an O to one of the ends thus created and a PPh₃ to the other. (Indeed, this process of erasing and replacing was used in a structure-drawing program to create these answers, shown below.) In cases where the desired product alkene is symmetrical, the two possible results of this operation are identical.

(a)

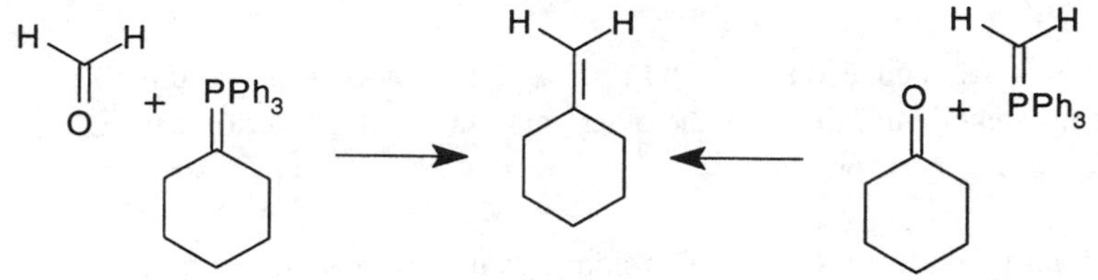

(b)

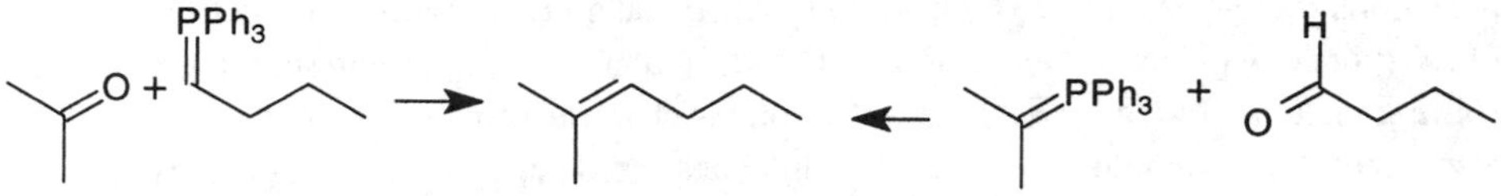

(c)

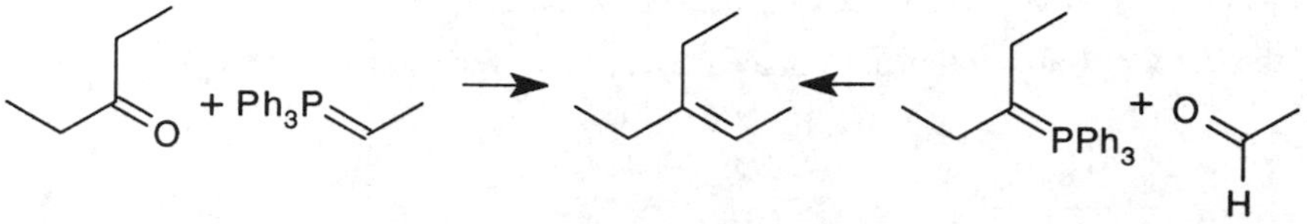

Ex 12.29 An exocyclic double bond is energetically disfavored over an endocyclic double bond, and thus dehydration of 1-methylcyclohexanol yields almost exclusively 1-methylcyclohexene (part *a*). Similarly, the trisubstituted alkene product in part *b* is favored over the disubstituted one. From the alcohol intermediate in part *c*, there is only one alkene that can be directly formed by acid-catalyzed dehydration.

(a)

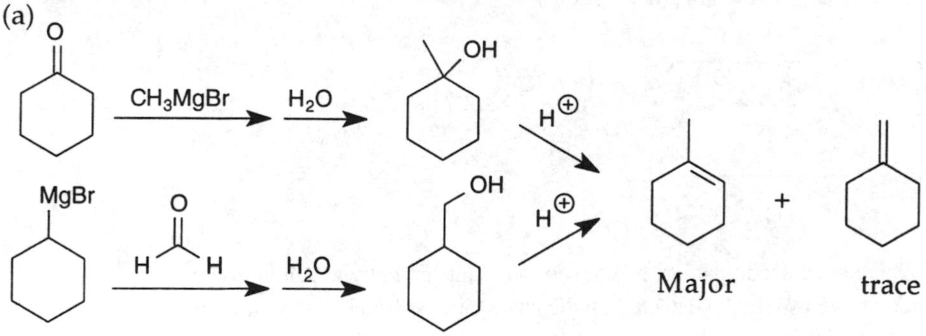

(b)

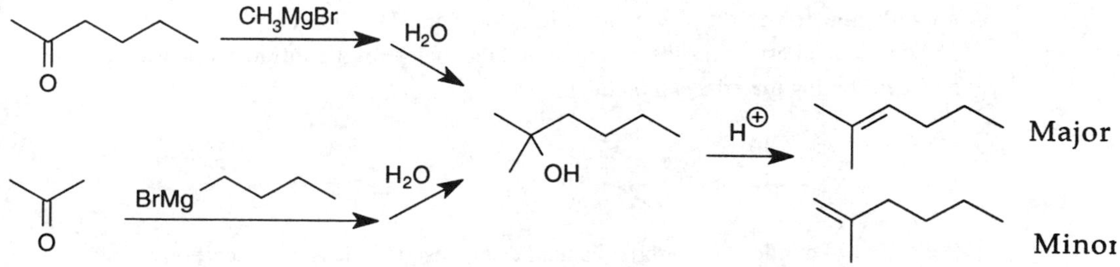

(c)

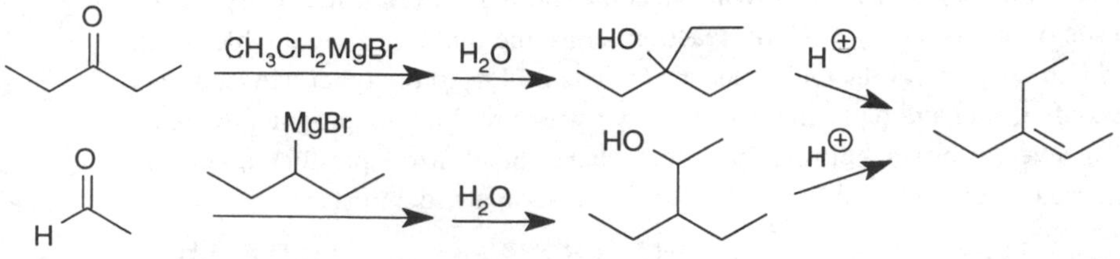

Formation of an alkene from an alcohol by acid catalyzed dehdration often results in a mixture of regioisomeric alkenes. On the other hand, the Wittig reaction can form only a single regioisomeric product.

Ex 12.30 This chapter has focused on the reactions of carbonyl groups. In reactions where the carbonyl group is replaced by a different functional group, the infrared spectrum of the product will lack the characteristic C=O stretching absorption present in the starting material. In many cases, new absorptions characteristic of the product will also appear, for example, the O—H stretching absorption of alcohols produced by reduction of aldehydes, ketones, and carboxylic acid derivatives.

(a) Here, the carbonyl absorption is replaced by the absorption of the O—H bond.

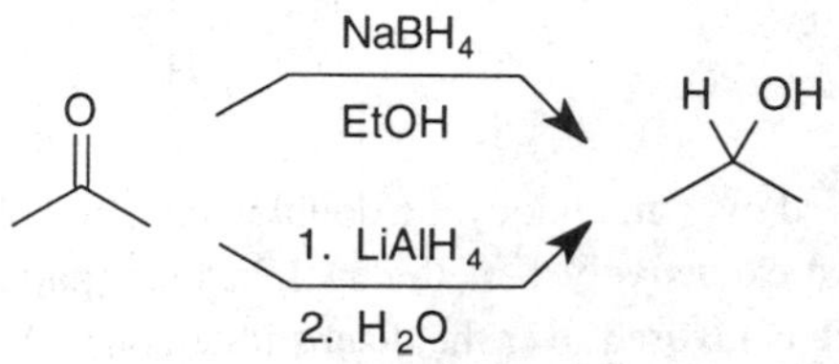

(b) In the formation of a ketal from a ketone, the C=O absorption is lost, replaced by C—O stretching absorptions.

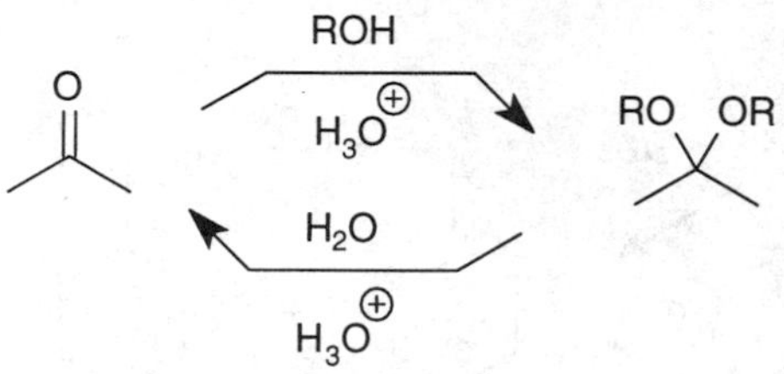

(c) The change that accompanies the conversion of one carboxylic acid derivative to another can usually be detected by the change in the frequency of the C=O absorptions—the greater the degree of overlap between the heteroatom and the π bond, the lower the frequency of absorption. The interconversion of carboxylic acid esters results in no significant change in the infrared spectra between starting material and product.

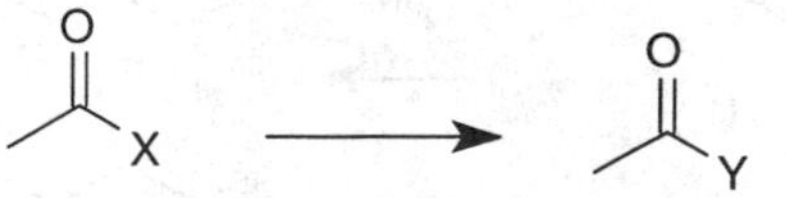

(d) Reduction of amides and carboxylic acid esters results in loss of the carbonyl absorption and appearance of either O—H or N—H absorptions.

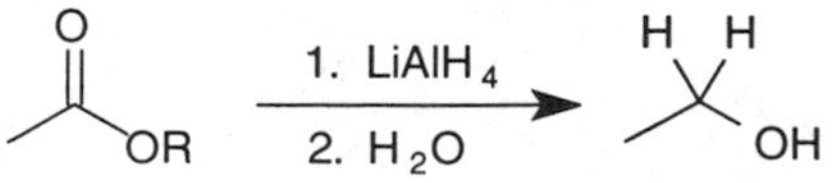

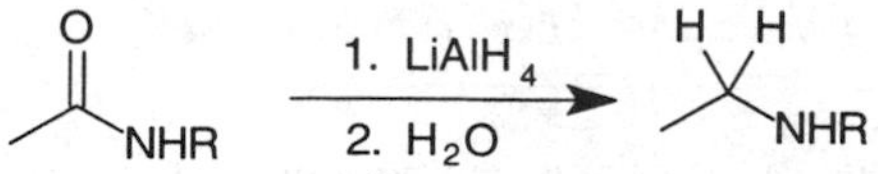

(e) Although the interconversion of various sulfonic acid derivatives is accompanied by changes similar to those for carboxylic acid derivatives, the specific absorptions have not been covered in this textbook.

Answers to Review Problems

Pr 12.1   (a)   Aldehydes are reduced by sodium borohydride to the corresponding alcohol: 1-pentanol.

(b)   Lithium aluminum hydride also reduces aldehydes: 1-pentanol.

(c)   Schiff base formation takes place upon treatment with amine:

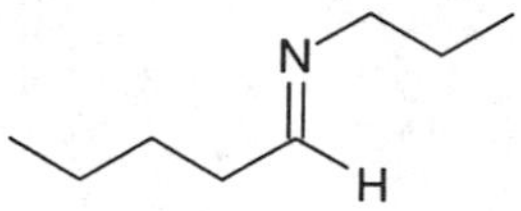

(d)   Phenylhydrazone formation results from the treatment of an aldehyde with phenyl hydrazine:

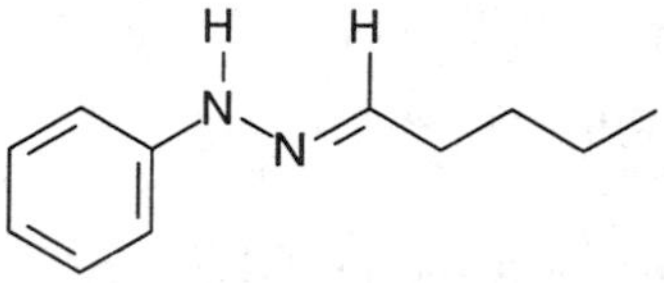

(e)   Amides do not react with aldehydes: no reaction.

(f)   Acetal formation is accomplished with a diol:

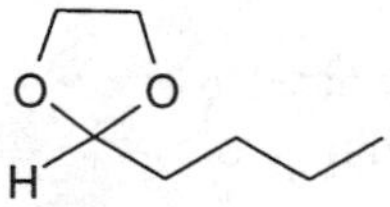

(g)   Sodium chloride is inert to aldehydes: no reaction.

(h)   Thionyl chloride attacks alcohols and acids, but not aldehydes, because they lack an OH group: no reaction.

(i)   Oximes are formed upon treatment of aldehydes with hydroxylamine:

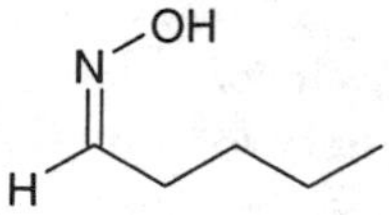

(j)   Because they are easily oxidized, aldehydes can be converted to carboxylic acids even with cold $KMnO_4$: pentanoic acid.

Pr 12.2   (a)   Esters are more difficult to reduce than aldehydes or ketones, and are inert to sodium borohydride: no reaction.

(b)   Lithium aluminum hydride is significantly more reactive as a hydride donor than is sodium borohydride, and reduction of an ester to an alcohol occurs readily: 1-pentanol and methanol.

(c)   Amides are prepared by the treatment of esters with amines: *N*-propyl-pentanoamide.

(d)   Sodium chloride is not sufficiently nucleophilic to attack an ester: no reaction.

Pr 12.3   (a)   Sodium borohydride is not sufficiently reactive to reduce an amide: no reaction.

         (b)   Lithium aluminum hydride reduces an amide to the corresponding amine in which the entire skeleton is retained: 1-pentylamine [$CH_3(CH_2)_4NH_2$].

         (c)   Amides are not easily converted to esters: no reaction.

         (d)   Amides resist reaction with amines: no reaction.

         (e)   Sodium chloride is not sufficiently nucleophilic to attack an amide: no reaction.

Pr 12.4   (a)   Acid chlorides are hydrolyzed to the corresponding acids upon treatment with water: acetic acid (CH_3CO_2H).

         (b)   Esters are formed from acid chlorides and alcohols:

             n-propyl acetate [$CH_3CO_2(CH_2)_2CH_3$].

         (c)   Amides are formed upon treatment with amines:

             N,N-dimethylacetamide [$CH_3CON(CH_3)_2$].

         (d)   The simplest amine, ammonia, produces a primary amide: acetamide (CH_3CONH_2).

         (e)   Friedel-Crafts acylation (Chapter 11) takes place with aromatic compounds in the presence of a Lewis acid: acetophenone (CH_3COPh).

         (f)   A thiol ester is formed by the reaction of an acid chloride with a thiol: $CH_3COSCH_2CH_3$.

         (g)   An anhydride is produced when an acid chloride reacts with a carboxylate anion: $CH_3CO_2COCH_3$.

         (h)   A phenyl ester is formed by the reaction of an acid chloride with phenol: CH_3CO_2Ph.

         (i)   Catalytic hydrogenation produces an aldehyde, which can be further reduced to an alcohol: CH_3CH_2OH. (With a catalyst poisoned by the addition of quinoline, the reduction stops at the aldehyde stage.)

Pr 12.5   (a)   Hydrolysis of an acid chloride is easily accomplished with water.

         (b)   Acid chlorides are formed by treatment of a carboxylic acid with thionyl chloride. Usually pyridine is also present to take up the HCl produced as a by-product.

         (c)   Esters can be converted to amides by treatment with ammonia under equilibrating conditions. High concentrations of ammonia are necessary to shift the equilibrium toward the amide.

         (d)   Transesterification is accomplished by shifting the equilibrium between the two esters by providing an acidic solution of the alcohol desired in the ester: here, this requires acidic ethanol.

         (e)   The preparation of amides from carboxylic acids involves activation (for example, by treatment with thionyl chloride to form the acid chloride), followed by nucleophilic substitution by ammonia to produce the amide.

         (f)   Hydrolysis of a nitrile is accomplished with aqueous acid.

Pr 12.6   (a)   This is a Cannizzaro reaction, involving the reaction of an aldehyde lacking α hydrogen atoms with hydroxide. It begins with hydroxide acting as a nucleophile, adding to the carbonyl carbon and forming a tetrahedral intermediate. Transfer of hydride from this intermediate to a second aldehyde effects simultaneously conversion of the second aldehyde to an alkoxide and oxidation of the first aldehyde to a carboxylic acid. Proton exchange completes the reaction. Neutralization of the reaction mixture produces the carboxylic acid.

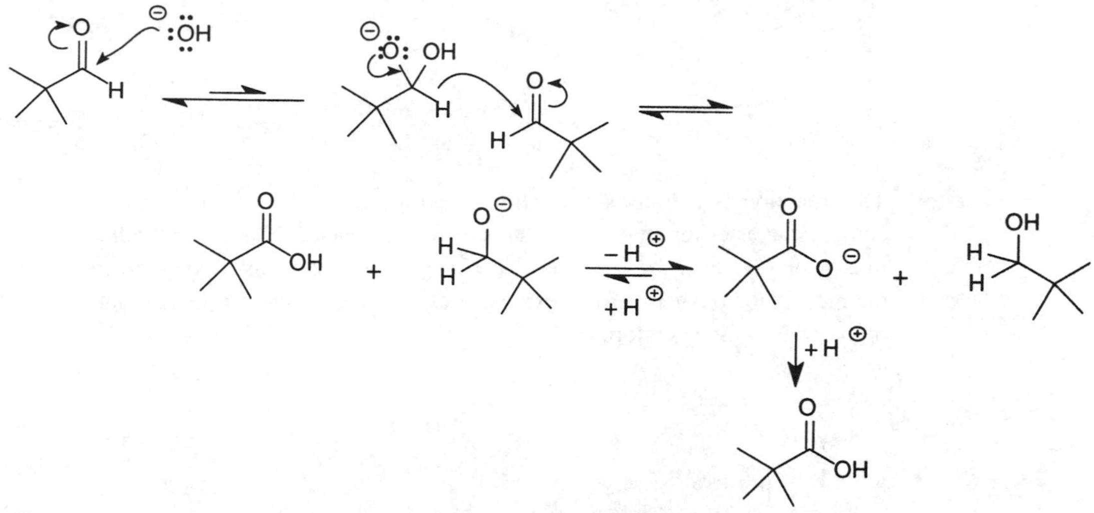

(b) This reaction requires enamine formation. In acid, the carbonyl group is protonated, allowing the nitrogen atom in the cyclic amine to attack the carbonyl carbon. After proton exchange, dehydration generates an iminium · ion. Deprotonation at the α position allows a shift of electron density that relieves the positive charge on nitrogen, forming an enamine.

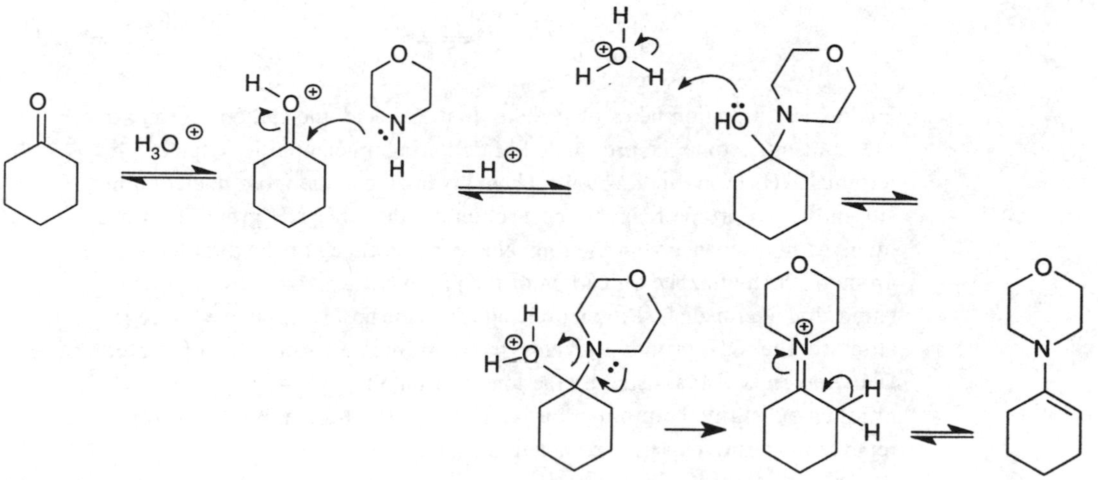

(c) This reaction is the hydrolysis of a ketal. In acid, one of the oxygen atoms of the ketal is protonated, converting it to a better leaving group. Loss of alcohol from this intermediate is facilitated by donation of lone-pair electron density from the other oxygen atom. The intermediate thus formed, an O-alkylated ketone, is electron deficient in the same way as a protonated ketone and reacts readily with water, forming a new C—O σ bond at the same time that the π bond is broken. This process is then repeated. Protonation of the ethereal oxygen, followed by use of the lone pair on the OH group to expel alcohol, produces a protonated carbonyl group. Neutralization gives the observed product.

(d) This reaction is a transesterification. Protonation of the carbonyl group activates an ester for attack by an alcohol, a weak nucleophile. A tetrahedral intermediate is thus formed. Proton exchange from the ethanolic oxygen to the methanolic oxygen permits expulsion of methanol as the carbonyl group of the product ester is reformed.

(e) Hydrazone formation takes place here. In dilute acid, the carbonyl group of the starting ketone is protonated, facilitating nucleophilic attack. The terminal NH_2 group in 2,4-dinitrophenylhydrazine is the more nucleophilic site in the hydrazone reagent because, unlike the other NH group, it is not involved in resonance with the ring. Nucleophilic attack by the terminal NH_2 group of the hydrazone on carbon of the protonated carbonyl group forms a tetrahedral intermediate. Proton exchange from the positively charged nitrogen atom to the OH group activates this intermediate for loss of water. Dehydration is also assisted by the lone pair on nitrogen. As water leaves, a nitrogen-substituted iminium ion is formed. Deprotonation on nitrogen relieves the positive charge, completing the reaction.

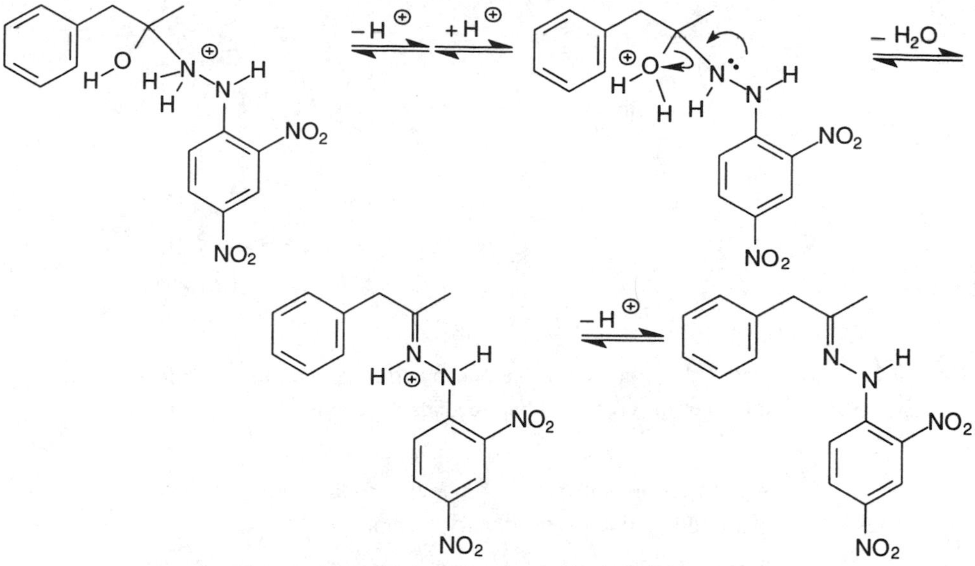

(f)   This is a reductive amination. It begins by formation of an imine by the addition of nitrogen as a nucleophile to the carbonyl carbon, facilitated by prior protonation. Loss of a proton from nitrogen and addition of a proton to oxygen convert the OH group to a good leaving group. Loss of water from this intermediate is facilitated by donation of lone-pair electron density from nitrogen, resulting in simultaneous formation of the π bond of an iminium ion. Nucleophilic delivery of a hydride to the carbon of the iminium ion from cyanoborohydride forms the carbon–hydrogen bond of the product amine after addition of water in the workup.

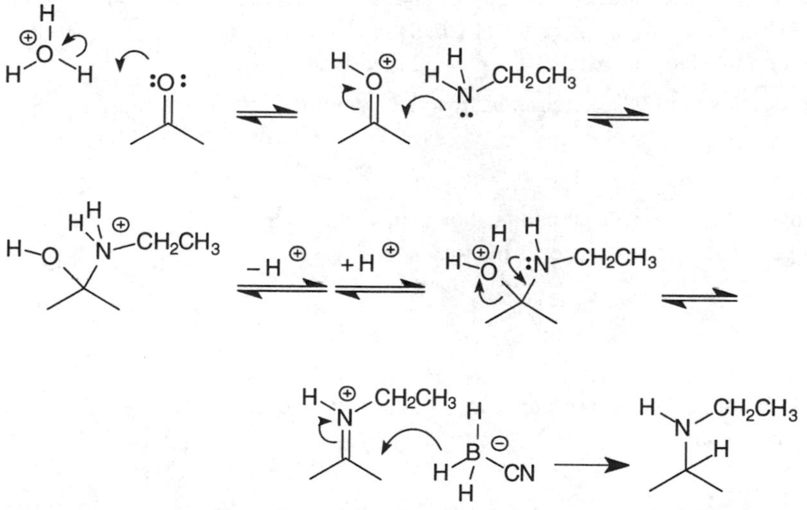

(g)   The formation of a sulfonamide from a sulfonyl chloride is a nucleophilic acyl substitution. The sulfonyl chloride is already highly active toward nucleophilic attack and requires no acid catalysis. The nitrogen atom of methylamine therefore attacks sulfur, forming a new S—N bond as electron density from the S$=$O π bond shifts onto oxygen. The positive charge on nitrogen is relieved by deprotonation and the S$=$O bond is reformed, expelling chloride ion and yielding product.

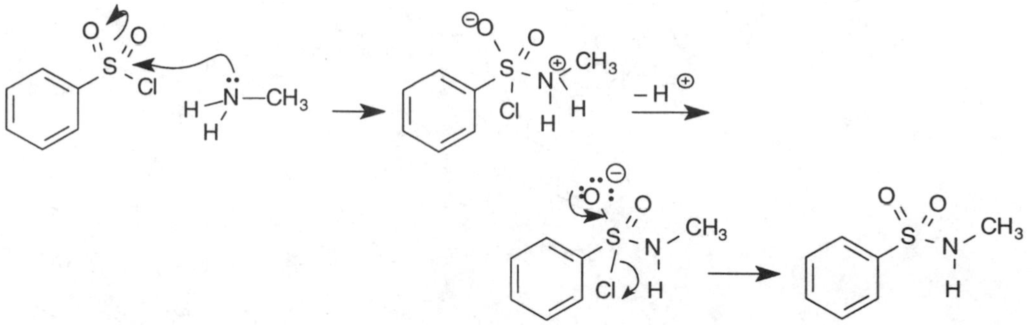

Pr 12.7

(a) The reduction of this ketone by lithium aluminum hydride produces the corresponding secondary alcohol.

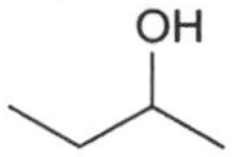

(b) Aldehydes are sufficiently active that sodium borohydride can effect their reduction to primary alcohols.

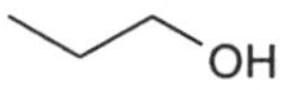

(c) Esters are reduced by lithium aluminum hydride to alcohols. The alcoholic group of the ester also appears in the product mixture as an alcohol.

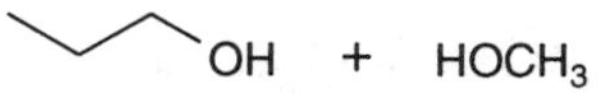

(d) Transesterification under acidic conditions converts a methyl ester to an ethyl ester.

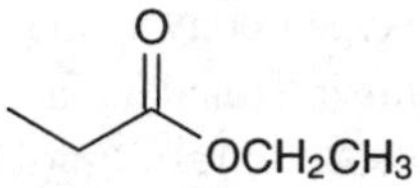

(e) Lithium aluminum hydride reduction of an amide forms the amine in which the amide C—N bond is retained in the product. [This differs from the result of LiAlH₄ reduction of esters (part *c*), in which the parallel C—O bond is cleaved.]

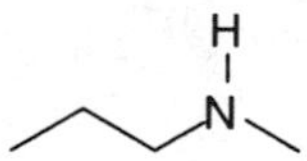

(f) A ketone is converted to a ketal upon treatment with acidic alcohol.

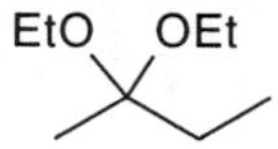

(g) Benzaldehyde undergoes a disproportionation in this Cannizzaro reaction upon treatment with aqueous hydroxide ion.

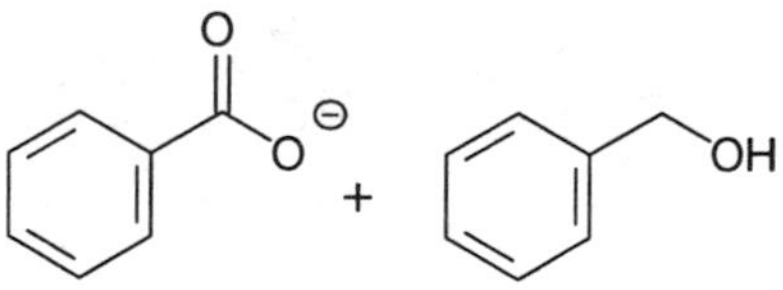

(h) Enamine formation takes place upon interaction of a ketone with a secondary amine.

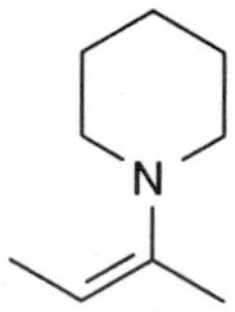

(i) Hydrolysis of a primary amide produces the parent carboxylic acid.

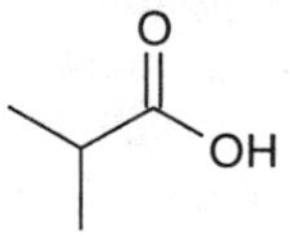

(j) The presence of a ketone does not interfere with the hydrolysis of a nitrile to a carboxylic acid.

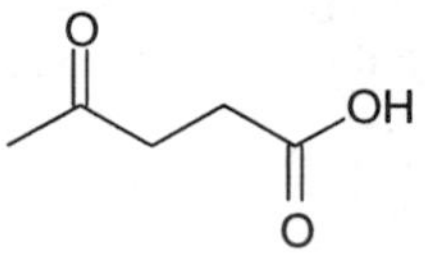

(k) Electrophilic aromatic sulfonation converts benzene to benzenesulfonic acid. This functional group is activated toward nucleophilic acyl substitution by treatment with $POCl_3$. Chloride is then displaced from the intermediate sulfonyl chloride to form the product sulfonamide.

(l) Sodium borohydride reduction of the ketone forms a secondary alcohol, which is converted to the corresponding chloride upon treatment with phosphorus oxychloride.

(m) A phenylhydrazone is formed by treatment of a ketone with phenylhydrazine.

(n) Aldehyde and ketone carbonyl groups are converted to imines (Schiff bases) upon treatment with primary amines.

(o) After conversion of the carboxylic acid to the acid chloride by treatment with thionyl chloride, nucleophilic substitution by an amine forms an amide.

(p) Reaction of an anhydride with an alcohol results in the formation of an ester.

(q) No reaction takes place with a tertiary amine. The ketone is recovered.

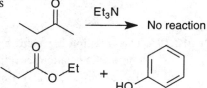

(r) Transesterification converts a phenyl ester to an ethyl ester. Because a lone pair of electrons of an oxygen atom attached to an aromatic ring is delocalized to the π system, phenyl esters are more reactive than alkyl esters.

(s) Friedel-Crafts alkylation involving attack on the electron-rich aromatic ring takes place at the *ortho* and *para* positions.

(t) Grignard addition on an ester takes place twice, yielding a tertiary alcohol in which two of the groups attached to the carbinol carbon derive from the Grignard reagent.

Pr 12.8 Ketal hydrolysis involves a sequence in which protonation of each alkoxy group oxygen atom leads to loss of the corresponding alcohol. Addition of water provides the oxygen atom that ultimately becomes the carbonyl group oxygen.

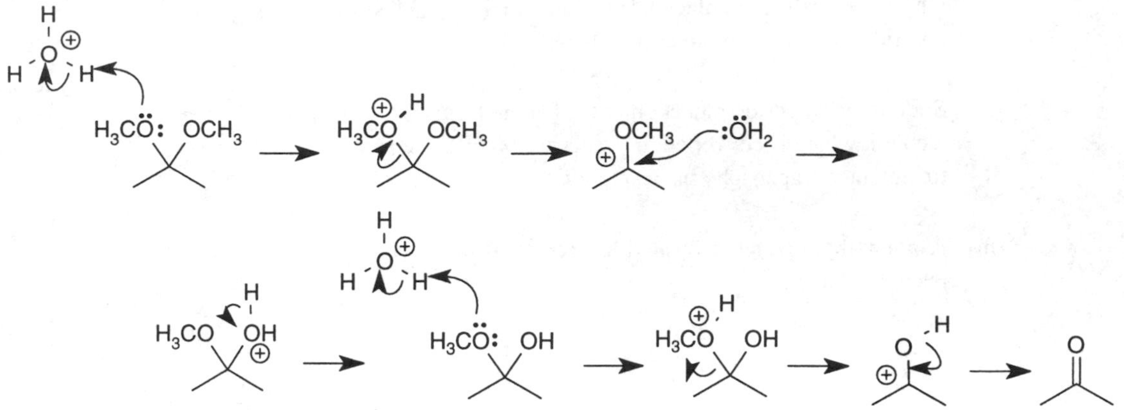

Pr 12.9 Structural features that stabilize the intermediate cation in ketal/acetal hydrolysis lower the activation energy and accelerate the reaction.

(a) The additional alkyl substituent of a ketone stabilizes the intermediate cation, and thus ketals hydrolyze more rapidly than comparable acetals.

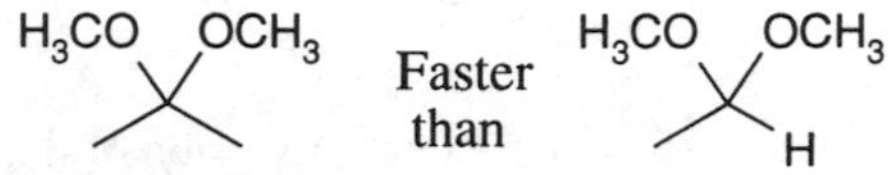

(b) The electron-withdrawing chlorine substituent destabilizes the intermediate cation, and thus hydrolysis of the unsubstituted ketal is faster.

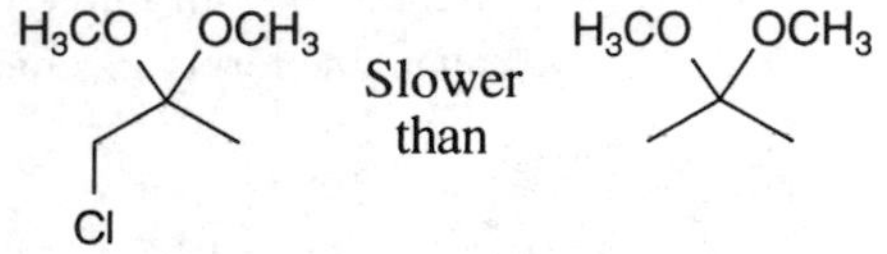

(c) The intermediate cation is stabilized by the attached aromatic ring to a greater extent than by a simple methyl group.

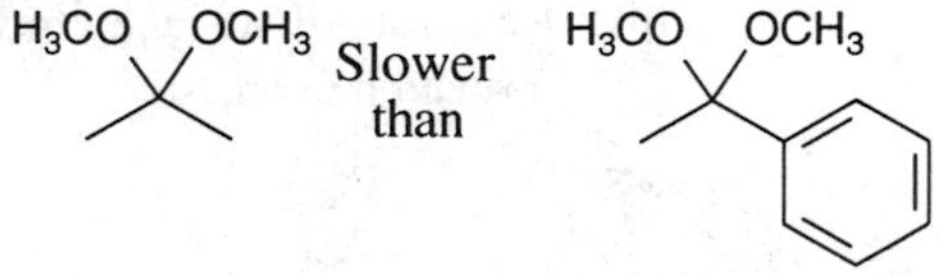

Pr 12.10 The rate-limiting step in acetal and ketal hydrolysis is the formation of the intermediate cation resulting from protonation and loss of an alcohol (or thiol in the case of thioketals). Because of orbital mismatch between sulfur and carbon, there is little resonance stabilization of the cationic intermediate formed from thioketals.

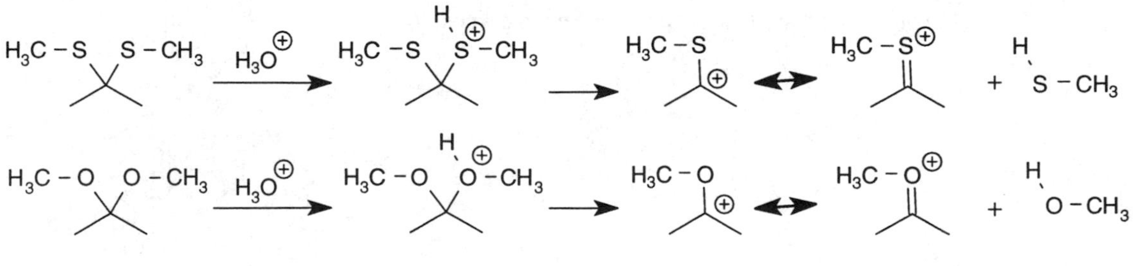

Important New Terms

Acetal: [RCH(OR)$_2$]; a functional group bearing an alkyl group, a hydrogen atom, and two alkoxy groups on one carbon atom produced in the acid-catalyzed alcoholysis of an aldehyde or a hemiacetal (12.3)

Alcoholysis: a reaction in which an alcohol displaces a leaving group or is added across a multiple bond (12.3)

Aluminate: a species containing an O—Al bond (12.2)

Base-induced reaction: a chemical conversion in which a base that is required for the reaction is consumed as product is formed (12.5)

Betaine: a zwitterionic species with a negatively charged atom and a positively charged atom that are separated by two atoms; name derived from a compound called betaine [$^-$O$_2$CCH$_2$N$^+$(CH$_3$)$_3$] isolated from beets; an intermediate in the Wittig reaction (12-7)

Borate: a species containing one or more O—B bonds (12.2)

Cannizzaro reaction: the conversion of an aldehyde lacking α-hydrogen atoms to equal amounts of the corresponding carboxylic acid and alcohol upon treatment with sodium or potassium hydroxide (12.3)

Complex metal hydride: a reagent in which hydride is bound to boron or aluminum and which is soluble in organic solvents, providing the equivalent of the hydride ion in nucleophilic reactions; most common members of this group are NaBH$_4$, LiAlH$_4$, and NaBH$_3$(CN) (12.2)

Complex metal hydride reduction: the use of a complex metal hydride to convert an aldehyde to the corresponding primary alcohol, a ketone to a secondary alcohol, an ester to a primary alcohol, an imine to an amine, or an amide to an amine (12.2)

Disproportionation: a reaction in which a species of intermediate oxidation level is converted to equal amounts of a more oxidized and a more reduced product (12.2)

Geminal diol: a functional group bearing two —OH substituents on the same carbon atom; *see also* **Hydrate** (12.3)

Hemiacetal: RCH(OR)(OH); a functional group bearing an alkyl group, a hydrogen atom, an alkoxy group, and a hydroxy group on one carbon atom; the product of the nucleophilic addition of an alcohol to an aldehyde (12.6)

Hemiketal: RRC(OR)(OH); a functional group bearing two alkyl groups, an alkoxy group, and a hydroxy group on one carbon atom ; the product of the nucleophilic addition of an alcohol to a ketone (12.3)

Hydrate: the product of nucleophilic addition of water to an aldehyde or ketone (12.3)

Hydrazone: R$_2$C=NNH$_2$; a condensation product of hydrazine (H$_2$NNH$_2$) with an aldehyde or ketone; often a highly colored solid used as a diagnostic test for the presence of a carbonyl group (12.4)

Imine: a family of compounds containing a C=N double bond (12.4)

Imine-enamine tautomerization: the process by which a proton is shifted from the α-carbon of an imine to the imine nitrogen, or from the N—H group of an enamine to the adjacent alkenyl carbon; a 1,3 shift of a proton in an imine or enamine (12.4)

Ketal: R$_2$C(OR)$_2$; a functional group bearing two alkyl groups and two alkoxy groups on one carbon atom; produced in the acid-catalyzed alcoholysis of a ketone or a hemiketal (12.3)

Nicotinamide adenine dinucleotide (NADH): a biological reducing agent that provides a hydride equivalent; a cofactor that effects the reduction of α-ketoacids in fatty acid biosynthesis (12.3)

Nucleophilic acyl substitution: *see* **Nucleophilic substitution** (12.5)

Nucleophilic addition: an addition reaction initiated by attack by an electron-rich reagent (a nucleophile) on a carbonyl compound or derivative (12.1)

Nucleophilic substitution (at an sp^2-hybridized center): a substitution reaction initiated by attack by an electron-rich reagent (a nucleophile) on a carboxylic acid derivative (12.5)

Oxime: R$_2$C=NOH; a condensation product of hydroxylamine (NH$_2$OH) with an aldehyde or ketone; often a highly colored solid used as a diagnostic test for the presence of a carbonyl group (12.4)

Phenylhydrazone: R$_2$C=NNHPh; a condensation product of phenylhydrazine (PhNHNH$_2$) with an aldehyde or ketone; often a highly colored solid used as a diagnostic test for the presence of a carbonyl group (12.4)

Phosphonium ylide: an α-deprotonated phosphonium salt; R$_3$P$^+$—($^-$CR$_2$) (12.7)

Phosphoric acid derivatives: a family of compounds containing the PO(OR)$_3$ group (12.6)

Reductive amination: the conversion of a carbonyl group to an amine through reduction of an intermediate imine (12.4)

Schiff base: an *N*-alkylated imine (R$_2$C=NR) (12.4)

Semicarbazone: $R_2C{=\!=}NNHC(O)NH_2$; condensation product of an aldehyde or ketone with semicarbazide [$H_2NNHC(O)NH_2$]; often a highly colored solid used as a diagnostic test for the presence of a carbonyl group (12.4)

Sulfonamides: RSO_2NR_2; a family of compounds containing the —SO_2NR_2 group (12.6)

Sulfonic acids: RSO_3H; a family of compounds containing the —SO_3H group (12.6)

Tetrahedral intermediate: an intermediate in nucleophilic addition and nucleophilic acyl substitution obtained upon covalent bond formation between an attacking nucleophile and a carbonyl carbon (12.5)

Tetravalent intermediate: *see* **Tetrahedral intermediate** (12.5)

Transesterification: the interconversion of one carboxylic acid ester to another (12.5)

Wittig reaction: reaction by which an aldehyde or ketone is converted into an alkene by condensation with a phosphonium ylide (12.7)

Substitution Alpha to Carbonyl Groups

13

Enolate Anions and Enols as Nucleophiles

Key Concepts

Formation of enolate anions and enols

Haloform reaction

Hell-Volhard-Zelinski reaction

Kinetic and thermodynamic formation of enolate anions

Alkylation of ketones and esters

Aldol and aldol condensation reactions

Conjugate addition to α,β-unsaturated carbonyl groups

Robinson ring annulation

Claisen and Dieckmann condensation reactions

Reformatsky reaction

Acetoacetic and malonic ester syntheses

Answers to Exercises

Ex 13.1 The conversion of a ketone to an enol is a rearrangement reaction that involves removal of a proton from carbon and addition of a proton to oxygen. The sequence of the two events is determined by the reaction conditions: in base, the proton is first removed from carbon, whereas in acid, the proton is first added to oxygen.

(a)

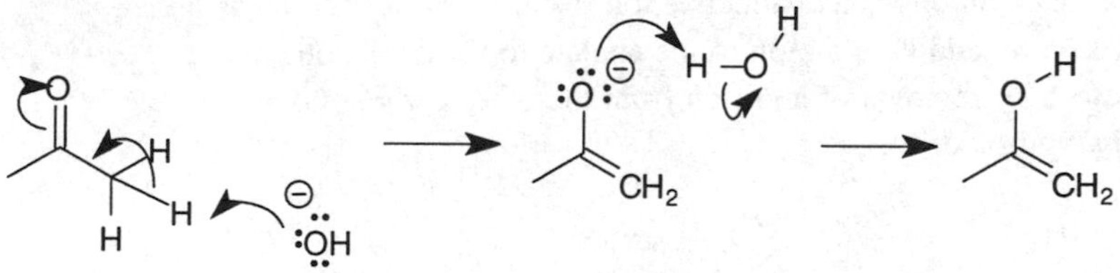

(b)

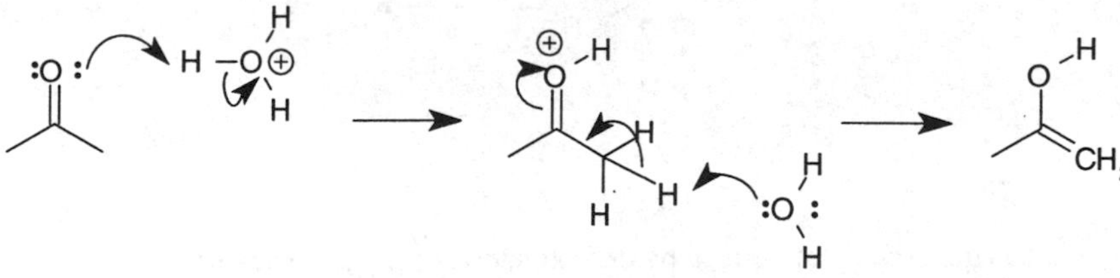

Ex 13.2 The formation of a bright yellow precipitate of iodoform constitutes a positive iodoform test, obtained for ketones bearing the CH_3CO functionality. This group is present in 2-pentanone and acetophenone, but not in 3-pentanone or pentanal. Although acetic acid has the key functional group, the acidity of the carboxylic acid group interferes with formation of the α enolate, which is needed for the iodination.

(a) negative (b) positive (c) negative (d) positive (e) negative

Ex 13.3 (a) Tautomerization of the α-proton in the acid bromide produces the enol shown. In this enol, appreciable electron density develops at the α position, making possible a nucleophilic attack on molecular bromine. (Bromine is present because PBr_3 is in equilibrium with phosphorus and Br_2.) After deprotonation,

253

an α-bromoacid bromide is formed. This product is converted to the acid by hydrolysis upon addition to water during work-up.

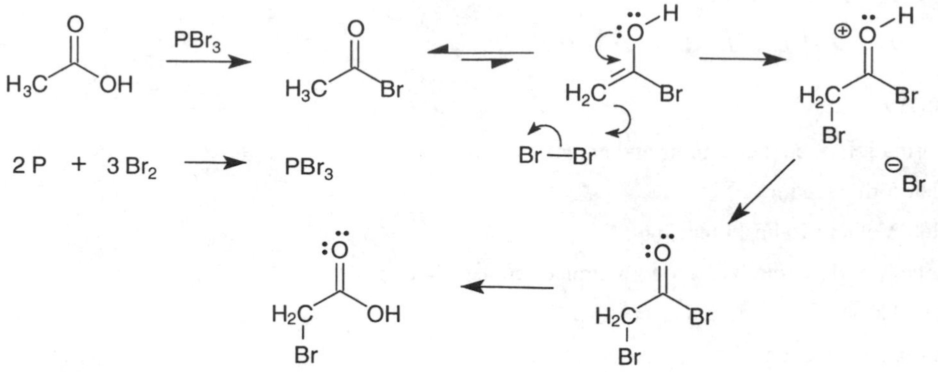

(b) The enolization of the α-bromoacid bromide is more difficult than is that of the starting acid bromide. This is the result of the presence of two bromine atoms as substituents on the double bond; both bromine atoms donate electron density by resonance to the already electron-rich π system. The equilibrium is therefore shifted to the left, as shown, and the predominant product is the mono-brominated acid.

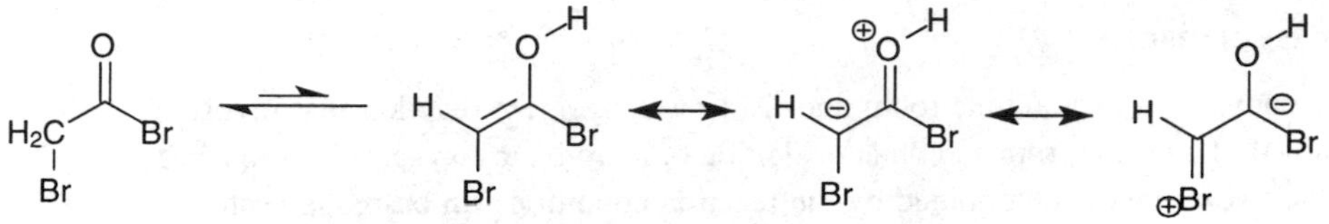

Ex 13.4 Interconversion of enolate anions is an isomerization reaction. Water serves as a catalyst for this reaction by adding a proton to the enolate to form a ketone and a hydroxide ion, followed by removal of a proton from the other carbon atom α to the carbonyl group by hydroxide ion.

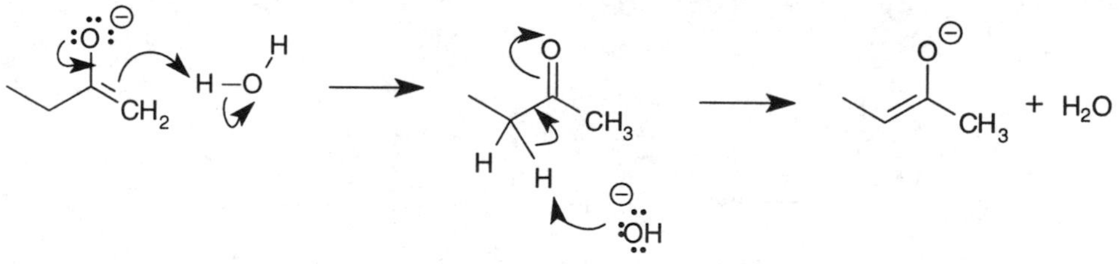

Ex 13.5 The kinetic enolate is generally that derived by deprotonation of the less substituted carbon atom α to the carbonyl group. In parts *a* and *d* of this Exercise, only a single enolate can be formed, because the starting carbonyl compounds are esters.

(a)

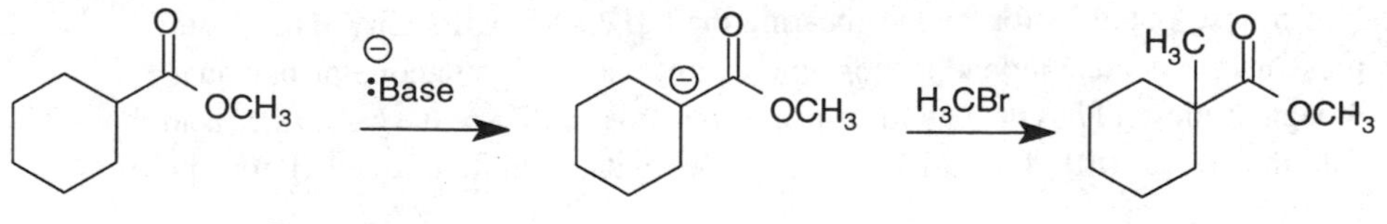

(b)

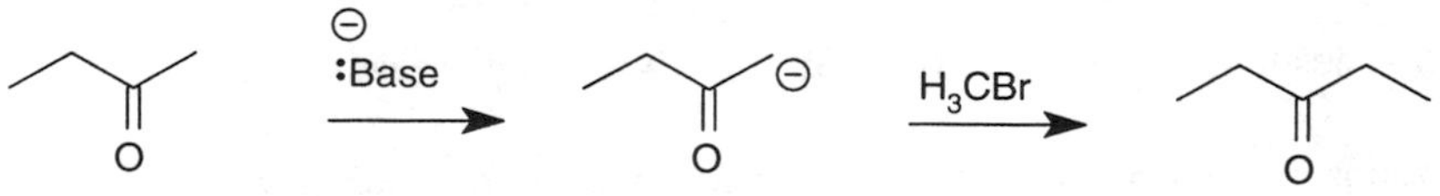

(c)

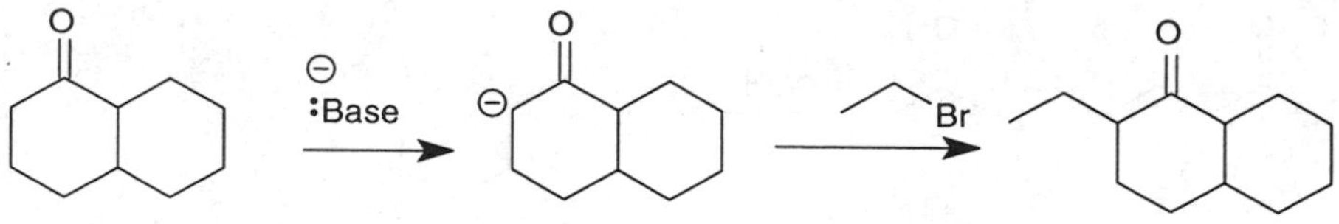

(d)

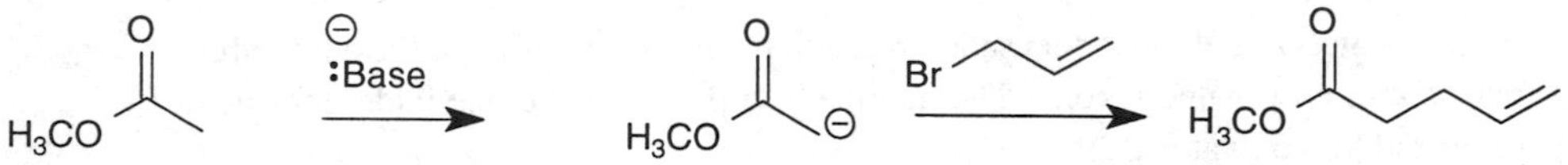

Ex 13.6 The position of the double bond in the aldol condensation reaction is determined by both kinetics and thermodynamics. Elimination of water from the aldol (β-hydroxyaldehyde) intermediate requires removal of a proton from a carbon atom adjacent to that bearing the hydroxyl group. The carbon α to the carbonyl group. is more acidic than the more remove carbon because the anion derived by deprotonation of the fomrer is stabilized by resonance delocalization. Further, the α,β-unsaturated aldehyde is stabilized by conjugation that is not present in the β,γ isomer.

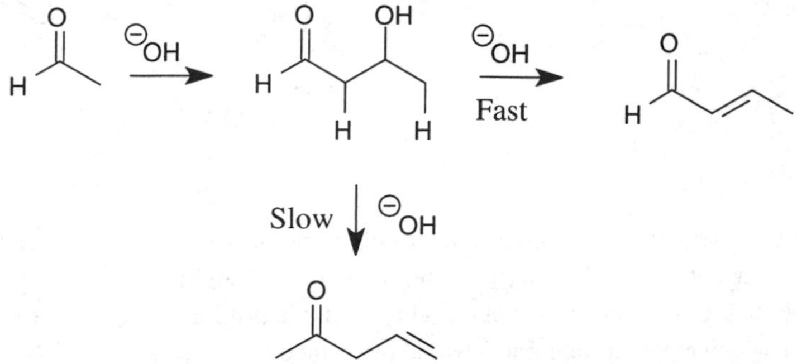

Ex 13.7 The intramolecular aldol condensation reaction generally forms only five- and six-membered-ring products. This criterion determines the product formed in parts *a, b,* and *d* of this Exercise. In part *c*, of the two possible products with six-membered rings, the one with the double bond at the bridgehead position is substantially destabilized by strain and is not formed.

(a)

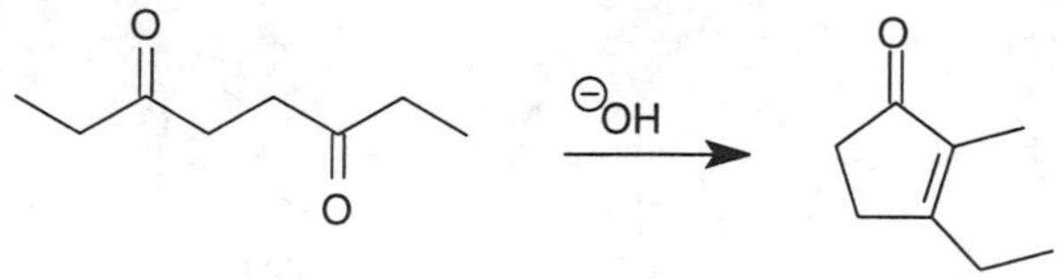

(b)

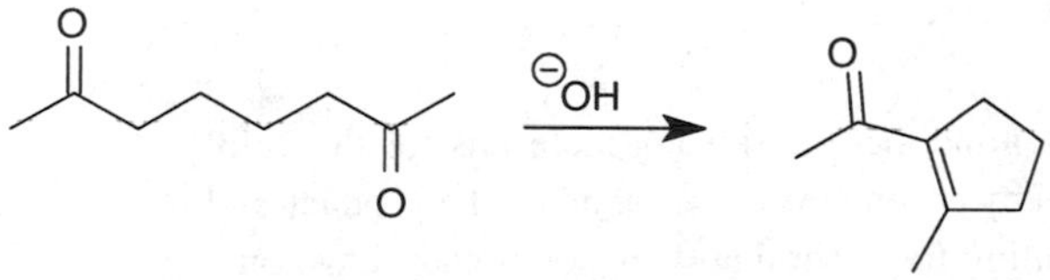

(c)

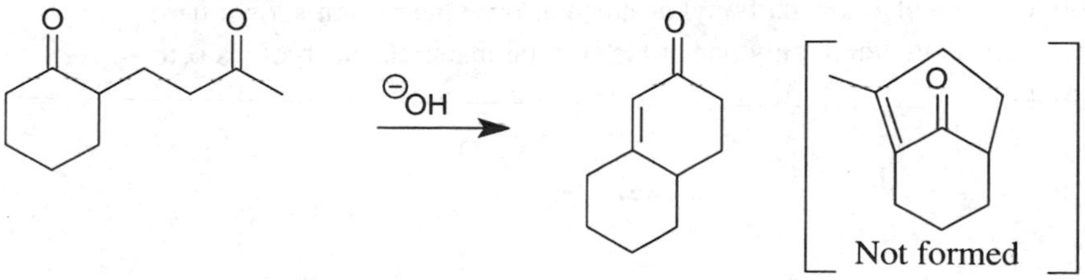

(d)

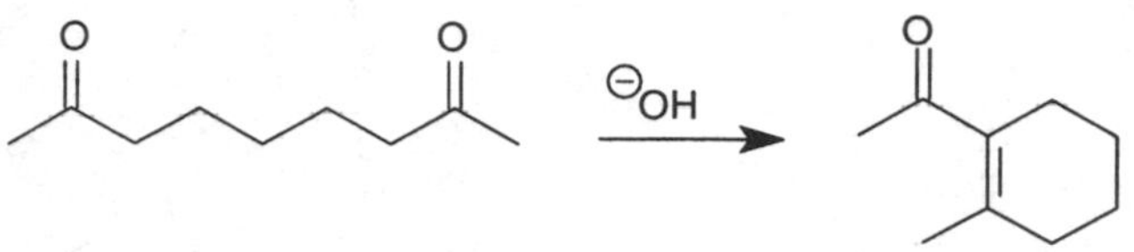

Ex 13.8 An acid-catalyzed aldol condensation reaction involves formation of the enol ,which reacts with protonated ketone. The intermediate β-hydroxyketone undergoes acid-catalyzed loss of water.

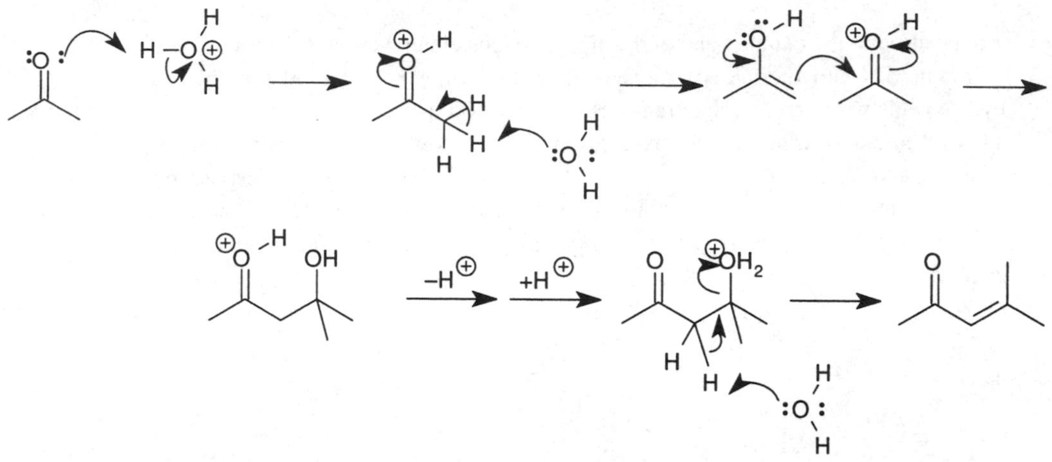

Ex 13.9 Both starting ketones are unsymmetrical. Therefore, four possible enolates are possible, each of which can react with either ketone, resulting in a total of eight products. Further, each product has at least one center of chirality, and one product has four centers of chirality for which eight diastereomers would be formed.

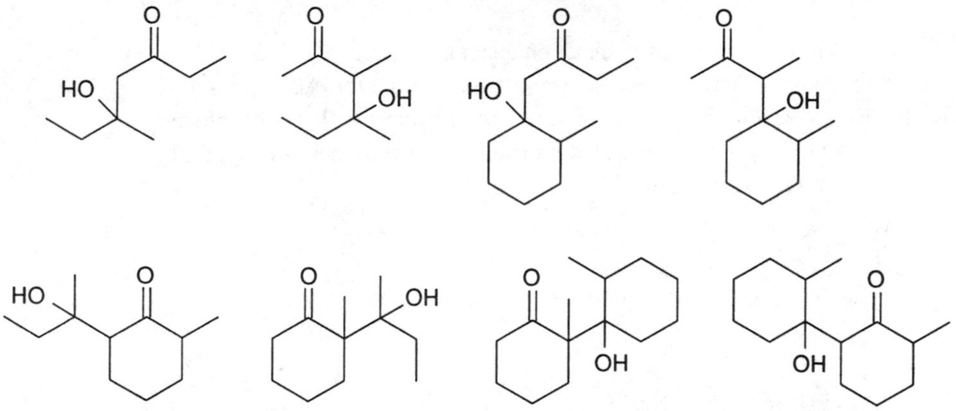

Ex 13.10 This exercise is easy if you keep in mind that the starting materials for the aldol condensation reaction can be derived by erasing the C=C bond of the product and adding a carbonyl group to the resulting fragment that does not already have one. For part *b*, both fragments have carbonyl groups but only the option shown here provides for a reaction in which only one of the starting materials has protons α to a carbonyl group.

(a)

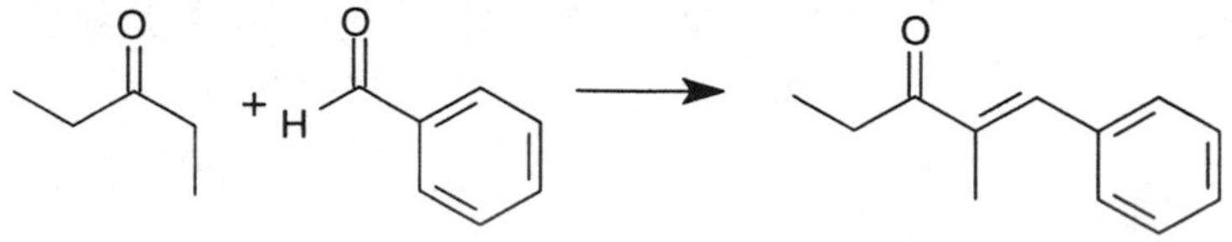

(b)

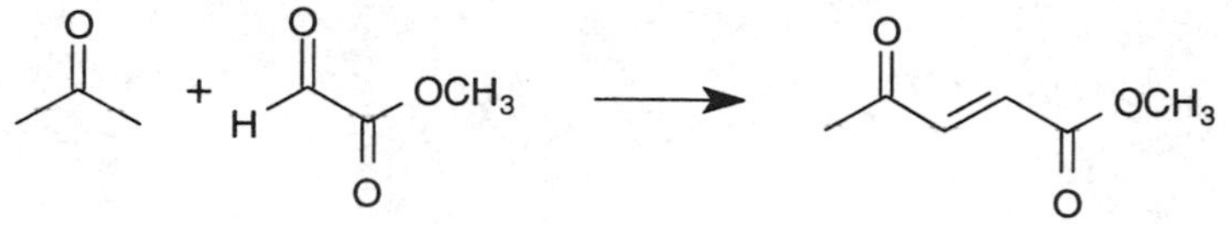

(c)

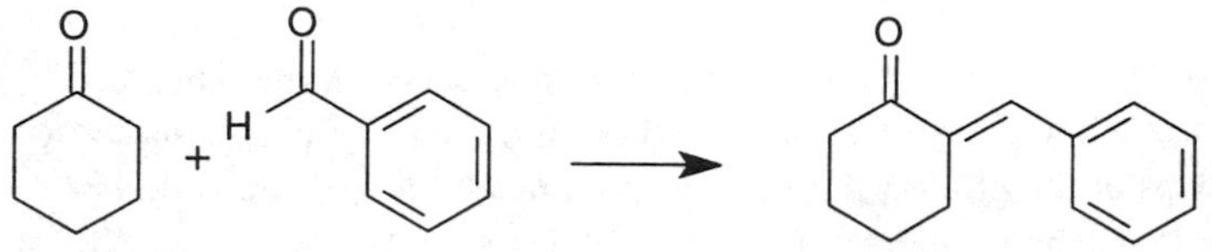

(d)

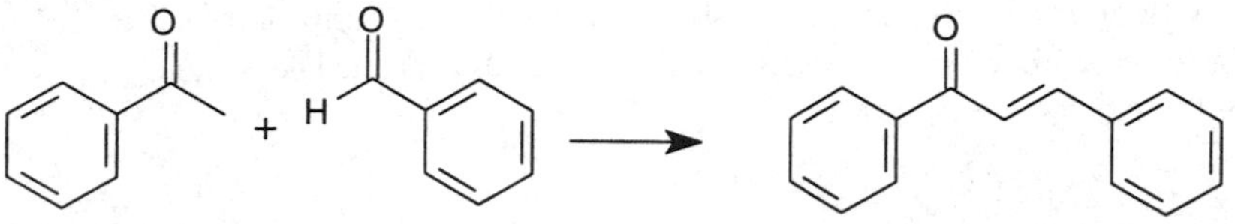

Ex 13.11 Both Grignard and alkyllithium reagents are charge-intensive nucleophiles that add in a 1,2 fashion, whereas cuprate reagents add 1,4.

(a)

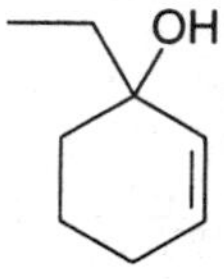

(b)

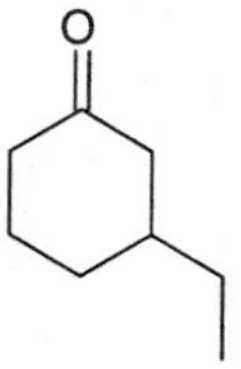

(c)

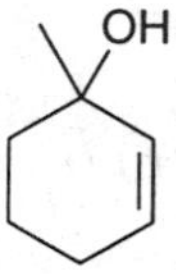

Ex 13.12 The three different enolate anions and the derived aldol cyclization products are shown below. **A** is the only product observed because **B** has a 4-membered ring and **C** is destabilized by steric interactions.

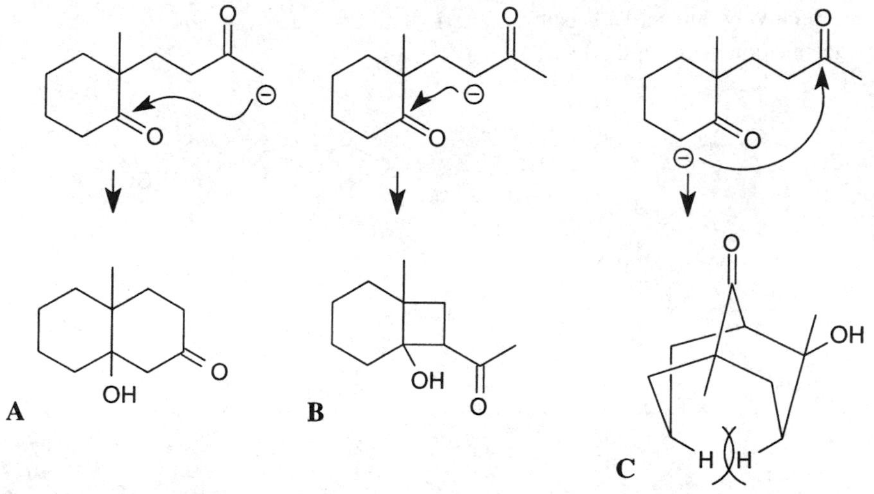

A

B

C

Ex 13.13 Each of these reactions is a Claisen condensation, forming the β-ketoesters shown. Because the reactions are conducted in ethanol as solvent, the methyl ester in part *c* and the phenyl ester in part *d* are converted to ethyl esters.

(a)

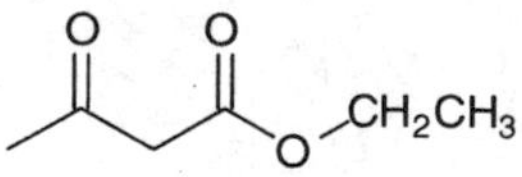

(b)

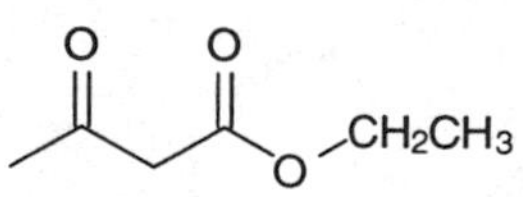

(c)

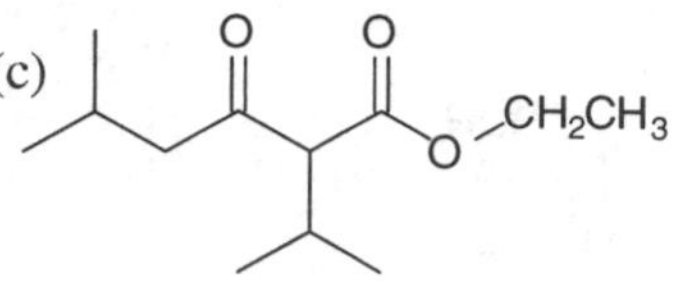

(d)

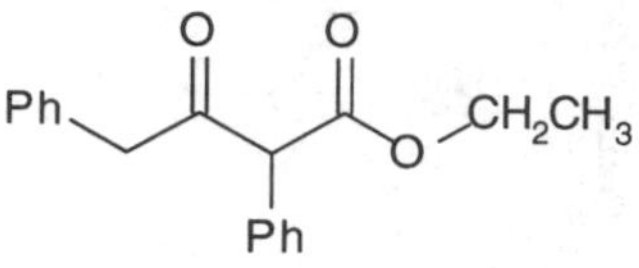

Ex 13.14 Deprotonation at C-2 produces an enolate anion that, after reaction with the other carbonyl group, results in a β-ketoester that does not have a proton on the carbon α to both carbonyl groups of the product β-ketoester. As a result, the ring closure is unfavorable. On the other hand, deprotonation at C-6 leads to an alternative β-ketoester with a relatively acidic C—H between the carbonyl groups. Deprotonation at this carbon provides the driving force that directs this Dieckmann cyclization only in the latter direction.

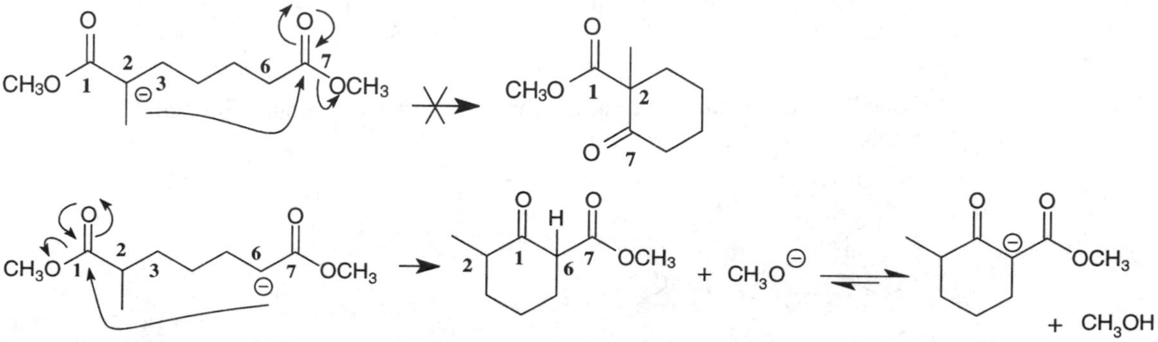

Ex 13.15 In each case, it is the enolate anion of methyl acetate that acts as a nucleophile because the carbonyl partners lack α-hydrogen atoms and thus cannot form enolate anions. The reactions with the carbonate, oxalate, and aryl esters are further facilitated by the enhanced reactivity of these carbonyl groups. On the other hand, pivalate esters are relatively unreactive because of steric hindrance; self-condensation of methyl acetate dominates in this case.

(a)

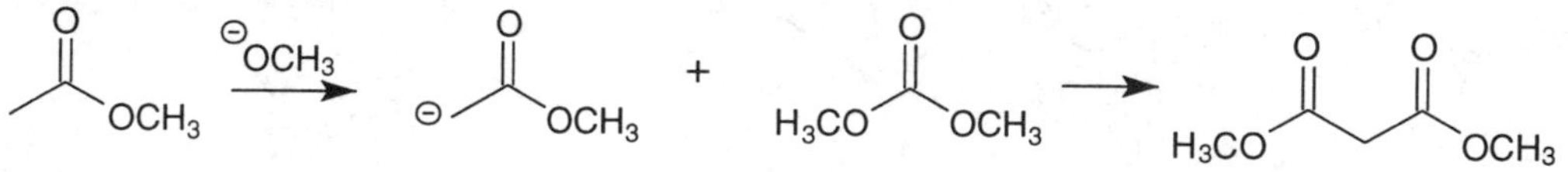

(b)

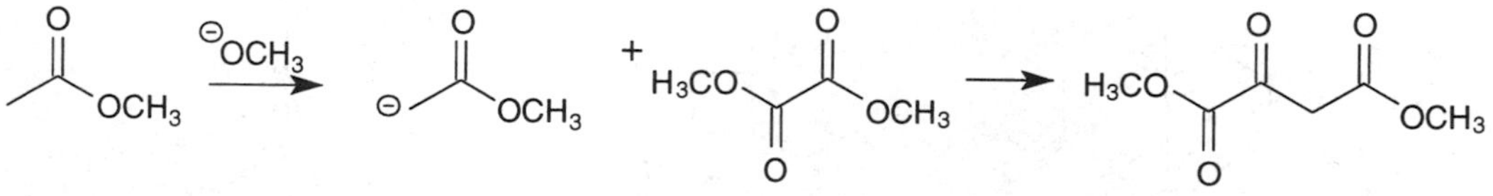

(c)

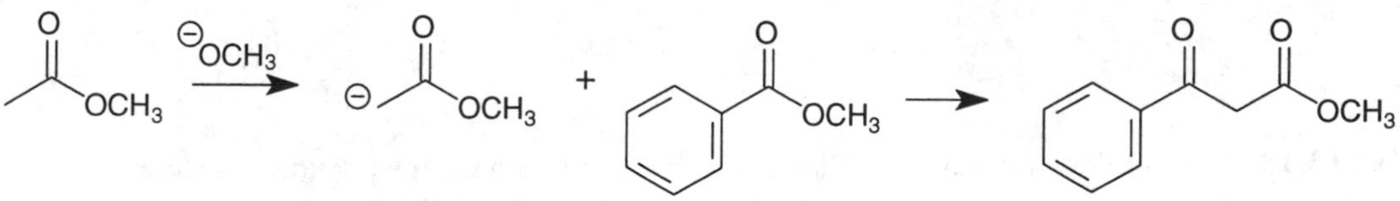

(d)

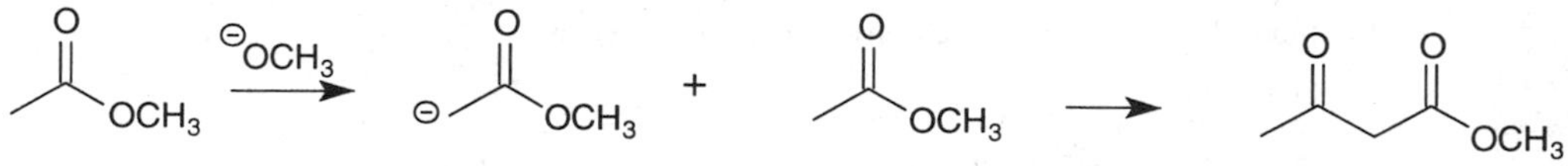

Ex 13.16 In part *a*, Claisen condensation of methyl propanoate yields the desired product. Parts *b* and *c* involve crossed Claisen condensations with methyl formate and dimethyl carbonate, respectively.

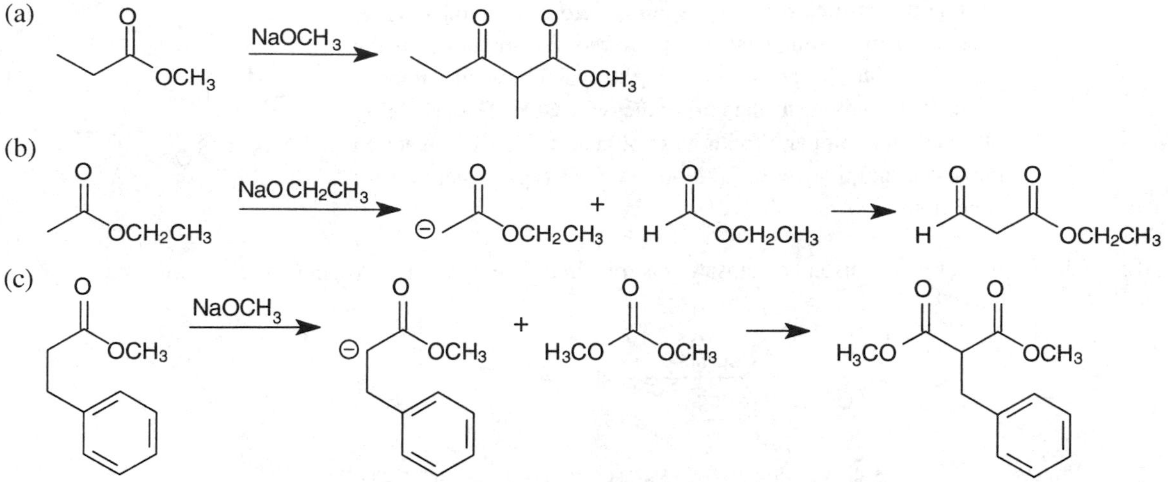

(a)

(b)

(c)

Ex 13.17 The Reformatsky reaction involves the formation of a zinc enolate from an α-bromoester, followed by reaction of this enolate with an aldehyde or ketone.

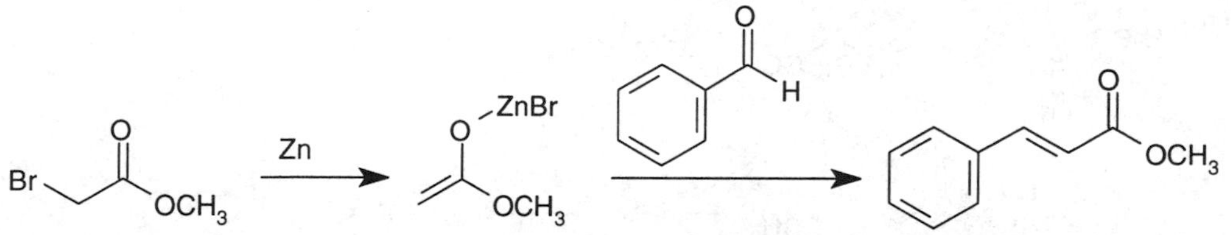

Ex 13.18 Hydrolysis of a carboxylic acid ester involves addition of hydroxide ion as a nucleophile to form a tetrahedral intermediate followed by loss of methoxide ion to form the carboxylic acid. Under alkaline conditions, the acid is rapidly converted to the carboxylate anion.

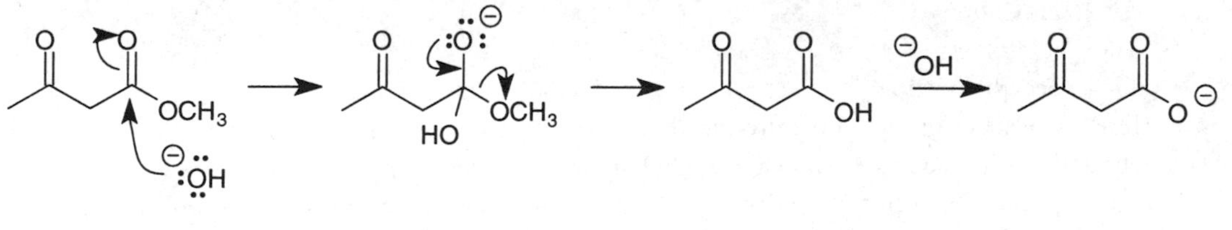

Ex 13.19 Under acidic conditions, tautomerization of an enol involves protonation on carbonand deprotonation of oxygen.

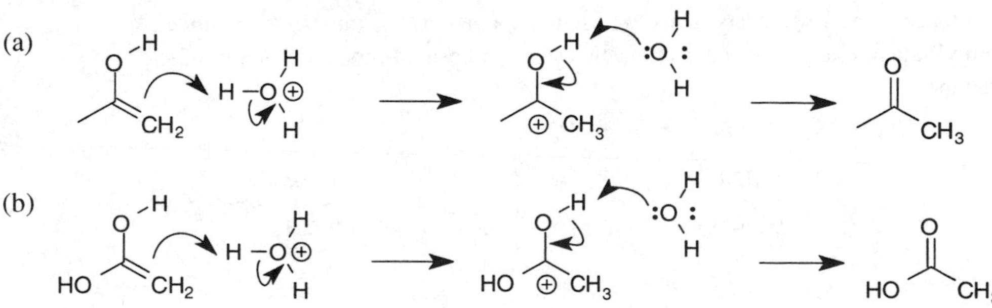

(a)

(b)

Ex 13.20 β-ketoacids readily undergo decarboxylation because a six-electron shift in a six-atom transition state permits the formation of two stable species, CO_2 and an enol. The analogous six-electron shift in an α-ketoacid would have to take place in a somewhat less preferred, five-membered-ring transition state. More importantly, this reaction would produce an unstable carbene as the co-product with CO_2. The instability of a carbene compared with an enol makes this latter reaction thermo-dynamically unfavorable under typical laboratory conditions.

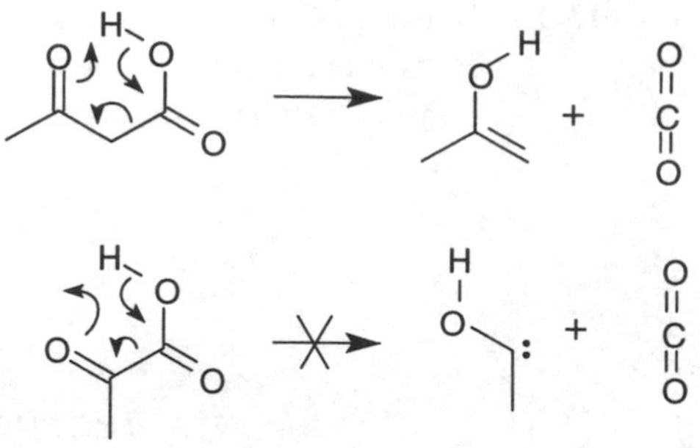

Ex 13.21 In part *a*, a Claisen condensation is required; in part *b*, a Dieckmann condensation is used.

(a)

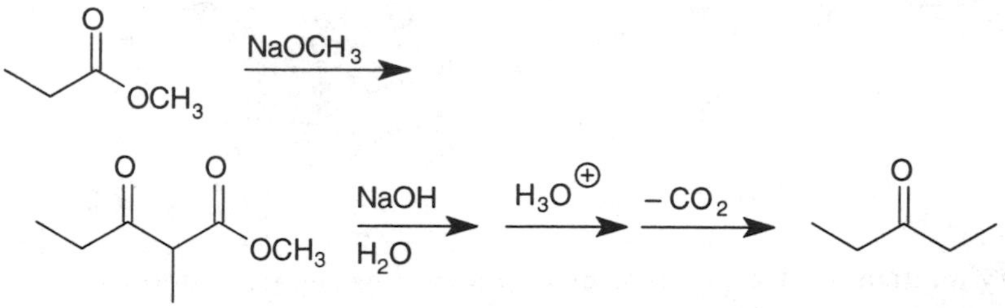

(b)

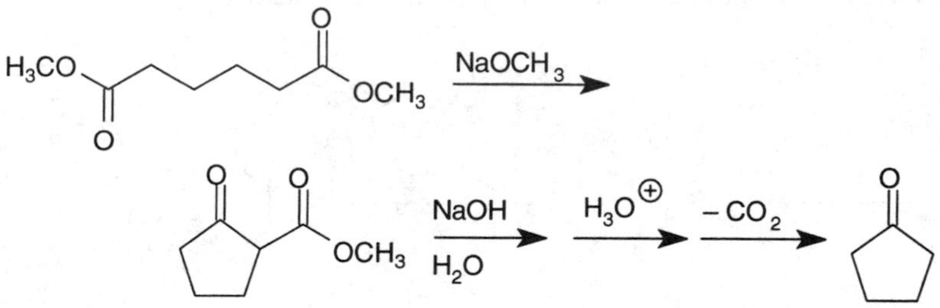

Ex 13.22 The significant steps of the malonic ester and acetoacetic ester reactions are the same:

1. deprotonation to form the enolate anion
2. alkylation
3. hydrolysis
4. decarboxylation

Here, two alkyl groups are introduced at the α position, so steps 1 and 2 are repeated. In the sequences shown, the ethyl group is attached first, followed by the methyl group. The opposite order of alkylation can be used, but the sequence in which the smaller group is introduced second is preferred. After alkylation, both intermediates are hydrolyzed and then undergo decarboxylation upon heating. A carboxylic acid results from the malonic ester synthesis and a methyl ketone from the acetoacetic ester synthesis. Thus, we use the malonic ester synthesis to produce 2-methylbutanoic acid and the acetoacetic ester synthesis to produce 3-methyl-2-pentanone.

Malonic Ester Synthesis

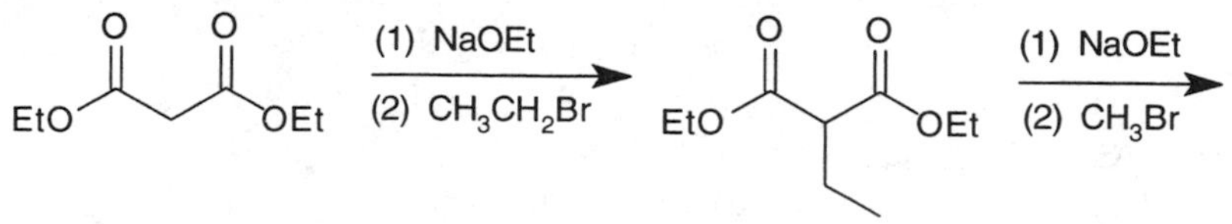

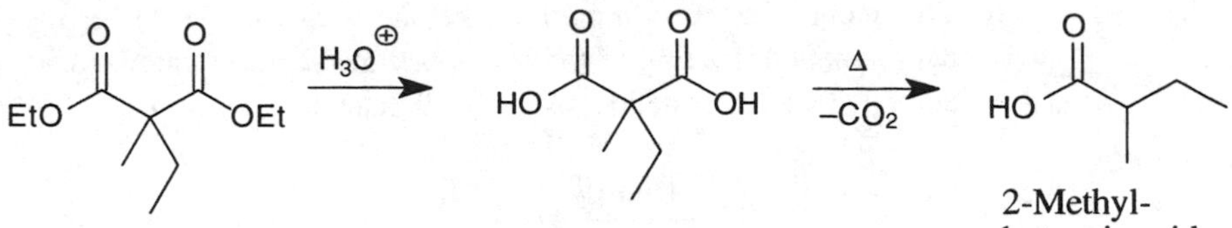

2-Methyl-
butanoic acid

Acetoacetic Ester Synthesis

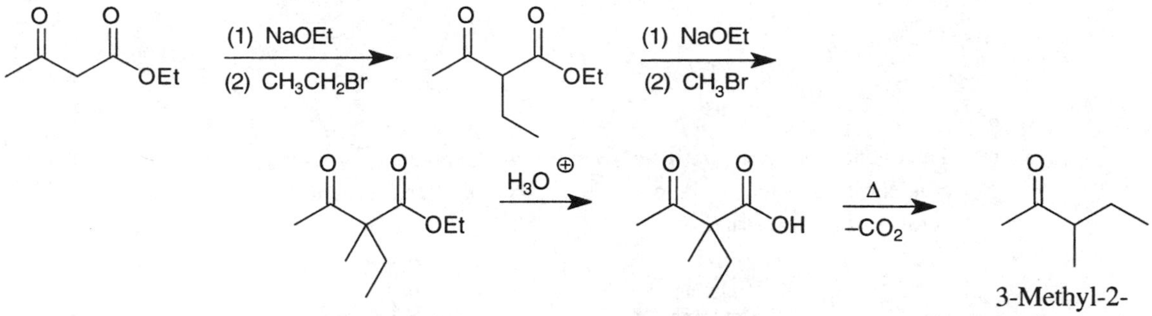

3-Methyl-2-
pentanone

Ex 13.23 Rings can be formed by both the malonic ester and acetoacetic ester syntheses by
the use of alkyl dihalides. In this Exercise, the product is a methyl ketone, so the
acetoacetic ester synthesis is the appropriate choice.

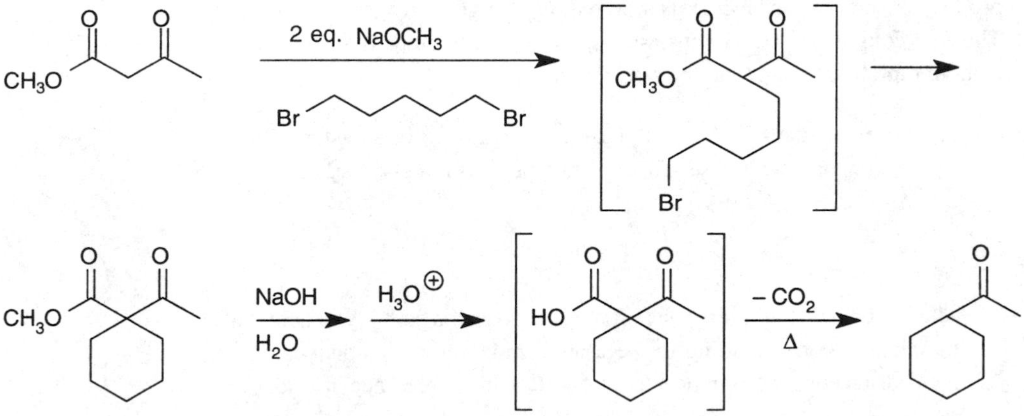

Answers to Review Problems

Pr 13.1 (a) Treatment of a methyl ketone with NaOH and I_2 result in the iodoform
reaction, forming (after acidification) the carboxylic acid and HCI_3.

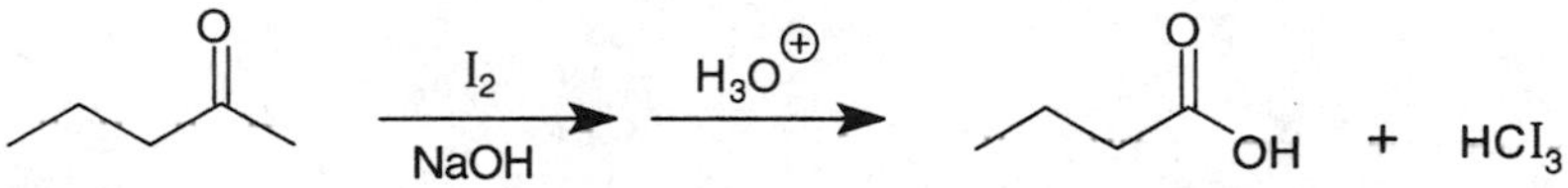

(b) Treatment of a ketone with NaOH at room temperature results in the aldol
reaction. Because 2-pentanone is unsymmetrical, two products result.

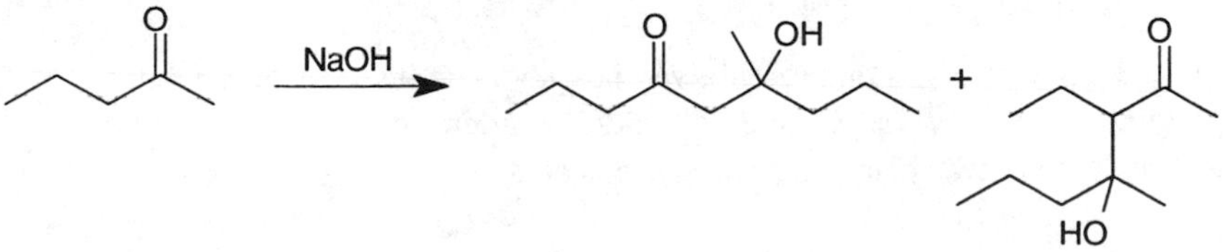

(c) Treatment of an unsymmetrical ketone with LiN(i-Pr)$_2$ results in kinetic deprotonation, leading to the less substituted enolate anion. Reaction of this anion with an alkyl halide results in alkylation.

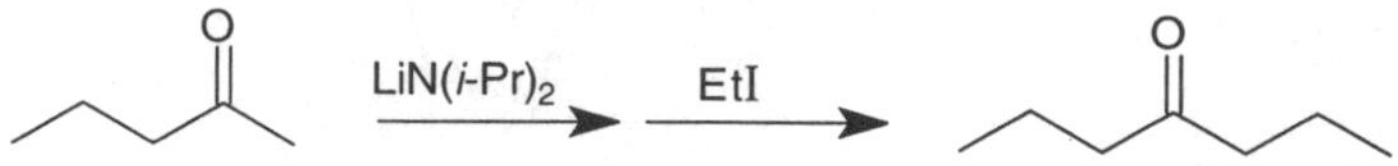

(d) Bromination of a ketone under acid conditions can be controlled to mono halogenation.

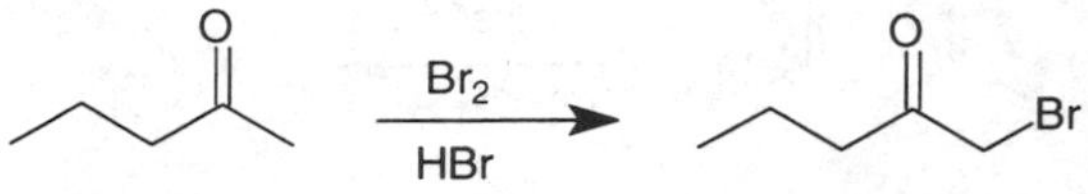

(e) Treatment of a ketone with NaOH at elevated temperatures results in an aldol condensation reaction, here producing two isomeric products because the ketone is unsymmetrical.

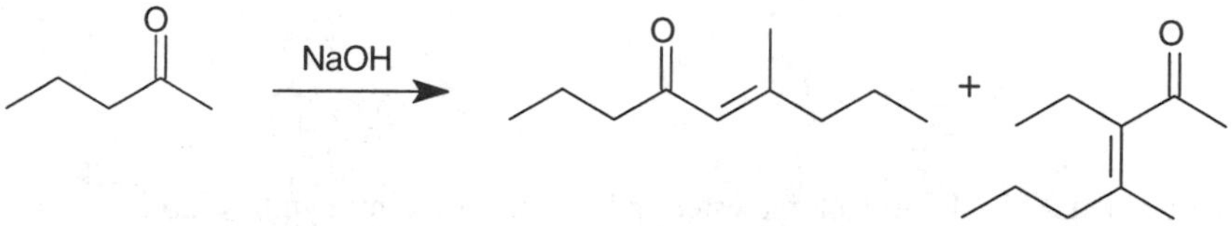

Pr 13.2 (a) Two equivalents of methyl Grignard are incorporated into the product skeleton when an ester is attacked by a Grignard reagent. The first equivalent effects substitution, producing a ketone that is more rapidly attacked than the starting material.

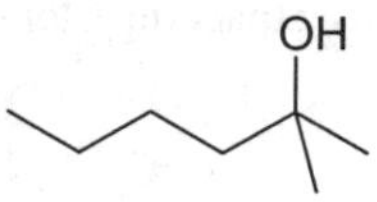

(b) A Claisen condensation takes place upon treatment of an ester with α hydrogen atoms with base. The resulting product is a β-ketoester.

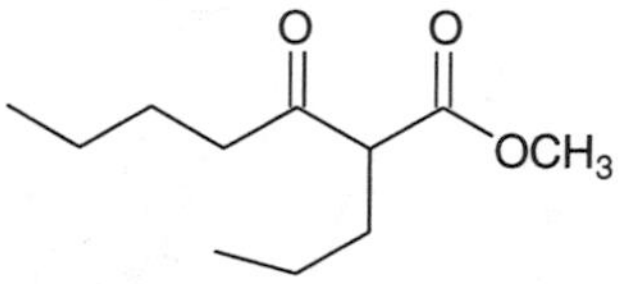

(c) The acid-catalyzed hydrolysis of an ester gives the carboxylic acid and the alcohol corresponding to the ester. Unlike the aldol condensation, the Claisen condensation does not take place in the presence of acid.

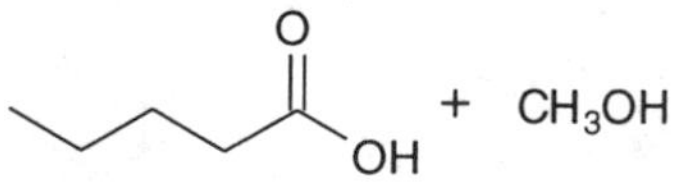

Pr 13.3 This question addresses the different reactivity of a simple ester and an α,β-unsaturated ester.

(a) Grignard reagents in the presence of Cu$^+$ react as do cuprate reagents, adding to the β-position.

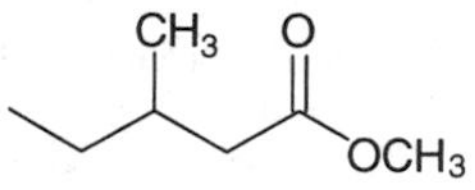

(b) Reaction of an α,β-unsaturated ester with NaOCH$_3$ results in no net reaction

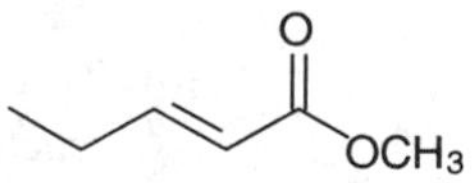

(c) The acid-catalyzed hydrolysis of an ester gives the carboxylic acid and the alcohol corresponding to the ester. Unlike the aldol condensation, the Claisen condensation does not take place in the presence of acid.

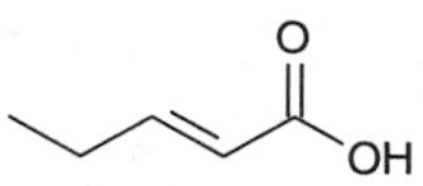

Pr 13.4   (a) H_3O^+. The product is the aldol condensation product of the starting material. If the product is formed in acid, dehydration can occur as soon as the condensation has taken place.

(b) $NaOCH_3$ This is the Claisen condensation product of the starting material.

(c) Reaction of the product from part *b* with NaOH in H_2O, followed by acid and heat (to effect decarboxyation).

(d) H_3O^+. This is an aldol condensation reaction.

Pr 13.5   In an acid-catalyzed aldol condensation reaction, dehydration of the initially formed β-hydroxyalcohol is usually rapid because of the stability of the α,β-conjugated double bond and proceeds at a rate comparable to or faster than the aldol reaction. Elimination of water from the aldol product derived from **A** takes place readily, but is blocked in the product derived from **B** where there is no α-hydrogen present.

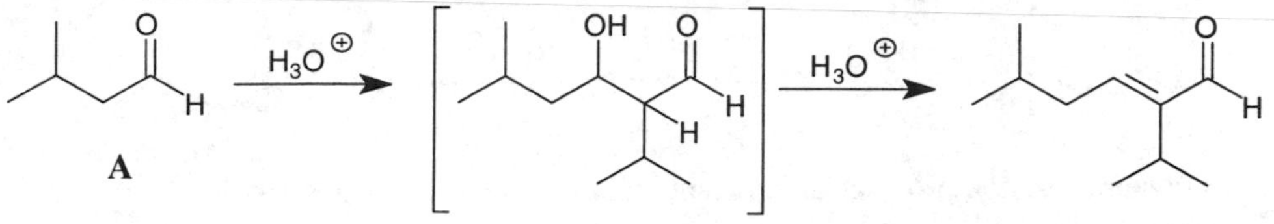

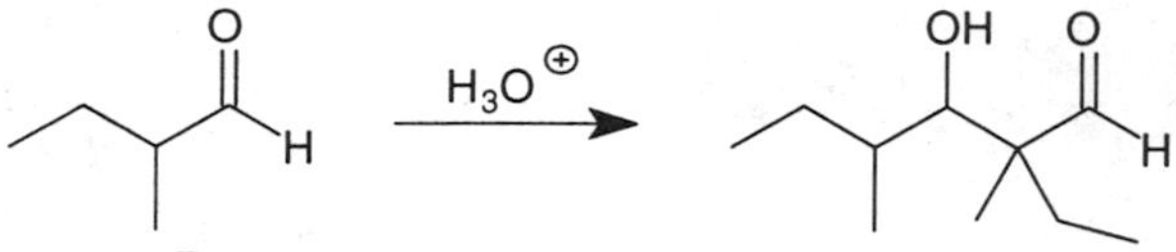

Pr 13.6   Several answers are possible. One possible route that uses the reactions described in this chapter is given here for each part.

(a) The desired product is the Claisen condensation product of an ester of propanoic acid.

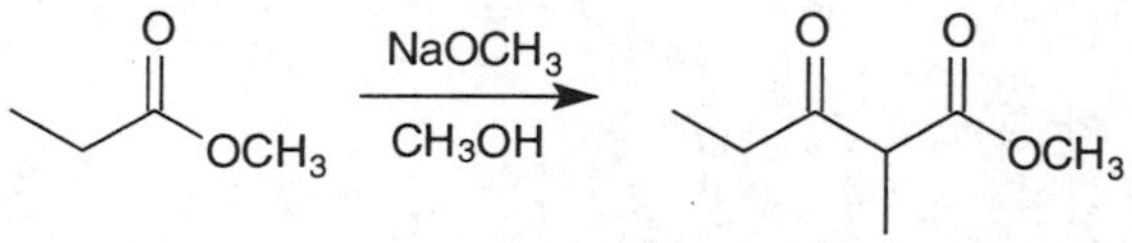

(b) The desired product is the Claisen condensation product from methyl butanoate, analogous to that produced in part *a*, but with an additional alkyl group at the α position. To follow strictly the constraints of the problem, the ester (containing five carbons) would have to be prepared by esterification of butanoic acid.

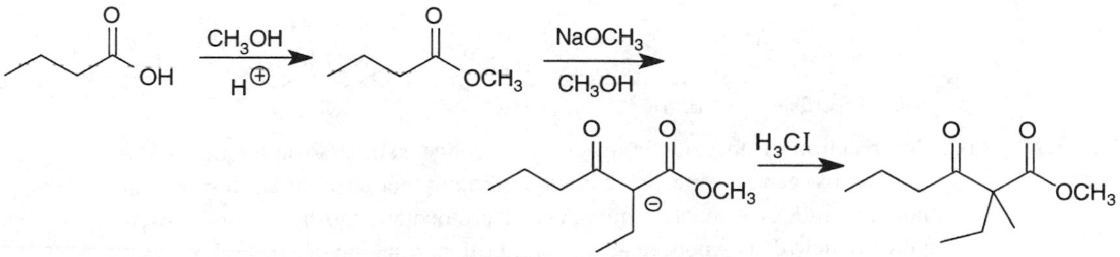

(c) Dehydration and decarboxylation of the β-ketoester product from part *b* gives the desired product.

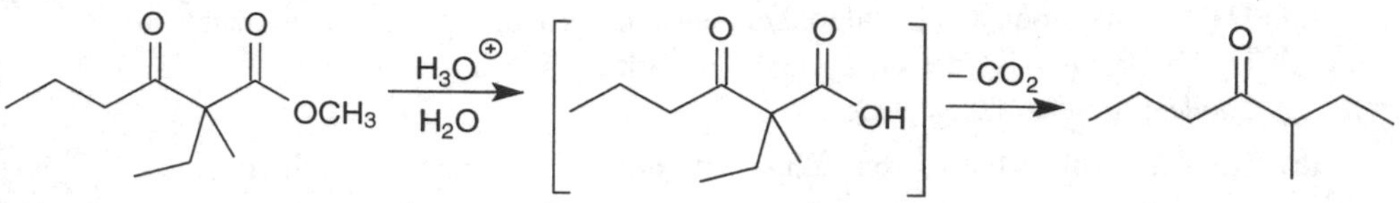

(d) Reduction of the ketone from part *c* gives the alcohol. Alternatively, the same alcohol can be prepared from a Grignard reaction between *s*-butylmagnesium bromide and butanal, as shown.

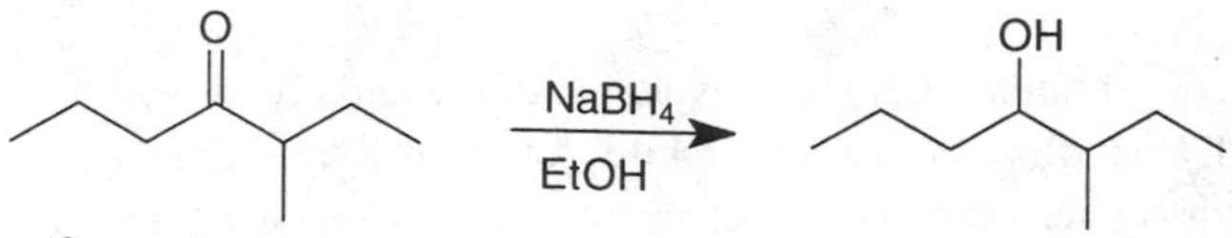

(from *c*)

Or:

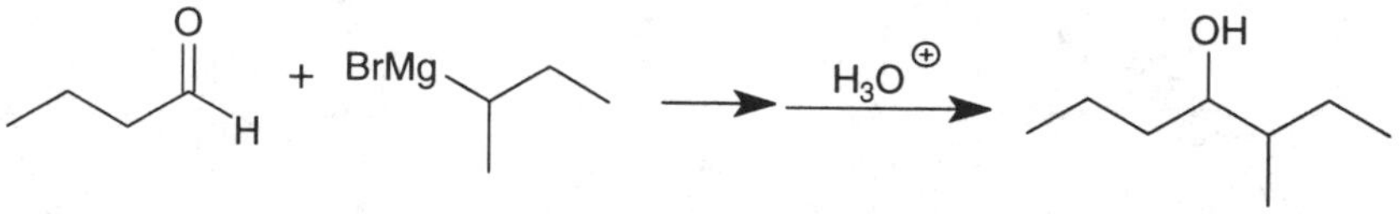

(e) A Grignard reaction provides an efficient method for preparing the desired product.

(f) Alkylation of the anion of methyl acetoacetic ester with *n*-butyl bromide gives a β-ketoester. Then hydrolysis and decarboxylation give the product.

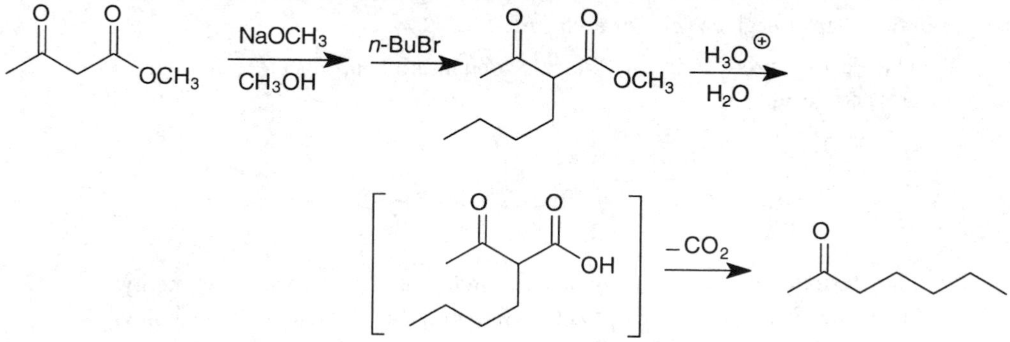

(g) A Grignard reaction of *n*-propylmagnesium bromide with an ester of acetic acid forms a tertiary alcohol. This is dehydrated under acidic conditions to form the product alkene through a Zaitsev elimination.

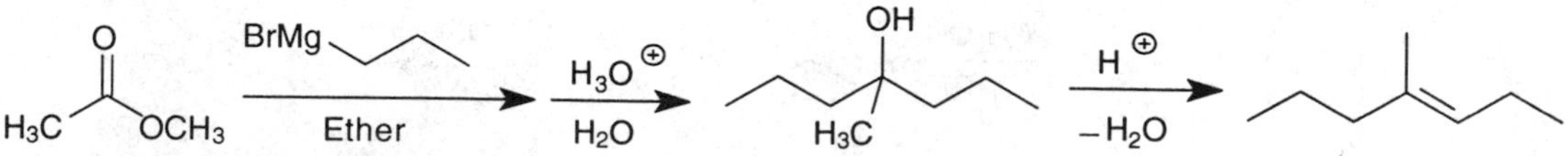

Pr 13.7 (a) This reaction is an acid-catalyzed aldol condensation. Protonation of the carbonyl oxygen activates the carbonyl group to tautomerization, forming an enol that attacks a second equivalent of protonated starting material. A β-hydroxyaldehyde is produced after deprotonation, resulting in a neutral product. Dehydration occurs by a pathway similar to that described in Chapter 9.

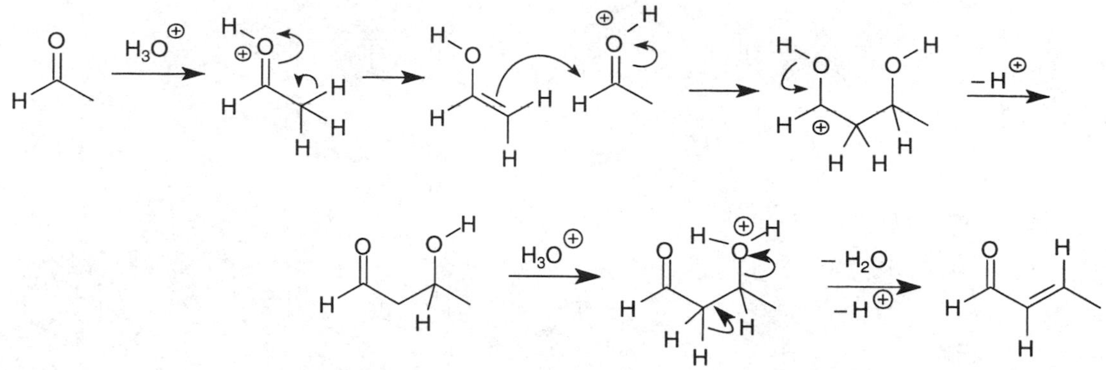

(b) Deprotonation at the α position of propanal generates an enolate anion, that adds, as a nucleophile, to another equivalent of aldehyde. The resulting alkoxide is trapped by protonation to form the β-hydroxyaldehyde.

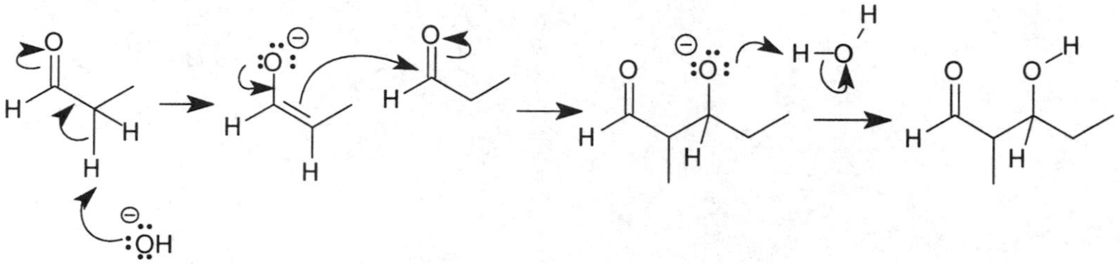

(c) Reaction with base effects deprotonation at the α position and generates an ester enolate anion, which attacks a second ester. The resulting tetrahedral alkoxide intermediate reforms the C=O bond with loss of ethoxide in this Claisen condensation.

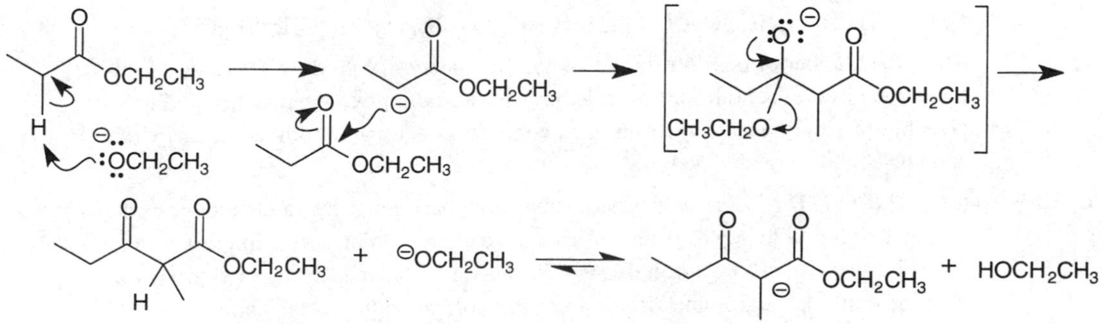

(d) The base-catalyzed Claisen condensation proceeds in the same manner as in the mechanism in part c. Treatment of the resulting β-ketoester with aqueous acid effects hydrolysis of the ester by nucleophilic acyl substitution. The resulting β-ketoacid is not stable at room temperature and undergoes concerted loss of CO_2 to form the enol. Proton tautomerization by protonation followed by deprotonation yields the observed ketone.

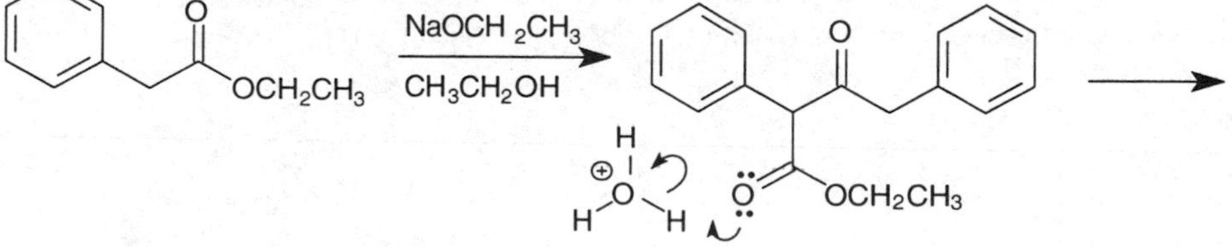

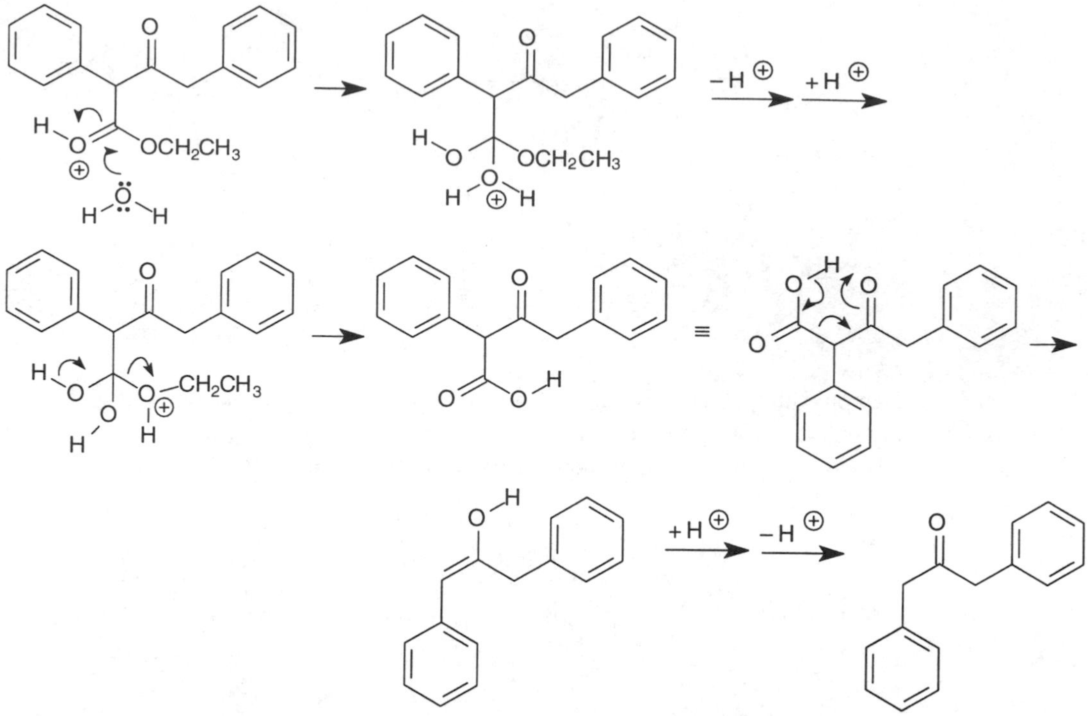

$-H^\oplus$ $+H^\oplus$

$+H^\oplus$ $-H^\oplus$

Pr 13.8 These are both examples of crossed aldol reactions in which one partner serves as nucleophile (after deprotonation by base) and the other as electrophile. In the first example, only one of the two starting aldehydes has acidic hydrogen atoms.

$C_6H_5CH=CHCHO + CH_3CH=CHCHO \rightarrow C_6H_5(CH=CH)_3CHO$

$C_6H_5CH_2CH=CHCHO + CH_3CH=CHCHO \rightarrow C_6H_5CH_2(CH=H)_3CHO$

Thus, the number of possible enolate anions (and hence of products) is reduced. In the second example, both starting aldehydes have acidic γ hydrogen atoms, and four possible products, resulting from each component acting as both as electrophile and nucleophile, are obtained.

Pr 13.9 (a) CH_3COSCH_3. Because of size mismatching between sulfur and carbon, back donation of a lone pair from sulfur to the carbonyl group is less important in a thiol ester than in a normal ester. Because this resonance interaction slows nucleophilic attack and destabilizes the corresponding ester enolate anion, Claisen condensations are slower with normal esters than with thiol esters.

(b) $C_6H_5CH_2CO_2CH_3$. $C_6H_5CO_2CH_3$ lacks an acidic α hydrogen and cannot form an ester enolate as required for a Claisen condensation. No such problem is encountered with $C_6H_5CH_2CO_2CH_3$ which smoothly condenses upon treatment with base.

(c) $CH_3CH_2CH_2CH_2CO_2CH_3$. As in part *b*, one este,r$((CH_3)_3CCO_2CH_3$, lacks an acidic α hydrogen and cannot form an ester enolate anion. In contrast, the condensation proceeds smoothly with $CH_3CH_2CH_2CH_2CO_2CH_3$.

Important New Terms

Acetoacetic ester: an α-acetylated derivative of an ester; $CH_3(CO)CH_2CO_2R$ (13.5)

Acetoacetic ester synthesis: a method for preparing an α–mono- or dialkylated derivative of a methyl ketone by sequentially alkylating an acetoacetic ester anion, hydrolyzing the alkylated ester, and decarboxylating of the resulting β-ketoacid (13.5)

Aldol: a β-hydroxyalcohol; a molecule containing both an aldehyde and an alcohol functional group (13.3)

Aldol condensation: the formation of an α,β-unsaturated aldehyde (or ketone) from two molecules of an aldehyde (or ketone) (13-3)

Aldol reaction: the formation of a β-hydroxyaldehyde (or ketone) from two molecules of an aldehyde (or ketone) (13-3)

Annulation: the formation of a ring on an existent ring (13.3)

β–Dicarbonyl compound: a functional group containing two carbonyl groups attached to a common atom (13.5)

Claisen condensation: a reaction producing a β-ketoester upon treatment of an ester with base (13.4)

Crossed aldol condensation: an aldol condensation between two different carbonyl compounds (13.3)

Crossed Claisen condensation: a Claisen condensation between two different esters (13.4)

Decarboxylation: the loss of CO_2, usually from a carboxylic acid; particularly easy from a β-ketocarboxylic acid (13.5)

Dieckmann condensation: an intramolecular variant of the Claisen condensation (13.4)

Enolization: keto-to-enol tautomerization; conversion of a ketone or aldehyde to its enol form (13.1)

Ester enolate anion: a resonance stabilized anionic species obtained by removal of a proton from the α-position of an ester (13.1)

Haloform reaction: the conversion of a methyl ketone to the corresponding carboxylic acid and haloform (CHX_3) upon treatment with aqueous base and dihalogen (13.1)

Hard: descriptor of a charge-intensive reagent; often applied in the description of nucleophiles, electrophiles, acids, and bases (13.3)

Hell–Volhard–Zelinski reaction: a method for the monobromination α to a carboxylic acid by treatment of a carboxylic acid bearing α hydrogen atoms with bromine in the presence of phosphorus tribromide (13.1)

β-Ketoacid: a functional group containing a keto group and a carboxylic acid attached to a common atom (13.5)

Iodoform test: a chemical color test for the presence of a $R(CO)CH_3$ functionality by treatment with aqueous base and iodine, evidenced by the formation of a yellow precipitate of CHI_3 (13.1)

Malonic ester: a diester in which both ester groups are bound to the same carbon atom; $CH_2(CO_2R)_2$ (13.5)

Malonic ester synthesis: a method for preparing mono- and dialkylated carboxylic acids by sequential alkylation of a malonic ester anion, hydrolysis of the alkylated diester, and decarboxylation of the resulting β-diacid (13.5)

Michael addition: a reaction in which a resonance-stabilized carbanion reacts with an α,β-enone in a conjugate addition (13.3)

Reformatsky reaction: a Claisen-like condensation of a preformed zinc ester enolate with a ketone or an aldehyde (13.4)

Robinson ring annulation: the use of an intramolecular aldol reaction to construct a six-membered ring fused to another ring (13.3)

Soft: descriptor of a charge-diffuse reagent; often applied in the description of nucleophiles, electrophiles, acids, and bases (13.3)

Skeletal–Rearrangement Reactions

14

Key Concepts

Cationic skeletal rearrangements by 1,2 shifts of alkyl groups:

Wagner-Meerwein and pinacol rearrangements

Rarity of anionic and radical skeletal rearrangements

Pericyclic reactions:

cycloadditions, sigmatropic shifts, electrocyclic reactions

Characteristics of pericyclic reactions

Skeletal rearrangements through [3,3] sigmatropic shifts:

the Cope and Claisen rearrangements

Synthesis of amides by the acid-catalyzed Beckmann rearrangement of oximes

Synthesis of amines by the Hofmann rearrangement of amides

Synthesis of esters by the Baeyer-Villiger oxidation of ketones

Structure and reactivity of diazoketones, ketenes, and isocyanates

Answers to Exercises

Ex 14.1 Addition of a proton from the acidic medium to the oxygen of an alcohol, followed by loss of water, produces a secondary carbocation. Migration of a methyl group from the adjacent carbon forms a more stable tertiary cation that leads to two products under the reaction conditions. Loss of a proton produces the most stable tetrasubstituted alkene, and trapping of the carbocation by water forms a tertiary alcohol.

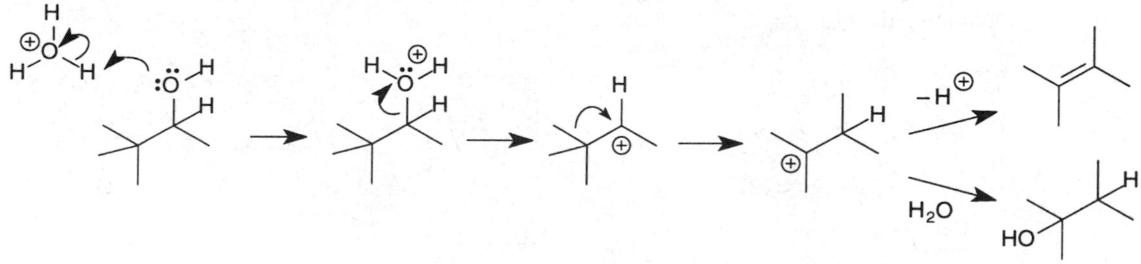

Ex 14.2 The pinacol rearrangement of a diol is initiated in the same way as the dehydration of the alcohol in Exercise 14.1. Loss of water from the protonated alcohol is accompanied by simultaneous migration of an adjacent alkyl group, forming a carbocation that is greatly stabilized by electron donation from a lone pair on oxygen. Loss of a proton completes the sequence, forming the product carbonyl compound (a ketone in *a* and an aldehyde in *b*).

(a)

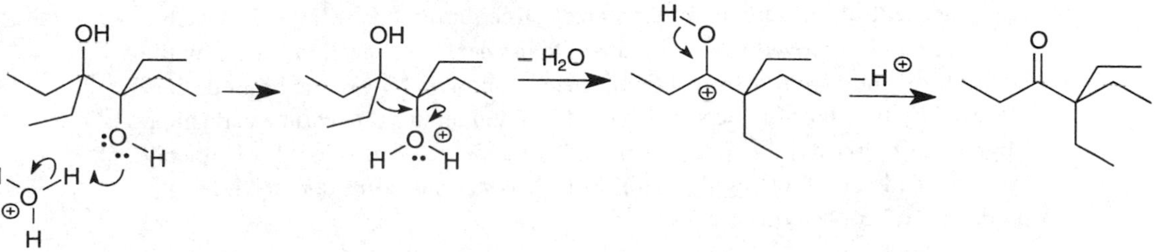

(b)

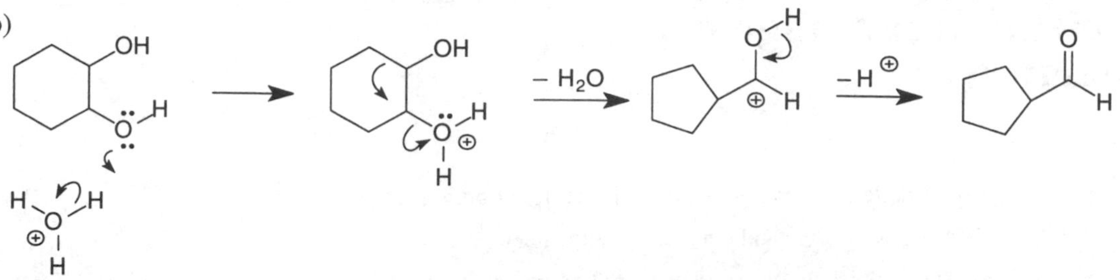

Observation of a pinacol product shows that carbon migrates in preference to hydrogen. Here, this migration accomplishes a contraction of a six-membered ring to a five-membered ring product. This represents a very valuable skeletal rearrangement because six-membered rings are readily prepared by the Diels-Alder reaction. However, this rearrangement is only possible because of the low ring strain associated with cyclopentanes: ring contraction of a five-membered ring to form a cyclobutane is usually not possible. On the other hand, ring expansion of a four- to a five-membered ring relieves the strain associated with the former and is favorable, as illustrated in Figure 14.1.

Ex 14.3   Benzil. All carbonyl groups are stabilized by contributions from zwitterionic resonance structures bearing negative charge on oxygen and positive charge on carbon. The presence of substituents that destabilize cationic centers reduces the importance of this resonance contributor and destabilizes the carbonyl group. In benzil, there are two immediately adjacent ketones; neither is as stable as the simple ketone in benzophenone. Thus, addition of a nucleophile to one of the ketones of benzil requires less energy to break the carbon–oxygen π bond than the same reaction of benzophenone. Furthermore, as one of the carbonyl groups of benzil is attacked by a nucleophile, the destabilizing effect on the remaining carbonyl group is reduced, further lowering the activation energy and thereby facilitating the reaction.

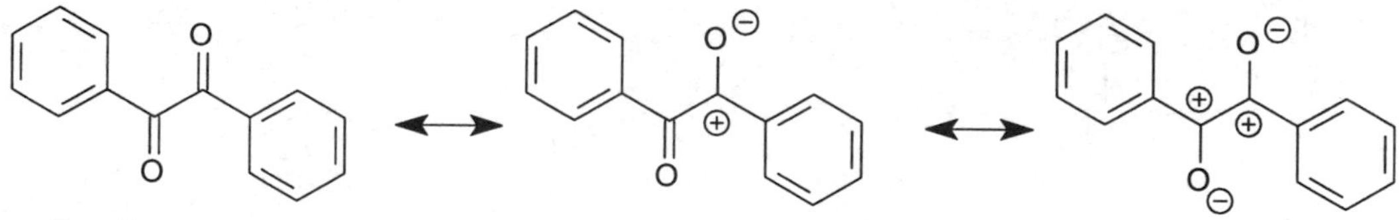

Benzil

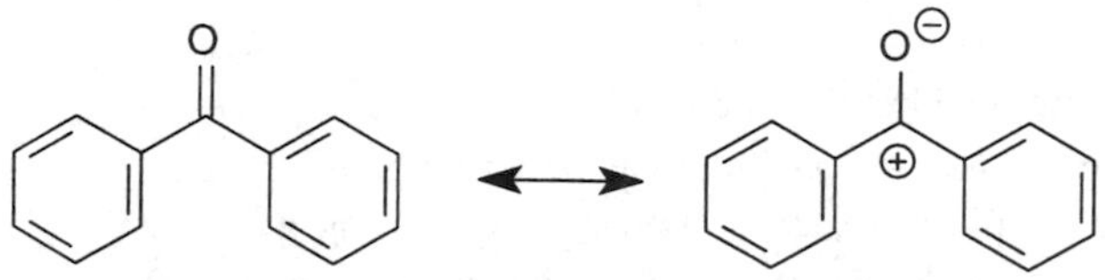

Benzophenone

Ex 14.4   Analyze the structures of the starting material and the product. Deduce that a rearrangement of the carbon skeleton takes place during this reaction in which a phenyl group has moved from one carbon to the next. The reaction begins by the addition of one electron from lithium metal followed by loss of bromide ion, followed by transfer of a second electron from lithium metal to form a carbanion. Migration of a phenyl group results in the formation of a carbanion that is doubly benzylic and therefore highly stabilized. Protonation of the anion yields the product, 1,1,2-triphenylethane.

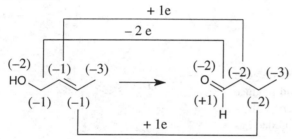

Ex 14.5 Examine the oxidation levels of all of the carbon atoms of both the starting allylic alcohol and the product aldehyde. (Assume that all hydrogens are +1 and all oxygens –2.) Then compare the corresponding atoms in starting material and product. C-1 undergoes a two-electron oxidation, going from a –1 to a +1 oxidation level. C-2 and C-3 each undergo a one-electron reduction. Thus, although several atoms change their oxidation levels, this reaction does not require either an oxidizing or a reducing reagent, because, overall, the oxidation is balanced by the reductions.

Ex 14.6 This rearrangement reaction is exothermic as written.

Bonds broken		Bonds made	
$C{=}C\ \pi$	63	$C{=}O\ \pi$	93
$O{-}H$	111	$C{-}H$	99
	174		192

$$\Delta H^{\circ} = 174 - 192 = -18\ \text{kcal/mole}$$

Ex 14.7 The *trans* isomer of 1,3,5-hexatriene cannot undergo an electrocyclic reaction. The transition state for this reaction has six carbon atoms in a cyclic array, and a *trans* double bond cannot be accommodated in such a small ring.

Ex 14.8 (a) Sigmatropic rearrangement (also called a [1,3] sigmatropic shift). Notice that only four electrons participate in this rearrangement and that the numbers of σ and π bonds in the product are the same as in the reactant.

(b) Cycloaddition reaction (Diels-Alder reaction).

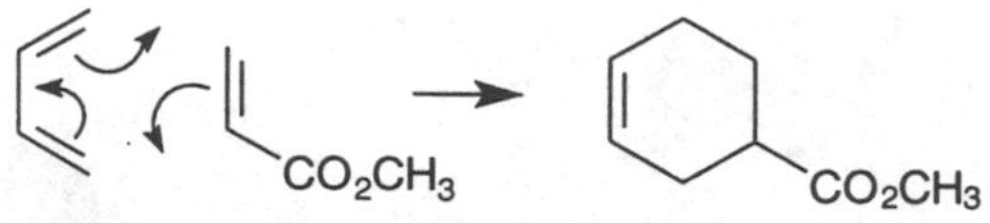

(c) Cycloreversion (the reverse of a cycloaddition, a retro Diels-Alder reaction).

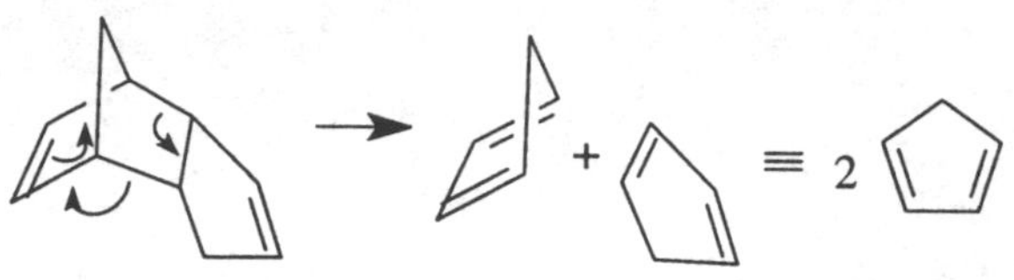

(d) Sigmatropic rearrangement (a degenerate re-arrangement except for the isotopic label). As in part a, the transition state for this rearrangement includes only four electrons and is a [1,3] sigmatropic shift.

(e) Electrocyclic reaction.

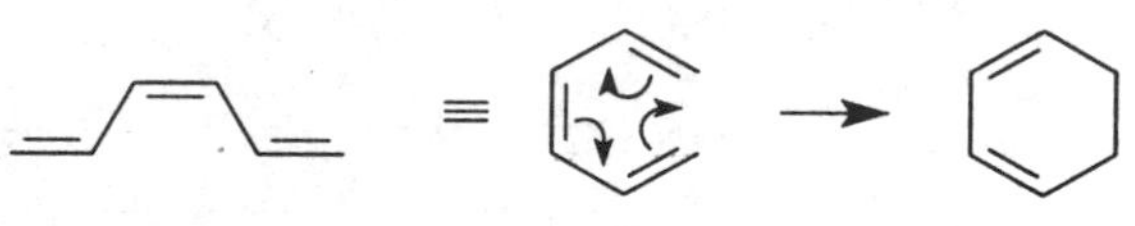

(f) Nonpericyclic reaction. Note that electrocyclic ring opening of the cyclobutene would produce 2-methyl-1,3-butadiene, an isomer with a different skeleton than the indicated product, as seen in part h.

(g) A rearrangement reaction (both starting material and product have the formula C_5H_8O).

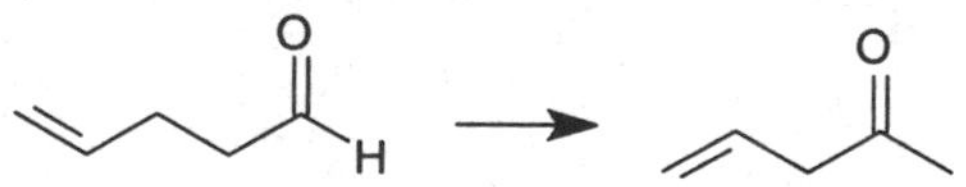

(h) Electrocyclic reaction.

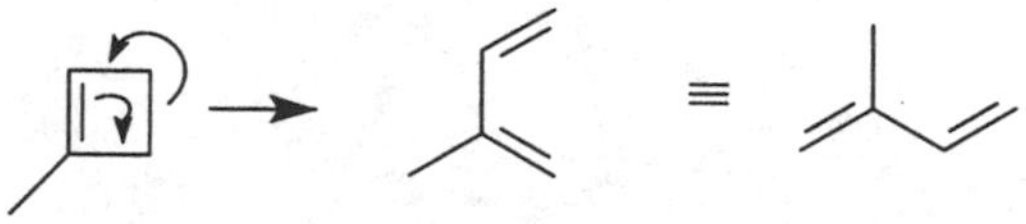

Ex 14.9 Overall, the formation of an oxime from a ketone, followed by a Beckmann rearrangement, effects the insertion of an NH group between the carbonyl carbon and an α carbon of the starting ketone, forming an amide (or lactam). Therefore, to identify the starting ketone that would produce the amides in this Exercise, it is only necessary to mentally remove the NH group and reconnect the carbon atoms.

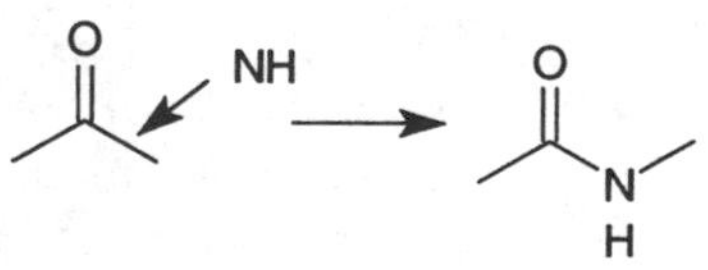

(a)

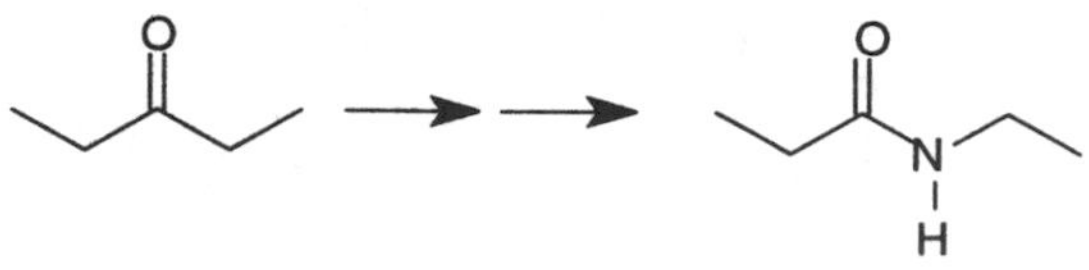

(b)

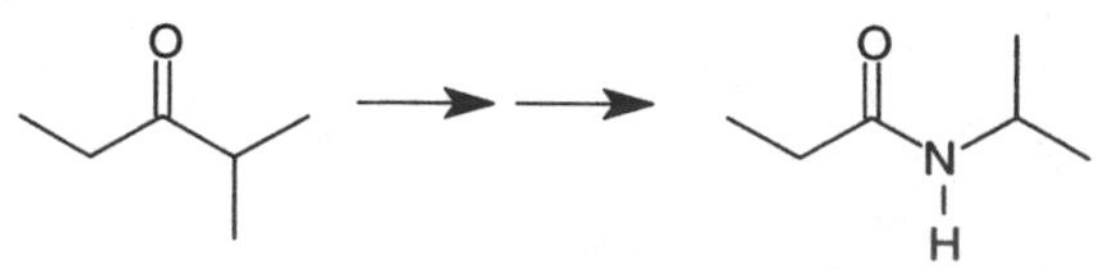

(c)

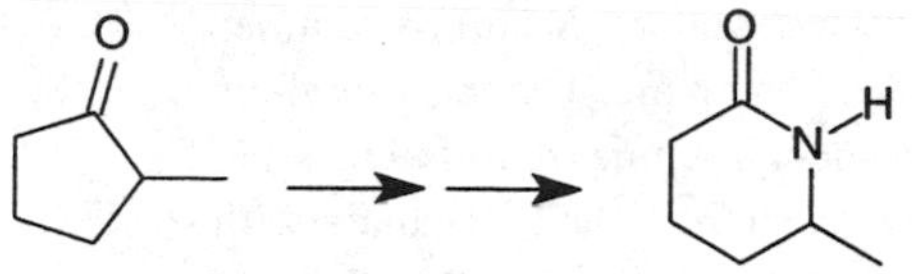

Ex 14.10 The overall transformation involves two major parts: the conversion of a ketone to an oxime; and the Beckmann rearrangement.

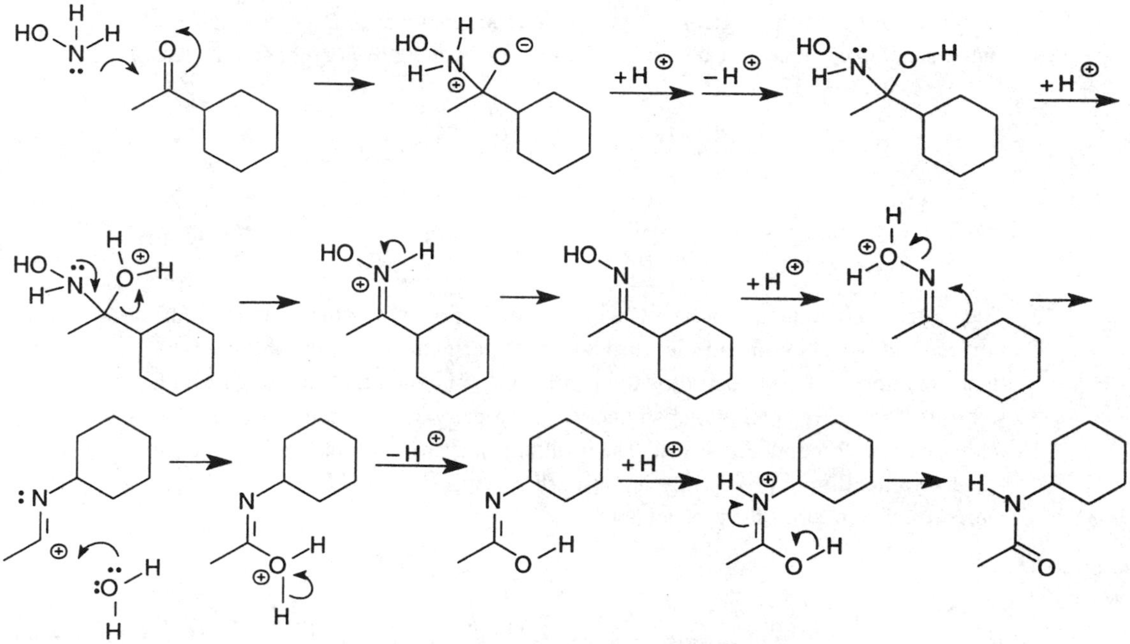

Ex 14.11 The slow (and therefore rate-limiting) step in the Beckmann rearrangement is that involving migration of carbon with simultaneous loss of water. The intermediate cation formed in this step is significantly stabilized by resonance donation of the lone pair of electrons on nitrogen. Ideally, then, this nitrilium ion intermediate will have a linear geometry, resembling an alkyne. Stabilization is greatly reduced when this intermediate cation is held in a nonlinear arrangement as part of a ring; the smaller the ring, the less stabilization is provided by resonance delocalization. Thus, the Beckmann rearrangement of cyclopentanone oxime is slower than that of cyclohexanone, and both rearrangements in these cyclic systems proceed more slowly than that of the oxime of an acyclic ketone.

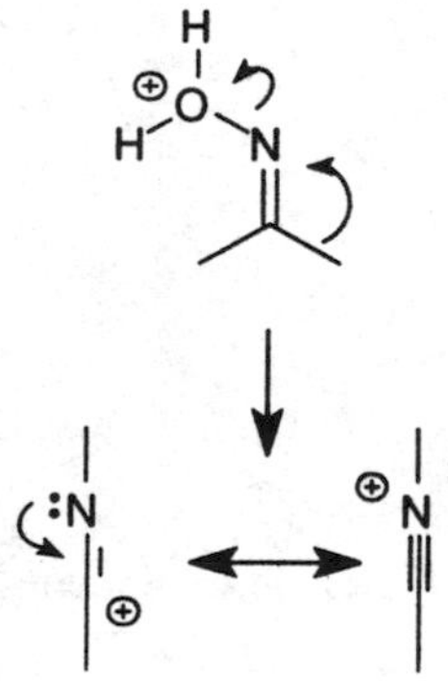

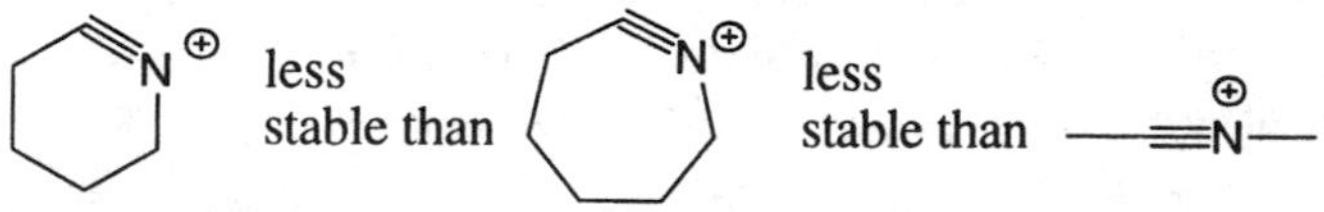

less stable than

less stable than

Ex 14.12 The isocyanate group is a relatively unstable functionality that undergoes rapid reaction with nucleophiles, including water, alcohols, amines, and carboxylic acids.

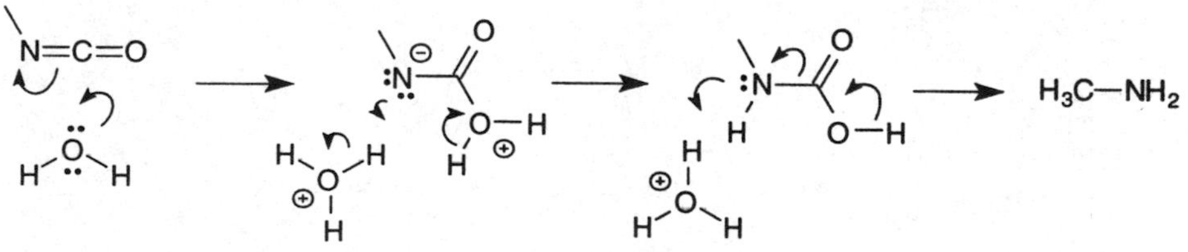

The first stage of the reaction involves addition of H–OH across the carbon–nitrogen π bond, forming the same intermediate, an *N*-carboxyamine (or carbamic acid anion) that we encountered in the Hofmann rearrangement. *N*-carboxyamines are not stable and rapidly lose carbon dioxide to form an amine. The most obvious toxic effect of methyl isocyanate is the release of methyl amine within the tissue. However, of greater biological consequence is the reaction of the isocyanate with biologically important alcohols and amines as nucleophiles (taking the place of water in the above reaction). The initially formed adducts from these nucleophiles, carbamates and ureas, respectively, are relatively stable. Where an alcohol or amine is part of a critical biological reagent such as an enzyme or DNA, the natural functions of these molecules are inhibited and even diverted toward other pathways.

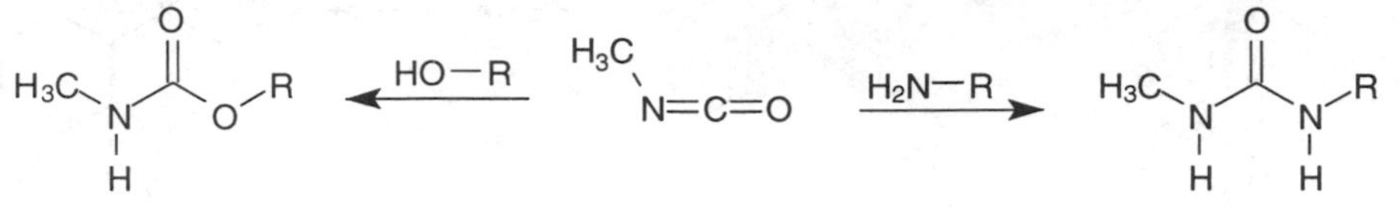

A carbamate   A urea

Ex 14.13 The Hofmann rearrangement begins with the conversion of a primary amide to the corresponding *N*-bromoamide through an intermediate anion obtained by deprotonation. A second deprotonation then forms an anion that rearranges, with the migration of the carbon substituent α to the carbonyl group to the nitrogen, taking place as bromide ion is lost. The resulting isocyanate is attacked by water in the basic medium, producing an *N*-carboxyamine as in Exercise 14.12. Decarboxylation forms the product amine.

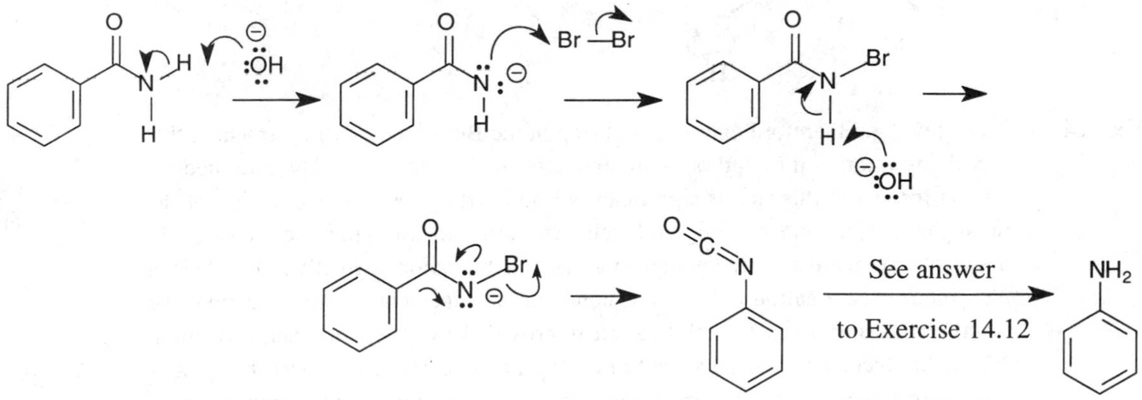

Ex 14.14 As with all synthesis problems, more than one route is possible. A reasonable one is suggested here. In each part of this Exercise, a Hofmann rearrangement is involved. In parts *a* and *b*, treatment of the starting amide with NaOH and Br$_2$ produces the desired product amine.

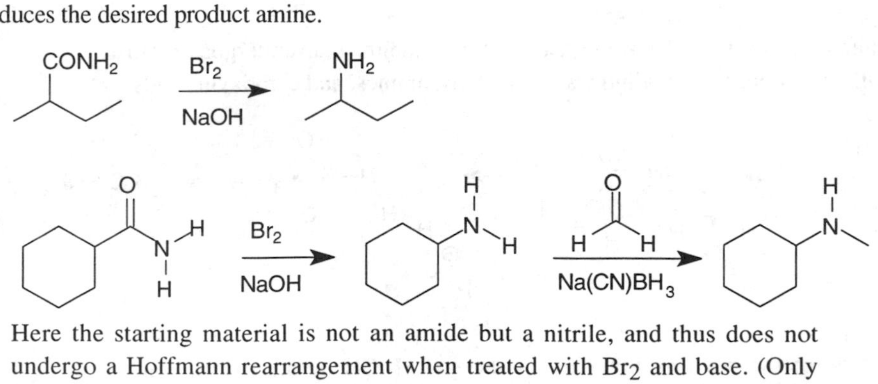

(c) Here the starting material is not an amide but a nitrile, and thus does not undergo a Hoffmann rearrangement when treated with Br$_2$ and base. (Only primary amides are substrates for this reaction.) Nonetheless, a nitrile can be hydrolyzed with aqueous base (or acid) to a primary amide. Treatment with NaOH and Br$_2$ then effects the rearrangement, producing the amine.

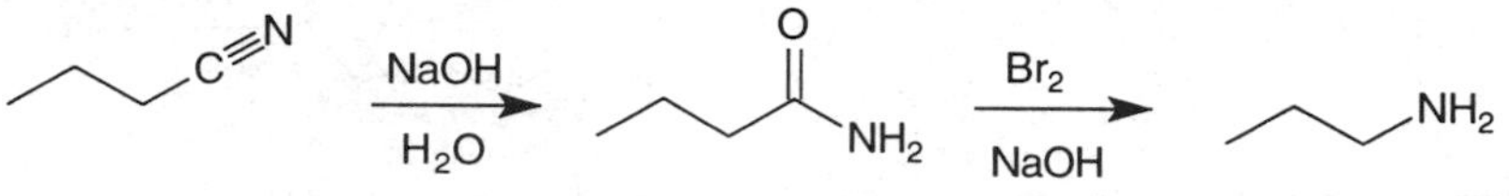

(d) Here the starting material is a primary amide, but the desired product is an amide rather than a primary amine that is the product of a Hofmann rearrangement. After the starting amide is treated with NaOH and Br$_2$, the primary amine can be converted to the desired acetamide derivative by reaction with acetic anhydride.

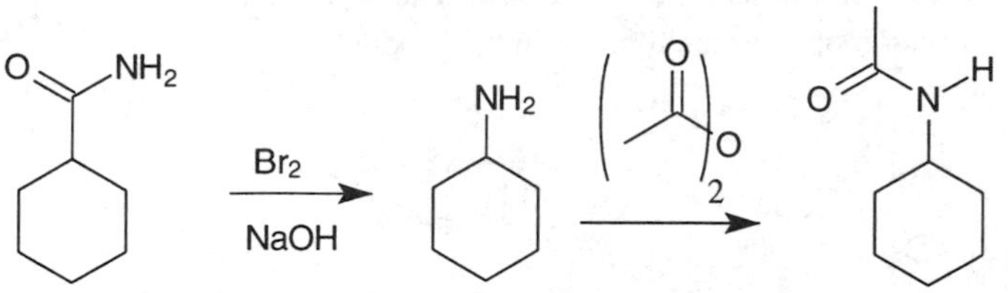

Ex 14.15 The conversion of acetic acid and hydrogen peroxide to peracetic acid and water is a nucleophilic acyl substitution.

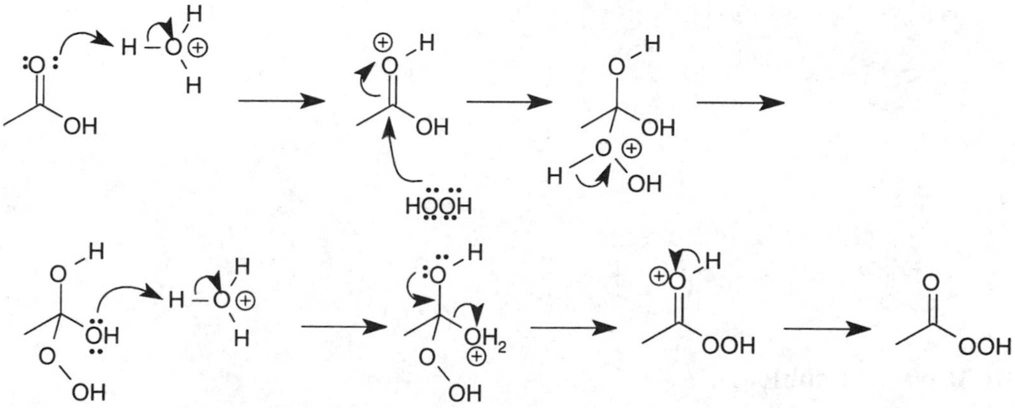

Ex 14.16 The Baeyer-Villiger oxidation converts a ketone to an ester or lactone by insertion of an oxygen atom between the carbonyl carbon and one of the α carbons. Thus, the reactant for a Baeyer-Villiger oxidation must be a ketone.

(a) The starting material is an alcohol, and oxidation with Cr^{6+} affords the ketone required for the Baeyer-Villiger oxidation.

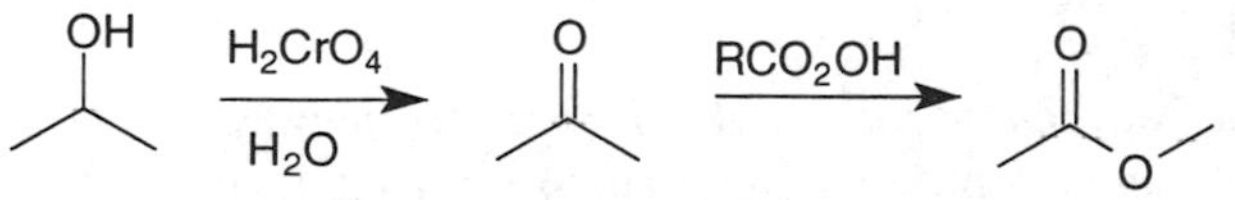

(b) The required ketone can be prepared from the starting alkene by a two-step sequence of hydroboration-oxidation, followed by oxidation of the resulting alcohol with Cr^{6+}. Baeyer-Villiger oxidation accomplishes the insertion of oxygen into the more highly substituted α carbon.

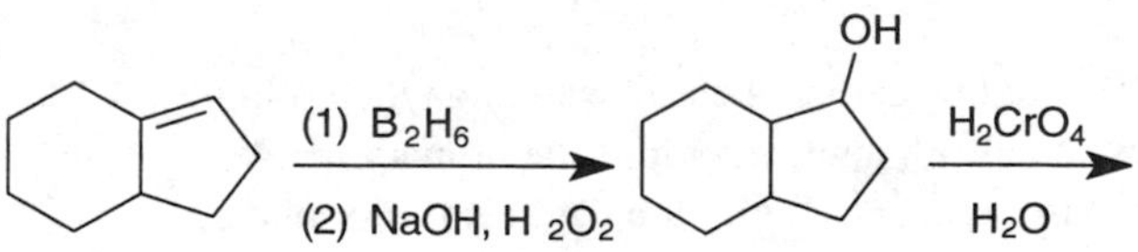

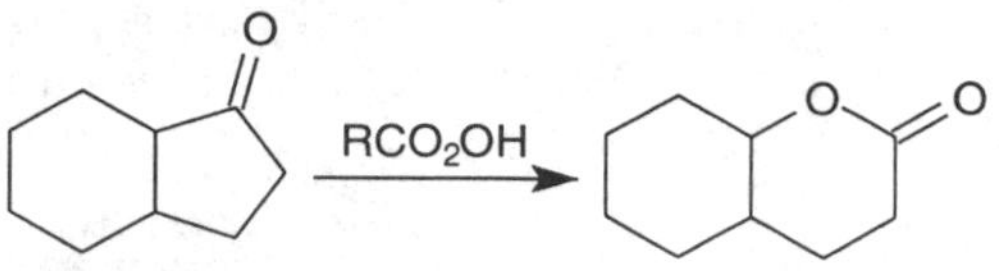

Ex 14.17 The Claisen rearrangement takes the form shown, where the product has both a carbonyl group and a double bond, fixed in positions relative to each other by the pericyclic nature of the reaction. Specific examples exist in which the R group is carbon, hydrogen, or oxygen so that, respectively, ketones, aldehydes, or carboxylate derivatives are formed. To find the starting materials for the products in this Exercise, we can mentally run the Claisen rearrangement backwards (keeping in mind that the Claisen rearrangement is exothermic because of the formation of a carbonyl group in the product).

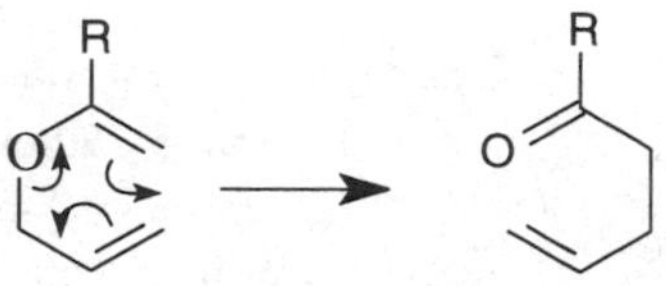

(a)

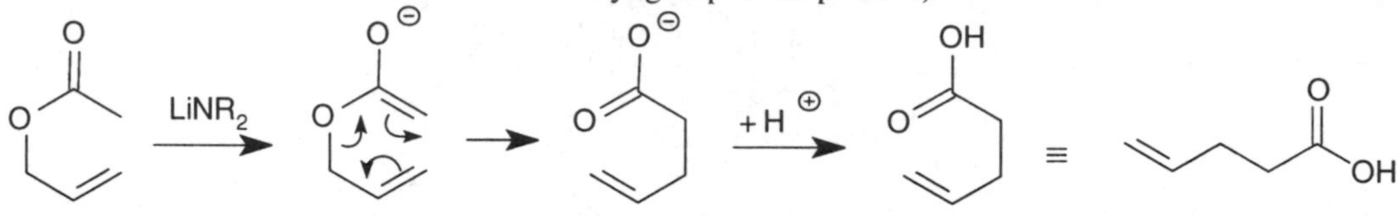

(b)

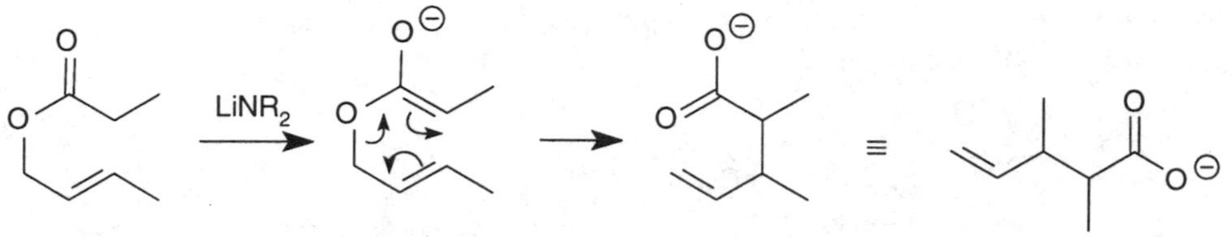

(c)

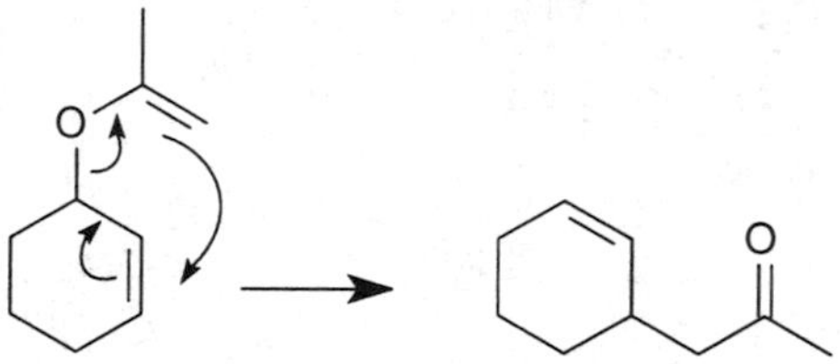

Answers to Review Problems

Pr 14.1 (a) This reaction is a pinacol-like rearrangement in which a methyl group migrates to form a more stable cationic intermediate. Loss of a proton produces 3,3-methylpropanal.

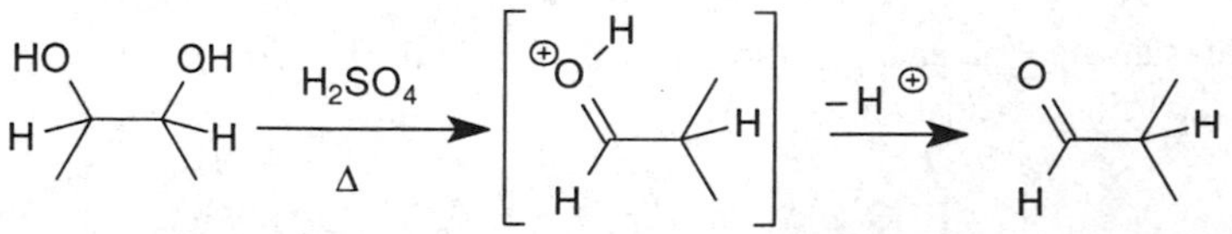

(b) Heating 1,5-dienes leads to a Cope rearrangement. In this case, the initially formed product of the [3,3] sigmatropic rearrangement is an enol that tautomerizes to the more stable aldehyde. The driving force for this reaction is the conversion of a carbon–carbon π bond to a carbon–oxygen π bond.

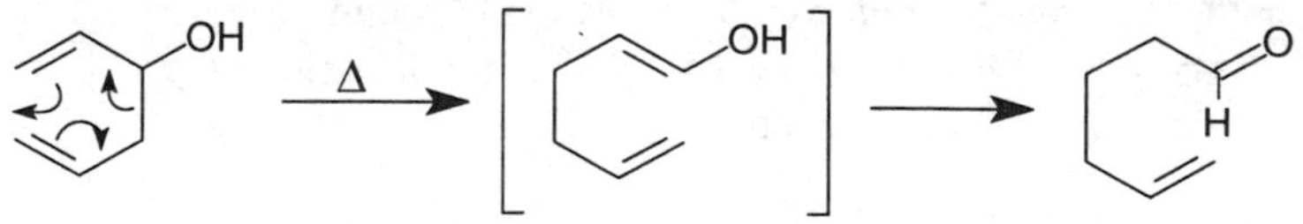

(c) In this sequence, treatment of the starting carboxylic acid with thionyl chloride forms the acid chloride, and reaction with ammonia forms the primary amide. Upon treatment with Br_2 and base, primary amides are degraded to primary amines by a Hofmann rearrangement through loss of the carbonyl carbon as carbon dioxide.

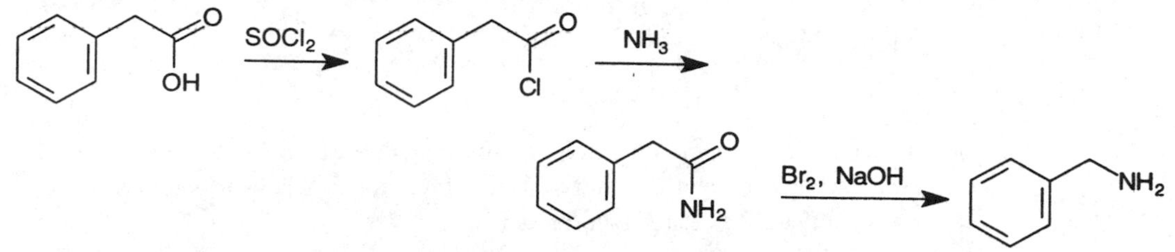

(d) Treatment of ketones with peracid leads to Baeyer-Villiger oxidation, with the insertion of an oxygen atom between the carbonyl carbon and one of the α carbons. Reaction of unsymmetrical ketones with peracids leads to insertion of the oxygen at the side of the more substituted α carbon.

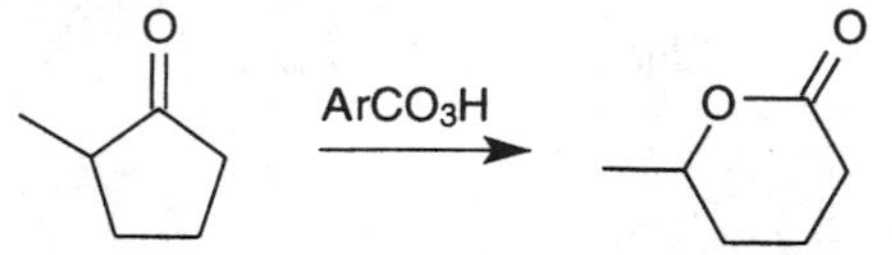

(e) In this sequence, a ketone is first converted to the oxime derivative, which then undergoes the Beckmann rearrangement upon treatment with H2SO4. As with the Baeyer-Villiger oxidation, the Beckmann rearrangement of an unsymmetrical oxime leads to insertion of the nitrogen between the carbonyl group carbon the more substituted a carbon.

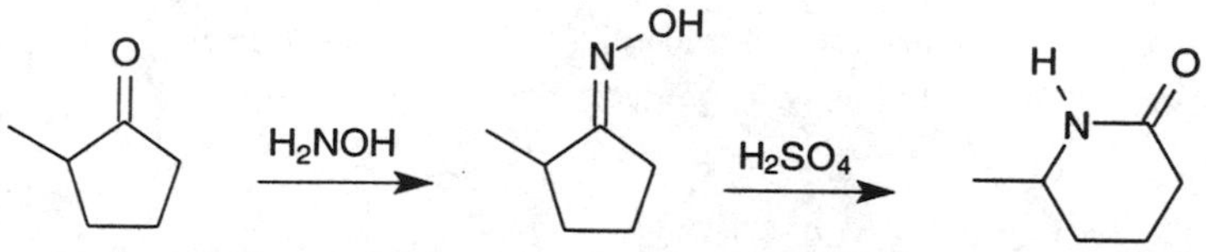

(f) This reaction is an example of a Diels-Alder (cycloaddition) reaction.

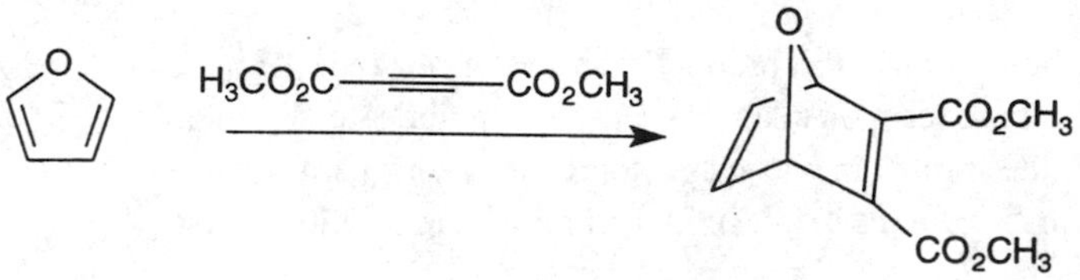

(g) In this example of the Claisen (or oxa-Cope) rearrangement, the initial product no longer has an aromatic system. However, proton tautomerization restores aromaticity in the final product.

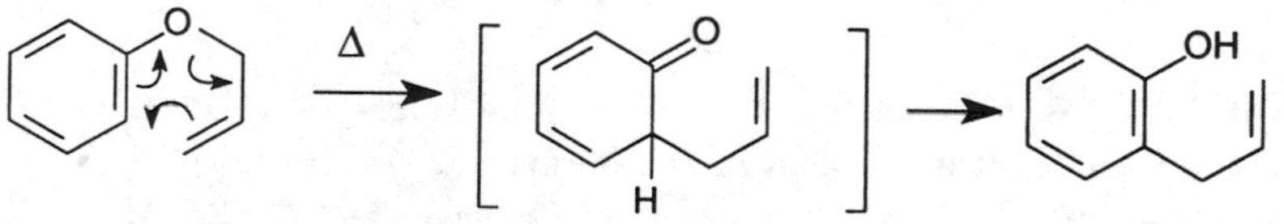

Pr 14.2 (a) Overall, this transformation requires the insertion of an NH group between the carbonyl carbon and an α carbon of the starting ketone. This is accomplished by reaction of the ketone with hydroxylamine and treatment of the resulting oxime with H_2SO_4 in a Beckmann rearrangement.

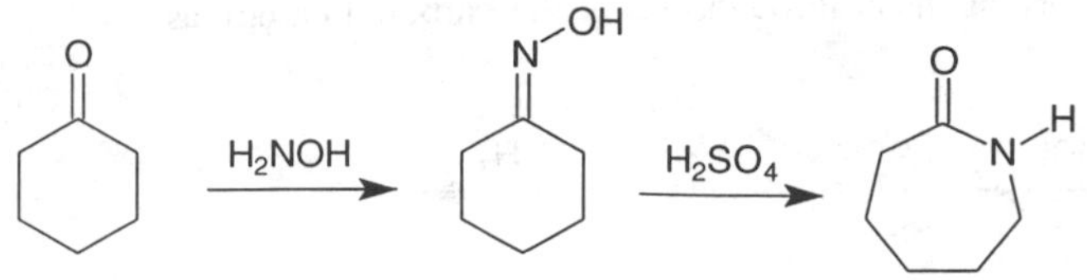

(b) The Baeyer-Villiger oxidation results in the insertion of an oxygen atom between a carbonyl carbon and an α carbon of a ketone. A peracid such as peracetic acid effects the desired transformation.

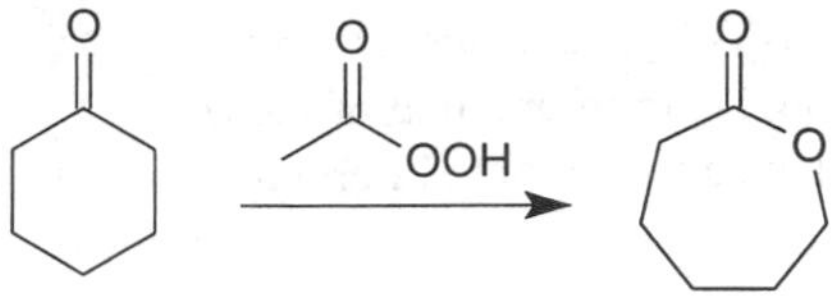

(c) Alcohols are conveniently prepared from ketones by reduction with $NaBH_4$ in ethanol.

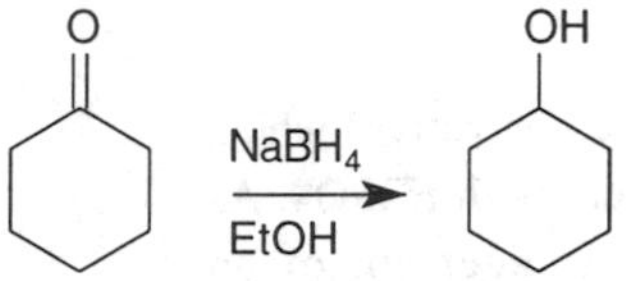

(d) Examination of the oxidation level of the altered carbon of the functional groups of the starting ketone (+2) and the product bromide (0) shows that a reduction is necessary. Hydride reduction to the alcohol (as in part *c*) followed by treatment with PBr_3 (or, alternatively, HBr) produces the desired bromide.

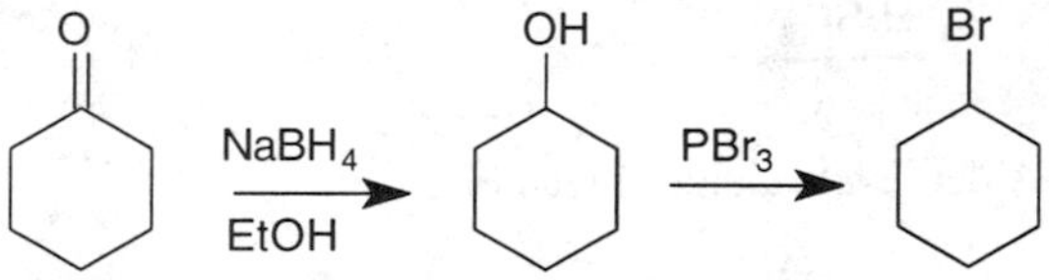

(e) There is one more carbon atom in the product than in the reactant. Thus, a carbon–carbon bond-forming reaction takes place at some point. A convenient sequence uses the bromide formed in part *d* as a source for a Grignard reagent treatment of which with CO_2 gives the desired acid upon acidification of the reaction mixture.

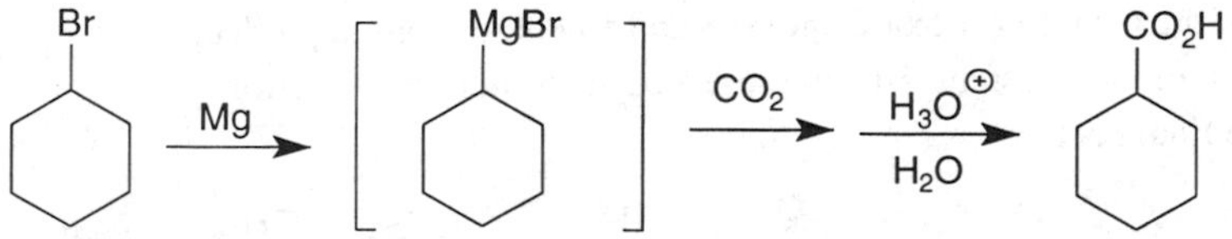

(f) The product acid has two more carbons than the starting ketone. Several different sequences accomplish this kind of transformation. Two use the bromide prepared in part *d*: (1) in a malonic ester synthesis, and (2) by conversion to a Grignard reagent followed by treatment with ethylene oxide and oxidation of the resulting primary alcohol.

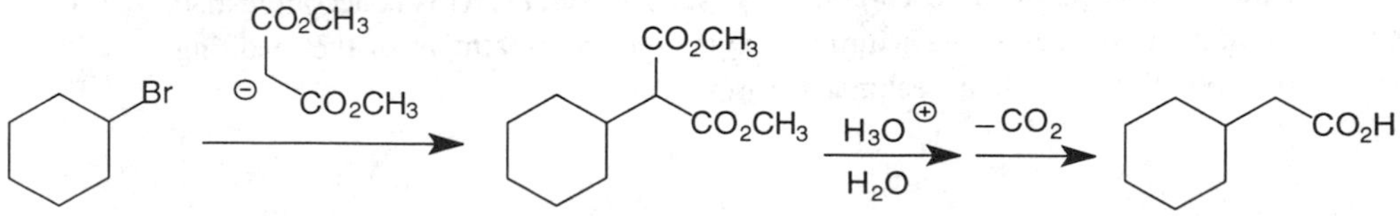

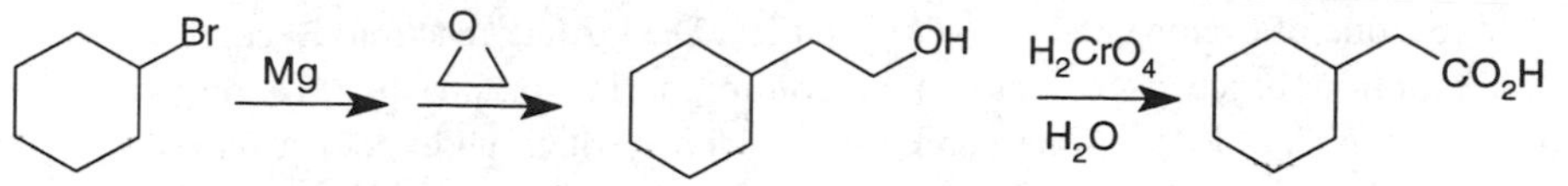

(g) Primary amines are best prepared either by reaction of an alkyl halide with a large excess of ammonia or by the Gabriel synthesis.

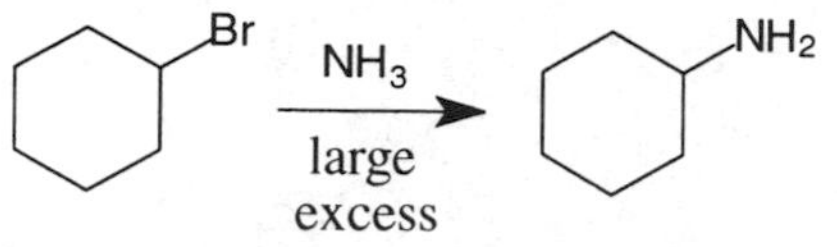

Pr 14.3   (a) The Gabriel synthesis adds the nitrogen of phthalimide to a carbon by nucleophilic substitution. Because only the anion of phthalimide is sufficiently reactive, the synthesis is an excellent way to prepare primary amines. Treatment of the phthalimide anion with *n*-butyl bromide yields *N*-butylphthalimide. After substitution, the amine is released by reaction of the substituted phthalimide with hydrazine.

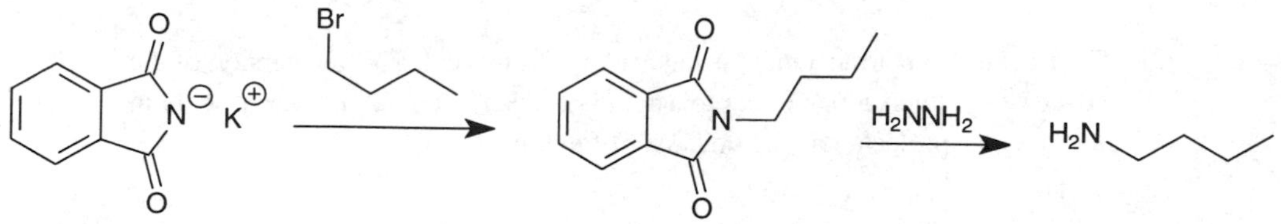

(b) The Hofmann rearrangement converts an unsubstituted carboxylic acid amide to the primary amine with one fewer carbon atom. Thus, treatment of pentanoic acid amide with Br_2 and NaOH produces butylamine.

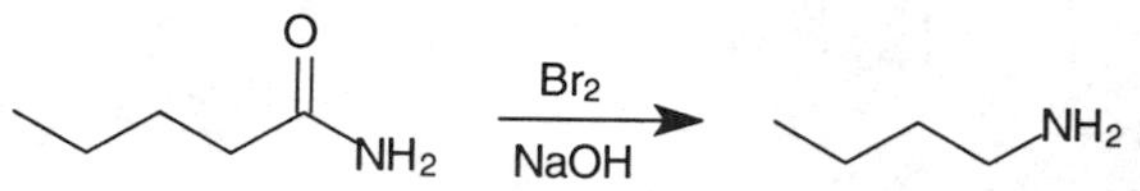

(c) The Beckmann rearrangement effects insertion of an NH group between the carbonyl carbon and one of the α carbons of a ketone. The resulting amide can be converted to an easily separated mixture of an amine and a carboxylic acid salt by hydrolysis with aqueous NaOH. By starting with the symmetrical ketone 5-nonanone, we ensure that the rearrangement produces the desired *n*-butylamine after hydrolysis.

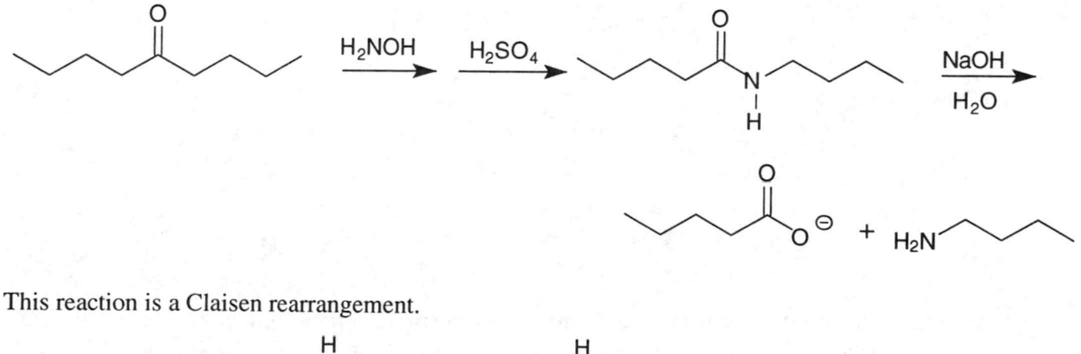

Pr 14.4   (a) This reaction is a Claisen rearrangement.

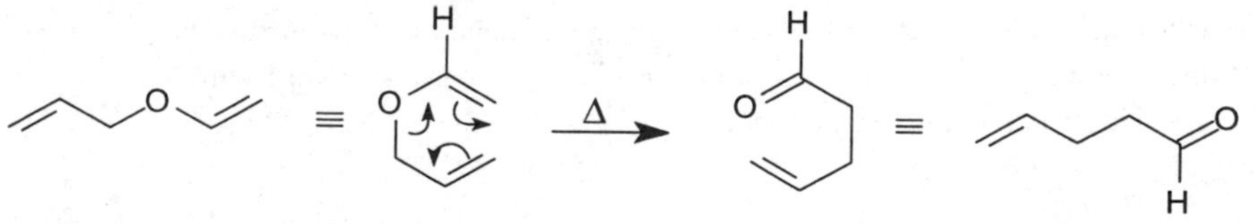

(b) The reaction of cyclopentadiene with itself is a Diels-Alder reaction. Because the reaction proceeds at room temperature, it is not possible to buy cyclopentadiene. Rather, one purchases the dimer which undergoes a retro Diels-Alder reaction to the monomer upon heating. (The diene is removed immediately by distillation and kept very cold.)

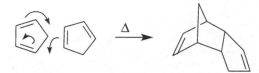

(c) The energy difference between starting material and product in this electrocyclic reaction is the result of the ring strain in the starting material. (The number of π and σ bonds is unchanged.)

(d) The Beckmann rearrangement is induced by strong acid. This conversion of the oxime of cyclohexanone to caprolactam is carried out on a very large scale in industry. (Caprolactam is the precursor of nylon.)

(e) The Baeyer-Villiger reaction converts a ketone to an ester (or a lactone, in the case of cyclic ketones).

(f) The reaction of a primary amide with base and Br$_2$ results in formation of an *N*-bromoamide, deprotonation of which induces a Hofmann rearrangement. The resulting isocyanate is attacked by water producing a carbamic acid which undergoes decarboxylation to an amine.

(g) The reaction of a 1,2-diol with acid induces a pinacol rearrangement by migration of a carbon substituent, leading to a ketone.

Pr 14.5 In the first step, hydroxide ion acts as a base to deprotonate phenol. (Phenols are considerably more acidic than water, pK_a 10 versus 16. Thus, this acid-base equilibrium lies to the side of the phenoxide ion.) Alkylation of the phenoxide ion by allylic bromide forms an allyl ether that, upon heating, undergoes a Claisen rearrangement. Notice that in the product of this pericyclic reaction the aromaticity of the original benzene ring has been destroyed. Proton tautomerization through deprotonation, followed by reprotonation, forms the product phenol.

Pr 14.6 **(a)** The starting diol shows infrared absorption bands for the OH groups at approximately 3600 cm^{-1}, whereas the product ketone has a strong band at 1710 cm^{-1}.

The four methyl groups in the starting diol are identical, whereas those in the product occur as three identical groups and one different group. This difference is clearly visible in both the proton and the carbon NMR spectra. Because the product has been dehydrated, its parent peak in the mass spectrum will be at 18 m/z units below that of the reactant.

(b) The two ketones in the starting material appear as hydroxyl and carboxylic acid functional groups in the product. The product shows infrared absorption bands for OH groups (at 3600 cm^{-1}). As in part a, the starting material is symmetrical.

$$\text{Ph} \overset{O}{\underset{O}{\diagdown}} \text{Ph} \xrightarrow{\text{NaOH}} \xrightarrow{\text{H}_3\text{O}^{\oplus}} \underset{\text{OH}}{\overset{\text{Ph}}{\underset{\text{Ph}}{\diagdown}}} \overset{O}{\diagdown} \text{OH}$$

Although the phenyl groups of the product are identical, the two remaining carbons are not and this change is evident from the more complex carbon NMR spectrum. In addition, there are two more protons in the product than in the starting material; the protons on oxygen in the product appear as a broad signal in the proton NMR spectrum. The product has two more hydrogen atoms and one more oxygen atom than the starting material; thus, the parent molecular ion of the product is 18 mass units higher than that of the starting material (210 versus 228).

(c) The infrared absorption bands for the carbonyl groups of ketones and amides are quite different (one band at 1715 versus two bands at 1680 and 1530 cm^{-1}). Furthermore, the amide product shows an N—H stretch at approximately 3400 cm^{-1} not present in the starting material. The starting ketone is symmetrical, whereas the product is not. Thus, the proton NMR spectrum of the starting material shows

$$\xrightarrow{\text{H}_2\text{NOH}} \xrightarrow{\text{H}_2\text{SO}_4}$$

one triplet and one quartet for the two identical ethyl groups. In the proton spectrum of the product, there are two triplets and two quartets. The proton on nitrogen appears as a distinct absorption. The carbon NMR spectrum of the product has more peaks than that of the reactant because of the change in symmetry, and the carbon attached to nitrogen absorbs considerably downfield from the other carbons in the product and from the carbons of the starting ketone. Because the Beckmann rearrangement effects a net insertion of an NH group, the mass of the molecular ion of the product increases from 86 to 101.

(d) The conversion from an amide to an amine is clear in the infrared spectra from the disappearance of the carbonyl absorptions of the starting material at 1680 and 1530 cm^{-1}. Changes in the proton NMR spectra

$$\overset{O}{\diagdown}\text{NH}_2 \xrightarrow{\text{Br}_2, \text{NaOH}} \diagdown\text{NH}_2$$

are subtle (both starting material and product have a simple propyl group and two hydrogens on nitrogen), although there are differences in chemical shifts. For example, the carbon adjacent to the carbonyl group becomes attached to nitrogen, resulting in a downfield shift of approximately 37 to 34 δ in the carbon NMR spectrum. The product has lost CO relative to the starting material, resulting in a decrease in mass of the molecular ion from 87 to 59.

(e) The Baeyer-Villiger oxidation converts the starting ketone to an ester by insertion of an oxygen between the carbonyl group and one of the α carbons. The carbonyl group absorption in the infrared spectrum

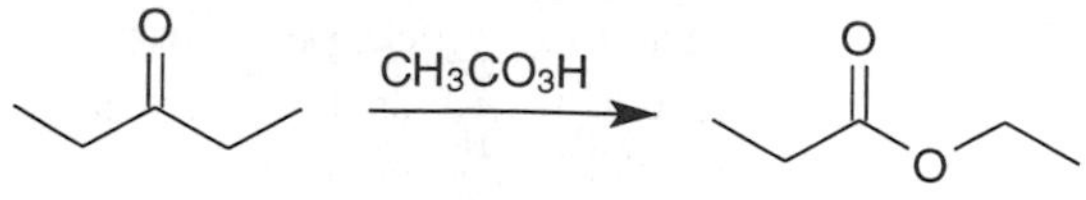

shifts from 1715 to 1735 cm^{-1}, and new C—O stretches are observed. Both the proton and carbon NMR spectra show dramatic changes resulting from the change from the symmetrical ketone with two identical ethyl groups to the product ester with two different ethyl groups. For example, 3-pentanone has three signals in the carbon NMR spectrum at 211, 35, and 8 δ. Ethyl propionate has five distinct signals, at 178, 60, 28, 27, and 14 δ. Because an oxygen atom has been added in the reaction, the mass of the product is 16 units higher than that of the starting material (102 versus 86).

(f) In this Claisen rearrangement, one of the carbon–carbon π bonds of the starting material is replaced by a carbonyl group. Thus, the product has a carbonyl absorption at 1715 cm^{-1} in its infrared spectrum that is not present in the starting material. Both the proton and carbon NMR spectra change significantly. In the starting material, there are five hydrogens on sp^2-hybridized carbons—in the product, there are only three. In the carbon NMR spectrum of the starting material, there are four signals at lower field than 121 δ, whereas in the product there are only three (the two alkenyl carbons and the carbonyl carbon). The mass does not change in this rearrangement.

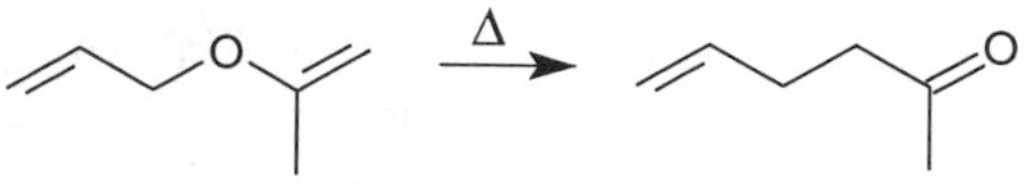

Pr 14.7 In the next chapter you will learn how the reactions from the first 14 chapters can be combined to prepare complex organic molecules from simple ones. It is important that you have a good working knowledge of these transformations. There are many possible answers to each part of this Problem: one acceptable pathway is given here. Use Table 14.2 as a guide in proposing others.

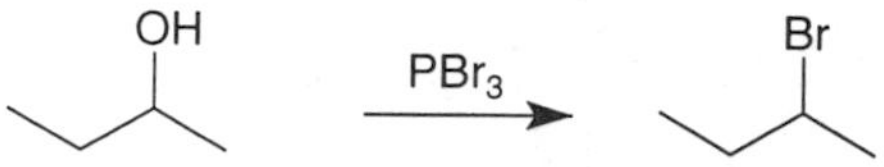

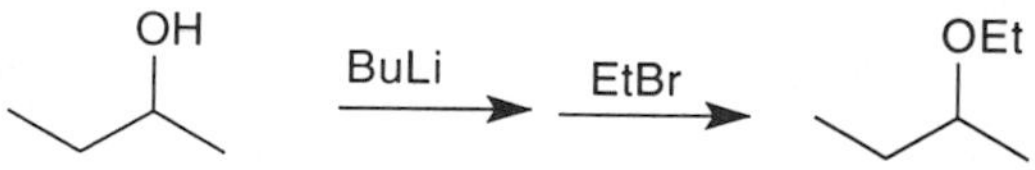

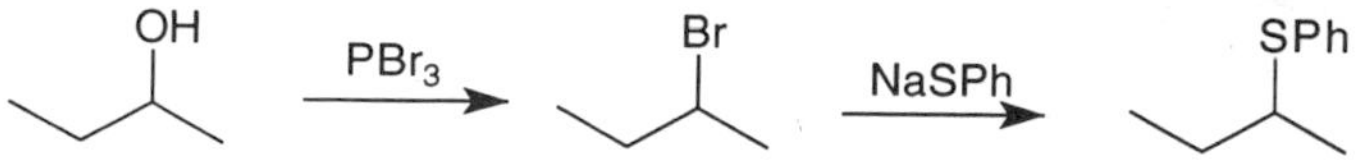

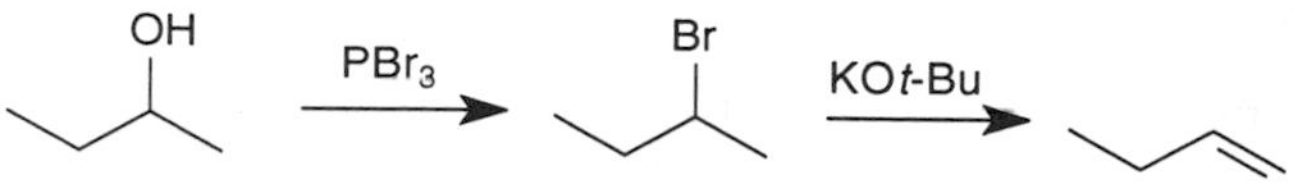

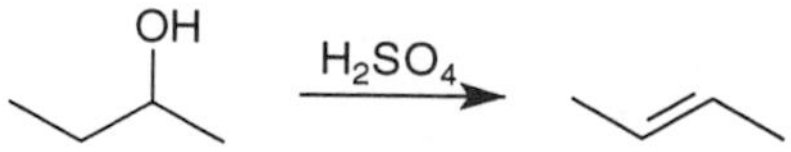

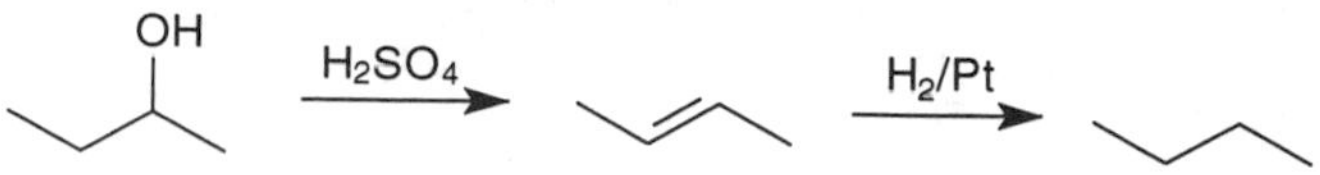

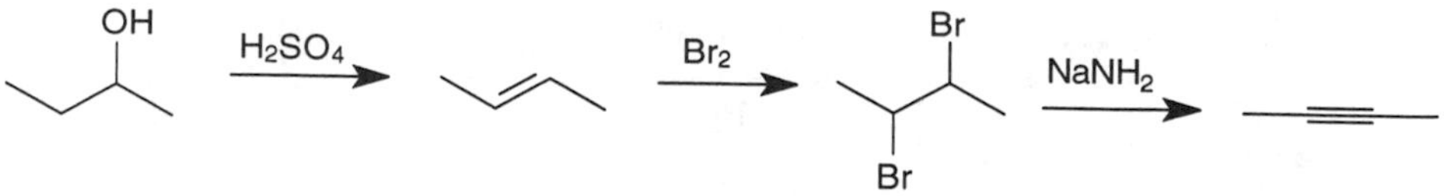

(h)

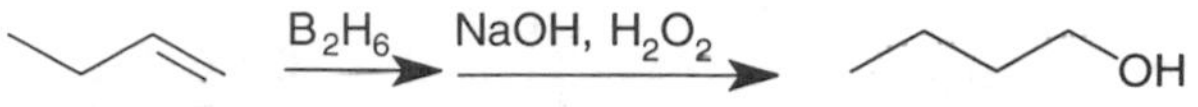
B2H6
NaOH, H2O2
OH

from part *d*

(i)

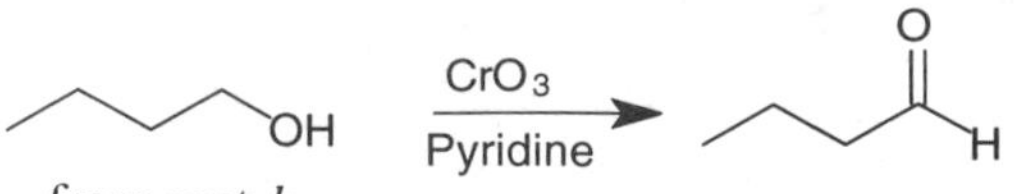
OH
CrO3
Pyridine
O
H

from part *h*

(j)

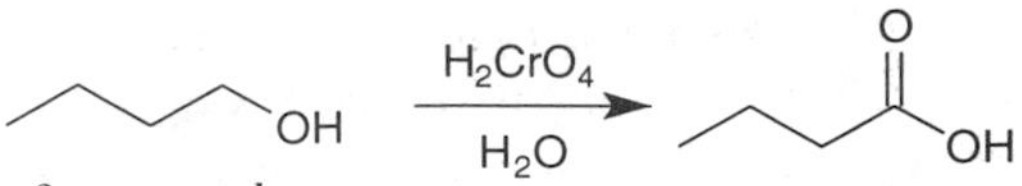
OH
H2CrO4
H2O
O
OH

from part *h*

(k)

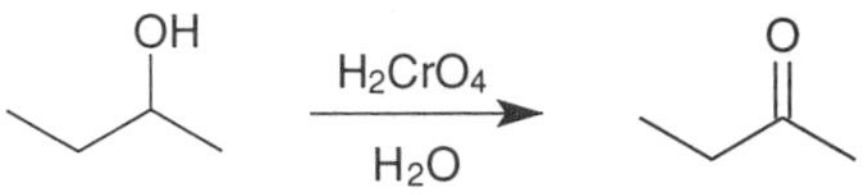
OH
H2CrO4
H2O
O

(l)

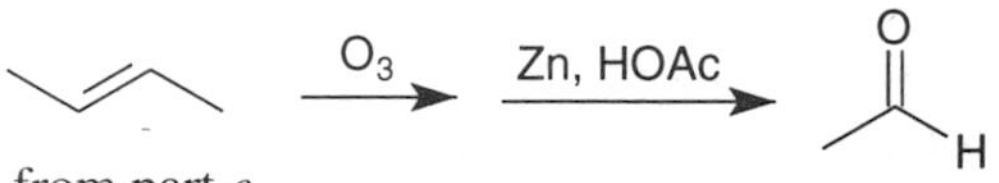
O3
Zn, HOAc
O
H

from part *e*

(m)

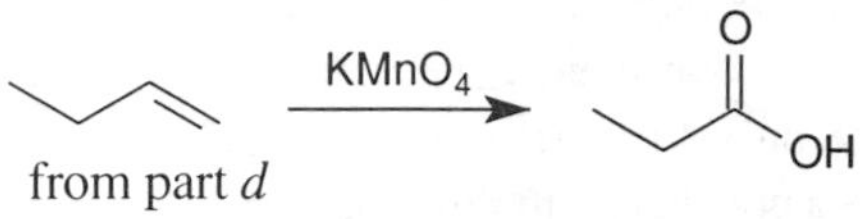
KMnO4
O
OH

from part *d*

(n)

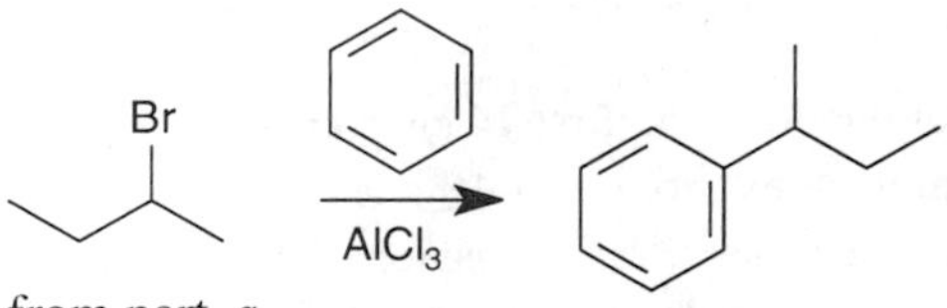
Br
AlCl3

from part *a*

(o)

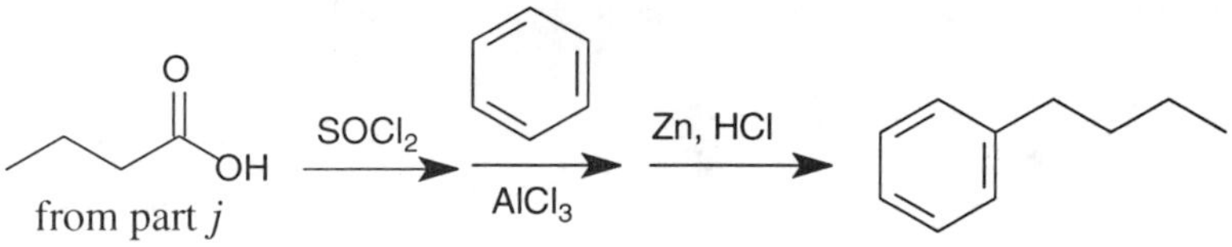
O
OH
SOCl2
AlCl3
Zn, HCl

from part *j*

(p)

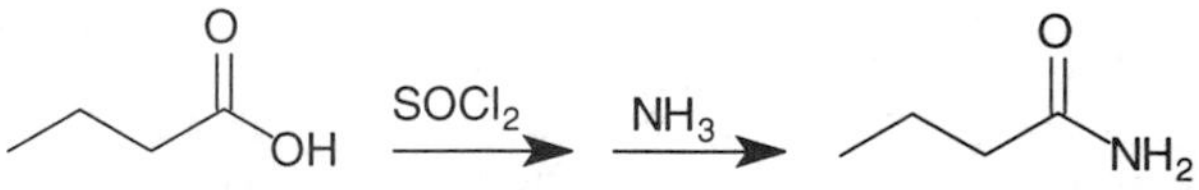
O
OH
SOCl2
NH3
O
NH2

from part *j*

(q)

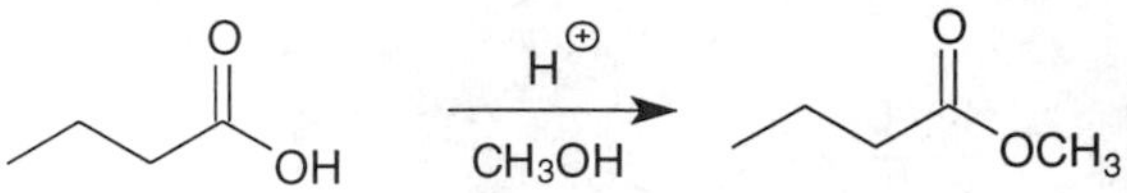
O
OH
H⊕
CH3OH
O
OCH3

from part *j*

(r)

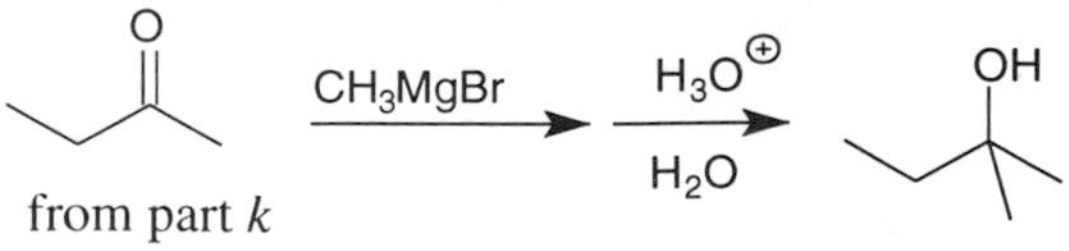
O
CH3MgBr
H3O⊕
H2O
OH

from part *k*

(s)

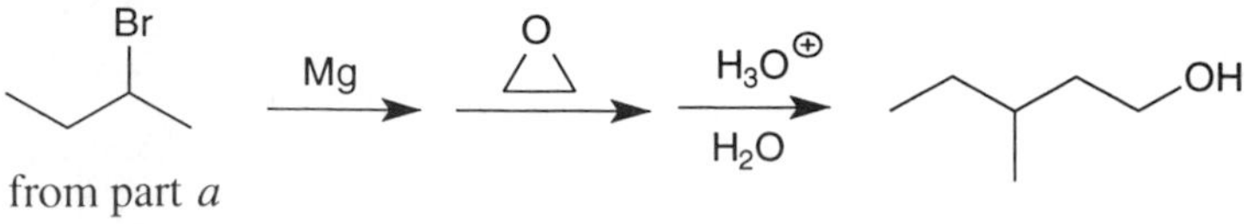
Br
Mg
O
H3O⊕
H2O
OH

from part *a*

(t) [structure: 2-bromobutane] from part *a* → Mg → CO_2 → $H_3O^{\oplus}$ / H_2O → [structure: 2-methylbutanoic acid]

Important New Terms

Baeyer-Villiger oxidation: the transformation of a ketone into an ester by reaction with a peracid; the net change is the insertion of an oxygen atom between the carbonyl carbon and an adjacent carbon (14.3)

Beckmann rearrangement: the transformation of the oxime of a ketone to an amide by reaction with a strong acid; the net change from the ketone is the insertion of an NH group between the carbonyl carbon and an adjacent carbon (14.2)

Benzilic acid rearrangement: the anionic skeletal rearrangement of an α-diketone to an α-hydroxyacid induced by treatment with aqueous hydroxide (14.1)

Claisen rearrangement: a [3,3] sigmatropic shift of a substituted allyl vinyl ether; a pericyclic reaction in which allyl vinyl ether is converted to a rearranged β,γ-enone; sometimes called an oxa-Cope rearrangement (14.3)

Cope rearrangement: a [3,3] sigmatropic shift; the process by which a new carbon–carbon σ bond is formed between C-1 and C-6 in a substituted 1,5-hexadiene at the same time that the bond between C-3 and C-4 is broken, with both π bonds shifting to take up new positions between different carbon atoms (14.1)

Cycloaddition reaction: a pericyclic reaction resulting from the combination of two separate π systems into a cyclic product (14.1)

Cycloreversion: a pericyclic reaction that is the inverse of a cycloaddition in which a cyclic molecule fragments into two or more smaller π systems (14.1)

Degenerate rearrangement: a skeletal rearrangement in which the breaking and forming of bonds lead to a product that is chemically identical to the reactant (14.1)

Electrocyclic reaction: a concerted, pericyclic, intramolecular, ring-forming reaction (14.1)

Hofmann rearrangement : the conversion of an amide to an amine containing one fewer carbon upon treatment with bromine in aqueous base (14.2)

Isocyanate: RN=C=O; an intermediate in the Hofmann rearrangement (14.2)

Lactone: a cyclic ester (14.3)

Nitrilium cation: a resonance-stabilized alkylated nitrile cation R-C≡N$^+$–R; encountered as an intermediate in the Beckmann rearrangement (14.2)

Pericyclic reaction: a concerted chemical conversion taking place through a transition state that can be described as a cyclic array of interacting orbitals (14.1)

Pericyclic rearrangement: a skeletal rearrangement proceeding through a concerted, pericyclic transition state (14.1)

Sigmatropic rearrangement: a skeletal rearrangement accomplished through the shift of a σ bond to the opposite end of a π system—for example, as in the Cope rearrangement; involves the migration of a group from one end of a π system to the other (14.1)

Sigmatropic shift: a pericyclic reaction in which a σ-bound substituent migrates from one end of a π system to the other (14.1)

Wagner–Meerwein rearrangement: a cationic rearrangement in which a carbon substituent participates in a 1,2 shift (14.1)

Multistep Syntheses

15

Key Concepts

Classification of synthetic transformations as carbon–carbon-bond forming, functional-group transformations, or oxidation–reduction reactions

Working backward in retrosynthetic analysis

Criteria for evaluating synthetic efficiency

Distinction between linear and convergent synthetic design

Recognizing functional group compatibility

Protecting groups for carbonyl compounds, alcohols, carboxylic acids, and amines

Reagents and conditions for protection and deprotection of various functional groups

Typical conditions for practical, large-scale synthesis

Answers to Exercises

Ex 15.1   These reactions represent examples of:

(a)   Reduction

(d)   Functional-group transformation

(b)   Carbon–carbon bond formation

(e)   Carbon–carbon bond formation

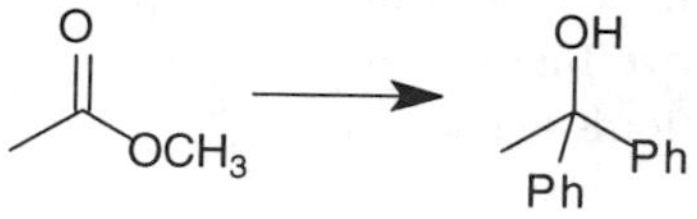

(c)   Functional-group transformation

(f)   Oxidation

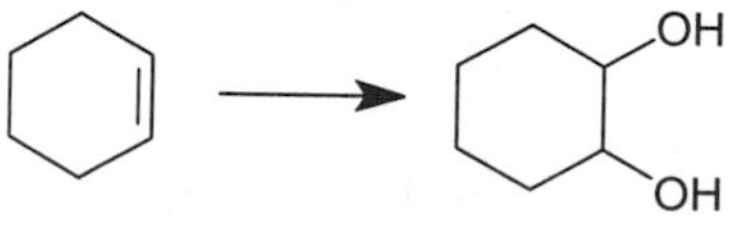

Ex 15.2   (a)   The starting material, acetone, has three carbons; the product amine has six. Therefore, synthesis of the product requires the construction of one new C—C bond. By dividing the product at the boldfaced bond, we obtain two three-carbon units. Although this Exercise does not require a synthetic route, one possibility is shown here.

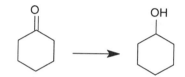

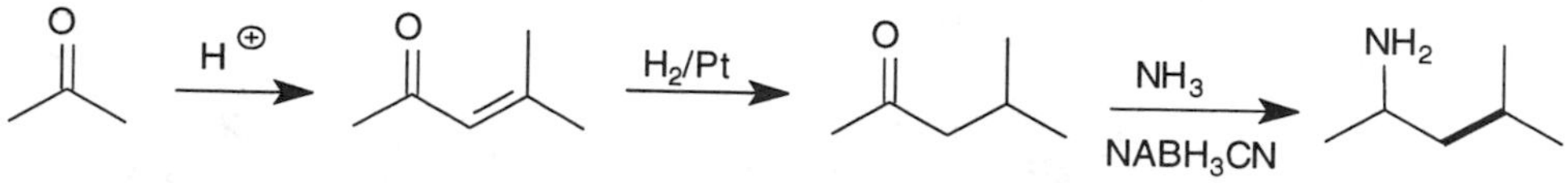

(b)   Because the starting material (methanol) has only one carbon atom, all three carbon–carbon bonds of the product (*t*-butanol) must be constructed. As before, a possible route is suggested here even though it is not required in the Exercise.

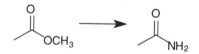

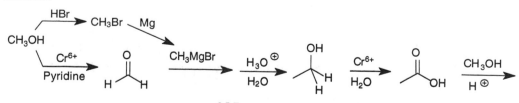

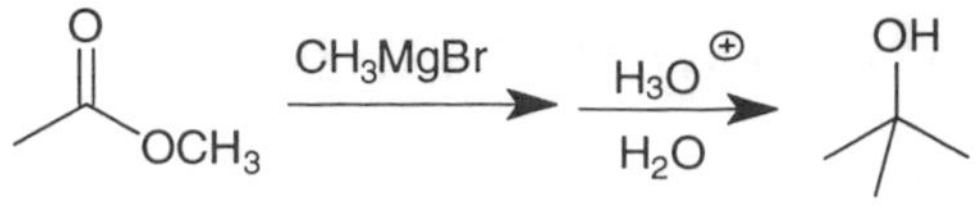

(c) Two new C—C bonds must be formed. The six-carbon product can be divided as shown with bold-faced bonds into three, two-carbon fragments. Again, possible syntheses that form these bonds are provided, although not required for the Exercise.

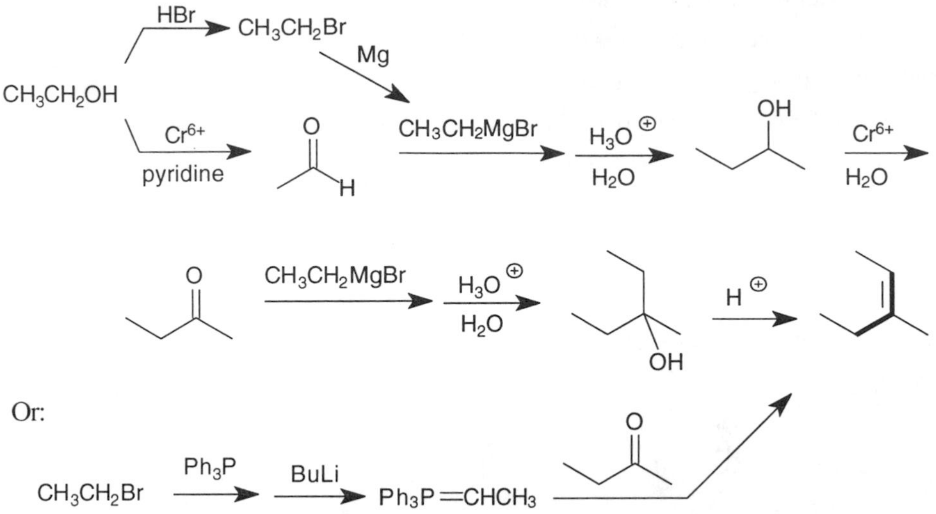

Or:

Ex 15.3 The most straightforward approach combines an unspecified one-carbon reagent with the three carbon atoms of propanal. By considering possible last steps that form carbon–carbon bonds *and* produce alcohols as products, we find that the reaction of methyl Grignard reagent with the starting aldehyde affords the desired product.

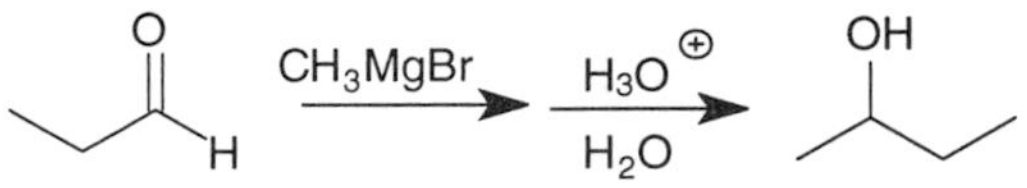

Ex 15.4 Combining two two-carbon units derived from ethanol with a one-carbon reagent leads to the five carbons present in the product. Two possible routes are shown.

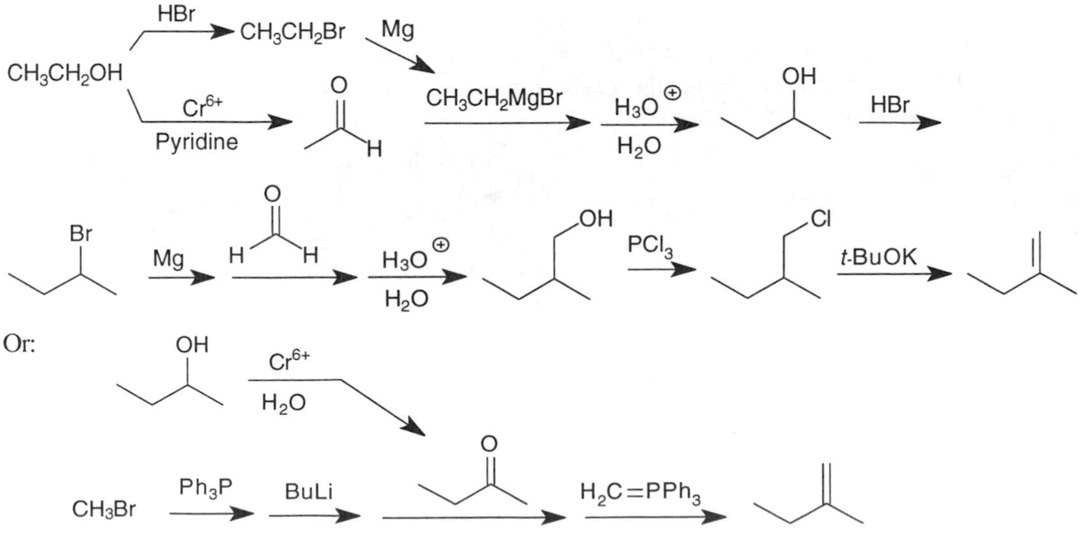

Or:

Ex 15.5 One possible set of reagents is illustrated for each step of this Exercise. Note that there are often several ways for effecting these transformations. Thus, for example, the conversion of 2-propanol to propene (shown in the upper sequence) can be accomplished either by acid-catalyzed dehydration, as shown, or by conversion of the alcohol to a tosylate ester or alkyl halide, followed by base-induced elimination (not shown). Thus, if your answer does not correspond to that shown, check with the textbook before you reject your answer.

Ex 15.6 The upper route in Exercise 15.5 involves several steps that are difficult to control and/or involve reagents that are difficult to handle.

(a) Elimination of water from 2-propanol, induced by heating with acid, leads to additional products other than the desired alkene. In addition, propene is a gas at room temperature and would be difficult to obtain in high yield with ordinary laboratory glassware.

(b) Anti-Markovnikov hydration of the alkene requires the use of diborane, a reagent that is somewhat awkward to use because it is a pyrophoric and toxic gas.

(c) Oxidation of 1-propanol to propanal requires nonaqueous conditions (Cr^{6+} in pyridine, for example) to prevent further oxidation to the carboxylic acid.

(d) The final step, reaction with methyl Grignard reagent or methyllithium, requires anhydrous conditions. The alternative (lower) route is not only shorter, but also involves more straightforward chemistry and more convenient reagents.

(e) Oxidation of 2-propanol to acetone can be carried out in water because further oxidation of ketones with Cr^{6+} is quite slow.

(f) Formation of the anion of acetone followed by alkylation with methyl iodide or methyl bromide forms the needed carbon–carbon bond.

(g) Reduction of the ketone with $NaBH_4$ completes the synthesis of 2-butanol.

Ex 15.7 The overall yield in a linear synthesis such as this one is the product of the yields in each of the individual steps; thus, $0.80 \times 0.80 \times 0.50 \times 1.00 \times 0.48 \times 0.30 = 0.046$ (4.6%). The molecular weight of sertraline is 306 and that of o-dichlorobenzene is 147.

$$\frac{10 \text{ g}}{\frac{306 \text{ g}}{\text{mole}}} = 0.046 \times \frac{X}{\frac{147 \text{ g}}{\text{mole}}} \; ; \qquad X = \frac{10 \text{ g}}{\frac{306 \text{ g}}{\text{mole}}} \times \frac{147 \text{ g}}{\text{mole}} \times \frac{1}{0.046} = 104 \text{ g}$$

Thus 104 g of starting material will be required to make 10 g of the final product.

(a) carbon-carbon bond formation

(b) carbon-carbon bond formation

(c) carbon-carbon bond cleavage

(d) reduction

(e) functional-group transformation followed by carbon-carbon bond formation

(f) functional-group transformation followed by reduction

Ex 15.8 This exercise asks us to examine the starting materials and reagents to see if some other, possibly faster reaction might take place.

(a) Treatment of the bromoketone with base results in formation of an enolate anion. However, intramolecular displacement of bromide ion by the enolate anion is faster than a bimolecular aldol condensation reaction. Both of these reactions are faster than either elimination of HBr to form an alkene or S_N2 displacement by hydroxide to form an alcohol.

(b) Dehydration of the alcohol functionality produces an equivalent of water, which can hydrolyze the ketal functional group. The unsaturated ketone will undergo proton tautomerization (from the α to the γ position), producing the conjugated ketone in competition with the desired product.

(c) As in part *a*, an intramolecular reaction, this time an esterification to form a lactone, is almost always faster than a bimolecular reaction. However, lactones are somewhat less stable than acyclic esters, and when methanol is used as solvent for the reaction, the desired methyl ester is the predominant product.

(d) Many things can go wrong with this seemingly straightforward electrophilic aromatic sulfonation. The benzylic alcohol functionality is not stable to the strongly acidic conditions required for the sulfonation (fuming sulfuric acid). The benzylic cation formed upon protonation and loss of water from the alcohol can itself act as an electrophile in a Friedel-Crafts alkylation (only the *para* product is shown, but *ortho* substitution as well as multiple substitution also occurs) and can also lose a proton to form an alkene. This alkene is styrene which will polymerize in the presence of a strong acid. Overall, then, this reaction is a mess!

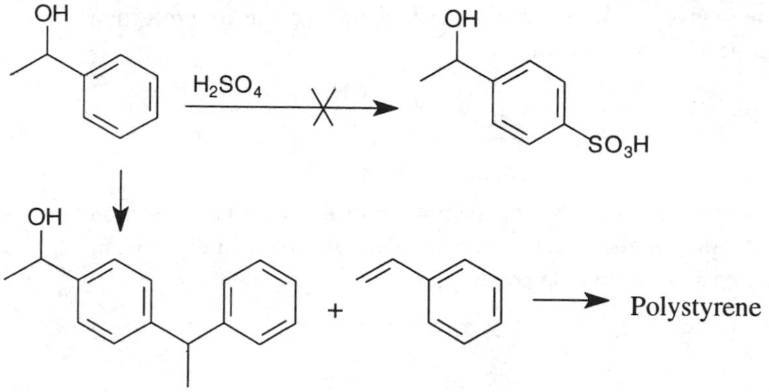

(e) Reduction of a ketone by catalytic hydrogenation is slower than reduction of an alkene. (To accomplish preferential reduction of the ketone as posed in the Exercise, we can use NaBH$_4$.)

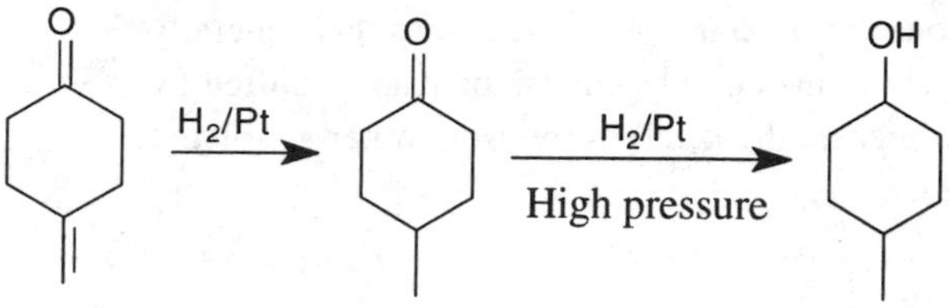

(f) Grignard reagents are not compatible with acidic functional groups. As the organometallic C—Mg bond is formed, it will undergo acid-base reaction with the alcohol, effecting net conversion to a carbon–hydrogen bond.

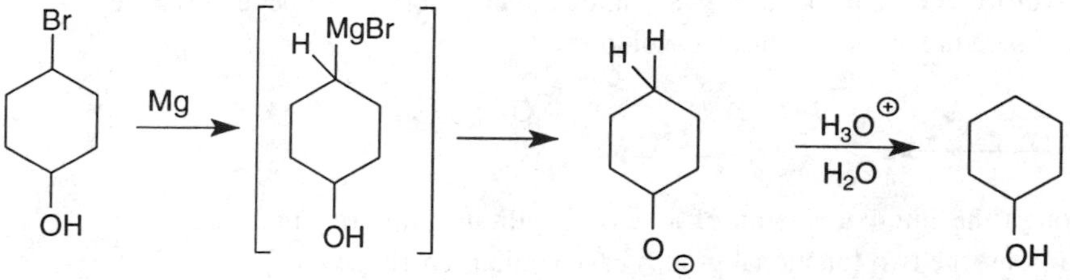

(g) Two different elimination products are formed in this reaction. The desired product is not stable to base and isomerizes to the more stable conjugated isomer. Both esters hydrolyze to form a mixture of regioisomeric carboxylate salts.

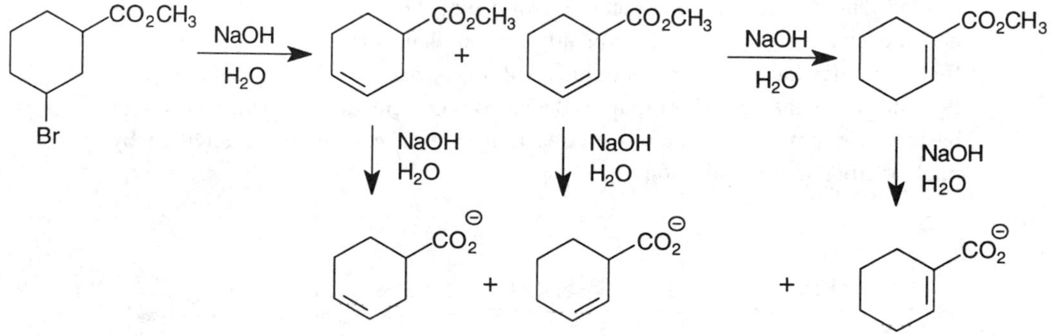

Ex 15.9 Using the three criteria given in the Exercise for evaluating reactions for protection of functional groups leads to the following:

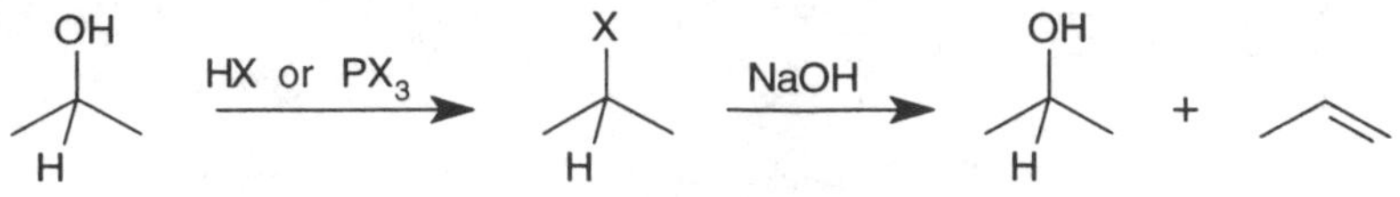

No. Conversion of an alcohol to an alkyl halide is a reasonably high yield reaction (typically >85%) by the reverse reaction almost always results in significant amounts of elimination and substitution products.

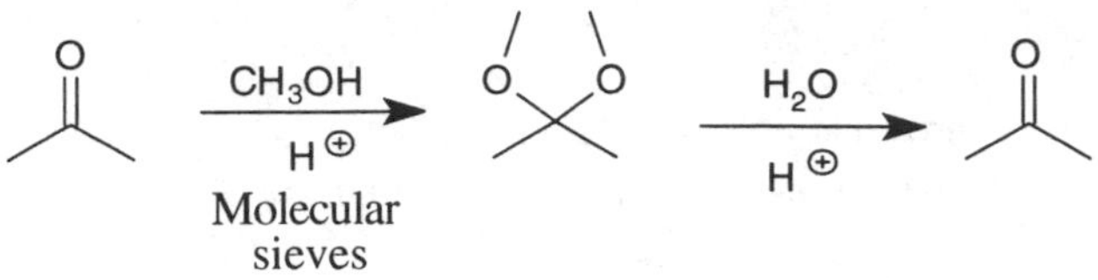

- -

Yes. The interconversion of a ketone and a ketal involves reactions that generally have few side reactions. The position of the equilibrium can be readily shifted by removing water as it is formed (to prepare the ketal) or by using water as solvent or co-solvent (to reform the ketone).

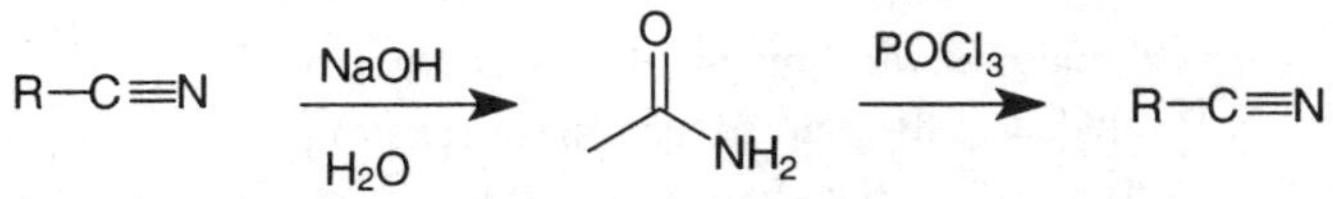

- -

No. The hydrolysis of a nitrile to form an amide requires highly basic conditions at elevated temperature, relatively harsh conditions.

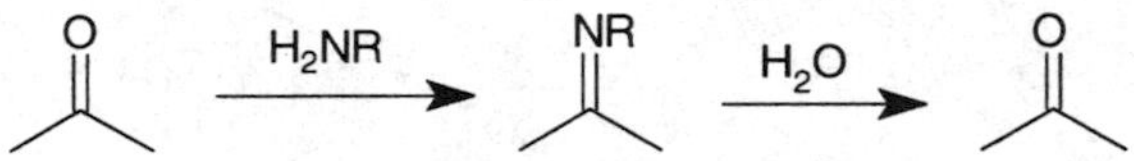

No. Although the interconversion of a ketone with an imine requires relatively mild conditions, the two functional groups have similar reactivities.

- -

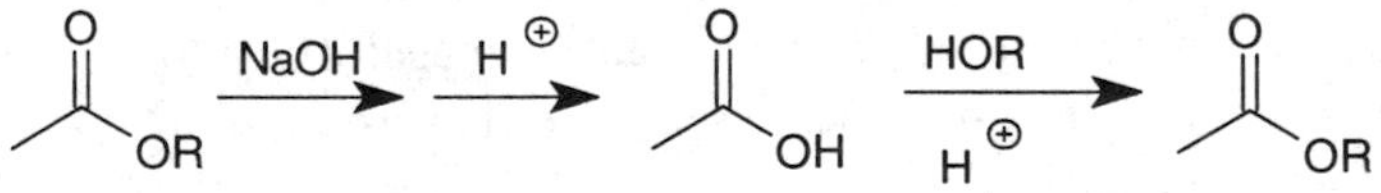

Maybe. The interconversion of an ester with an acid occurs under mild conditions and the yields are quite high, typically greater than 90%. However, these two functionalities have similar reactivities with many reagents. Two exceptions are: the ester lacks the acid OH group of the carboxylic acid; and the acid reacts with basic nucleophiles in an acid-base reaction (proton exchange) rather than by nucleophilic, acyl substitution.

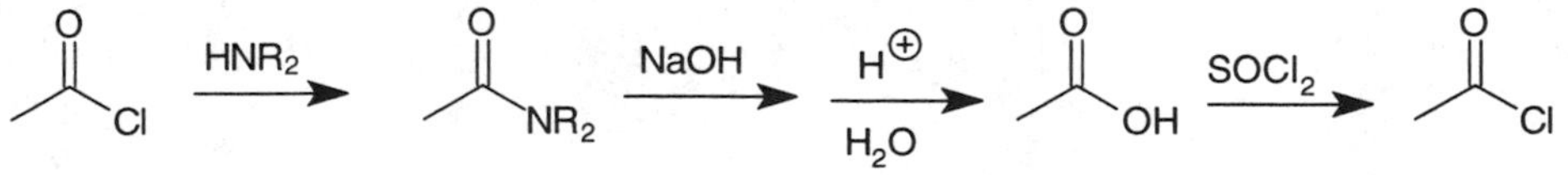

- -

No. An acid chloride is such a reactive functionality that it reacts more rapidly with all reagents than does a carboxylic acid amide. Further, the amide is not a good protecting group for an acid chloride because the hydrolysis of an amide to a carboxylic acid (an essential part of reformation of the acid chloride) is very slow, even with concentrated base at elevated temperatures.

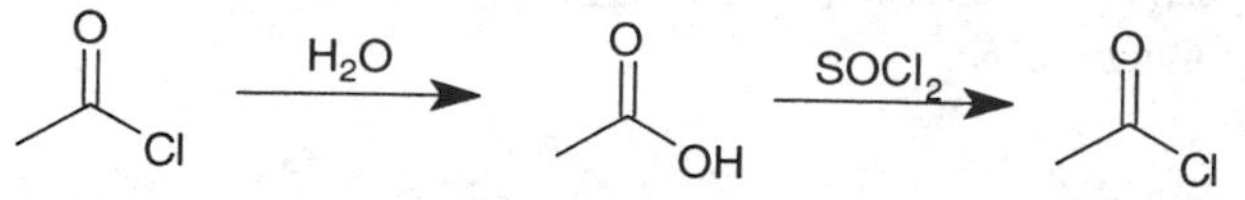

No. For the reasons discussed above, an acid chloride is not an appropriate protecting group for a carboxylic acid. In the reverse sense, the carboxylic acid is less reactive than the acid chloride toward nucleophilic, acyl substitution, but it reacts as or more rapidly by an acid-base reaction with most nucleophiles.

Ex 15.10 Benzylic ethers are unusually easy to hydrolyze because of the stability of the intermediate benzylic cation. Thus, the carbon-oxygen bond to this benzylic carbon is broken in the reaction rather than the carbon-oxygen bond to the CH_3 group, which would result in an unstable, methyl cation. If the second R group is also benzylic, both bonds cleave with comparable facility.

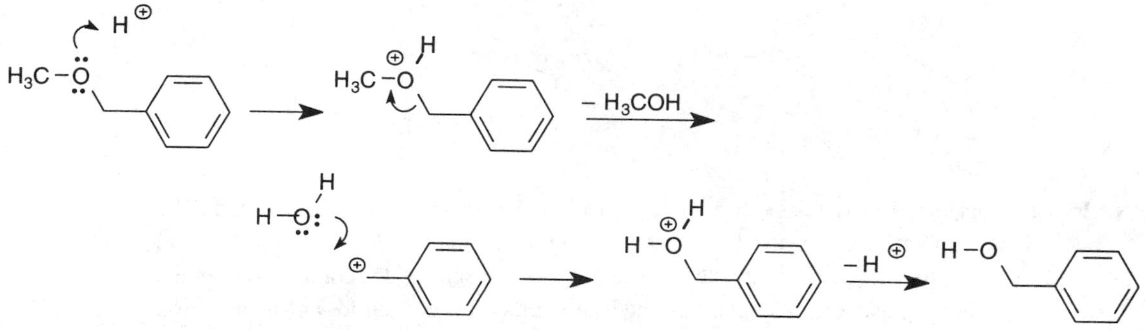

Ex 15.11 The formation of an ester from the reaction of an alkene with a carboxylic acid is an example of an electrophilic addition. It begins by protonation of the alkene to form, in this case, a tertiary cation that is then trapped by the carboxylic acid. (The carbonyl oxygen of the carboxylic acid is shown here as the nucleophilic oxygen. Can you think of a reason why the two oxygen atoms differ in nucleophilicity?)

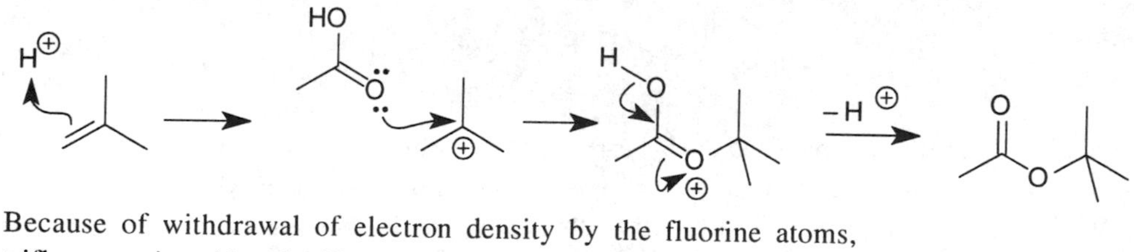

Ex 15.12 Because of withdrawal of electron density by the fluorine atoms, trifluoroacetic acid and trifluoroacetate ion are each less nucleophilic than the corresponding acetic acid species. Without a good nucleophile present, the *t*-butyl cation loses a proton to form isobutylene.

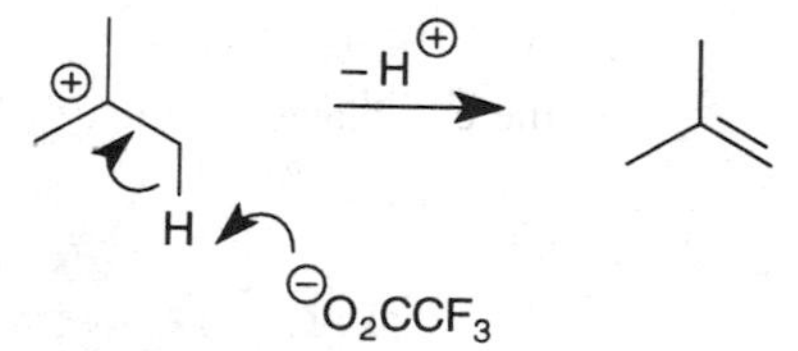

Ex 15.13 All three functional groups have in common the first three types of resonance structures shown below. The uncharged ones on the left are the most important. The second structures are the least important because they have both charge separation and one fewer bond. The third structures have similar stabilities in urethanes and amides, but the analogous resonance structure in esters contribute less significantly to the hybrid because positive charge is placed on the more electronegative oxygen atom rather than on nitrogen. The urethane has a fourth

contributing resonance structure, shown at the right. The urethane thus has the most negative charge on the carbonyl oxygen; the ester has the least. The amide has a larger fraction of partial positive charge on nitrogen than does the urethane, because there are three competing resonance structures that place negative charge on oxygen for the amide and only two for the urethane.

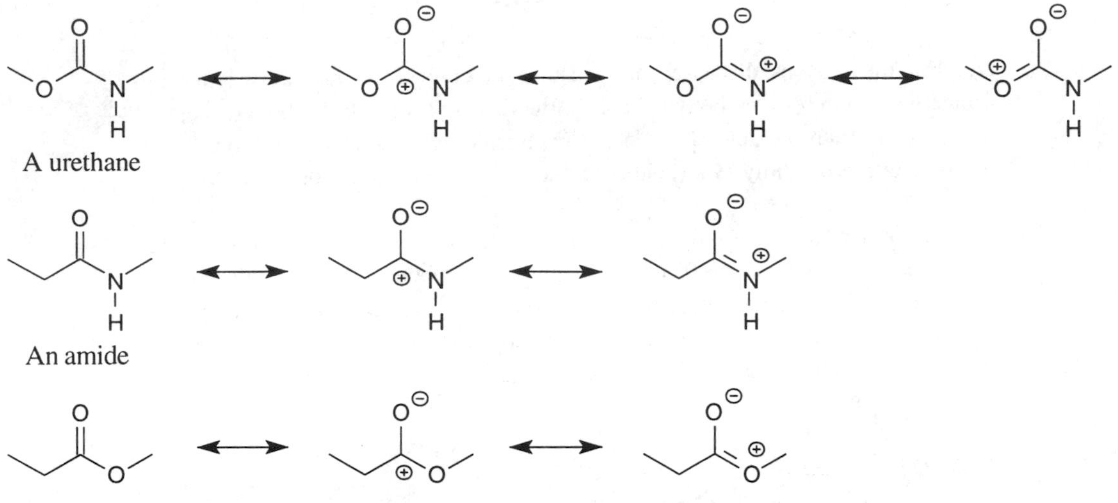

A urethane

An amide

An ester

Ex 15.14 This reaction is faster in an acidic medium where nitrogen is protonated. The neutral amine is a better leaving group than the amide ion (RNH⁻), which would be the leaving group in the absence of protonation. Indeed, loss of a neutral amine from the positively charged intermediate is much faster than loss of a carbanion from a simple carboxylic acid.

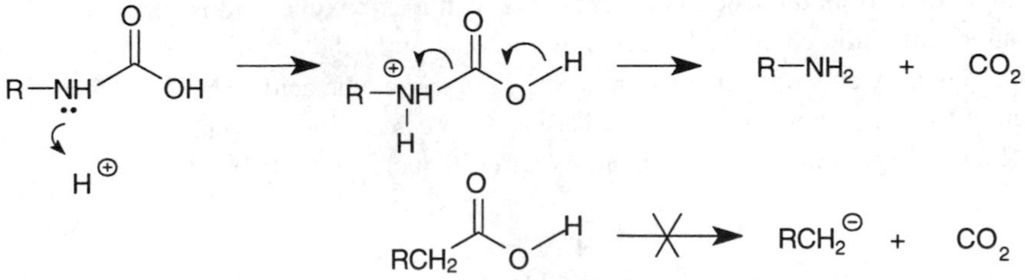

Ex 15.15 The reaction of a nitrile with base and diethyl carbonate begins by formation of the anion of the nitrile by α deprotonation. The resulting anion then acts as a nucleophile, attacking the carbonyl carbon of diethyl carbonate. Loss of ethoxide ion from the resulting tetrahedral intermediate results in net nucleophilic acyl substitution, forming the product nitrile ester.

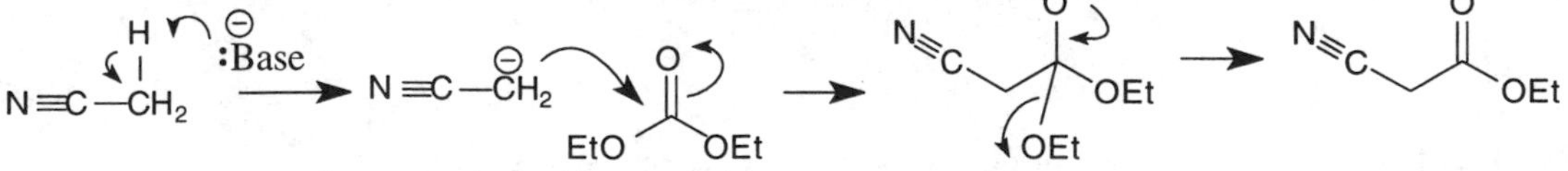

Ex 15.16 As we saw in Exercise 15.15, the reaction of a nucleophile with dimethyl carbonate effects nucleophilic acyl substitution, whereas with dimethyl sulfate the reaction is an S_N2 displacement. These two reactions follow different pathways because of the stability of the leaving groups formed. The mesylate ion formed by S_N2 displacement is considerably more stable than the methoxide ion lost in

nucleophilic acyl substitution. and the C—O bond in dimethyl sulfate is more polarized and weaker than that in dimethyl carbonate.

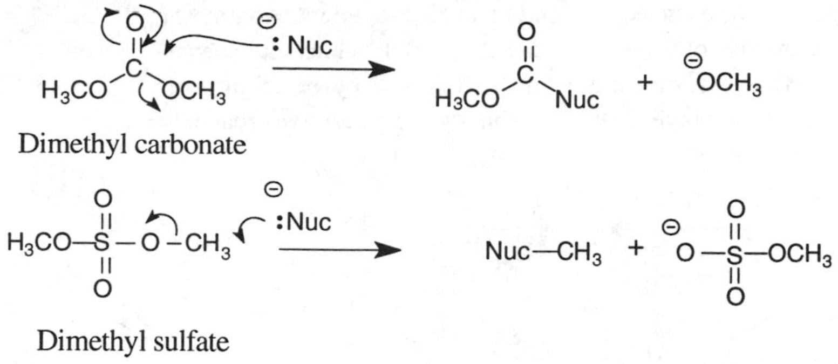

Dimethyl carbonate

Dimethyl sulfate

Ex 15.17 After the formation of the amide, the alkylation is limited to the introduction of a single methyl group, whereas methylation of a primary amine will also produce significant amounts of the di- and trialkylamine. Furthermore, the amines in question are substituted anilines that, because of resonance delocalization of lone-pair electron density, are less reactive than alkyl amines toward electrophiles. In the case of the amide, a more reactive nucleophile can be easily formed by deprotonation, generating a resonance-stabilized anion.

Answers to Review Problems

Pr 15.1
  (a) Simultaneous oxidation and reduction (no net change)

  (b) Functional group transformation

  (c) Oxidation

  (d) Reduction

  (e) Carbon–carbon bond formation

  (f) Functional-group transformation followed by reduction

  (g) Functional-group transformation

Pr 15.2
  (a) $H_2O + H_3O^+$

  (b) CH_3NH_2

  (c) $KMnO_4$, H_2O; or H_2CrO_4, H_2O

  (d) (1) $LiAlH_4$, ether (2) H_2O; or H_2/Pt, 2000 psi, 100 °C

  (e) $NaOCH_3$

  (f) CH_3NH_2, $NaBH_3CN$, H_2O

  (g) $(CH_3)_2NH$ (large excess for efficient conversion of the alkyl halide)

Pr 15.3 Because both methanol *and* ethanol must be incorporated into the product, analyze the problem by conceptually dissecting a two-carbon unit from the product. This reveals two carbon atoms that are not connected. Thus, one route would involve two molecules of methanol and one of ethanol:

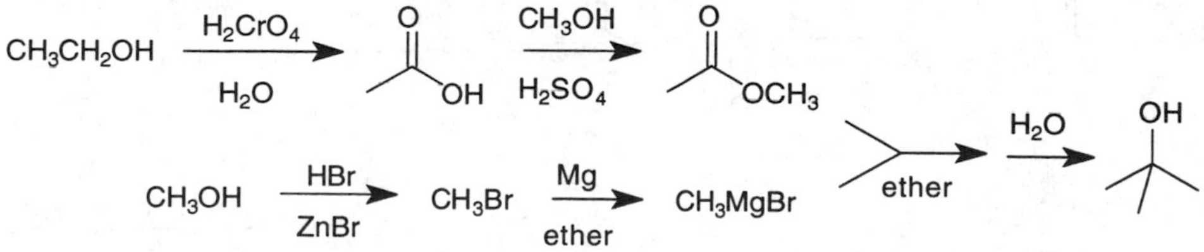

Pr 15.4 Dissect the product into two two-carbon units because ethanol is the sole source of carbon. To form the one additional carbon–carbon bond in the product, ethanol must be converted separately to a nucleophile and to an electrophile. Oxidation of ethanol with Cr^{6+} in the absence of water forms the electrophilic aldehyde ethanal, whereas reaction with HBr to form bromoethane, followed by reaction with lithium, forms the nucleophilic alkyllithium. Combination of these two reagents produces the desired product, 2-butanol.

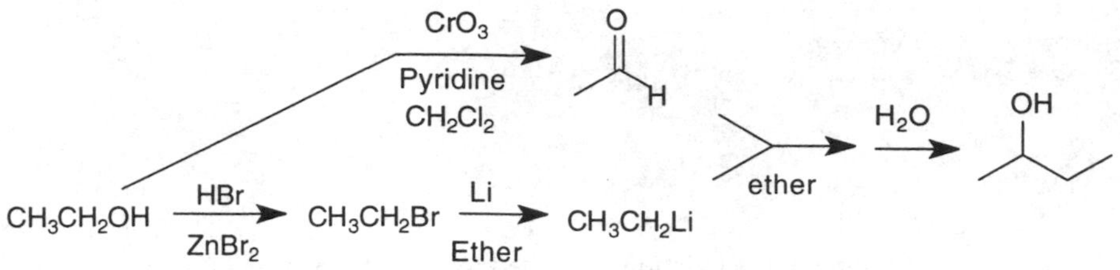

Pr 15.5 (a) In this transformation, the more reactive ketone can be protected from reaction with a Grignard reagent by conversion to a ketal. After formation of the tertiary alcohol by reaction with methyl Grignard, the ketal is hydrolyzed, reforming the ketone group.

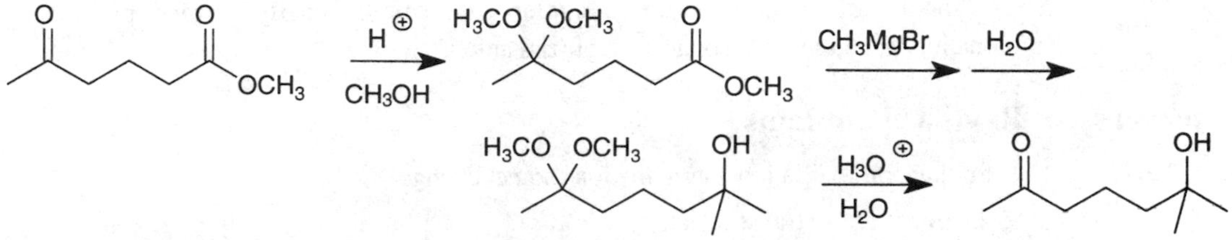

(b) The alcohol functionality must be changed to one that is compatible with a Grignard reagent. A benzyl ether is a suitable protecting group.

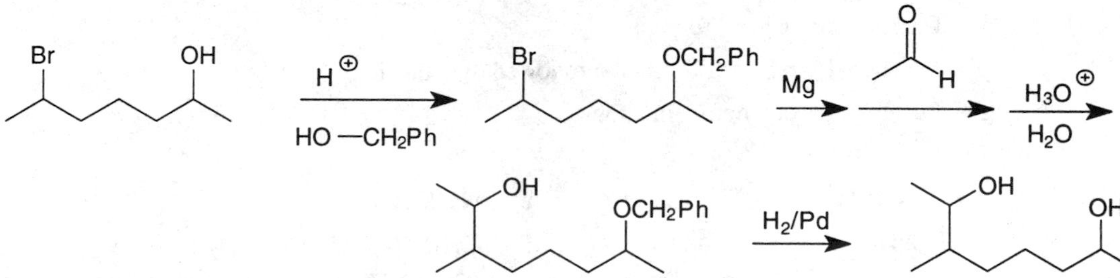

(c) Attempts to convert the ester functional group of an aminoester to an amide results instead in the production of polymer. The amino group must first be protected so that the nitrogen is less nucleophilic. Protection as an amide (by reaction with a carboxylic acid chloride) is adequate here because a tertiary amide is much more difficult to hydrolyze than a secondary amide. (We will see a better protecting group for an amine in Chapter 18.)

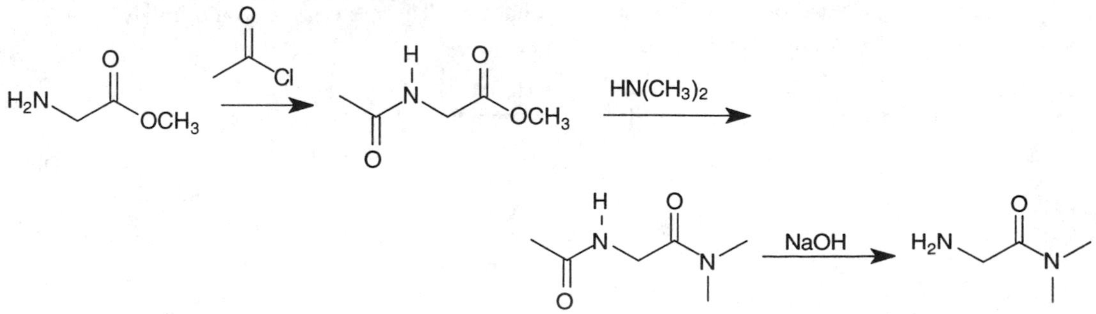

Pr 15.6 These reactions were discussed in detail in Chapter 12.

Ketal formation:

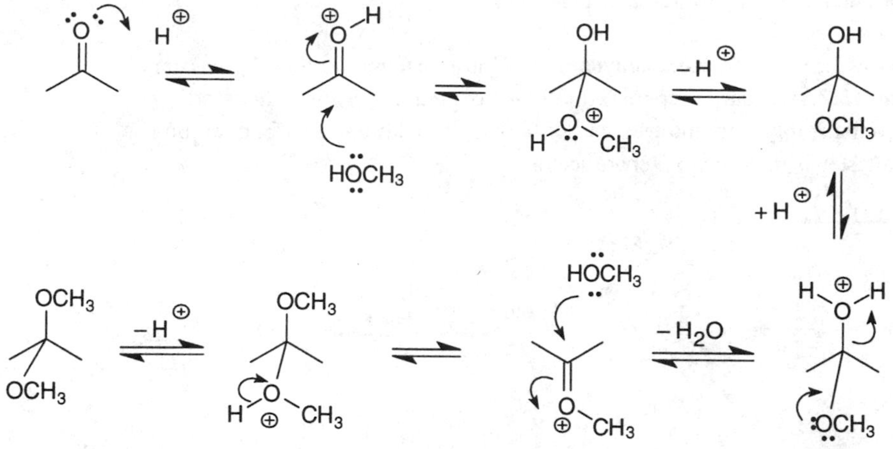

Ketal hydrolysis:

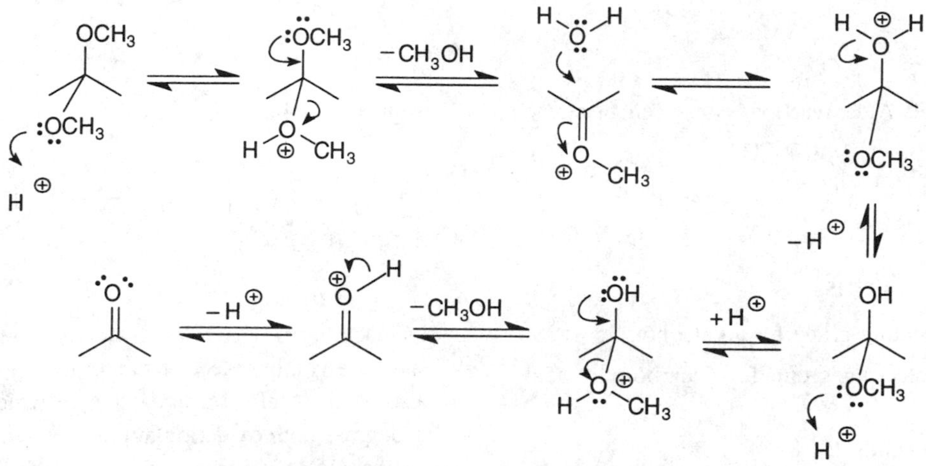

Pr 15.7 As in most syntheses, many routes are possible. Because the product has a *cis* double bond, we can use a Wittig reaction in the synthesis. Because the product has five carbons and our carbon sources contain one and two carbons, it is reasonable to condense a three-carbon ylide with a two-carbon aldehyde (as shown here) or a two-carbon ylide with a three-carbon aldehyde (not shown).

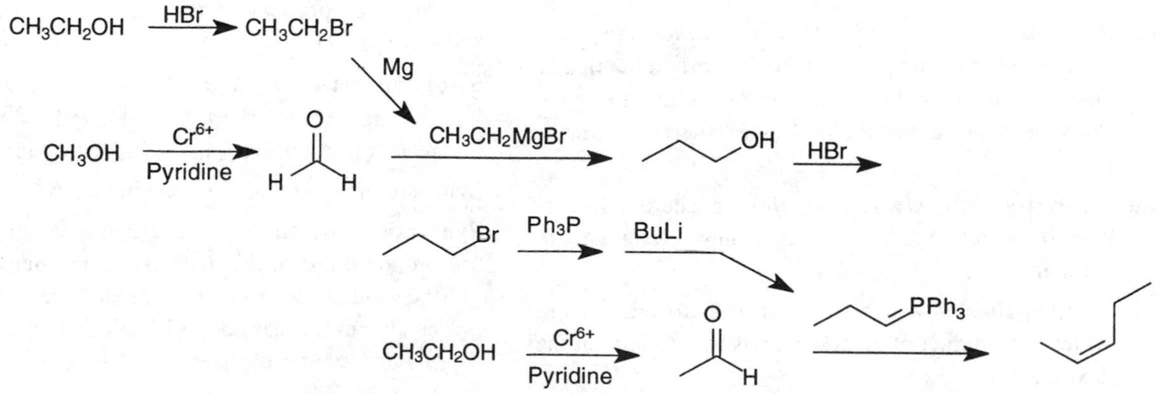

Pr 15.8   The product, *cis*-2-pentene, has a boiling point of 37 °C. Therefore, it is difficult to handle and to isolate from typical reaction solvents, such as ethyl ether (b.p. 35 °C), dichloromethane (b.p. 40 °C), and THF (b.p. 67 °C).

Pr 15.9   By substituting acetone for acetaldehyde in the Wittig reaction used in the answer to Problem 15.7, we can prepare 2-methyl-2-pentene. Because the starting materials contain only one and two carbon atoms, an additional carbon–carbon bond-forming step is required to prepare acetone.

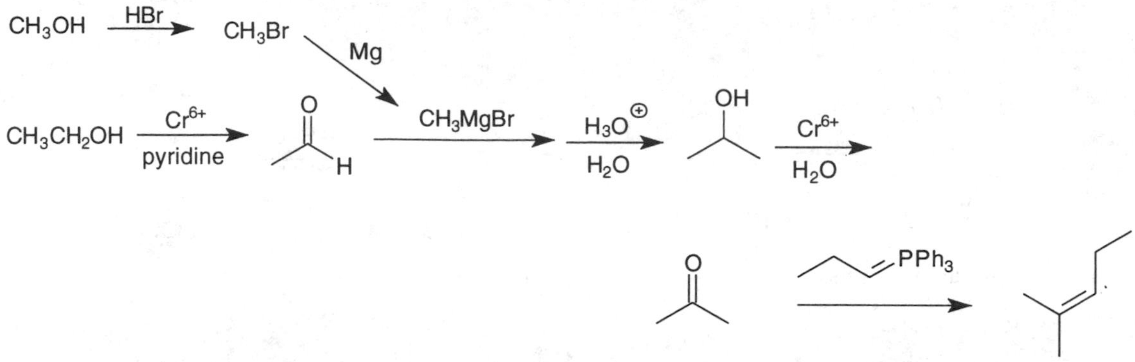

Pr 15.10   The Diels-Alder reaction is an excellent way to make substituted cyclohexenes.

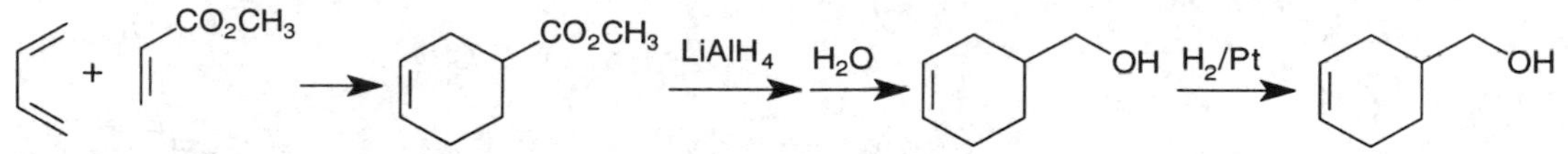

Important New Terms

Benzyl ether: a protecting group for an alcohol (15.8)

t-Butyl ester: a protecting group for a carboxylic acid (15.8)

Carbamate: *see* **Urethane** (15.8)

Carbon–carbon bond-forming reaction: a chemical transformation in which two previously unconnected carbon atoms become covalently bound (15.1)

Convergent synthesis: a branched synthesis in which two or more synthetic intermediates react with each other (15.6)

Functional-group compatibility: descriptor of a reagent or reaction that is sufficiently chemically selective so that only the desired functional group (of the several present in the molecule) interacts with the reagent (15.7)

Functional-group transformation: a chemical reaction in which one functional group is changed to another (15.1)

Linear synthesis: a sequence of transformations in which the product of one reaction is the reactant in the next reaction (15.6)

Oxidation–reduction reaction: a chemical transformation in which the oxidation level of a reactant and

its reaction partner are equivalently changed, with one substrate gaining electrons and the other losing them; also used to refer to a reaction in which a substrate undergoes both oxidation at one atom and reduction at another (15.1)

Pharmaceuticals: biologically active compounds sold by a drug company; may be synthetic, semi-synthetic, or obtained from natural sources (15.7)

Retrosynthetic analysis: a method for planning an organic synthesis by working backward, step-by-step, from the product to the possible starting materials (15.1, 15.2)

Semi-synthetic: a naturally occurring (or cultured) material that is chemically altered, sometimes in a relatively minor way, in the laboratory (15.7)

Synthetic: prepared in the laboratory (15.7)

Synthetic efficiency: the evaluation of the utility of a proposed synthesis; depends on the number of steps, the yield of each step, the ease and safety of the reaction conditions, the ease of purification of intermediates, and the cost of starting materials, reagents, and personnel time (15.6)

Urethane: a protecting group for an amine (15.8)

Polymeric Materials

Key Concepts

    Mechanisms of cationic, anionic, and radical polymerization reactions

    Mechanisms of condensation polymerizations

    Distinguishing physical properties of linear and branched polymers

    Contrast between addition and condensation polymerizations

    Consequences of living polymerization techniques

    Methods for crosslinking: vulcanization and difunctional monomers

    Structures, stabilities, and synthesis of synthetic polymers: polyols, poly(ethylene glycol)s, polyethers, polyacetals, polyesters, polyurethanes

    Structures of biological polymers: polysaccharides, peptides, proteins, nucleic acids

    Cyclization equilibria between open-chain and hemiacetal forms of carbohydrates

    Stereochemistry of linkages between saccharide units

    Effect of the stereochemistry of anomeric linkages on the structures and physical properties of sugars

    Intramolecular hydrogen bonding in peptides and proteins: α-helix formation

    Intermolecular hydrogen bonding in peptides and proteins: β-pleated sheet formation

    Primary, secondary, tertiary, and quaternary structure of proteins

    Stereochemistry of polymers with chiral centers: preparing atactic, isotactic, and syndiotactic polymers

    Ziegler-Natta polymerization

Answers to Exercises

Ex 16.1   The cyclic bis-lactone of lactic acid is shown below at the right. The mechanism for its formation from the dimer of lactic acid is shown below. The steps in this process begin with activation of the carbonyl group of the carboxylic acid by protonation, followed by nucleophilic attack by the hydroxyl group α to the other carboxylic acid, to form a tetrahedral intermediate.

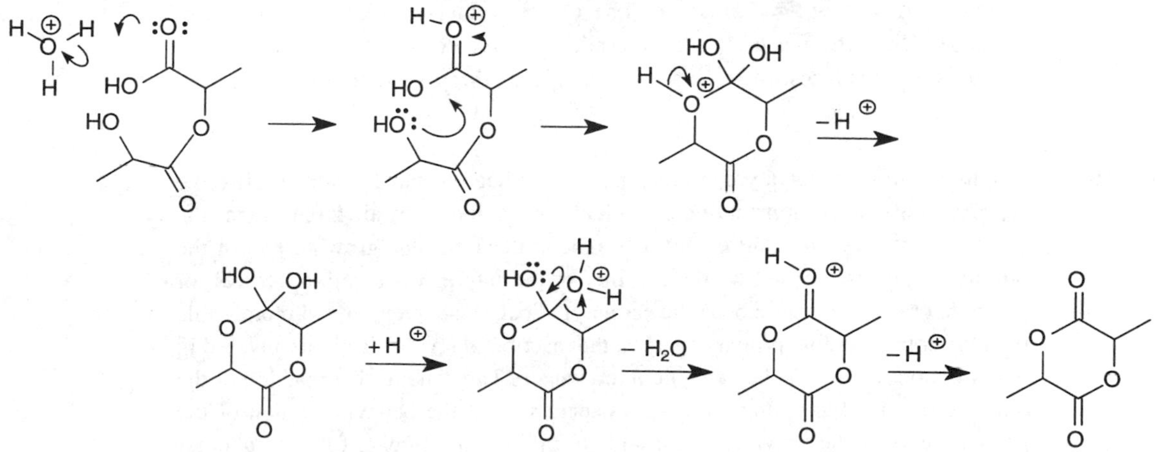

bis-Lactone

Ex 16.2   Radical terminations attained by combination of a polymer chain radical with an initiating radical X· or with another polymeric radical are both highly exothermic, so the relative probability of termination by each of these competing pathways depends on the ratio of available X· (rate is proportional to the product [X·] [R·]) and of polymer radicals (rate is proportional to the product [R·] [R·]. The initiating radical X· is produced by thermal or photochemical activation of the initiator, and as the initiation reaction continues, progressively higher concentrations of X· become available. These radicals (X·), however, are consumed when they become chemically bound to the end of a polymer chain as the polymerization begins. Thus, the available concentration of X· depends on the relative rates of initiation and of attack on ethylene. The rate of termination by combination of radical X· with the radical end of a polymer chain remains relatively constant, because the increase in the concentration of chains compensates for the decrease in concentration of X·. As the initiation proceeds, however, the number of polymer radical chains does continue to increase until termination reactions begin to dominate. Because the termination by X· is first order in [R·], whereas the termination by radical combination with itself is second order in [R·], the latter process becomes more favorable as the reaction proceeds.

Ex 16.3 Termination of the growth of a polymer chain requires only that the high reactivity character of a free radical be lost. This occurs in several ways, including the loss of a hydrogen atom from the radical to form an alkene, as shown below at the right. Concurrently, the hydrogen is transferred to another radical, as shown below at the left. Thus, by this process two radicals disappear, and two chains are terminated. These pathways consume radicals in the same way as in radical combination, as discussed in the answer to Exercise 16.2, in which two radicals become joined by a covalent bond.

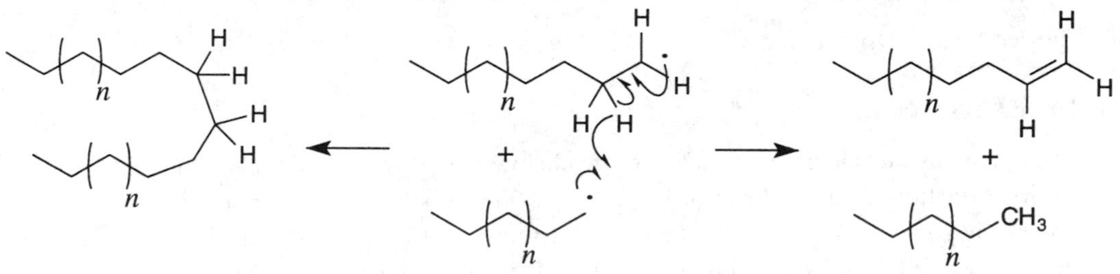

The net bonding change for hydrogen transfer between two carbon radicals is the formation of a new C=C π bond (one C—H bond is broken and another is formed). Thus, the reaction is exothermic by about 63 kcal/mole. The radical combination produces a C—C σ bond, a process that is exothermic by approximately 83 kcal/mole.

Ex 16.4 Polymerization of polystyrene takes place in a head-to-tail fashion in all three types of polymerization: cationic, radical, and anionic. In all three cases, the regiochemistry of the additions of a styrene molecule to the growing end of the polymer chain is dictated by the stability of the intermediate cation, radical, or anion that results. Because benzylic cations, radicals, and anions are all more stable than the corresponding primary species, the reaction of styrene is highly favored in the direction that forms the benzylic intermediate. Thus, the further reaction of the reactive intermediate (almost) always occurs so that the benzylic carbon of the growing end of the polymer chain adds to the end of styrene, forming another benzylic intermediate.

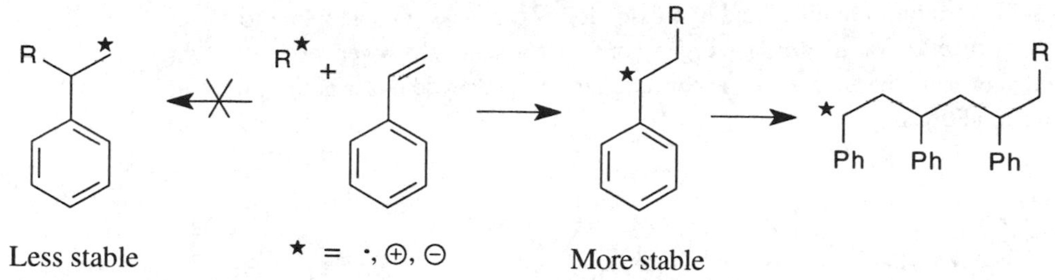

Less stable $\star = \cdot, \oplus, \ominus$ More stable

Ex 16.5 (a) The shortest repeat unit for Teflon is $-CF_2-$, not $-CF_2CF_2-$.

(b) The direction of polymerization cannot be determined by examining a central portion of a polymer, because the resulting polymer has an alternating structure in which CH_2 units alternate with a substituted carbon, with no indication whether the bond to a given CH_2 unit was present initially or formed during the polymerization. For example, it is not possible to determine from the repeat unit alone whether one of the carbon–carbon bonds emanating from the carbon bearing the $-CN$ group in Orlon was formed during polymerization or if it was present in the monomer.

Ex 16.6 Shown below are segments of all-*trans*- and all-*cis*-polyisoprene. Because each of these chains has a regular, repeating structure, two or more polymer chains can approach each other more closely than in a polymer with random *cis* and *trans* double bonds. The stronger van der Waals attractive interactions that result causes these regular polymers to be more viscous than one with random *cis* and *trans* geometries about the double bonds.

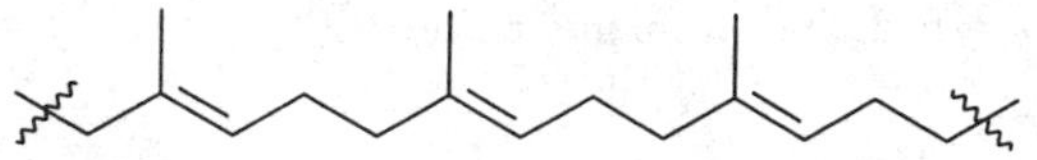

All-*trans*-polyisoprene

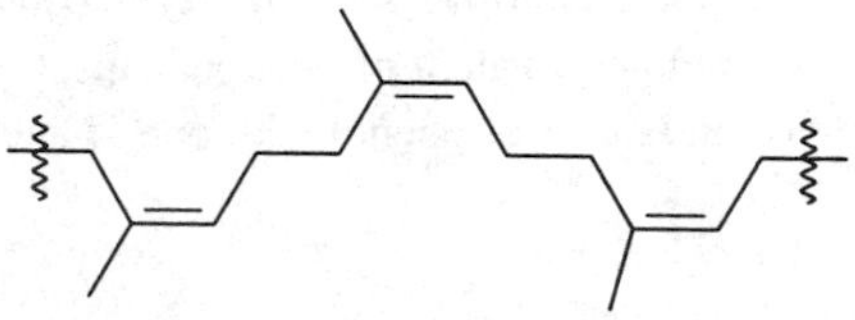

All-*cis*-polyisoprene

Ex 16.7 The difference in water solubility between poly(vinyl alcohol) and poly(ethylene glycol) results from the different functional groups present in each. Both contain oxygen functional groups and therefore act as hydrogen bond acceptors. However, the hydroxyl groups of poly(vinyl alcohol) also act as hydrogen bond donors, and because of these additional interactions, this polymer is more soluble in water. Likewise, poly(vinyl alcohol) can hydrogen bond with itself, and these intermolecular interactions increase the attractive interactions between polymer chains [as compared with poly(ethylene glycol)], reducing solubility in hydrocarbon solvents. These additional interactions contribute to a higher viscosity.

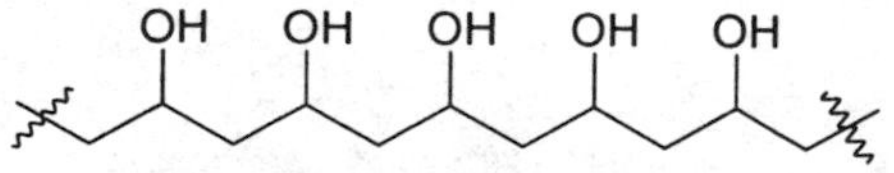

Poly(vinyl alcohol)

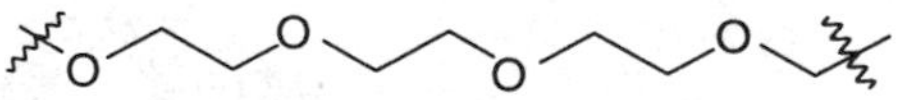

Poly(ethylene glycol)

Ex 16.8 Trioxane is a cyclic triacetal. Therefore, it undergoes reactions typical of acetals, including acid-catalyzed hydrolysis to release the carbonyl compound. The slowest step is the acid-catalyzed cleavage of one of the carbon–oxygen bonds that results in an acyclic diol after loss of a proton. This diol then undergoes relatively rapid, sequential loss of one-carbon units from one of the ends until it is completely converted to formaldehyde. This first step cannot be accelerated by base, as there is no acidic proton to remove, and hydroxide ion is too weak a nucleophile to effect an S_N2 reaction on a carbon–oxygen bond of trioxane. However, after the first bond is broken, the acyclic diol is converted to formaldehyde under both acidic and basic conditions.

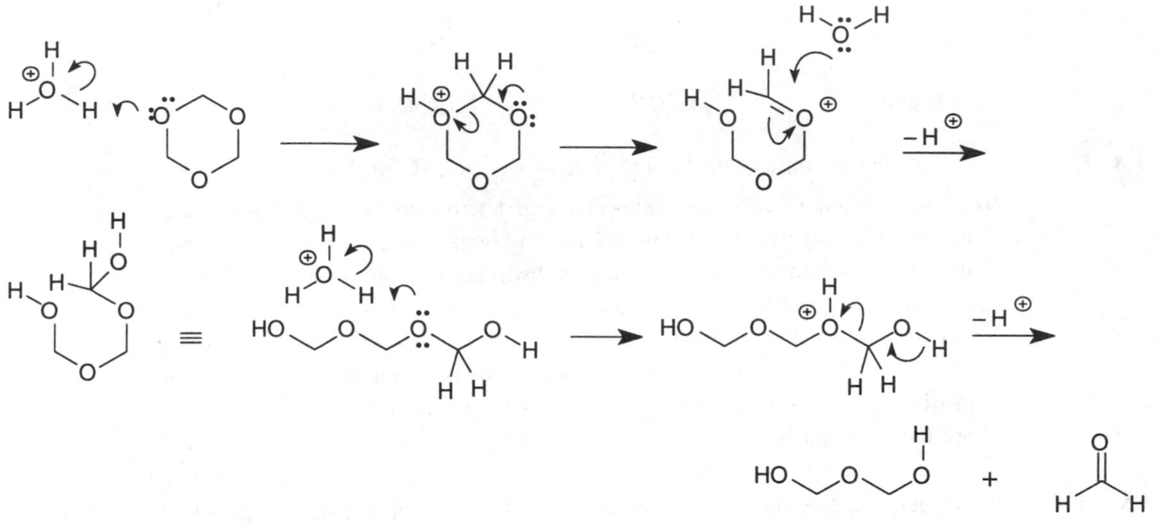

Ex 16.9   The reaction of acetone with phenol under acidic conditions resembles a Friedel-Crafts alkylation. In the first step, protonated acetone acts as an electrophile, resulting in the formation of a diol. (Recall that the OH group of phenol is a strongly activating, *ortho, para* director.) This benzylic alcohol is readily converted under acidic conditions to a benzylic cation that acts as an electrophile in attacking the second equivalent of phenol. Thus, bisphenol A is derived from the reaction of two equivalents of phenol with one of acetone.

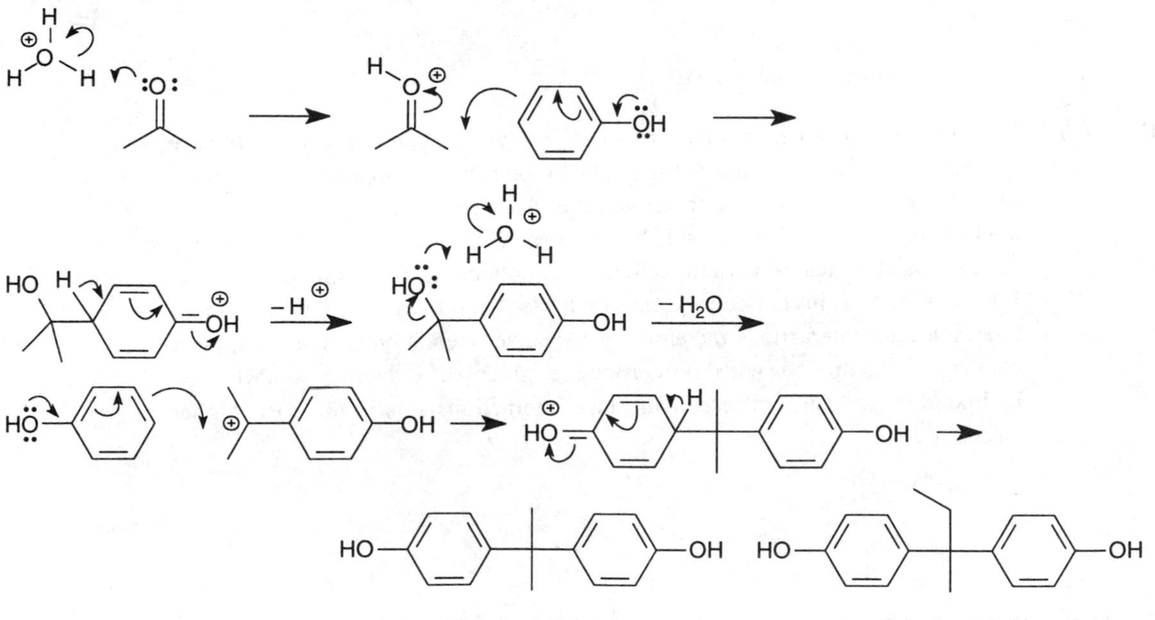

Bisphenol A   Bisphenol B

Bisphenol B is almost identical to bisphenol A. It is derived from *butanone* and phenol and thus has one more carbon.

Ex 16.10   The formation of polymer from bisphenol A and both dimethyl and diphenyl carbonate are transesterification reactions. Representing bisphenol A schematically as shown below simplifies illustration of the mechanism without removing any of the details.

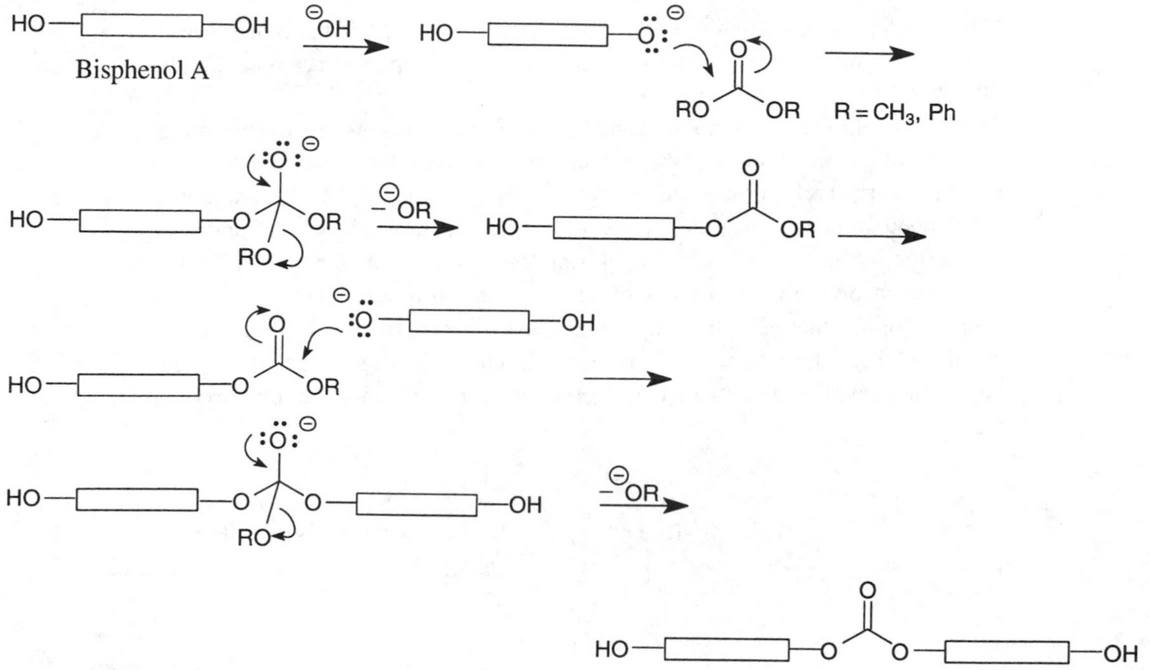

Shown above are two stages in the polymerization that produces a diester with two phenolic OH groups. Each of these reacts with the starting carbonate ester to continue the polymer chain by mechanisms identical in detail to those shown above.

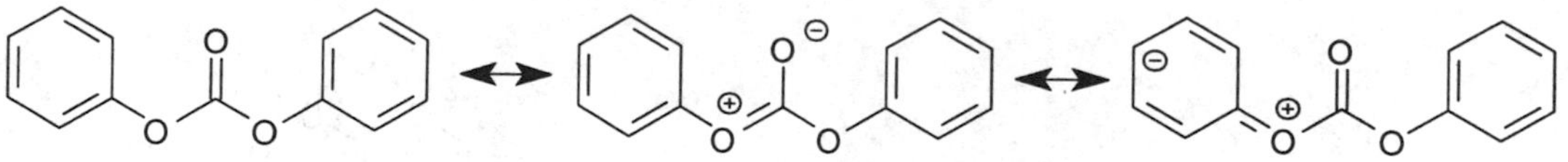

Transesterification of dimethyl carbonate is slower than that of diphenyl carbonate because there is a greater degree of stabilization of the carbonyl group in the former. The lone pairs of electrons on oxygen in diphenyl carbonate are partially delocalized to the aromatic rings. As a result, there is reduced delocalization of these electrons to the carbonyl π system and thus, less stabilization that must be overcome during attack by the anionic nucleophile in the rate-determining step (formation of the tetrahedral intermediate).

Ex 16.11   Replacement of the hydroxyl group of a hemiacetal by an alkoxy group to form an acetal is an S_N1 reaction, involving cleavage of the carbon–oxygen bond after protonation of the OH group. The resulting cation is then attacked by the alcohol in the second stage to form the C—OR bond in the product. Although there are two different carbon–oxygen bonds in the hemiacetal, cleavage of only one of these bonds is productive in proceeding to the acetal.

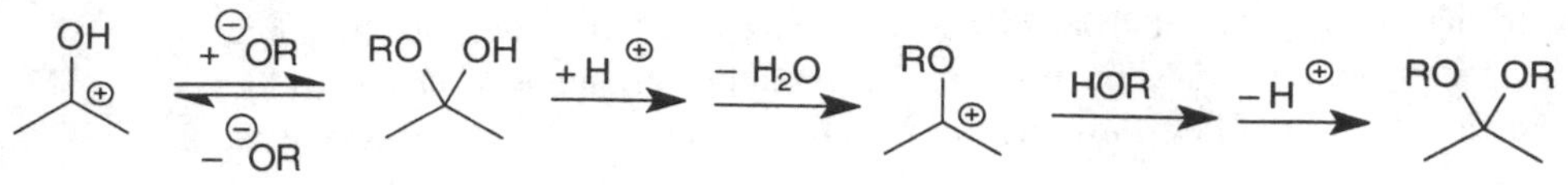

However, in a cyclic hemiacetal such as found in carbohydrates like glucose, cleavage of one bond results in loss of the hydroxyl group, whereas breaking the other bond opens the ring.

Protonation of the OH group, followed by loss of water, results in a cyclic cation. Addition of the alcohol to this cation then leads to the cyclic acetal after loss of a proton. Alternatively, protonation of the ring oxygen, followed by cleavage of a C—O bond in the ring, leads to a protonated, acyclic hydroxylaldehyde. Capture of this cation by the alcohol and loss of a proton produces an acyclic hemiacetal. Upon protonation of the OH group of this hemiacetal followed by loss of water, a cation is formed that is captured by the remote hydroxyl group, reforming the six-membered ring. All of the steps of carbon–oxygen bond cleavage are acid-catalyzed: that is, they involve the initial addition of a proton from the medium to an oxygen atom.

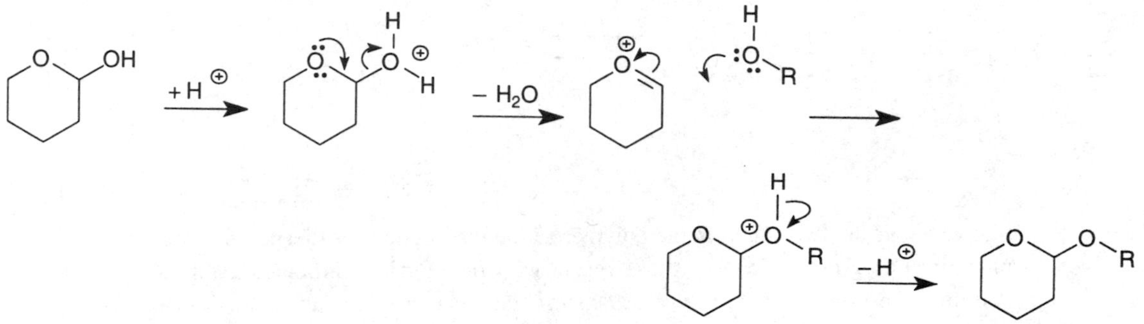

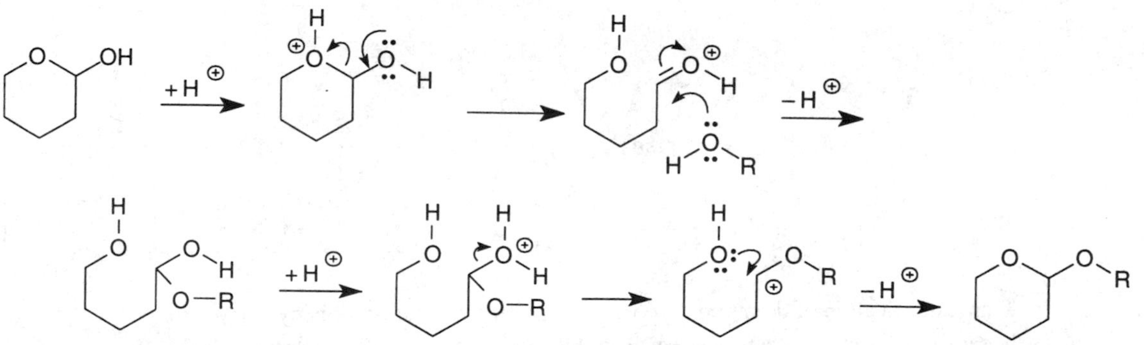

Because both routes give the same product, they cannot be distinguished by product analysis and other details that might indicate which is faster have not yet been examined. How then should we decide which option to choose as "the" mechanism for this reaction? There is a principle followed in science known as Ockham's razor (named for Sir William of Ockham, an English scholastic who rejected the reality of universal concepts). This principle states that we should use the simplest explanation among all reasonable possibilities until data compel us to conclude that this answer is not correct. In this case, the first sequence that does *not* involve opening and closing of the ring is preferred on philosophical grounds.

Ex 16.12 This reaction is a difficult one, involving nucleophilic attack of an amine on a carbonyl carbon of a carboxylic acid. Under neutral conditions, the fastest reaction between acetic acid and methylamine is, instead, proton transfer.

$$CH_3CO_2H \quad + \quad CH_3NH_2 \quad \rightleftharpoons \quad CH_3CO_2^{\ominus} \quad + \quad CH_3\overset{\oplus}{N}H_3$$

The resulting carboxylate and ammonium ions are both ill-suited for nucleophilic acyl substitution. The carboxylate anion is a poor electrophile because of its negative charge, and the ammonium ion lacks an unshared pair of electrons needed for bond formation. Nucleophilic acyl substitution is generally accelerated in the presence of acid where the carbonyl group is protonated and, therefore, activated. Conversely, under basic conditions, deprotonation converts the nucleophile to a more reactive species. The reaction of acetic acid with methylamine poses a dilemma—on the one hand, as the pH is decreased (increasing acidity), the quantity of free amine decreases to a vanishingly small amount before significant quantities of the acid are protonated. On the other hand, in the presence of a base sufficiently strong to effect deprotonation of the ammonium ion, the carboxylic acid is converted completely to the carboxylate, a very unreactive species for nucleophilic acyl substitution. It is possible to form an amide by heating an ammonium salt of a carboxylic acid to high temperature (>200 °C), but this process is practical only in special cases.

Ex 16.13 The Beckmann rearrangement of the oxime of cyclohexanone follows the general pathway discussed in Chapter 14, as shown below. (The intermediate, ring-expanded cation is not twisted as shown here. This depiction is used only so that a correspondence can be more readily seen with its precursor, the protonated oxime.)

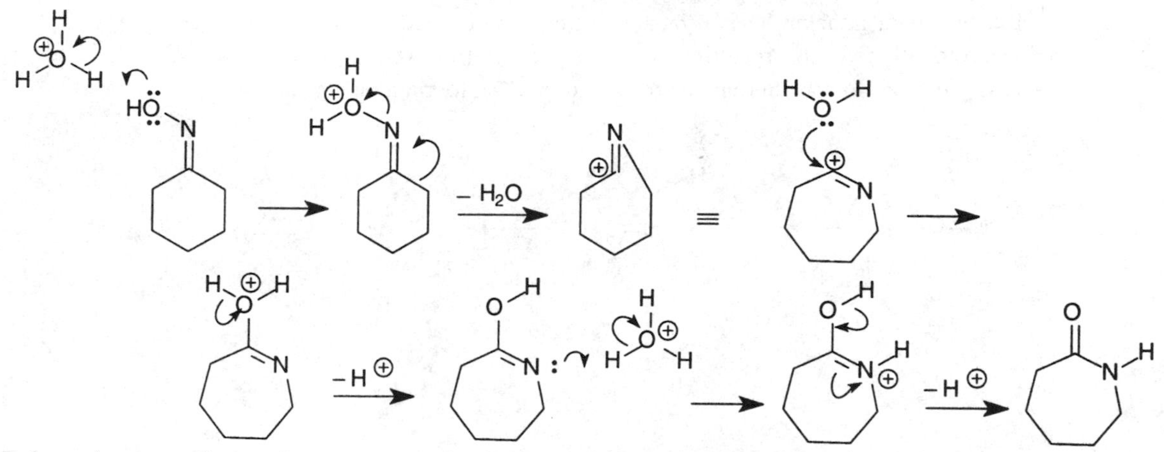

Polymerization of caprolactam occurs by nucleophilic acyl substitution of the carbonyl group. To start, an external nucleophile reacts with the carbonyl group, forming a tetrahedral intermediate. The carbonyl group π bond is then reformed as the carbon–nitrogen bond is broken. The resulting amine then serves as the nucleophile for nucleophilic acyl substitution of another molecule of caprolactam. As each caprolactam is added, a new amide bond is formed, as well as a free amino group that reacts with yet another molecule of caprolactam.

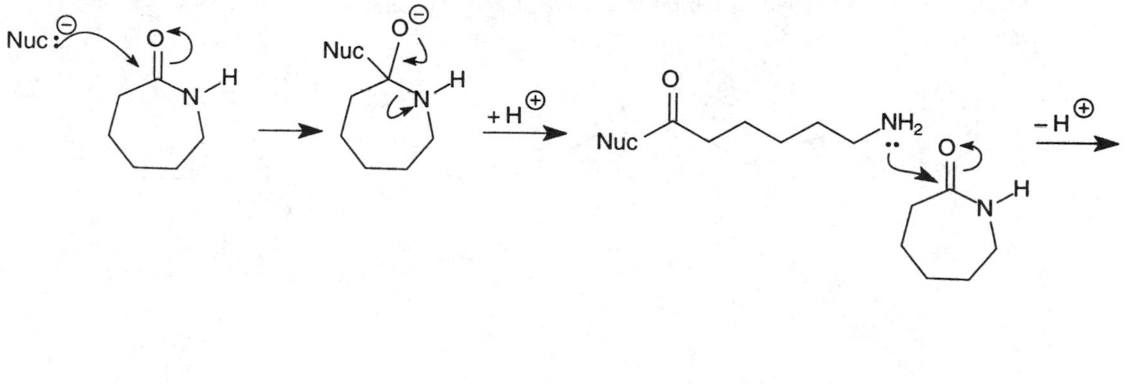

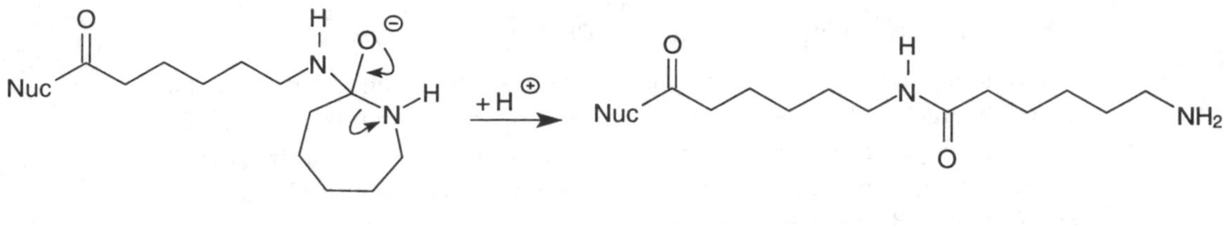

Ex 16.14 The reaction of an isocyanate with water begins by protonation at nitrogen, followed by addition of water to carbon, an electrophilic addition across the C=O π bond that forms an *N*-carboxyamine. Further protonation on nitrogen then sets the stage for loss of carbon dioxide, producing the free amine as a proton is transferred to a weak base in the medium.

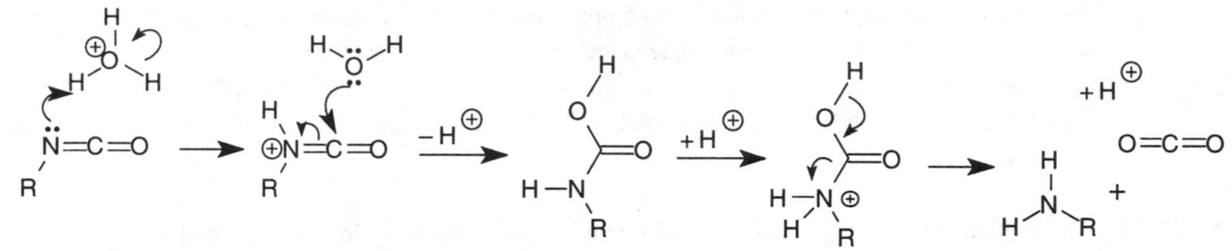

As is often the case, the distinction between catalysis by acid and base is only the order of proton additions and removals. The reaction of an isocyanate with water under basic conditions begins with the deprotonation of water to form hydroxide ion that then attacks the isocyanate as nucleophile, adding to the π system. Deprotonation of the resulting *N*-carboxy acid, followed by protonation at nitrogen, leads to a zwitterion that readily loses CO_2, forming the free amine.

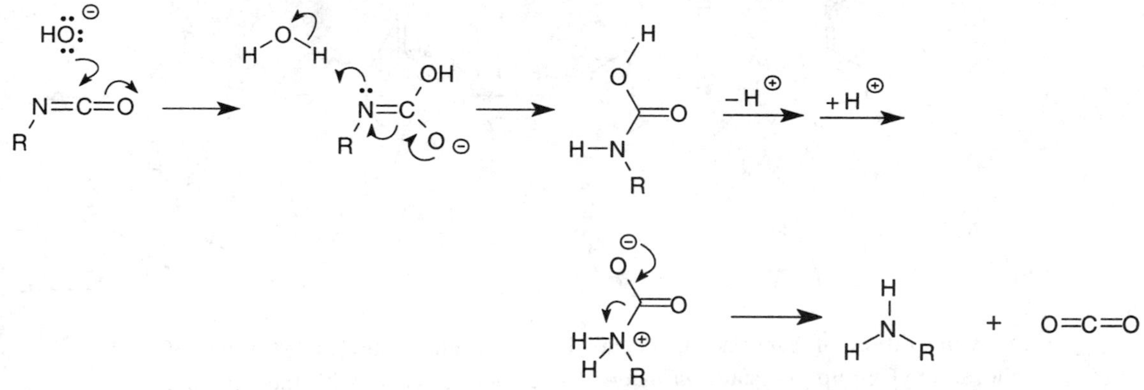

Ex 16.15 Formaldehyde is activated by complexation of the Lewis acid with a lone pair of electrons on oxygen. (The Lewis acid is represented here as L+.) The resulting cation effects electrophilic aromatic substitution by addition to the π system of phenol. Loss of a proton reforms the aromatic system, and exchange of a proton for the Lewis acid produces the product alcohol and regenerates the Lewis acid catalyst.

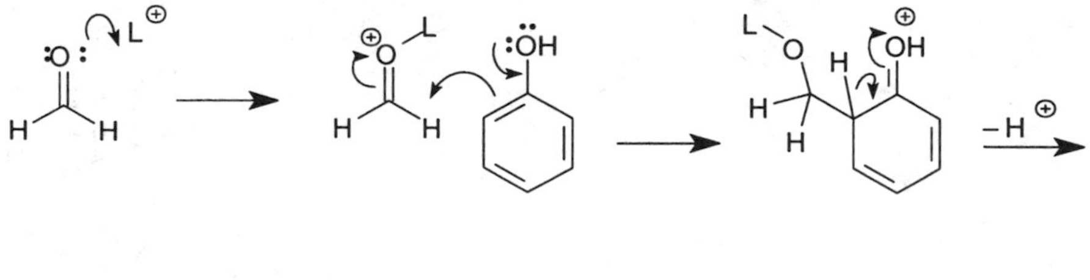

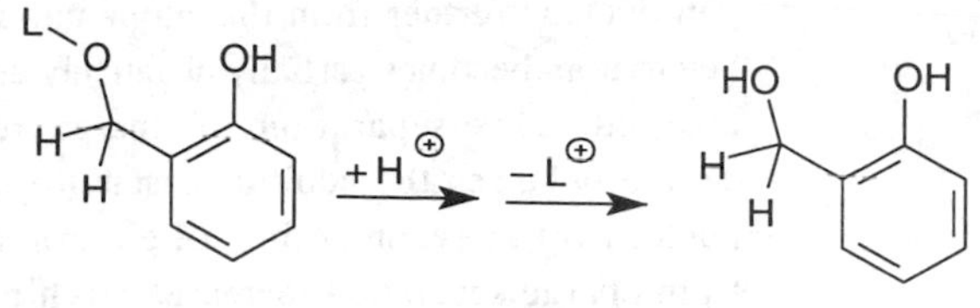

Substitution takes place *ortho* and *para* to the hydroxyl group, because the hydroxyl group of phenol is a strong electron-releasing group. (Only *ortho* substitution is shown). When the electrophile reacts with phenol at the *ortho* (and *para*) positions, the intermediate cation (structure at the right of the first line) is stabilized by delocalization of lone-pair electron density from oxygen. No such stabilization is possible, were the attack to occur at the *meta* positions.

Ex 16.16 All three carbon atoms of epichlorohydrin are electrophilic sites, because each bears a leaving group. Based on our knowledge of S_N2 reactions, the less substituted, primary ring carbon of the epoxide is more reactive toward nucleophiles than is the other. However, the relative reactivity of the epoxide carbons and the primary alkyl chloride can be determined only by measurement of the rates of reaction for these separate processes. Reaction of methoxide ion at the chlorine-bearing carbon produces the simple substitution product in which the labeled carbon is one of the epoxide carbons.

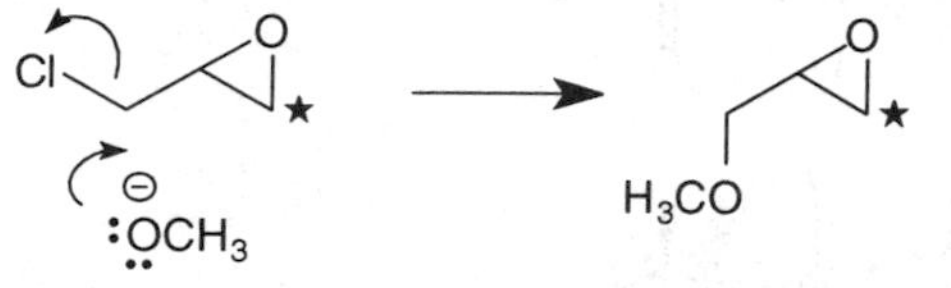

Alternatively, attack of the nucleophile with the less substituted carbon of the epoxide results in ring opening. Intramolecular attack of the alkoxide ion on the carbon bearing chlorine results in the substitution product in which the labeled carbon is bonded to the nucleophile.

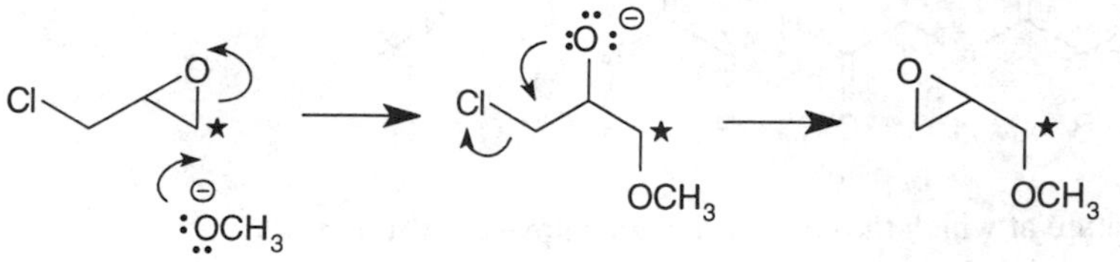

Ex 16.17 Remember that the two ends of a polymer chain are different. Thus, any atom along the chain is a center of chirality if it has two additional substituents that are different. Thus, of all the polymers shown in Figure 16.1, only Teflon does not have stereoisomers.

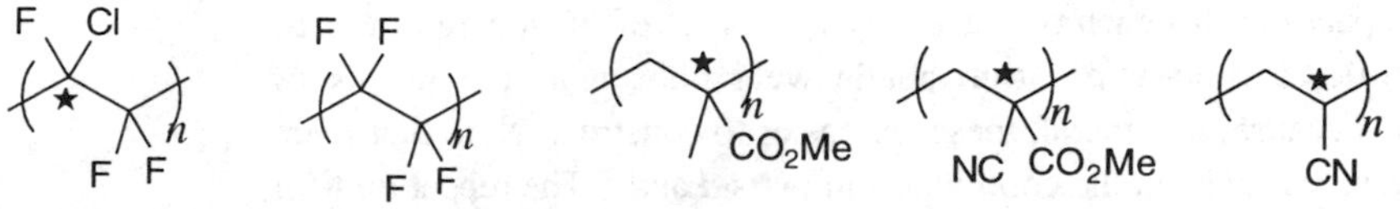

| Kel-F | Teflon | Plexiglas, Lucite | Instant Glue | Orlon |

Ex 16.18 In theory, the most stable position for a hydrogen atom between two identical atoms might indeed be in the middle. However, the types of hydrogen bonds that are important in organic chemistry are those between a hydrogen bound to one neutral heteroatom and another such atom. If we start at one extreme in which the hydrogen interacts only with one heteroatom and move the hydro-

gen nucleus further from this atom and closer to the other, we find that the first heteroatom becomes partially negatively charged and the second partially positively charged. This separation of charge represents an increase in energy that counterbalances the additional stability gained by association of the hydrogen nucleus with a second center of electron density. The loss of stability that results from charge separation increases much more rapidly as the hydrogen position is shifted, and, as a result, the actual shift in position (and the resulting amount of charge separation) is quite small.

$$X\text{—}H\text{------}X \longrightarrow {}^{\ominus}X\text{------}H\text{—}X^{\oplus}$$

The fundamental principle that determines the relative position of atoms in molecules is the repulsive interactions between electrons. A linear arrangement of a hydrogen bond provides the maximum spatial separation between the two pairs of electrons associated with the hydrogen atom.

Ex 16.19 Shown below are three parallel strands of a peptide derived from an α-amino acid, hydrogen bound to one another. Notice that there are far fewer hydrogen bonds (half as many) than when the strands are antiparallel (Figure 16.7).

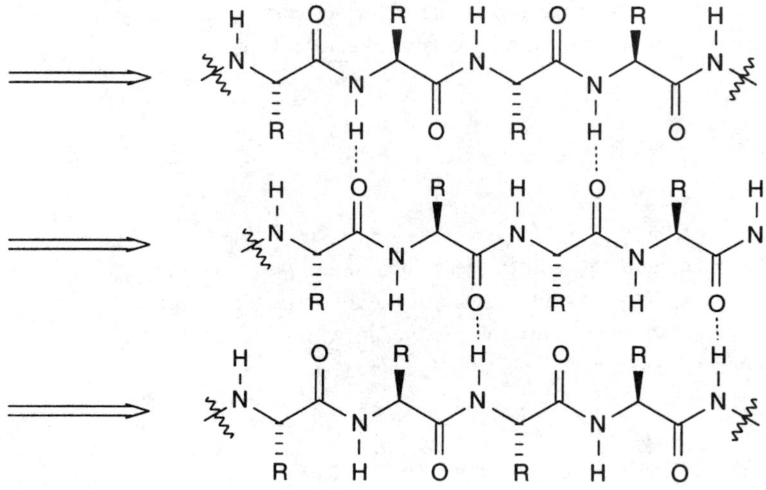

Ex 16.20 To determine the temperature at which the entropy and enthalpy contributions are equal, we set = $T\Delta S°$ and solve for T.

$$T = \frac{\Delta H°}{\Delta S°} = \frac{(-95 \text{ kcal / mole})}{(-285 \text{ cal /mole·K})} \times \frac{1 \text{ cal}}{1000 \text{ ckal}} = 333 \text{ K} = 60 °C.$$

Answers to Review Problems

Pr 16.1 The minimum repeat unit for each of these polymers is set off in square brackets. Wherever possible, carbon-heteroatom bonds were chosen as the points of "disconnections" because, in general, these are easier to construct. Note that there are two different repeat units in the copolymers in parts *d* and *f*. The repeat unit for these polymers could also be represented as a combination of the two units shown, that is, by removing the central pair of brackets).

(a)

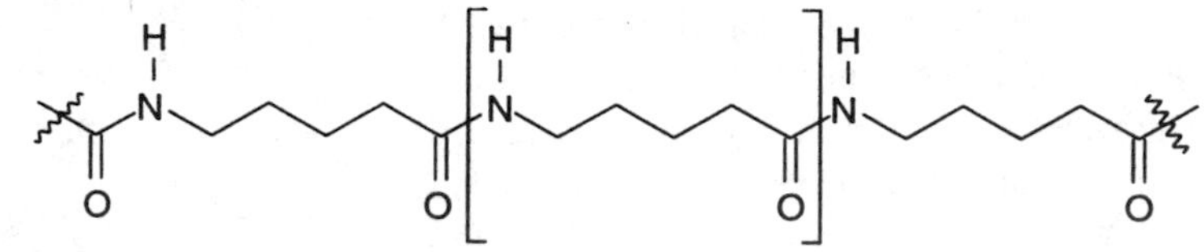

(b)

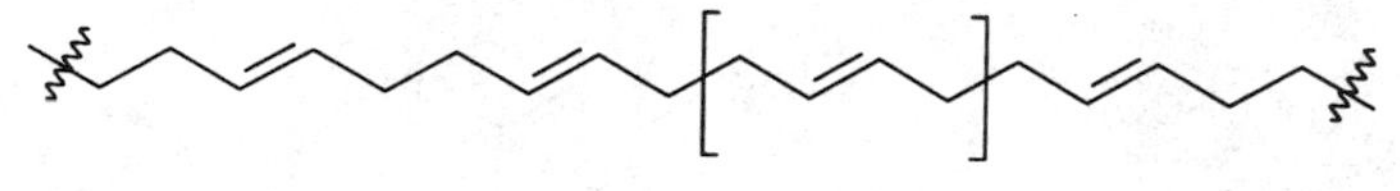

(c)

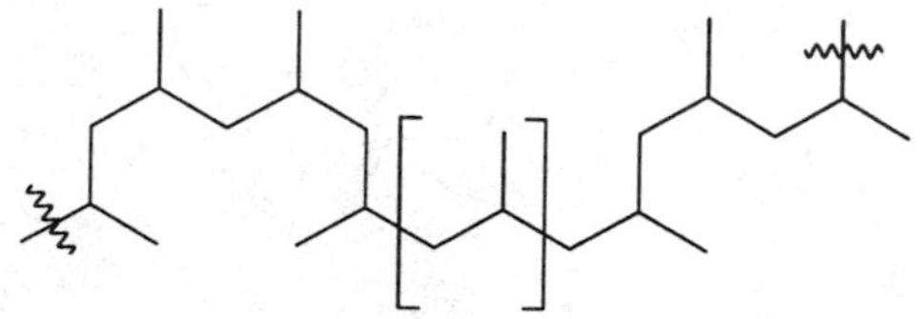

(d)

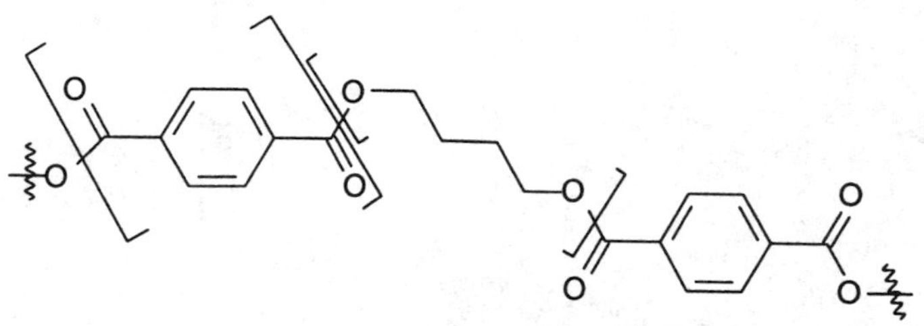

(e)

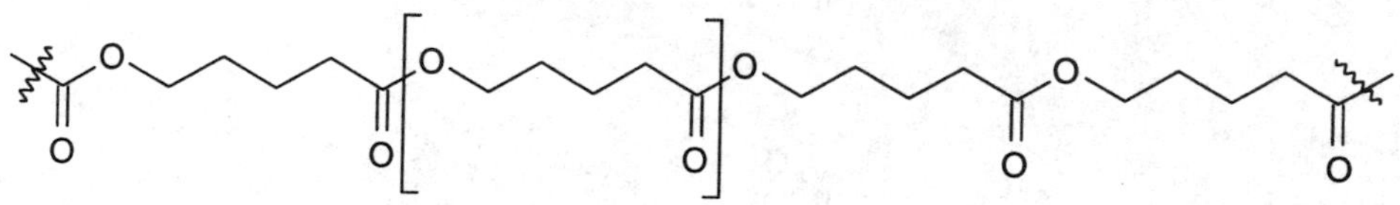

(f)

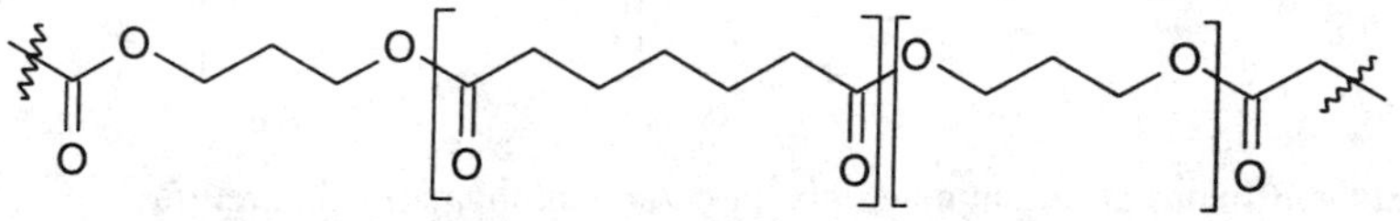

Pr 16.2 A reasonable precursor monomer for each of the polymers in Problem 16.1 is shown below. In the case of parts *d* and *f*, the polymer is best prepared from a mixture of two monomers.

(a)

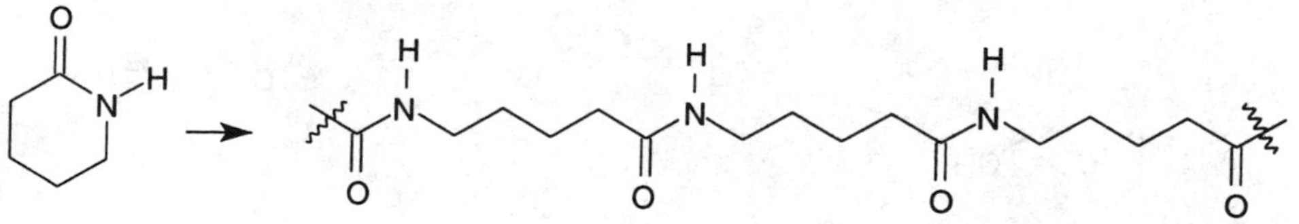

(b)

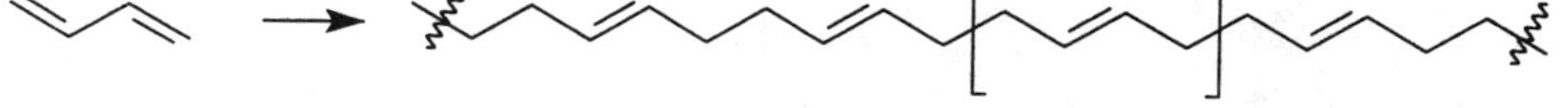

(c)

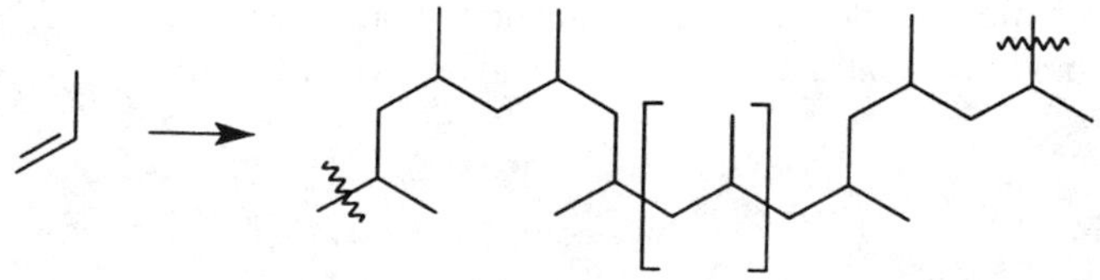

(d)

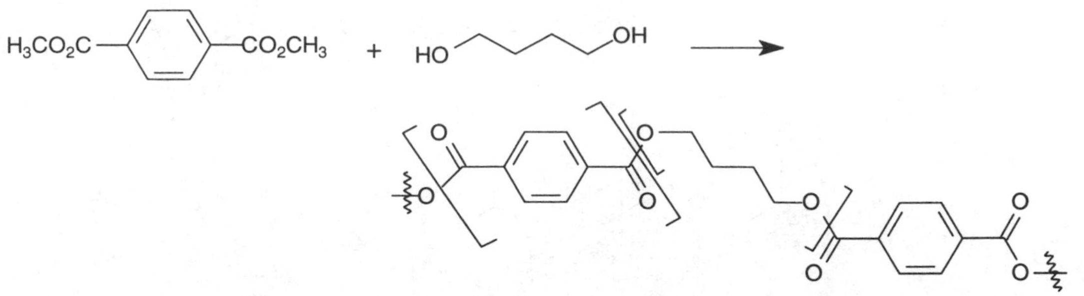

(e)

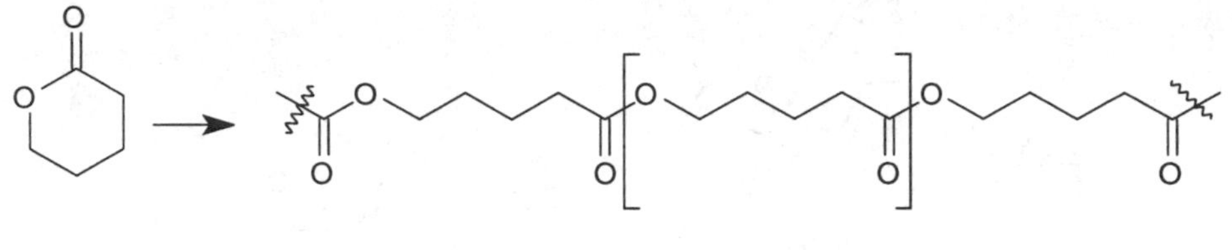

(f)

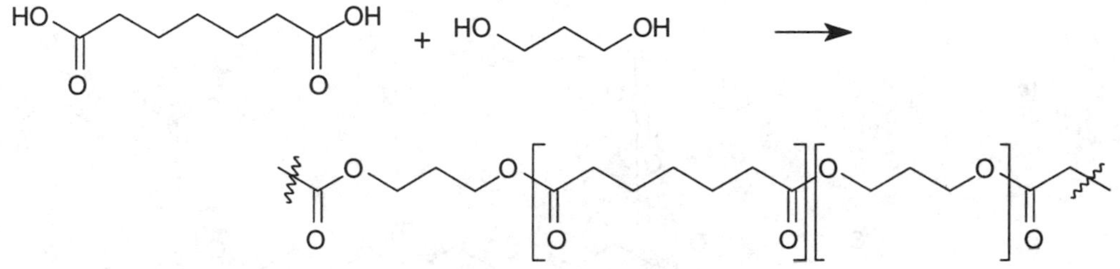

Pr 16.3 Both Celcon and Deldrin are excellent materials for certain applications that require a plastic that is strong yet resilient and '"slippery". (Uses include moving parts such as gears and pipe fittings.) Both materials are more stable than paraformaldehyde because the hemiacetal functional group of the latter is not present. Hemiacetals undergo rapid cleavage in the presence of both acids and bases.

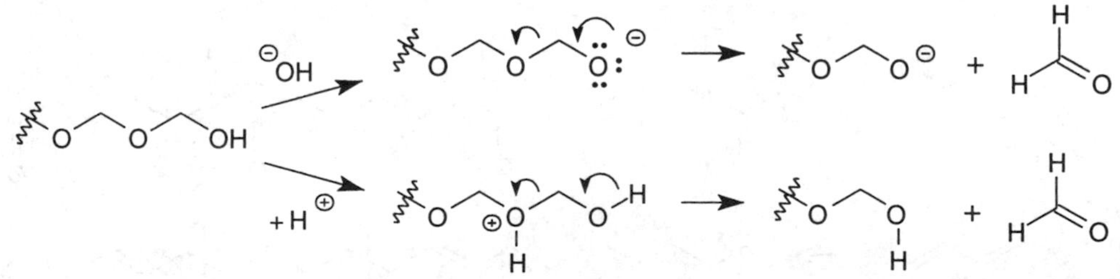

By contrast, acetals are much more stable. There are no acidic protons in an acetal that can be removed by base, in contrast to the acidic hydroxyl group of a hemiacetal.

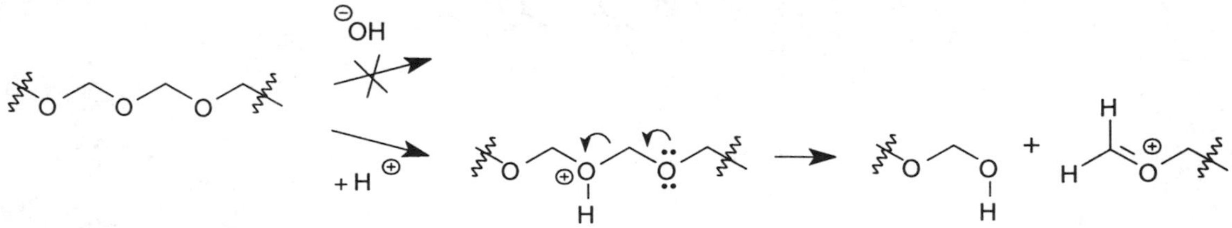

Furthermore, the acid-catalyzed cleavage of an acetal requires the formation of a discrete cationic intermediate rather than the neutral aldehyde produced by acid-catalyzed decomposition of a hemiacetal. Reaction of a hemiacetal with acid leads to bond cleavage and simultaneous loss of a proton so that no intermediate carbocation is formed.

Deldrin, the polymer that is capped on the ends with acetate groups, is susceptible to hydrolysis with base, removing the end group and leaving a hemiacetal that can be further degraded by base. Although Celcon does have free hydroxyl groups at the end, these are not part of a hemiacetal and thus this plastic is considerably more stable to alkaline conditions than is Deldrin.

Deldrin

Celcon

Pr 16.4 The reaction of ethylene oxide with base is a bimolecular reaction that is catalyzed by base, with hydroxide ion as the attacking nucleophile. However, there is an alternate reaction that involves deprotonation of ethylene glycol to an alkoxide ion that is comparable in nucleophilicity to hydroxide ion. As the reaction proceeds and the concentration of product builds, more of the reaction involves reaction of the anion of ethylene glycol with ethylene oxide, forming di(ethylene glycol). Because the formation of ethylene glycol and di(ethylene glycol), as well as higher oligomers, are bimolecular reactions, their rates depend on the concentrations of both the active nucleophile and ethylene oxide.

As the concentration of the products increases, the rates of formation of di(ethylene glycol) also increases. With a high concentration of water, the relative concentration of hydroxide ion is greater than that of the anion of ethylene glycol: thus, the formation of the higher members of the series is reduced under these conditions. However, ethylene glycol is miscible with water and separation of excess water from the product requires distillation of the latter, adding greatly to the cost of producing large quantities of this diol. In practice, the water concentration is kept low and the oligomers are marketed in the form of automobile antifreeze, for which purpose they are as good or better than ethylene glycol.

Pr 16.5 Bisphenol-F is the condensation product of two equivalents of phenol with one of formaldehyde and is similar to both bisphenol-A and bisphenol-B. The mechanism of its formation is essentially identical to that for bisphenol-A provided in the answer to Exercise 16.9, substituting formaldehyde for acetone.

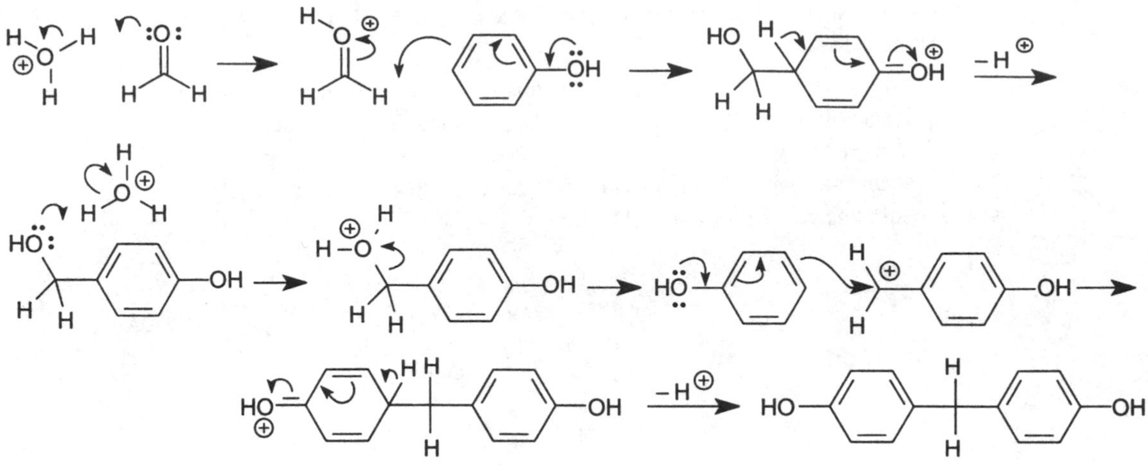

Bisphenol F

Pr 16.6 The oxidation of an alkyl side chain of an aromatic compound is conveniently carried out in the laboratory with $KMnO_4$. This reagent is not a practical alternative for use on a large scale because substantial quantities of the reduction product, MnO_2, are produced. (The oxidation of each methyl group represents a change in oxidation level from -3 to $+3$. Reduction of $KMnO_4$ to MnO_2 represents a change in oxidation level of Mn from $+6$ to $+4$, and thus three equivalents of $KMnO_4$ are required for each methyl group. For fun, calculate how much MnO_2 would be produced upon oxidation of 1,000,000 pounds of *p*-xylene, keeping in mind that there are two methyl groups to be oxidized.)

Pr 16.7 The principal difference between polyamides made from α- and β-amino acids is the position of functional groups along the chain. As shown in the answer to Exercise 16.19, the N—H group of an amino acid residue of a peptide derived from α-amino acids points to the same side as the carbonyl group. Because there is one more carbon atom between the acid and amine of β-amino acids, these groups point to opposite sides of the resulting polymer, as shown below. Nonetheless, two adjacent chains of such a peptide are associated through just as many hydrogen bonds as one derived from α-amino acids. There are fewer atoms per residue in the latter, however, and on a weight-adjusted basis, peptides derived from α-amino acids have stronger mutual attractions.

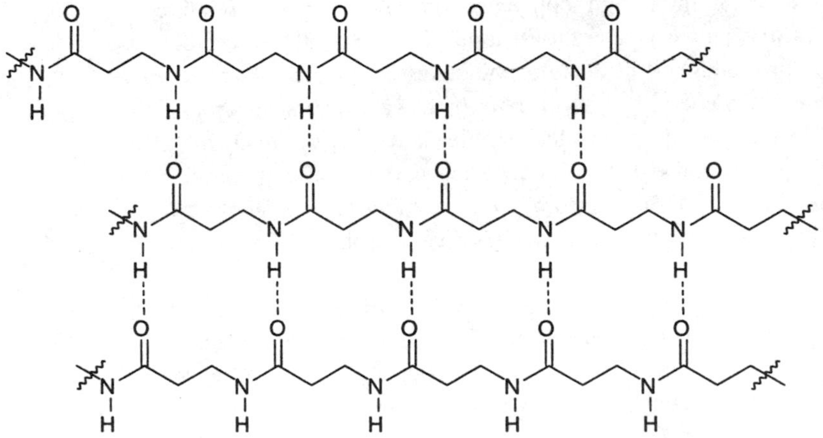

Pr 16.8   This mechanism begins with nucleophilic attack by methylamine on the anhydride carbonyl group. The tetrahedral intermediate thus generated then reforms an amide carbonyl group, while expelling the carboxylate anion. Because this anion is a carbamic acid, it undergoes rapid decarboxylation, producing a free amine. This amine then acts as a nucleophile to initiate the same sequence of bond making and bond breaking on the next N-carbonic anhydride. The dipeptide thus produced also has a free amino end that can participate in further chain extension by the same sequence. Because a small molecule (CO_2) is lost with each sequential step, this is a condensation polymerization.

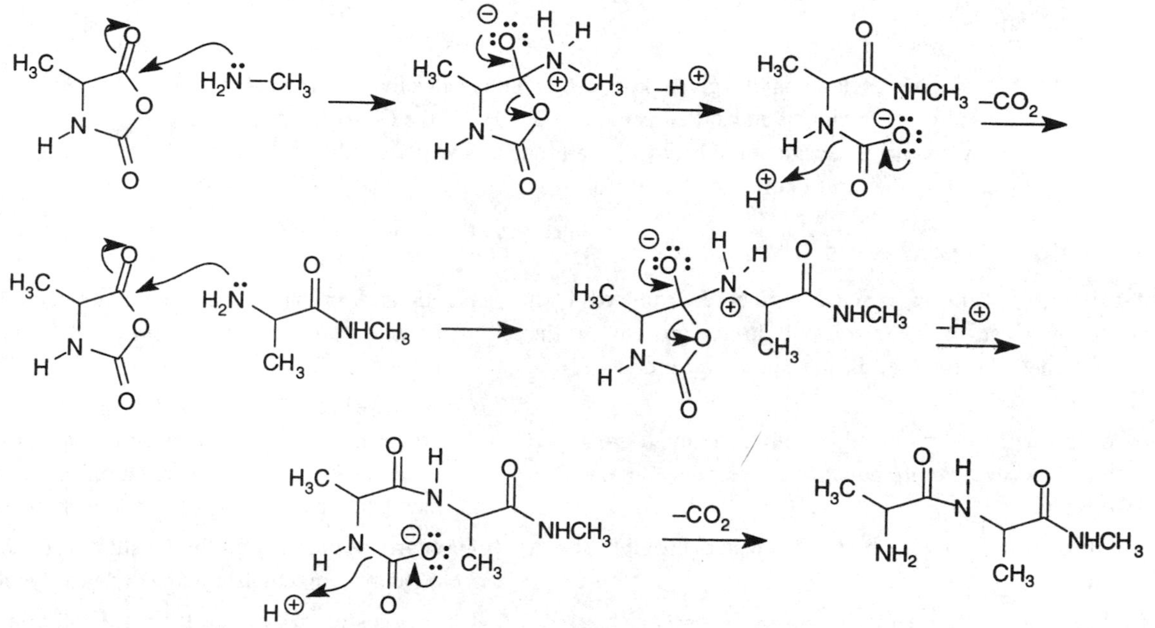

Pr 16.9   Both products are Diels-Alder adducts, obtained when one of the reactants acts as a diene (with two conjugated double bonds participating in the π-interaction) and the other reactant acts as a dienophile (with one double bond so participating). Diels-Alder reactions take place through a concerted cyclic transition state involving six electrons. The conversion of one Diels-Alder product into the other is an electrocyclic reaction discussed in Chapter 14, the Cope rearrangement.

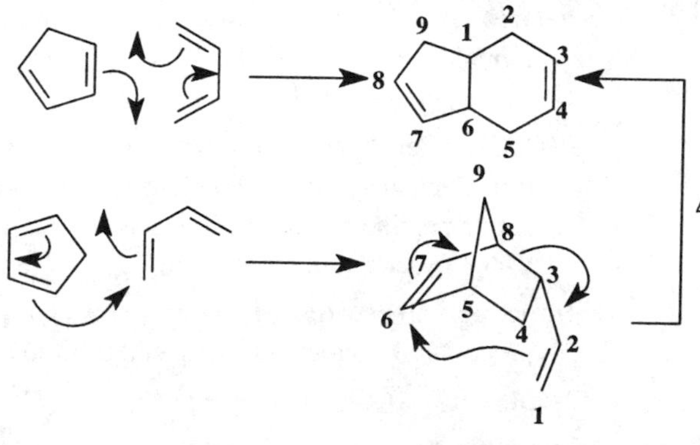

5-Vinylnorbornene

Pr 16.10   Both of the double bonds in 5-vinylnorbornene can be attacked to form a secondary radical. However, the double bond in the five-membered ring of 5-vinylnorbornene is strained. (The C—C=C bond angle is constrained to approximately 108° by the bicyclic structure.) Relief of this ring strain accompanies attack on this double bond by a radical, resulting in a faster reaction.

Polymerization reactions generally slow as they proceed, not only because the monomer concentration is constantly decreasing but also because large molecules move more slowly. The endocyclic double bond becomes incorporated to the growing polystyrene polymer because it has reactivity comparable to that of styrene. On the other hand, the other double bond is less reactive but ultimately reacts when the concentration of styrene drops to quite low levels. At this point, two quite large polymer chains are joined, to make one very high molecular weight polymer.

Pr 16.11   Often, the acceleration of a given process by light is indicative of a radical process, possibly producing a radical that is trapped by O_2. Of the C—H bonds present in polystyrene, cleavage at the benzylic site produces the most stable radical, and is therefore the most likely site of oxidation.

Important New Terms

Addition polymer: a macromolecule produced in a polymerization in which all atoms present in the monomer are retained in the polymeric product (16.3, 16.4)

α-Amino acid: a compound in which an amino group and a carboxylic acid are attached to the same carbon atom (16.5)

Amylopectin: a highly branched, water-insoluble starch (16.5)

Amylose: *see* **Starch** (16.5)

Anionic polymerization: the formation of a polymer by a process in which the growing end is a carbanion (16.4)

Atactic: stereochemical designator of a polymer with random orientation of groups at centers of chirality (16.7)

Branched polymer: a macromolecule in which chemical bonds interconnect chains, forming a complex, three-dimensional network (16.2)

Carbamate: *see* **Urethane** (16.5)

Carbohydrate: a polyhydroxylated aldehyde or ketone with the molecular formula $C_m(H_2O)_n$ (16.5)

Carbowax: a synthetic poly(ethylene glycol) (16.4)

Cationic polymerization: the formation of a polymer by a process in which the growing end is a carbocation (16.4)

Cellulose: a water-soluble biopolymer containing 3000–5000 glucose units connected exclusively by β linkages (16.5)

Cellulose acetate: an optically transparent polymer obtained by treating cellulose with acetic anhydride, thus converting many of the polysaccharide hydroxyl groups to acetate esters (16.5)

Condensation polymer: a macromolecule produced in a polymerization in which a small-molecule by-product is formed (16.5)

Cross-linking: the covalent interconnections between polymer chains from which a three-dimensional network results; the process in which a bifunctional molecule is incorporated in two separate polymer chains (16.2, 16.6)

Dacron: a commercial polyester produced by linking dimethyl terephthalate with ethylene glycol (16.5)

Epoxy resin: a structurally rigid material obtained by crosslinking a diol with epichlorohydrin (16.6)

Ethylene glycol: $HOCH_2CH_2OH$ (16.4)

Glass: a polymer based on a three-dimensional network of tetrahedrally arranged silicon atoms linked by oxygen (16.4)

Glycol: a 1,2- or 1,3-diol (16.4)

α-Helix: a right-handed spiraling structure imposed by intramolecular hydrogen bonding between groups along a single peptide chain (16.7)

β-Helix: a left-handed spiraling structure imposed by intramolecular hydrogen bonding between groups along a single peptide chain; not found with naturally occurring α-amino acids (16.7)

Ionic polymerization: the formation of a polymer by a process in which the growing end is an ion (16.4)

Isoprene: 2–methylbutadiene (16.4)

Isotactic: stereochemical designator of a polymer in which all groups at centers of chirality along the chain point in the same direction (16.7)

Lactam: a cyclic amide (16.5)

Linear polymer: a macromolecule in which the monomer units are attached end-to-end (16.2)

Living polymer: a macromolecule in which the end of the chain is chemically reactive but in which two such ends will not react with each other; often applied to anionic, cationic, and organometallic polymerizations (16.4)

Macromolecule: *see* **Polymer** (16.1)

Monomer: the chemical precursor of a polymer (16.1)

Neoprene: poly(2-chlorobutadiene) (16.4)

Nylon 6: a polyamide formed in the ring-opening polymerization of caprolactam (16.5)

Nylon 66: a polyamide formed in the cross reaction between adipic acid and 1,6-diaminohexane (16.5)

Peptide: a polyamide composed of 2–10 or fewer α-amino acid residues (sometimes used interchangeably with polypeptide) (16.5, 16.7)

Peptide bond: an amide linkage (16.5)

Plastics: polymers that can be heated and molded while relatively soft; from the Greek *plastikos*: fit to be molded (16.4)

β-Pleated sheet: a folded, sheetlike structure imposed by intermolecular hydrogen bonding between peptide chains (16.7)

Pleating: the deviation from a planar arrangement in the hydrogen bonded structure of two intermolecularly associated peptide chains to avoid steric interaction of the alkyl groups at the α-position (16.7)

Plexiglas: $-[CH_2C(CH_3)(CO_2CH_3)]_n-$; poly(methyl methacrylate) (16.4)

Polyacetal: $-(CHRO)_n-$ (16.4)

Polyamide: polymer in which the repeat units are joined by an amide linkage (16.5)

Polycarbonate: $-(ROCO_2)_n-$ (16.5)

Polyester: a polymer in which the repeat units are joined by an ester linkage (16.1, 16.5)

Polyether: a polymer in which the repeat units are joined by an ether linkage (16.4)

Poly(ethylene glycol): $-(CH_2CH_2O)_n-$; condensation polymer from ethylene glycol (16.4)

Polymer: a large molecule composed of many repeating subunits; from the Greek *polumeres*: having many parts (16.1)

Polymerization: the process of linking monomer units to a polymeric matrix (16.1)

Polypeptide: a polyamide derived from α-amino acids, specifically composed of 10–100 α-amino acids (sometimes used interchangeably with peptide) (16.5, 16.7)

Polysaccharide: a polyacetal formed by condensation of a hemiacetal group of one sugar unit with an alcohol group of another sugar unit, taking place with the loss of water (16.5)

Polystyrene: $-[CH_2CH(Ph)]_n-$ (16.4)

Polyurethane: $-(OCONH)_2-$ polymer in which the repeat units are joined by a urethane (carbamate) linkage (16.5)

Poly(vinyl alcohol): $-[CH_2CH(OH)]_n-$ (16.4)

Poly(vinyl chloride): $-[CH_2CH(Cl)]_n-$ (16.4)

Primary structure of a peptide or protein: the sequence of amino acid units along a peptide or protein chain (16.7)

Protein: a poly(α-amino acid) composed of more than 100 α-amino acids (16.7)

Quaternary structure of a peptide or protein: clusters formed as several large polypeptide or protein units join together to form a functional object (16.7)

Radical polymerization: polymerization initiated by a radical and in which the chain-carrying step is a radical (16.4)

Repeat unit: the segment of atoms and groups that is encountered sequentially as one moves along a polymer chain (16.1)

Resin: a highly viscous polymeric glass (16.6)

Ring-opening polymerization: a polymerization reaction in which the driving force for bond formation between repeat units is supplied by relief of ring strain in a monomer (16.4)

Rubber: naturally occurring poly(2-methylbutadiene) (16.4)

Saccharide: *see* **Carbohydrate** (16.5)

Secondary structure of a peptide or protein: a complex three-dimensional structure describing local organization of chain segments such as α-helices and β-pleated sheets (16.7)

Silk: a protein containing high fractions of glycine and alanine (16.5)

Starch: a water-soluble biopolymer containing as many as 4000 glucose units connected by α linkages (16.5)

Sugar: *see* **Carbohydrate** (16.5)

Syndiotactic: stereochemical designator of a polymer in which the alkyl groups at centers of asymmetry in a polymer chain point alternately in one direction and in the opposite direction (16.7)

Teflon: $-(CF_2)_n-$ (16.4)

Tertiary structure of a peptide or protein: the three-dimensional description of how the β-pleated sheets and α-helices are spatially dispersed; describes protein folding (16.7)

Urethane: a carbonyl group bound on one side to the nitrogen of an amine and on the other side to the oxygen of an alcohol (16.5)

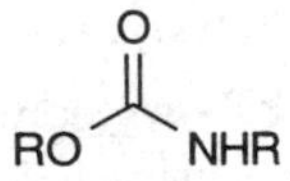

$$RO \overset{\displaystyle O}{\underset{}{\|}} NHR$$

Vulcanization: the cross-linking of a polymer by heating it with sulfur (16.4)

Wool: a structurally complex, naturally occurring protein heavily cross-linked with sulfur–sulfur bonds (16.5)

Ziegler–Natta catalyst: an organometallic polymerization initiator that produces isotactic polypropylene (16.7)